Reservoir Quality Prediction in Sandstones and Carbonates

Edited by

J.A. Kupecz
J.G. Gluyas
and
S. Bloch

AAPG Memoir 69

Published by
The American Association of Petroleum Geologists
Tulsa, Oklahoma, U.S.A. 74101

ISBN: 0-89181-349-7

Reservor quality prediction in sandstones and carbonates / edited by
    J. A. Kupecz . . . [et al.].
        p.   cm. -- (AAPG memoir ; 69)
    Includes bibliographical references (p.   ) and index.
    ISBN 0-89181-349-7 (alk. paper)
    1. Petroleum--Migration. 2. Carbonate reservoirs. 3. Sandstone.
    4. Petroleum reserves--Forecasting. 5. Petroleum--Prospecting.
    I. Kupecz, Julie A. II. Series.
    TN870.52.R47 1998
    622' . 1828--dc21   97-48365
                                                   CIP

Association Editor: Neil F. Hurley
Science Director: Richard Steinmetz
Publications Manager: Kenneth M. Wolgemuth
Special Projects Editor: Anne H. Thomas
Production: Custom Editorial Productions, Inc., Cincinnati, Ohio

This book and other AAPG publications are available from:

The AAPG Bookstore
P.O. Box 979
Tulsa, OK 74101-0979
Telephone: (918) 584-2555;
or (800) 364-AAPG (U.S.A.—book orders only)
Fax: (918) 584-2652
or (800) 898-2274 (U.S.A.—book orders only)

Geological Society Publishing House
Unit 7, Brassmill
Enterprise Centre
Brassmill Lane
Bath BA1 3JN
United Kingdom
Tel 1225-445046
Fax 1225-442836

Australian Mineral Foundation
AMF Bookshop
63 Conyngham Street
Glenside, South Australia 5065
Australia
Tel (08) 379-0444
Fax (08) 379-4634

Canadian Society of Petroleum Geologists
#505, 206 7th Avenue S.W.
Calgary, Alberta T2P 0W7
Canada
Tel (403) 264-5610

About the Editors

Julie A. Kupecz

Julie A. Kupecz received degrees in geology from the University of Oklahoma (B.S., 1979), Colorado School of Mines (M.S., 1984), and the University of Texas at Austin (Ph.D., 1989). Her professional experience includes geological research, petroleum exploration and reservoir development, and reservoir characterization. She has been employed by Cities Service Co., the University of Texas Bureau of Economic Geology, and ARCO (ARCO Alaska and ARCO Exploration and Production Technology). She is president of Kupecz and Associates, Ltd., and is currently a Senior Research Advisor for Intevep, S. A., Los Teques, Venezuela.

Julie's research interests include carbonate sedimentology, sequence stratigraphy, and diagenesis, with emphasis on the integration of techniques for characterization and prediction of reservoir quality. She is an Associate Editor of the *Journal of Sedimentary Research*, and has been a co-convenor of AAPG, SEPM, and IAS Annual Meeting sessions on porosity prediction; carbonate and evaporite depositional systems; outcrop studies of carbonates and evaporites; emerging plays/development in North America; and carbonate sedimentology of Latin American basins.

Jon G. Gluyas

Jon G. Gluyas completed a B.Sc. degree in geology at Sheffield University, England, and a Ph.D. on shale geochemistry and diagenesis at Liverpool University, England. He joined BP Exploration in Aberdeen in 1981. He was transferred to BP's Research Centre in 1984 to work on devising methods for accurate prediction of reservoir quality. Between 1987 and 1991 Jon worked in production geoscience around the "BP world." Jon returned to R&D at BP Sunbury (London) in mid-1991. For the following two years, he led a large research team on reservoir quality prediction. In late 1993, he moved to Venezuela to lead the appraisal and production effort on a field reactivation project. In mid-1996, Jon joined London-based independent company Monument Oil and Gas. He has written 25 scientific papers on sediment diagenesis, exploration and production geology, and has presented papers at international meetings, including those organized by AAPG and the Royal Society.

Sal Bloch

Sal Bloch is a Senior Research Associate at Texaco's E&P Technology Department. His previous employers include Norsk Hydro, ARCO, and Oklahoma Geological Survey. Sal received his M.S. degree in geology from Wroclaw University in Poland and his Ph.D. from George Washington University. His primary research and technical service focus for the past 15 years has been sandstone diagenesis and reservoir quality prediction.

AAPG
Wishes to thank the following
for their generous contributions
to

Reservoir Quality Prediction in Sandstones and Carbonates

◆

ARCO Exploration and Production Technology

◆

Exxon Production Research Company

◆

R.P. Major

◆

Eric Mountjoy
McGill University

◆

Norsk Hydro ASA Exploration and Production

◆

Texaco E&P Technology

◆

Contributions are applied against the production costs of publication, thus directly reducing the book's purchase price and making the volume available to a greater audience.

Table of Contents

Reservoir Quality Prediction in Sandstones and Carbonates: An Overview

Julie A. Kupecz
Intevep, S.A.
Los Teques, Venezuela
and
Kupecz and Associates, Ltd.
Denver, Colorado, U.S.A.

Jon Gluyas
Monument Oil and Gas
London, United Kingdom

Salman Bloch
Texaco E&P Technology Department
Houston, Texas, U.S.A

INTRODUCTION

The accurate prediction of reservoir quality is, and will continue to be, a key challenge for hydrocarbon exploration and development. Prediction is a logical and critically important extension of the description and interpretation of geological processes. However, in spite of the profusion of publications on sandstone and carbonate diagenesis, relatively few articles illustrate the application of such studies to reservoir quality prediction. This Memoir represents the first attempt to compile worldwide case studies covering some predictive aspects of both siliciclastic and carbonate reservoir characteristics. We have attempted here to focus on the variability due to diagenetic effects in sandstones and carbonates, rather than on sedimentological effects, i.e., the presence or absence of a given reservoir. The chapters cover the spectrum of stages in the exploration-exploitation cycle (Table 1).

The importance of reservoir quality in pay evaluation has been illustrated by Rose (1987), who analyzed an unnamed company's exploration results over a 1-year period. Of 87 wildcat wells drilled, 27 were discoveries (31% success rate); incorrect predictions of the presence of adequate reservoir rocks were made in 40% of the dry holes. Importantly, the geologists believed that reservoir quality was the primary uncertainty in 79% of the unsuccessful wells. Similarly, a comparison of predrill predictions with postdrill results by Shell (Sluijk and Parker, 1984) indicated that reservoir quality was seriously overestimated, whereas hydrocarbon charge and retention predictions were more accurate. Although these statistics do not clearly separate drilling failure due to lack of potential reservoir from the lack of adequate reservoir quality, it seems that although explorers are aware of the significance of reservoir quality prediction, generation of predictive models continues to be a formidable task.

Accurate prediction of reservoir quality is needed throughout the entire "life cycle" of a reservoir (Sneider, 1990). Proper assessment of reservoir quality must be continually refined, from prior to exploratory drilling, to discovery, during appraisal and development drilling, and throughout reservoir management. At the *Exploration Stage*, the main challenge is to assess and predict the reservoir facies, its geometry, and its distribution; reservoir porosity and permeability for use in petroleum reserves calculations; seismic characteristics; and migration pathways. In this Memoir, papers by Brown, Ehrlich et al., Evans et al., Gluyas, Gluyas and Cade, Gluyas and Witton, Primmer et al., Ramm et al., Sombra and Chang, Tobin, and Zempolich and Hardie address various aspects of the assessment process.

At the *Appraisal, Planning, and Development Stages*, it is necessary to understand and predict reservoir porosity, permeability, and reservoir distribution to

Table 1. Overview of Chapters in This Memoir.

Author	Stage in Exploration-Exploitation Cycle	Location/Basin	Reservoir Age	Lithology	Data/ Methodology	Summary of Chapter
Brown	Exploration	North Dakota, Williston Basin, U.S.A.	Mississippian	Carbonates (limestone, dolomite, argillaceous carbonate)	Wireline logs, cuttings descriptions, temperature, numerical regression	Determination of influence of carbonate mineralogy, shale content, and fabric on loss of porosity with burial.
Cabrera-G, Arestad, Dagdelen, and Davis	Development	Western Canada Sedimentary Basin	Devonian	Carbonate (dolomite), evaporite, shale	Seismic	Porosity prediction from multicomponent seismic data via geostatistical methods.
Cavallo and Smosna	Development	West Virginia, U.S.A.; Appalachian Basin	Mississippian	Carbonate	Formation Microscanner (FMS) logs	FMS logs with sidewall core, integrated into depositional model for ooid shoals. Used to predict optimal location for development wells.
Erlich, Bowers, Riggert, and Prince	Exploration; Development	Examples from Thailand (Pattani Basin), Oklahoma, U.S.A. (Cherokee Basin)	Miocene; Permian–Late Carboniferous, respectively.	Sandstone	Petrographic Image Analysis (PIA), mercury porosimetry	Integration of PIA and porosity to understand variations in permeability.
Evans, Cade, and Bryant	Variable; Overview of permeability prediction	N/A	N/A	Sandstone	Modeling of empirical data (porosity, lithology)	Modeling effects of geological processes that affect permeability (burial, cementation) to calculate changes in permeability.
Gluyas	Exploration	Norwegian Central Graben	Late Jurassic	Sandstones	Petrography; porosity, permeability data	Risking of porosity evolution models for predrill porosity prediction.
Gluyas and Cade	Exploration	Worldwide published data	Permian to Pleistocene	Sandstones (quartz, feldspar)	Integration of experimental, petrographic, and porosity data (worldwide)	Porosity–depth relationship for prediction in uncemented sandstones gives maximum porosity baseline to compare cement volumes and cemented ss porosity.
Gluyas and Witton	Exploration	Southern Red Sea, offshore Yemen	Miocene	Sandstone	Petrography, burial and thermal history, provenance, depositional environment	Case study of predrill reservoir quality prediction.
Love, Strohmenger, Woronow, and Rochenbauch	Development	N. Germany; Southern Zechstein Basin	Permian	Carbonate (dolomite; calcitized dolomite)	Statistics; neural networks; core; well logs; structural data; geochemistry	Statistical relationships of geological data for prediction of predrill reservoir quality.

Author	Application	Location	Age	Lithology	Data / Methods	Key points
Major and Holtz	Development; Reservoir Management	Permian Basin, west Texas and SE New Mexico, U.S.A.	Permian	Carbonate (dolomite)	Petrography, well logs, capillary pressure data cores; well-logs; porosity and permeability data; production history	Determination of "flow units" controlled by depositional facies and diagenetic alteration; quantification of bypassed oil in low-permeability flow units and heterogeneous flow units.
Mountjoy and Marquez	Development	Western Canada Sedimentary Basin	Devonian	Carbonate (dolomite, limestone)	Petrography	Controls of depositional facies and diagenesis on pore systems and reservoir continuity; effects of dolomitization on pore types and reservoir character; comparison of reservoir characteristics of limestone vs. dolomite at depth.
Primmer, Cade, Evans, Gluyas, Hopkins, Oxtoby, Smalley, Warren, and Worden	Exploration base	Worldwide data	Variable; predominantly Mesozoic and younger	Sandstones	Depositional environment, composition, maximum burial time; fluid inclusions, stable isotopes, and organic maturation where available	Subdivision into five "styles" of diagenesis via relationship between detrital composition, burial depth, temperature, cement type.
Ramm	Exploration	Norwegian Central Graben	Late Jurassic	Sandstones	Petrography; fluid inclusions	Porosity prediction by prediction of composition, texture, and microquartz coatings that inhibit quartz cementation.
Smosna and Bruner	Exploration	Pennsylvania, U.S.A.; Appalachian Basin	Devonian	Sandstones (litharenites and sublitharenites)	Petrography	Prediction of reservoir potential of range of depositional facies.
Sombra and Chang	Exploration	Brazil: Santos, Campos, Espiritu Santo, Cumuruxatiba, Reconcavo, Sergipe, Alagoas, and Potiguar basins	Late Jurassic–Tertiary	Sandstones	Petrography; porosity vs. depth	Time Depth Index (TDI) to quantify influence of burial history on porosity evolution.
Tobin	Exploration	Examples from China, Myanmar, Turkey	Triassic; Paleocene–Eocene; Jurassic, respectively	Sandstones, carbonates, respectively	Outcrop	Decision Tree to classify outcrop for risk assessment.
Zempolich and Hardie	Exploration	Venetian Alps, Italy	Middle Jurassic	Carbonate (limestone, dolomite)	Outcrop; petrography; geochemistry	Field mapping of dolomite distribution for information on size and distribution of dolomite bodies and evidence for fluid pathways. Study of progressive textural modification for prediction of reservoir-grade porosity, permeability.

determine the location and optimal number of development wells, as well as to estimate economic production cutoff values, hydrocarbon pore volumes, recoverable reserves, and production rates (Sneider, 1990). By understanding controls on the degree of reservoir heterogeneity and distribution of flow units, a more accurate understanding and predictability of interwell connectivity and fluid-flow pathways can be gained (Tyler et al., 1984; Ebanks, 1990; Kerans et al, 1994; Stoudt and Harris, 1995; Tinker, 1996). Studies at the development scale in this Memoir are provided by Love et al., Smosna and Bruner, and Cavallo and Smosna. Prediction of permeability is addressed by Evans et al., Gluyas and Witton, and Erlich at al. The evolution of permeability during diagenesis is addressed in this Memoir by Zempolich and Hardie, and Mountjoy and Marquez.

At the *Reservoir Management Stage,* predictability of diagenetic patterns that control reservoir quality is used to identify bypassed and uncontacted pay, and in tertiary recovery planning and modification. Identification of bypassed oil and quantification of remaining hydrocarbons is addressed in this volume by Major and Holz.

COMPARISON OF SANDSTONES AND CARBONATES: REASONS FOR SIMILARITIES AND DIFFERENCES IN PREDICTIVE APPROACHES

There are some similarities and many differences between siliciclastics and carbonates, both in their depositional characteristics and in the way in which they respond to physical and chemical conditions during burial and lithification. Clearly, the total of the depositional and diagenetic effects control the final "reservoir-quality" product. In the following discussion, we compare both similarities and differences between sandstones and carbonate rocks under the guise of three headings: depositional controls, diagenetic controls, and resultant pore types.

Depositional Processes and Controls on Reservoir Quality Prediction

In contrast to siliciclastics, the generation and deposition of most carbonates is controlled by biological activity (~90%; Moore, 1989); sand generation and deposition is much less influenced by life. The significance of biological control on carbonate accumulation is that thickness and depositional properties of carbonates can form independently of allochthonous sediment supply. Certain prerequisites must be met for carbonates to form (e.g., temperature, light, salinity, and the availability of nutrients), which will control their geographical location as well as their environments of deposition. As a result, most carbonates are limited to shallow, tropical marine depositional settings. Adding complexity to reservoir quality prediction is that carbonate-producing organisms have evolved through time (e.g., Wilson, 1975; James, 1978).

In contrast, sand is derived mainly from erosion of a parent source and is transported to its site of deposition by physical processes. Physical parameters of sandstones (grain size, sorting, roundness, etc.) are used to understand and predict depositional processes and environments in which they were deposited. Some carbonate depositional environments are also strongly influenced by hydrologic controls, and resulting facies will have similar depositional characteristics to siliciclastic sandstones (e.g., bars, shoals, beaches, dunes, tidal flats, tidal channels, tidal deltas, and basin-margin sediment gravity flow deposits; Scholle et al., 1983, and references therein).

The similarities and differences between carbonate and siliciclastic sedimentology are reflected in similar, yet contrasting, concepts of sequence stratigraphy. The concepts of carbonate sequence stratigraphy are summarized by Sarg (1988), Schlager (1992), and Handford and Loucks (1993) and can be compared to sandstone sequence stratigraphy (e.g., Mitchum, 1977; Mitchum et al., 1977; Vail et al., 1977; Posamentier et al., 1988; Van Wagoner et al., 1988, 1990; among others). Large-scale stratal geometries of siliciclastic sediments (onlap, downlap, toplap, etc.) are also the fundamental geometries of carbonate depositional sequences. The relative volumetric importance of different systems tracts, however, is different for sands vs. carbonates.

Siliciclastics are controlled by physical sediment supply. During relative highstand of sea level, most coarse-grained clastics are "trapped" in fluvial systems and are not deposited in marine settings. During relative lowstands of sea level, coarse-grained sediments are able to bypass the shelf to be deposited in basinal marine settings. Therefore, lowstand systems tracts (LST) generally contain the most volumetrically abundant deposits of coarse-grained siliciclastics in petroleum basins. In contrast, the most significant factor for carbonate deposition is the inundation of shallow carbonate platforms (Sarg, 1988; Schlager, 1992; Handford and Loucks, 1993). As a result, during relative highstands of sea level, carbonates will be able to generate and accumulate the most significant quantities of sediment, varying according to relative rates of sediment production, accumulation, and sea level rise (Sarg, 1988). Therefore, highstand systems tract (HST) deposits are generally the most volumetrically significant for carbonates. During relative sea level lowstands, carbonate deposition is generally geographically and volumetrically restricted and less significant, although allochthonous slope-derived material and autochthonous deposits may be locally important.

The fundamental differences between the way in which carbonates and siliciclastics accumulate and are eroded and redeposited during a highstand–lowstand cycle have a major effect on the evolution of reservoir quality. Typically, sands deposited during highstands will suffer erosion and redeposition down systems tract as sea level falls, but the modification of the sediment is dominantly physical rather than chemical. Highstand carbonate deposits are unlikely to suffer the same fate. Exposure during sea level fall will be dominated by dissolution and reprecipitation rather than physical reworking of sediment. Depending on the climate, time, and magnitude of exposure, karstification, dolomitization, and evaporite precipitation can occur, all of which will result in a profound modification of reservoir quality.

In summary, differences in depositional controls, depositional and sequence stratigraphic settings, and sequence stratigraphic concepts between sandstones and carbonates necessitate that approaches to facies- and reservoir-quality prediction in sandstones vs. carbonates, although fundamentally similar, must also be specific and characteristically different.

Mineralogy, Diagenesis, and Reservoir-Quality Modification

Mineralogy

The second fundamental difference between carbonates and sandstones is mineralogy and the way in which the mineralogy both responds to and, indeed, controls diagenesis. Mineralogy of sandstones, although variable, commonly consists of grains that are chemically stable in the near-surface depositional environment. Although dissolution of feldspars and lithic fragments can be locally important (Heald and Larese, 1973; Milliken et al., 1989; Milliken, 1992; Bloch and Franks, 1993; among others), changes in porosity and permeability are not generally sufficient to significantly improve the overall quality of a reservoir (Bloch, 1994).

Carbonate sediments, in contrast, are composed of a small variety of minerals that are highly susceptible to chemical alteration, recrystallization, and dissolution (e.g., aragonite, Mg-calcite, calcite, and dolomite of varying stoichiometry). The effects of carbonate mineral instability on reservoir quality may be accentuated by the tendency of highstand carbonate systems to be exposed during falling sea level. The water:rock ratio during meteoric flushing and repeated seawater inundation is clearly much larger than that likely to be experienced during burial conditions. Consequently, there is significant potential for diagenetic modification before and throughout burial, often with multiple diagenetic events superimposed, and a continual modification of reservoir quality.

Meteoric Diagenesis

Subaerial exposure, meteoric diagenesis, and subsequent porosity evolution in carbonates have been addressed by Saller et al. (1994) and Budd et al. (1995). Among the most significant factors that determine the magnitude of carbonate porosity redistribution are the following: mineralogy, existing pore networks, depositional facies and stratigraphy, climate, the reactive potential of the groundwater, duration of exposure, hydrologic systems, size and topography of the exposed area, magnitude of base-level change, and tectonic setting. Exposure of carbonates can be manifest in two important diagenetic processes, karstification and meteoric cementation, with significant

redistribution of porosity and permeability taking place from the time of exposure throughout burial.

Studies of modern and ancient carbonate rocks subjected to exposure and meteoric diagenesis have documented the variability of the cementation process and its variable effectiveness. Enos and Sawatsky (1981) documented the high but variable nature of initial porosity of modern carbonate sediments (values ranging from 40% to 78%), and inferred that early diagenetic processes are responsible for the significant loss of preburial porosity (~20% loss in porosity) in analogous facies of nearby Pleistocene rocks. Budd et al. (1993) estimated that precompaction meteoric cements account for 3–37 vol. % in grainstones. However, Halley and Beach (1979) and Scholle and Halley (1985), based on studies of Holocene and Pleistocene sediments of Florida and the Bahamas, have claimed that porosity loss is slight during mineralogical stabilization, and that secondary porosity developed during early cementation preserves the overall magnitude of preburial porosity. These examples highlight the problem of uncertainty in preburial porosity prediction in carbonates.

Meteoric diagenesis in sandstones is a controversial topic. Much of the controversy has focused on the generation of secondary porosity. The complexity of the processes involved precludes any *a priori* assumptions as to the quantitative importance, or even presence, of secondary and enhanced porosity associated with meteoric diagenesis (Bloch, 1994). Furthermore, identification and quantification of secondary porosity often rely on subjective criteria. Even when positive evidence exists, such as partially dissolved grains and/or cements, it may be difficult to prove a meteoric origin for mineral dissolution. Giles and Marshall (1986), in a review of secondary porosity in sandstones, made a plausible case for the involvement of meteoric water dissolution in some settings. More recently, Emery et al. (1990) have furnished strong evidence using a combination of wireline log, core analysis, thin section, isotope geochemical, and seismic acoustic impedance data to highlight meteoric water dissolution of sandstones beneath an unconformity. The possibility that meteoric water can penetrate deep into a basin and still influence the course of diagenesis has been demonstrated from analysis of the oxygen and hydrogen/deuterium isotope ratios in authigenic minerals (Gluyas et al., 1997).

Marine Diagenesis

Active marine cementation, the occlusion of porosity, and the modification of pore types in various modern carbonate marine depositional settings have been documented by many workers (Bathurst, 1975, and references therein). Attesting to its economic importance, the significance of marine cementation in ancient carbonate reefs and buildups has been documented in a vast number of studies (e.g., Playford, 1980; and in books edited by Bebout and Loucks, 1977; Toomey, 1981; Schneidermann and Harris, 1985; Schroeder and Purser, 1986; and Monty et al., 1995; among others). The variability and magnitude of marine diagenetic effects on reservoir quality in carbonates are illustrated by Walls and Burrowes (1985), who documented that 15% to 70% of total porosity in Devonian reefs of Canada has been occluded by marine cement. Kerans et al. (1986) estimated that in Devonian reefs of the Canning Basin, Australia, radiaxial and microcrystalline marine cements each locally comprise 20–50% of the reef by volume.

There is no well-defined division of sandstone diagenesis into marine vs. nonmarine. Admittedly, meteoric water-influenced mineral dissolution has been much investigated because of the potential effect on reservoir quality improvement, as discussed above. However, near-surface precipitation processes can occur in a variety of environments (fluvial, marine, evaporitic, etc.). Carbonates, sulfates, and possibly halite tend to be the most important. These cements, however, rarely completely destroy the pore system in a large sand body. Moreover, because it is common for such cements (particularly carbonate) to form concretions, layers, or irregular masses, the effect on reservoir quality is often best represented as a reduction in the net (petroleum) pay thickness of a reservoir rather than the average effect on porosity (Bjørkum and Walderhaug, 1990). The diagenetic processes controlling these near-surface reactions are relatively well understood, and commonly involve bacterial destruction of organic matter in oxic, suboxic, and anoxic pore waters (Berner, 1980). However, although the process is well understood, methods are as yet unavailable for predicting the volume of syndepositional/early diagenetic cements in sandstones awaiting the drill bit.

Burial Diagenesis

Numerous diagenetic studies have documented that abundant cementation of carbonates occurs in the burial realm, which reduces or occludes any remaining porosity. The use of cathodoluminescence stratigraphy (e.g., Meyers, 1991; among others) has been shown to be an extremely useful tool for identifying and correlating generations of cement. Cathodoluminescence techniques have allowed workers to correlate phases of cementation to geochemical environments (e.g., meteoric, marine, burial) and then to estimate volume of cement precipitated during the various diagenetic phases. Grover and Read (1983) concluded that major, but variable, cementation has occurred under burial conditions in the Middle Ordovician of Virginia (U.S.A.), with 3–45 vol. % of cement during shallow burial (≤3 km) and 50–95% during deep burial. Meyers and Lohmann (1985), in their study of the Mississippian limestones of New Mexico (U.S.A.), estimated that approximately 60% of total cement was related to shallow-burial, marine phreatic processes, while approximately 40% was related to burial deeper than 1 km. Dorobek (1987) estimated that approximately 32% of the total cement in the Silurian–Devonian Helderberg Group of the central Appalachians (U.S.A.), was precipitated during shallow burial, with cementation by deep burial fluids occluding all remaining porosity. Using chemical, isotopic, and petrographic analysis, Prezbindowski (1985) estimated that 14 vol. % cement in the Cretaceous Stuart City reefs of Texas (U.S.A.) was due to marine cementation, 7 vol. % to near-surface, meteoric cementation, and 9 vol. % as the result of burial cementation.

Burial diagenesis and its effects on the quality of petroleum reservoirs is a much-researched topic. The range of minerals that can reduce the quality of a reservoir is large: quartz, carbonate minerals, clays, zeolites, and others (Primmer et al., this volume). The application of quantitative petrographic, geochemical, and isotopic analyses to authigenic minerals during the past decade has allowed scientists to date minerals, determine the temperature of precipitation, and characterize the pore waters from which precipitation occurred (e.g., Emery and Robinson, 1993; Williams et al., 1997). When such data are coupled with analyses of thermal and burial history information, powerful descriptions of diagenetic process have emerged (Glasmann et al., 1989; Kupecz and Land, 1991; Robinson and Gluyas, 1992; Hogg et al., 1993; Walderhaug, 1994). However, some key questions remain unanswered (e.g., there appears to be too little connate water in sediments to redistribute the observed cement volumes in the time available to the process). Essentially, there is insufficient knowledge at present to determine the controls (source/transport/precipitation of solutes) on the diagenetic evolution of sandstones. As for transport itself, there are advocates of lateral fluid flow, advection, and diffusion as the major harbingers of cementing fluids. This paucity of quantitative knowledge means that process-based predictive methodologies are few, and empiricism remains the prime tool for prediction of reservoir quality.

Dolomitization

Dolomitization can occur during essentially synsedimentary replacement or cementation of precursor carbonate and can continue throughout the burial realm. A spectrum of environments have been proposed by many (summarized by Land, 1980, 1982, 1985, 1986; Morrow, 1982, among others). Work in recent years has highlighted the fact that nonstoichiometric dolomites are susceptible to recrystallization (e.g., Kupecz et al., 1993), and that recrystallization is commonly associated with a progressive increase in crystal size (Kupecz and Land, 1994). The significance of dolomitization for reservoir quality is that an increase in crystal size (either during dolomitization of a micrite-dominated precursor or during dolomite recrystallization) and/or the rearrangement of touching pore space is generally associated with increased permeability (Lucia et al., 1995; Zempolich and Hardie, this volume). Because of the complexity of the dolomitization process and the potential for continued dolomite modification, prediction of reservoir quality will have inherent uncertainties.

Variability in Pore Types and Reservoir Quality Prediction

Pore types and their distribution are fundamentally different in sandstones and carbonates (e.g., Choquette and Pray, 1970, their table 1). The dominant primary pore type in sandstones is interparticle, regardless of depositional setting, with the pore diameter and pore-throat size a function of grain size and sorting (e.g., Evans et al., this volume). Cementation by quartz (a solid grain coating) and mechanical compaction will reduce pore and pore-throat dimensions, but the pore types remain essentially the same. The process of compaction or quartz cementation can proceed to low porosity levels without altering the relationship between porosity and permeability. Only when cementation proceeds to the point where pore coordination number declines (i.e., pore throats are being closed off) is there a major change in the poroperm relationship, with permeability falling to very low levels. Typically for a clean quartzose, medium-grained sandstone, porosity can be reduced to ~10% before the poroperm relationship declines. The porosity threshold will be higher for finer grained and more poorly sorted sands. Disruption of the pore network can occur at much higher porosity levels, where a mineral plugs pores randomly or creates "furry" microporous grain coats. Typically, carbonate minerals or clusters of kaolinite platelets plug pores, while chlorite and illite are common as clay coats with much trapped microporosity. Grain dissolution may result in moldic and micromoldic porosity.

Carbonate primary pore types are highly variable, with their shapes and sizes having little relation to energy, grain size, or sorting. Diagenetic modification of carbonate pore types adds additional complexity, with the resulting "ultimate" pore type varying widely (Choquette and Pray, 1970). Pores in carbonate rocks can range in size from <1 μ to caverns >100 m in diameter, and may be juxtaposed within the same rock unit. The complexity of porosity in carbonates is the result of many factors, which include the variable dimensions of sedimentary carbonate particles, the variability of skeletal pores, partial to total occlusion of pores by internal sediment or cement, creation of secondary pores [fabric selective or fabric independent, and of highly variable dimensions (e.g., breccias)], dolomitization, and recrystallization (e.g., Murray, 1960; Choquette and Pray, 1970). Because of the combination of biological and physical depositional processes, and diagenetic overprint of metastable chemical deposits, buried carbonates tend to have a greater heterogeneity of porosity and permeability than do buried sandstones and, as a result, generally have a greater uncertainty in prediction of average porosity.

PRESENT AND EMERGING METHODOLOGIES OF RESERVOIR QUALITY PREDICTION

Current geological approaches to predict porosity and permeability in reservoirs prior to drilling range between theoretical chemical models and purely empirical models (Byrnes, 1994). Regardless of the approach, to be useful from a practical point of view, a predictive technique must meet a number of criteria (Bloch and Helmold, 1995):

1. Sufficient accuracy must be achieved from a limited number of input parameters that can be estimated prior to drilling;
2. Prediction must be possible for a wide range of lithologies occurring in different geologic settings;
3. Permeability should be predicted independently of porosity to reduce the margin of error;
4. Although current understanding of processes responsible for porosity preservation, destruction, and enhancement is limited, the predictive model should at least implicitly account for the most important processes that take place during sediment burial;
5. For production and exploration purposes, the approach should be applicable on the reservoir scale, field scale, and subbasin scale. Basin-scale predictions are adequate for basin modeling, but not for the drilling of specific targets; and
6. The technique should be flexible, so that when it is not adequate by itself, reasonable accuracy can still be achieved by using it with another approach.

Choice of approach depends upon the type of anticipated reservoir rock and the amount of information available. In mature areas where cores and logs provide a calibration data set, the empirical approaches may prove best. This is especially true with field development prediction. In undrilled basins or targets, some aspect of theoretical relationships must be used, because there are no empirical data. In some cases, the uncertainty of the prediction will be large. This uncertainty should be related along with the predictive value so the value of the prediction can be correctly assessed.

Sandstones

Process-Oriented Models

Process-oriented models (or chemical reaction path models) do not meet some of the above criteria (most notably the first criterion). Such models are useful in simulating formation of some cements and diagenetic sequences in simple compositional systems (Bruton, 1985; Harrison, 1989; Harrison and Tempel, 1993), but are not yet capable of quantifying changes in porosity and permeability (Surdam and Crossey, 1987; Schmoker and Gautier, 1988; Meshri, 1989; Harrison and Tempel, 1993). The limitations of these models include the following: (1) uncertainties in thermodynamic and kinetic data used in the reaction path calculations (Surdam and Crossey, 1987; Meshri, 1989; Harrison and Tempel, 1993), (2) inaccuracies in paleohydrologic reconstructions, (3) inability to quantify mass transfer processes and the effect of these processes on reservoir quality (Harrison and Tempel, 1993), and (4) lack of feedback between compactional porosity loss and mineral reactions (Harrison and

Tempel, 1993). Despite their limitations, chemical reaction path models are useful, as they attempt to explain mechanistically what is occurring during porosity evolution and, thus, are helpful in identifying critical issues for further scientific studies of porosity evolution (Waples and Kamata, 1993).

Empirical Models

By contrast, empirical techniques can be a powerful predictive tool, but their effectiveness is to a large extent a function of availability and quality of calibration data sets. Reservoir quality prediction is no exception to the general rule that the fewer the calibration data, the less certain the prediction. The statement of Waples et al. (1992, p. 47), that maturity models "are simply too weak at present to allow us to carry out highly meaningful modeling unless our input is constrained by measured data" is also true of predicting reservoir quality.

In frontier areas, where data are sparse or not available, only comparative analogs can be used. If surface outcrops are available, the approach proposed by Tobin (this volume) can significantly assist in assessing potential subsurface porosity and permeability. Where some subsurface data are available, compaction models (Pittman and Larese, 1991; Gluyas and Cade, this volume), the relationship of porosity vs. vitrinite reflectance (Schmoker and Gauthier, 1988; Schmoker and Hester, 1990), or the predictive model of Scherer (1987) can be utilized for sandstones. If the prospective reservoir is expected to be quartz rich (quartz arenite, subarkose, sublitharenite) the "Exemplar" model (Lander et al., 1995) can be an effective tool for predrill porosity evaluation (Lander and Walderhaug, 1997). "Exemplar" is based on empirically calculated precipitation rates of quartz cement in quartz-rich sandstones (Walderhaug, 1994) ranging in age from Ordovician to Plio–Pleistocene (Lander et al., 1995). Significant progress in predicting quartz cementation rates with a minimum of basin-specific information has been recently reported by Bjørkum et al. (in press). Each of these approaches has its limitations and strengths and cannot be used indiscriminantly. The applicability of some of these models to reservoir-quality assessment in frontier basins was discussed by Bloch and Helmold (1995).

In mature basins, where calibration data sets are often available, cement presence in the calibration samples is the determining factor in choosing the predictive approach (Bloch and Helmold, 1995; Primmer et al., this volume). Weakly cemented sandstones display "global" trends in reservoir quality, as first published by Scherer (1987). If cement in all or most of the samples does not exceed 5–10%, multiple regression analysis can an effective predictive tool (Scherer, 1987; Bloch, 1991; Byrnes and Wilson, 1991). In uncemented or weakly cemented quartz-rich sandstones, the relationship between porosity and effective stress derived by Gluyas and Cade (this volume) can be very useful. Significant progress in prediction of reservoir quality of quartz-poor sandstones was made by Wilson and Byrnes (1988). Wilson and Byrnes generated a series of

proprietary linear regression functions for the prediction of porosity, permeability, and irreducible water saturation in lithic sandstones. The functions were based on a petrophysical and petrographic study of >500 samples representing a diverse suite of ductile- and volcanic-rich sandstones from various U.S. basins. Samples ranged in depth from 550 to 6460 m (1800 to 21,200 ft) and in age from Early Cretaceous through Miocene. The porosity function was able to predict porosity within a standard deviation of 1.9–2.2%.

Sandstones containing significant amounts of cements appear to have predictable diagenetic styles (Primmer et al., this volume). In such sandstones, several scenarios exist for porosity prediction. In many quartzose sandstones, quartz cementation is related to depth or burial history [e.g., Middle Cambrian sandstones of the peri-Baltic area (Brangulis, 1985); Mississippian Kekiktuk sandstone of the North Slope of Alaska (Bloch et al., 1990); Middle Jurassic sandstones of the North Sea and Haltenbanken area offshore Norway (Bjørlykke et al., 1986, 1992; Bloch et al., 1986; Ehrenberg, 1990; Giles et al., 1992; Ramm, 1992; Wilson, 1994]. Although many pay- and basin-specific predictive relationships have been developed for quartzose sandstones, at this time only Exemplar appears to provide a more general predictive tool (Lander and Walderhaug, 1997).

Where cementation is not directly related to burial history, a satisfactory predictive model for samples with a wide range of cement content can be obtained by grouping the data into two or more subsets and developing a predictive model for each subset (Bloch and Helmold, 1995). If controls on the distribution of cement cannot be quantified, a qualitative (high-low) assessment is usually possible. Even in rocks with a complex diagenetic history, reservoir quality is frequently related to simple parameters, such as grain size (for a given provenance and burial history). For example, in the Norphlet Formation, stylolitization (not just intergranular pressure dissolution) and quartz cementation have been shown to be affected by grain size (Thomas et al., 1993). As noted by Taylor and Soule (1993, p. 1554) for the North Bellridge field (California), "despite the important effects of diagenesis, reservoir quality is still a function of the change in grain size associated with depositional processes." Usually the relationship of grain size and permeability is not expressed as a simple correlation. Rather, in many reservoirs, sandstones coarser than a certain grain size are characterized by permeabilities exceeding a cutoff value (Bloch and McGowen, 1994). This relationship allows assessment of reservoir quality based on a facies model, assuming a depositional facies control of sand texture.

Future Trends

Although significant progress in reservoir quality prediction has been made in the last decade, there is clearly a need for methodologies that are both more general ("global") and more accurate. The emphasis of effect-oriented/empirical modeling will be on expert systems, hybrid process-effect approaches, nonlinear multivariate regression analyses, possibility analysis, and neural networks (Wood and Byrnes, 1994).

Future activities in process-oriented/geochemical modeling will be focused on: (1) code development (recoding that makes programs "more user-friendly, more transportable between various operating systems, and better suited to a modern coding environment"), (2) improvement of mass transfer algorithms, and (3) development of a universal and robust, easily updatable database for minerals and aqueous species (Wood and Byrnes, 1994, p. 395). Most importantly, the quantitative effects of subsurface rock-fluid interaction on porosity/permeability and the significance of local vs. allochthonous cement sources need to be better understood.

Carbonates

In spite of the complexities of carbonate systems, advances in our ability to predict reservoir quality in advance of drilling have been made. Current successes, because of the complexities discussed above, have been with empirical approaches and three-dimensional reservoir models.

Process-Oriented Models

Process-oriented studies and models in carbonates are very useful in our understanding of the mechanisms and complexities of aragonite, calcite, and dolomite precipitation and dissolution, and their interaction with various diagenetic fluids. Back and Hanshaw (1971), Kharaka and Barnes (1973), Berner (1975), Parkhurst et al. (1980), Matthews and Froelich (1987), Banner and Hanson (1990), Dewers and Ortoleva (1990, 1994), Dreybrodt (1990), Quinn and Matthews (1990), and Kaufman (1994) have studied various aspects of process-oriented modeling of carbonates and diagenetic fluids. Most of the models calculate geochemical parameters of the water and rock during reactions, without directly addressing changes in porosity and its distribution. Although these models provide vast amounts of information and have furthered our understanding of carbonate diagenesis, because of the complexity of the chemical systems and because diagenetic environments change during progressive burial of carbonates, none of these models can effectively simulate reservoir quality evolution of shelf limestones or dolomites.

Empirical Models

Empirical techniques have been shown to be a powerful tool for the prediction of reservoir quality in carbonates. Different approaches must be used depending on the amount of subsurface data and whether outcrop analogs are present. In frontier areas, where analogous outcrops are present, the methods of Tobin (this volume), as discussed in the sandstone section, offer a viable technique to predict reservoir quality. Tobin uses examples from both sandstone and carbonate outcrops.

In mature areas with extensive data sets, even given the potential for variability in preburial porosity,

empirical studies clearly document the decrease in porosity of carbonates with burial depth (Scholle, 1977, 1978, 1981; Schmoker and Halley, 1982; Halley and Schmoker, 1983; Schmoker and Hester, 1983; Schmoker, 1984; Schmoker et al., 1985; Amthor et al., 1994; Brown, this volume). These empirical studies can be subdivided into two main groups: those of pelagic limestones composed of low-Mg calcite; and limestones and dolomites interpreted to have been deposited in shallow marine depositional environments. The subdivision, as acknowledged by researchers (e.g., Scholle, 1981), is mainly for reasons of depositional complexity and diagenetic potential. Data from the low-Mg calcite pelagic limestones (Scholle, 1977, 1978, 1981) have simpler diagenetic histories and, as a result, have significantly less scatter in the data than in shallow marine counterparts. Pelagic carbonates are relatively stable, with no significant preburial porosity modification, and more predictable facies trends. The result is that changes in porosity in pelagic carbonates are most affected by mechanical and chemical compaction during burial (Scholle, 1977, 1978, 1981). Prediction of porosity requires the understanding of the maximum burial depth and the pore-water chemistry (Scholle, 1977).

Scatter in the data from shallow marine carbonates is interpreted as being due to early diagenetic variations in preburial porosity (Halley and Schmoker, 1983; Schmoker, 1984; Schmoker et al., 1985), which suggests that specific predictions of reservoir porosity may not be possible. These studies show that porosity is related to burial pressure, temperature and time, and lithology (limestone, dolomite, and shale content). Depositional fabrics (e.g., mudstone, wackestone, packstone, grainstone) do not display significant differences in average porosity, even though they do differ in the range in porosity values (Brown, this volume).

A different approach is presented by Love et al. (this volume), using statistical methods in data-intensive areas to allow the predrill prediction of reservoir quality. The authors analyze detailed geological data with a neural network predictive technique.

Additional examples of empirical predictions of carbonate reservoir quality are provided by integrated studies using a combination of stratigraphy, structural geology, petrophysics, seismic reflection data, production data, and numerical methods. The predictions were verified as successes or nonsuccesses by subsequent drilling (Maureau and van Wijhe, 1979; Serna, 1984; Beliveau and Payne, 1991). The strength of these studies is in the analysis of successes and failures.

Studies integrating geological and petrophysical data have proven very useful for reservoir characterization and detailed infill drilling. By integrating detailed analyses of depositional facies, facies tracts, sequence stratigraphy (especially at the parasequence level), diagenesis, pore types, porosity, permeability, capillary pressure, and saturation data, workers have been able to predict reservoir quality, reservoir performance, and bypassed pay. Studies include those by Aufricht and Koepf, (1957), Keith and Pittman (1983),

Bebout et al. (1987), Lucia and Conti (1987), Alger et al. (1989), Lucia et al. (1992a, b), Lucia (1993, 1995), Kerans et al. (1994), Martin et al. (1997), and Major and Holtz (this volume), among others. Incorporation of data into three-dimensional visualization models allows for reservoir quality prediction based on empirical correlations. Excellent examples of this methodology are presented Eisenberg et al. (1994), Kerans et al. (1994), Lucia et al. (1995), Tinker and Mruk (1995), Weber et al. (1995), and Tinker (1996).

Future Trends

Because of the complexity of carbonates (their extensive postdepositional modification, pore types, and reservoir-quality distribution), empirical predictions appear to be the only feasible way to realistically predict predrill reservoir quality. Future studies of predrill reservoir-quality prediction in carbonates are expected to continue to focus on the integration of detailed studies of subsurface cores and/or outcrop analog facies, detailed analysis of diagenesis, petrophysical analyses (particularly pore and pore-throat-type distribution, saturation, and capillary pressure data), production data, fluid-flow modeling, and reservoir simulation. By using three-dimensional modeling, all detailed variables can be mapped prior to drilling. As mentioned above, examples of this methodology are presented by Eisenberg et al. (1994), Kerans et al. (1994), Lucia et al. (1995), Tinker and Mruk (1995), Weber et al. (1995), and Tinker (1996). However, future studies must also include substantiation by subsequent drilling, and discussions of successes and failures of reservoir quality prediction.

OVERVIEW OF MEMOIR

The Memoir consists of 17 chapters emphasizing either reservoir-quality prediction techniques or exploration and exploitation case studies. Because of the diversity of papers, Table 1 is provided to help the reader gain an overview of the individual papers, including information on location, reservoir age, reservoir mineralogy, stage in the exploration cycle, tools used, and techniques used.

We have subdivided the chapters into two groups, those that address approaches to reservoir quality prediction and those that represent specific case studies. As a result, the chapters are not strictly subdivided by "sandstone" and "carbonate" examples. We hope that this approach serves to "cross-pollinate" ideas among workers in the field.

Approaches to Reservoir Quality Prediction

Tobin

Tobin shows how data obtained from sandstone and carbonate outcrop exposures can be used to evaluate subsurface porosity and permeability in potential reservoirs. His approach, based on a systematic decision-tree analysis, can be very useful in exploration risk assessment, particularly in frontier basins with limited or no subsurface information. Case studies from China, Myanmar, and Turkey illustrate the proposed procedure.

Gluyas and Cade

Gluyas and Cade present a new equation for compactional porosity reduction as a function of depth for uncemented, clean, ductile-grain-poor sandstones under hydrostatic pressure. The equation is based on field and experimental data. A modification of the equation relates porosity to effective stress, rather than to depth, and thus can be used to predict porosity in overpressured sands in which overpressure is relatively "early." This technique provides a convenient way to predict porosity in uncemented sands or to provide an upper limit on porosity in sandstones expected to contain authigenic cements. This technique, tested against a global data set, has an accuracy of ±2.5 porosity units at 95% confidence limits.

Brown

Brown addresses the influence of carbonate mineralogy, fabric, and shale content on the rate of porosity loss with burial. Because of the availability of modern well log suites, the Mississippian of the U.S. Williston Basin is used as a study area. Porosity data obtained at consistent intervals [10 ft (3 m)] help eliminate sampling bias, thus allowing an understanding of basin-scale porosity-loss mechanisms. Brown concludes that porosity is selectively preserved in dolomites (vs. limestones) at similar burial conditions, and that porosity decreases with increasing temperature. Cementation is a more important factor in loss of carbonate porosity than is mechanical compaction.

Love, Strohmenger, Woronow, and Rockenbauch

Love et al. present a statistical approach to the predrill prediction of reservoir quality. The authors stress that this methodology can be applied to both carbonate and siliciclastic reservoirs, and illustrate their techniques with a study of the Permian Zechstein carbonates of the Southern Zechstein Basin of northern Germany. A three-dimensional distribution of reservoir attributes is obtained by integrating geological data (facies, mineralogy, porosity, permeability, well logs, geochemistry) for 287 wells and applying a statistical analysis of these data. Because of the complexity of the spatial distribution of porosity and permeability, a neural network predictive technique is proven to be more effective than linear regression.

Primmer, Cade, Evans, Gluyas, Hopkins, Oxtoby, Smalley, Warren, and Warden

Based on an analysis of a "global" data set, Primmer et al. conclude that chemical diagenesis impacts sandstones through five predictable diagenetic "styles": (1) quartz, commonly with lesser amounts of diagenetic clays, and late ferroan carbonate; (2) clay minerals (illite or kaolinite) with lesser amounts of quartz (or zeolite) and late carbonate; (3) early grain-coating clays that may inhibit quartz cementation during deeper burial; (4) early evaporite or carbonate cements, and (5) zeolites, often in association with chlorite and/or smectite and late nonferroan carbonates.

The chemical diagenetic styles are a function of detrital mineralogy, depositional environments, and burial histories. Once the chemical diagenetic style is predicted, a "most likely" value of cement abundance can be estimated. This value is then subtracted from porosity values obtained from compaction curves or equations (e.g., Gluyas and Cade, see above).

Sombra and Chang

Sombra and Chang emphasize the correlation between a parameter they term "the time-depth index" (TDI) and porosity. The TDI-porosity relationship for three lithological types of reservoirs was established for Upper Jurassic to Tertiary sandstones of the Brazilian continental margin. Their approach involves (1) integration of the area enclosed between the time-depth axes and the burial history curve of a sandstone body (TDI) and (2) correlation of the integrated "TDI" with the porosity of the corresponding sandstone. The porosity of a lithologically similar sandstone can then be predicted prior to drilling if information on its burial history TDI is available. This technique can be useful when vitrinite reflectance data are not available to calibrate the vitrinite reflectance-porosity relationship in formations in which such relationship exists.

Evans, Cade, and Bryant

Evans et al. discuss permeability prediction based on a combination of empirical and modeling techniques. This approach can be used in both frontier and data-rich areas. The main difficulty in applying it is posed by the limitations in predicting variations in geologic factors that are used to predict permeability. Evans et al. demonstrate that, provided the input data are accurate, the permeability modeling technique commonly is able to predict permeability to within half an order of magnitude.

Ehrlich, Bowers, Riggert, and Prince

Ehrlich et al. apply petrographic image analysis to detailed porosity analysis to equate porosity elements to variations in permeability. This approach can be used to predict the highest permeability possible in a reservoir as a function of depth or basin location for a particular fabric. The concept is applied to investigations of Miocene sandstones of the Satun Field in the Pattani basin (Gulf of Thailand) and Upper Carboniferous sandstones from the Cherokee basin (Oklahoma).

Cabrera-Garzón, Arestad, Dagdelen, and Davis

Seismic reflection data from the Devonian Nisku dolomites of Joffrey Field, Western Canada Sedimentary Basin, were used by Cabrera-Garzón et al. for reservoir quality prediction. Geostatistical simulation of porosity distribution within the field was obtained through the analysis of P- and S-wave travel times from multicomponent (3D, 3C) seismic reflection data, integrated with porosity, permeability, and petrographic information from cores. Correlation of porosity and Vp/Vs allows prediction of the three-dimensional distribution of porosity.

Zempolich and Hardie

Using the Jurassic of the Venetian Alps of Italy as their study area, Zempolich and Hardie utilize detailed field relationships, supplemented with geochemistry, to better understand and predict the geometries, distribution, timing and mechanism of formation of potential dolomite reservoirs. They further use petrography to constrain the evolution of reservoir-quality dolomites. The authors conclude that reservoir-grade porosity is initiated by the replacement of limestone by dolomite, but that reservoir-grade permeability is created later, through the progressive recrystallization of the replacement dolomite.

Case Studies

Gluyas and Witton

The diagenetic sequence encountered in Miocene sandstones by a wildcat well in the southern Red Sea was nearly identical to that predicted prior to drilling. However, predrill assessment of the abundance of authigenic cements was too conservative. Early halite, although expected, formed a "killer" cement that plugged the entire porosity in the target sandstone. This work shows that with minimal data, reasonably accurate diagenetic predictions can be made.

Ramm, Forsberg, and Jahren

High porosity (>20%) in deeply buried (>4000 m) Upper Jurassic sandstones of the Norwegian Central Graben is interpreted to have been preserved by microquartz coats. These coats inhibit precipitation of pore-filling syntaxial quartz overgrowths during deeper burial. Microquartz appears to occur within isochronous layers and has most likely been sourced by syndepositional volcanic glass or sponge spicules.

Gluyas

Unlike Ramm et al., Gluyas attributes differences in porosity in Upper Jurassic sandstones of the Norwegian Central Graben to the competition of quartz cementation and oil emplacement ("race for space"). High porosity at deep burial depths is interpreted to be the result of retardation of quartz cementation by petroleum emplacement rather than by the presence of microquartz coats. This philosophy was used to predict the porosity of the reservoir in a prospect a few kilometers from existing data. Three porosity models were constructed to represent cases of cementation before, during, and after oil emplacement. The most likely outcome was predicted to be synchronous cementation and oil emplacement; thus, the porosity was estimated accordingly. Once drilled, the prospect was found not to contain oil but water; however, the core porosity of the sand was identical to that for the model in which cementation predated oil emplacement. Perhaps the oil will arrive shortly!

Cavallo and Smosna

Cavallo and Smosna present a case study of a reservoir at the development stage, the Mississippian Greenbrier Limestone of the U.S. Appalachian Basin, West Virginia. This study integrates Formation Microscanner (FMS) logs into an analysis and drilling program of an ooid shoal complex. By calibrating facies characteristics with the log response and integrating dip information from the logs, the authors illustrate reservoir quality prediction at the development scale.

Major and Holtz

Reservoir quality prediction at the development and reservoir management stages is presented by Major and Holtz. This study of the Permian San Andres Formation, West Texas (U.S.A.) Permian Basin illustrates the importance of reservoir quality prediction in a mature basin. Major and Holtz determine that flow units are controlled by a combination of depositional facies and subsequent diagenetic alteration, and are able to quantify the amount of bypassed oil in both low-permeability and heterogeneous flow units.

Mountjoy and Marquez

Detailed petrographic studies of the Devonian Leduc Formation of the Western Canada Sedimentary Basin are presented by Mountjoy and Marquez. Reservoir character of the dolomites is complex and can be observed at different scales. The distribution of pore types is controlled by original depositional facies, whereas the distribution of permeability is more a function of diagenetic processes, especially dolomitization. Mountjoy and Marquez compare dolomites and limestones at variable burial depths, and illustrate that dolomites have higher porosity and permeability than limestones at similar depths, because the dolomites are more resistant to pressure solution.

Smosna and Bruner

The content of shale and phyllite rock fragments in the Devonian Lock Haven Formation of the Appalachian Basin (U.S.A.) is controlled by depositional environments. The best reservoir quality occurs in depositional facies characterized by an intermediate labile grain content (distributary mouth bar and shelf). In those sandstones, secondary (lithmoldic) porosity enhances primary porosity. By contrast, sandstones with a low content of lithic grains (barrier island) have low lithmoldic and total porosity. Porosity in sandstones with a high abundance of lithic rock fragments (fluvial) was lost early due to compaction, thus preventing subsequent generation of lithmoldic porosity.

ACKNOWLEDGMENTS

We would like to extend our sincere thanks to the following individuals who dedicated their time and effort, and shared their expertise, toward improving the quality of the manuscripts in this Memoir: John Aggatt (Lincolnshire, England), John Bell (Bogota, Colombia), Mike Bowman (London, England), Andrew Brayshaw (Anchorage, Alaska, U.S.A.), Sean Brennan (Lawrence, Kansas, U.S.A.), Alton Brown (Plano, Texas, U.S.A.), Steve Bryant (Milan, Italy), Charles Curtis (Manchester, England), Martin Emery (Dallas, Texas, U.S.A.), Paul Enos (Lawrence, Kansas, U.S.A.), Laura Foulk (Denver, Colorado, U.S.A.), Steven Franks (Plano, Texas, U.S.A.), Mitch Harris (La Habra, California, U.S.A.), Richard Heaton (Edinburgh, Scotland), Andrew Horbury (London, England), Neil Hurley (Denver, Colorado, U.S.A.), Kerry Inman (Houston, Texas, U.S.A.), Nev Jones (Caracas, Venezuela), Marek Kacewicz (Plano, Texas, U.S.A.), Rob Kendall (Houston, Texas, U.S.A.), Andy Leonard (Aberdeen, Scotland), Bob Loucks (Plano, Texas, U.S.A.), Jerry Lucia (Austin, Texas, U.S.A.), Rick Major (Austin, Texas, U.S.A.), Jim Markello (Dallas, Texas, U.S.A.), Pascual Marquez (Maturin, Venezuela), Malcolm McClure (London, England), Mark Osborne (Durham, England), Jackie Platt (London, England), David Roberts (London, England), Jim Schmoker (Denver, Colorado, U.S.A.), Per Svela (Stavanger, Norway), Dick Swarbrick (Durham, England), Pete Turner (Birmingham, England), and Bill Zempolich (Dallas, Texas, U.S.A.). The photomicrographs on the dust cover were taken by Mark Hopkins (London, England). Comments by Alton Brown, Dick Larese, Mike Wilson, and Neil Hurley improved the introduction to the Memoir. We also acknowledge the diligent work of the AAPG editorial staff, including Kevin Biddle, Neil Hurley, Ken Wolgemuth, and Anne Thomas.

REFERENCES

Alger, R.P., D.L. Luffel, and R.B. Truman, 1989, New unified method of integrating core capillary pressure data with well logs: Society of Petroleum Formation Evaluation, v. 4, p. 145–152.

Amthor, J.E., E.W. Mountjoy, and H.G. Machel, 1994, Regional-scale porosity and permeability variations in Upper Devonian Leduc buildups: implications for reservoir development and prediction in carbonates: AAPG Bulletin, v. 78, p. 1541–1559.

Aufricht, W.R., and E.H. Koepf, 1957, The interpretation of capillary pressure data from carbonate reservoirs: Transactions of the American Institute of Mining, Metallurgical, and Petroleum Engineers, v. 210, p. 402–405.

Back, W., and B.B. Hanshaw, 1971, Rates of physical and chemical processes in a carbonate aquifer: Advances in Chemistry, v. 106, p. 77–93.

Banner, J.L., and G.N. Hanson, 1990, Calculation of simultaneous isotopic and trace element variations during water-rock interaction with applications to carbonate diagenesis: Geochimica et Cosmochimica Acta, v. 54, p. 3123–3137.

Bathurst, R.G.C., 1975, Carbonate sediments and their diagenesis: Developments in Sedimentology 12: New York, Elsevier, 658 p.

Bebout, D.G., and R.G. Loucks, eds., 1977, Cretaceous carbonates of Texas and Mexico, applications to subsurface exploration: University of Texas Bureau of Economic Geology Report of Investigations 89, 332 p.

Bebout, D.G., F.J. Lucia, C.F. Hocott, G.E. Fogg, and G.W. Vander Stoep, 1987, Characterization of the Grayburg reservoir, University Lands Dune field, Crane County, Texas: University of Texas at Austin Bureau of Economic Geology Report of Investigations 168, 98 p.

Beliveau, D., and D.A. Payne, 1991, Analysis of waterflood response of a naturally fractured reservoir: Society of Petroleum Engineers 22946, p. 603–613.

Berner, R.A., 1975, Diagenetic models of dissolved species in the interstitial waters of compacting sediments: American Journal of Science, v. 275, p. 88–96.

Berner, R.A., 1980, Early diagenesis: a theoretical approach: Princeton, New Jersey, Princeton University Press, 241 p.

Bjørkum, A.A., and O. Walderhaug, 1990, Geometrical arrangement of calcite cementation within shallow marine sandstones: Earth Science Reviews, v. 29, p. 145–161.

Bjørkum, P.A., E.H. Oelkers, P.N. Nadeau, O. Walderhaug, and W.M. Murphy, in press, Porosity prediction in quartz-rich sandstones as a function of time, temperature, depth, stylolite frequency, and the presence of hydrocarbons: AAPG Bulletin, May, 1988.

Bjørlykke, K., P. Aaagard, H. Dypvik, D.S. Hastings, and A.S. Harper, 1986, Diagenesis and reservoir properties of Jurassic sandstones from the Haltenbanken area, offshore mid-Norway, *in* A.M. Spencer, ed., Habitat of hydrocarbons on the Norwegian continental shelf: Norwegian Petroleum Society, p. 275–286.

Bjørlykke, K., T. Nedkvitne, M. Ramm, and G.C. Saigal, 1992, Diagenetic processes in the Brent Group (Middle Jurassic) reservoirs of the North Sea: an overview, *in* A.C. Morton, R.S. Haszeldine, M.R. Giles, and S. Brown, eds., Geology of the Brent Group: Geological Society Special Publication 61, p. 263–287.

Bloch, S., 1991, Empirical prediction of porosity and permeability in sandstones: AAPG Bulletin, v. 75, p. 1145–1160.

Bloch, S., 1994, Secondary porosity in sandstones: significance, origin, relationship to subaerial unconformities, and effect on predrill reservoir quality prediction, *in* M.D. Wilson, ed., Reservoir quality assessment and prediction in clastic rocks: SEPM Short Course 30, p. 137–159.

Bloch, S., and S.G. Franks, 1993, Preservation of shallow plagioclase dissolution porosity during burial and aluminum mass balance: AAPG Bulletin, v. 77, p. 1488–1501.

Bloch, S., and K.P. Helmold, 1995, Approaches to predicting reservoir quality in sandstones: AAPG Bulletin, v. 79, p. 97–115.

Bloch, S., and J.H. McGowen, 1994, Influence of depositional environment on reservoir quality prediction, *in* M.D. Wilson, ed., Reservoir quality assessment and prediction in clastic rocks: SEPM Short Course 30, p. 41–57.

Bloch, S., J.H. McGowen, J.R. Duncan, and D.W. Brizzolara, 1990, Porosity prediction, prior to drilling, in sandstones of the Kekiktuk Formation (Mississippian), North Slope of Alaska: AAPG Bulletin, v. 74, p. 1371–1385.

Bloch, S., R.K. Suchecki, J.R. Duncan, and K. Bjørlykke, 1986, Porosity prediction in quartz-rich sandstones: Middle Jurassic, Haltenbanken area, offshore central Norway (abs.): AAPG Bulletin, v. 70, p. 567.

Brangulis, A.P., 1985, Vend i kembriy Latvii: stratigrafiya, litologiya i kollektorskiye svoystva (The Vendian and Cambrian of Latvia: stratigraphy, lithology, and reservoir quality) (in Russian): Riga, Department of Natural Gas of the USSR, 134 p.

Brown, A., this volume, Porosity variation in carbonates as a function of depth: Mississippian Madison Group, Williston Basin, *in* J.A. Kupecz, J. Gluyas, and S. Bloch, eds., Reservoir quality prediction in sandstones and carbonates: AAPG Memoir 69, p. 29–46.

Bruton, C.J., 1985, Predicting mineral dissolution and cementation during burial: synthetic diagenetic sequences (abs.): SEPM Gulf Coast Section Program With Abstracts, v. 6, p. 2–3.

Budd, D.A., U. Hammes, and H.L. Vacher, 1993, Calcite cementation in the upper Floridan aquifer: a modern example for confined-aquifer cementation models?: Geology, v. 21, p. 33–36.

Budd, D.A., A.H. Saller, and P.M. Harris, eds., 1995, Unconformities and porosity in carbonate strata: AAPG Memoir 63, 313 p.

Byrnes, A.P., 1994, Empirical methods of reservoir quality prediction, *in* M.D. Wilson, ed., Reservoir quality assessment and prediction in clastic rocks: SEPM Short Course 30, p. 9–21.

Byrnes, A.P., and M.D. Wilson, 1991, Aspects of porosity prediction using multivariate linear regression (abs.): AAPG Bulletin, v. 75, p. 548.

Choquette, P.W., and L.C. Pray, 1970, Geologic nomenclature and classification of porosity in sedimentary carbonates: AAPG Bulletin, v. 54, p. 207–250.

Dewers, T., and P. Ortoleva, 1990, Interaction of reaction, mass transport, and rock deformation during diagenesis: mathematical modeling of intergranular pressure solution, stylolites, and differential compaction/cementation, *in* I.D. Meshri and P.J. Ortoleva, eds., Prediction of reservoir quality through chemical modeling: AAPG Memoir 49, p. 147–160.

Dewers, T., and P. Ortoleva, 1994, Formation of stylolites, marl/limestone alternations, and dissolution (clay) seams by unstable chemical compaction of argillaceous carbonates, *in* K.H. Wolf and G.V. Chilingarian, eds., Diagenesis IV: Elsevier, New York, Developments in Sedimentology 51, 155–216.

Dorobek, S.L., 1987, Petrography, geochemistry, and origin of burial diagenetic facies, Siluro–Devonian Helderberg Group (carbonate rocks), Central Appalachians: AAPG Bulletin, v. 71, p. 492–514.

Dreybrodt, W., 1990, The role of dissolution kinetics in the development of karst aquifers in limestone: a model simulation of karst evolution: Journal of Geology, v. 98, p. 639–655.

Ebanks, W.J., 1990, Geology of the San Andres reservoir, Mallet lease, Slaughter field, Hockley County, Texas: implications for reservoir engineering projects, *in* D.G. Bebout and P.M. Harris, eds., Geologic and engineering approaches in evaluation of San

Andres/Grayburg hydrocarbon reservoirs—Permian Basin: University of Texas Bureau of Economic Geology Publication, p. 75–85.

Ehrenberg, S.N., 1990, Relationship between diagenesis and reservoir quality in sandstones of the Garn Formation, Haltenbanken, mid-Norwegian continental shelf: AAPG Bulletin, v. 74, p. 1538–1558.

Ehrlich, R., et al., this volume, Detecting permeability gradients in sandstone complexes—quantifying the effect of diagenesis on fabric, in J.A. Kupecz, J. Gluyas, and S. Bloch, eds., Reservoir quality prediction in sandstones and carbonates: AAPG Memoir 69, p. 103–114.

Eisenberg, R.A., P.M. Harris, C.W. Grant, D.J. Goggin, and F.J. Conner, 1994, Modeling reservoir heterogeneity within outer ramp carbonate facies using an outcrop analog, San Andres Formation of the Permian Basin: AAPG Bulletin, v. 78, p. 1337–1359.

Emery, D., K.J. Myers, and R. Young, 1990, Ancient subaerial exposure and freshwater leaching in sandstones: Geology 18, p. 1178–1181

Emery, D., and A.G. Robinson, eds., 1993, Inorganic geochemistry: applications to petroleum geology: London, Blackwell Scientific Publications, 254 p.

Enos, P., and L.H. Sawatsky, 1981, Pore networks in Holocene carbonate sediments: Journal of Sedimentary Petrology, v. 51, p. 961–985.

Evans, J., C. Cade, and S. Bryant, this volume, A geological approach to permeability prediction in clastic reservoirs, in J.A. Kupecz, J. Gluyas, and S. Bloch, eds., Reservoir quality prediction in sandstones and carbonates: AAPG Memoir 69, p. 91–102.

Giles, M.R., and J.D. Marshall, 1986, Constraints on the development of secondary porosity in the subsurface: re-evaluation of process: Marine and Petroleum Geology 7, p. 378–397.

Giles, M.R., S. Stevenson, S.V. Martin, S.J.C. Cannon, P.J. Hamilton, J.D. Marshall, and G.M. Samways, 1992, The reservoir properties and diagenesis of the Brent Group: a regional perspective, in AC. Morton, R.S. Haszeldine, M.R. Giles, and S. Brown, eds., Geology of the Brent Group: Geological Society Special Publication 61, p. 289–327.

Glasmann, J.R., R.A. Clark, S. Larter, N.A. Briedis, and P.D. Lundegard, 1989, Diagenesis and hydrocarbon accumulation, Brent Sandstone (Jurassic), Bergen area, North Sea: AAPG Bulletin, v. 73, p. 1341–1360.

Gluyas, J.G., this volume, Poroperm prediction for reserves growth exploration: Ula Trend, Norwegian North Sea, in J.A. Kupecz, J. Gluyas, and S. Bloch, eds., Reservoir quality prediction in sandstones and carbonates: AAPG Memoir 69, p. 201–210.

Gluyas, J., and C.A. Cade, this volume, Prediction of porosity in compacted sands, in J.A. Kupecz, J. Gluyas, and S. Bloch, eds., Reservoir quality prediction in sandstones and carbonates: AAPG Memoir 69, p. 19–28.

Gluyas, J.G., and T. Witton, this volume, Poroperm prediction for wildcat exploration prospects: Miocene Epoch, Southern Red Sea, in J.A. Kupecz, J. Gluyas, and S. Bloch, eds., Reservoir quality prediction in sandstones and carbonates: AAPG Memoir 69, p. 163–176.

Gluyas, J.G., A.G. Robinson, and T.P. Primmer, 1997, Rotliegend sandstone diagenesis: a tale of two waters, in J. Hendry, P. Carey, J. Parnell, A. Ruffel, and R. Worden, eds., Geofluids II 1997: Belfast, The Queen's University of Belfast, p. 291–294.

Grover, G., Jr., and F.J. Read, 1983, Paleoaquifer and deep burial related cements defined by regional cathodoluminescent patterns, Middle Ordovician carbonates, Virginia: AAPG Bulletin, v. 67, p. 1275–1303.

Halley, R.B., and D.K. Beach, 1979, Porosity preservation and early freshwater diagenesis of marine carbonate sands (abs.): AAPG Bulletin, v. 63, p. 460.

Halley, R.B., and J.W. Schmoker, 1983, High-porosity Cenozoic carbonate rocks of South Florida: progressive loss of porosity with depth: AAPG Bulletin, v. 67, p. 191–200.

Handford, C.R., and R.G. Loucks, 1993, Carbonate depositional sequences and systems tracts—responses of carbonate platforms to relative sea level changes, in R.G. Loucks and J.F. Sarg, eds., Carbonate sequence stratigraphy: recent developments and applications: AAPG Memoir 57, p. 1–41.

Harrison, W.J., 1989, Modeling fluid/rock interactions in sedimentary basins, in T. A. Cross, ed., Quantitative dynamic stratigraphy: New York, Prentice Hall, p. 195–231.

Harrison, W.J., and R.N. Tempel, 1993, Diagenetic pathways in sedimentary basins, in A.D. Horbury and A.G. Robinson, eds., Diagenesis and basin development: AAPG Studies in Geology 36, p. 69–86.

Heald, M.T., and R.E. Larese, 1973, The significance of the solution of feldspar in porosity development: Journal of Sedimentary Petrology, v. 43, p. 458–460.

Hogg, A.J.C., P.J. Hamilton, and R.M. Macintyre, 1993, Mapping diagenetic fluid flow within a reservoir: K-Ar dating in the Alwyn area (UK North Sea): Marine and Petroleum Geology 10, p. 279–294.

James, N.P., 1978, Facies models: reefs: Geoscience Canada, v. 5, p. 16–26.

Kaufman, J., 1994, Numerical models of fluid flow in carbonate platforms: implications for dolomitization: Journal of Sedimentary Research, v. A64, p. 128–139.

Keith, B.D., and E.D. Pittman, 1983, Bimodal porosity in oolitic reservoir—effect on productivity and log response, Rodessa limestone (Lower Cretaceous), East Texas Basin: AAPG Bulletin, v. 67, p. 1391–1399.

Kerans, C., N.F. Hurley, and P.E. Playford, 1986, Marine diagenesis in Devonian reef complexes of the Canning Basin, western Australia, in J.H. Schroeder and B.H. Purser, eds., Reef diagenesis: New York, Springer-Verlag, p. 357–380.

Kerans, C., F.J. Lucia, and R.K. Senger, 1994, Integrated characterization of carbonate ramp reservoirs using Permian San Andres Formation outcrop analogs: AAPG Bulletin, v. 78, p. 181–216.

Kharaka, Y.K., and I. Barnes, 1973, SOLMINEQ: a solution-mineral equilibrium computation: Springfield, Virginia, National Technical Information Service Report PB 214-897, 82 p.

Kupecz, J.A., and L.S. Land, 1991, Late-stage dolomitization of the Lower Ordovician Ellenburger Group, west Texas: Journal of Sedimentary Petrology, v. 61, p. 551–574.

Kupecz, J.A., and L.S. Land, 1994, Progressive recrystallization and stabilization of early-stage dolomite: Lower Ordovician Ellenburger Group, West Texas, *in* B. Purser, M. Tucker, and D. Zenger, eds., Dolomites, a volume in honour of Dolomieu: IAS Special Publication 21, p. 255–279.

Kupecz, J.A., I.P. Montañez, and G. Gao, 1993, Recrystallization of dolomite with time, *in* R. Rezak and D.L. Lavoie, eds., Carbonate microfabrics, frontiers in sedimentology: New York, Springer-Verlag, p. 187–194.

Land, L.S., 1980, The isotopic and trace element geochemistry of dolomite: the state of the art, *in* D.H. Zenger, J.B. Dunham, and R.L. Ethington, eds., Concepts and models of dolomitization: SEPM Special Publication 28, p. 87–110.

Land, L.S., 1982, Introduction to dolomites and dolomitization: dolomites and dolomitization school: AAPG Course Notes, 29 p.

Land, L.S., 1985, The origin of massive dolomite: Journal of Geological Education, v. 33, p. 112–125.

Land, L.S., 1986, Environments of limestone and dolomite diagenesis; some geochemical considerations, *in* J. Warme and K. Shanley, eds., Carbonate depositional environments, modern and ancient, Part 5: diagenesis I: Colorado School of Mines Quarterly, v. 81, no. 4, p. 26–41.

Lander, R.H., and O. Walderhaug, 1997, An empirically calibrated model for sandstone reservoir quality prediction (abs.): Program of the 1997 Annual Convention of the AAPG, Dallas.

Lander, R.H., O. Walderhaug, A. Lyon, and A. Andersen, 1995, Reservoir quality prediction through simulation of compaction and quartz cementation (abs.): Program of the 1995 Annual Convention of the AAPG, Houston, p. 53A.

Love, K.M., C. Strohmenger, A. Woronow, and K. Rockenbauch, this volume, Predicting reservoir quality using linear regression models and neural networks, *in* J.A. Kupecz, J. Gluyas, and S. Bloch, eds., Reservoir quality prediction in sandstones and carbonates: AAPG Memoir 69, p. 47–60.

Lucia, F.J., 1993, Carbonate reservoir models: facies, diagenesis, and flow characterization, *in* D. Morton-Thompson and A.M. Woods, eds., Development geology reference manual: AAPG Methods in Exploration 10, p. 269–274.

Lucia, F.J., 1995, Lower Paleozoic cavern development, collapse, and dolomitization, Franklin Mountains, El Paso, Texas, *in* D.A. Budd, A.H. Saller, and P.M. Harris, eds., Unconformities and porosity in carbonate strata: AAPG Memoir 63, p. 279–300.

Lucia, F.J., and R.D. Conti, 1987, Rock fabric, permeability, and log relationships in an upward-shoaling, vuggy carbonate sequence: University of Texas at Austin Bureau of Economic Geology Geological Circular 87-5, 22 p.

Lucia, F.J., C. Kerans, and R.K. Senger, 1992a, Defining flow units in dolomitized carbonate-ramp reservoirs: Society of Petroleum Engineers, APE 24702, p. 399–406.

Lucia, F.J., C. Kerans, and G.W. Vander Stoep, 1992b, Characterization of a karsted, high-energy, ramp-margin carbonate reservoir: Taylor-Link West San Andres unit, Pecos County, Texas: University of Texas at Austin Bureau of Economic Geology Report of Investigations 208, 46 p.

Lucia, F.J., C. Kerans and F.P. Wang, 1995, Fluid-flow characterization of dolomitized carbonate ramp reservoirs: San Andres Formation (Permian) of Seminole field and Algerita escarpment, Permian Basin, Texas and New Mexico, *in* E.L. Stoudt and P.M. Harris, eds., Hydrocarbon reservoir characterization: SEPM Short Course 34, p. 129–153.

Major, R.P., and M.H. Holtz, this volume, Predicting reservoir quality at the development scale: methods for quantifying remaining hydrocarbon resource in diagenetically complex carbonate reservoirs, *in* J.A. Kupecz, J. Gluyas, and S. Bloch, eds., Reservoir quality prediction in sandstones and carbonates: AAPG Memoir 69, p. 231–248.

Martin, A.J., S.T. Solomon, and D.J. Hartmann, 1997, Characterization of petrophysical flow units in carbonate reservoirs: AAPG Bulletin, v. 81, p. 734–759.

Matthews, R.K., and X. Froelich, 1987, Forward modeling of bank-margin carbonate diagenesis: Geology, v. 15, p. 673–676.

Maureau, G.T.F.R., and D.H. van Wijhe, 1979, The prediction of porosity in the Permian (Zechstein 2) carbonate of eastern Netherlands using seismic data: Geophysics, v. 44, p. 1502–1517.

Meshri, I.D., 1989, On prediction of reservoir quality through chemical modeling (abs.): AAPG Bulletin, v. 73, p. 390–391.

Meyers, W.J., 1991, Calcite cement stratigraphy: an overview, *in* C.E. Barker and O.C. Kopp, eds., Luminescence microscopy and spectroscopy: qualitative and quantitative applications: SEPM Short Course 25, p. 133–148.

Meyers, W.J., and K.C. Lohmann, 1985, Isotope geochemistry of regionally extensive calcite cement zones and marine components in Mississippian limestones, New Mexico, *in* N. Schneidermann and P.M. Harris, eds., Carbonate cements: SEPM Special Publication 36, p. 223–239.

Milliken, K.L., 1992, Chemical behavior of detrital feldspars in mudrocks versus sandstones, Frio Formation (Oligocene), South Texas: Journal of Sedimentary Petrology, v. 62, p. 790–801.

Milliken, K.L., E.F. McBride, and L.S. Land, 1989, Numerical assessment of dissolution versus replacement in the subsurface destruction of detrital feldspars, Oligocene Frio Formation, south Texas: Journal of Sedimentary Petrology, v. 59, p. 740–757.

Mitchum, R.M., 1977, Seismic stratigraphy and global changes of sea level, part I: glossary of terms used in seismic stratigraphy, *in* C.W. Payton, ed., Seismic stratigraphy applications to hydrocarbon exploration: AAPG Memoir 26, p. 205–212.

Mitchum, R.M., P.R. Vail, and S. Thompson III, 1977, Seismic stratigraphy and global changes of sea level, Part II: the depositional sequence as a basic unit for stratigraphic analysis, *in* C.W. Payton, ed., Seismic stratigraphy applications to hydrocarbon exploration: AAPG Memoir 26, p. 53–62.

Monty, C.L.V., D.W.J. Bosence, P.H. Bridges, and B.R. Pratt, 1995, Carbonate mud-mounds, their origin and evolution: IAS Special Publication 23, 537 p.

Moore, C.H., 1989, Carbonate diagenesis and porosity: Developments in Sedimentology 46: New York, Elsevier, 338 p.

Morrow, D.W., 1982, Diagenesis 1. Dolomite—Part 2. Dolomitization models and ancient dolostones: Geoscience Canada, v. 9, p. 95–110.

Murray, R.C., 1960, Origin of porosity in carbonate rocks: Journal of Sedimentary Petrology, v. 30, p. 59–84.

Parkhurst, D.L., D.C. Thorstenson, and N. Plummer, 1980, PHREEQUE: a computer program for geochemical calculations: USGS Water Resources Investigational Report 80-96, 210 p.

Pittman, E.D., and R.E. Larese, 1991, Compaction of lithic sands: experimental results and applications: AAPG Bulletin, v. 75, p. 1279–1299.

Playford, P.E., 1980, Devonian "Great Barrier Reef" of Canning Basin, western Australia: AAPG Bulletin, v. 64, p. 814–840.

Posamentier, H.W., M.T. Jervey, and P.R. Vail, 1988, Eustatic controls on clastic deposition I—sequences and systems tracts models, *in* C.K. Wilgus, B.S. Hastings, C.G.St.C. Kendall, H.W. Posamentier, C.A. Ross, and J.C. Van Wagoner, eds., Sea-level changes: An integrated approach: SEPM Special Publication 42, p. 125–154.

Prezbindowksi, D.R., 1985, Burial cementation—is it important? A case study, Stuart City reef trend, south central Texas, *in* N. Schneidermann and P.M. Harris, eds., Carbonate cements: SEPM Special Publication 36, p. 241–264.

Primmer, T.J., C.A. Cade, J. Evans, J.G. Gluyas, M.S. Hopkins, N.H. Oxtoby, P.C. Smalley, E.A. Warren, and R.H. Worden, this volume, Global patterns in sandstone diagenesis: their application to reservoir quality prediction for petroleum exploration, *in* J.A. Kupecz, J. Gluyas, and S. Bloch, eds., Reservoir quality prediction in sandstones and carbonates: AAPG Memoir 69, p. 61–78.

Quinn, T.M., and R.K. Matthews, 1990, Post-Miocene diagenetic and eustatic history of Enewetak Atoll: model and data comparison: Geology, v. 18, p. 942–945.

Ramm, M., 1992, Porosity-depth trends in reservoir sandstones: theoretical models related to Jurassic sandstones, offshore Norway: Marine and Petroleum Geology, v. 9, p. 563–567.

Ramm, M., A.W. Forsberg, and J.S. Jahren, this volume, Porosity-depth trends in deeply buried Upper Jurassic reservoirs in the Norwegian Central Graben: an example of porosity preservation beneath the normal economic basement by grain-coating microquartz, *in* J.A. Kupecz, J. Gluyas, and S. Bloch, eds.,

Reservoir quality prediction in sandstones and carbonates: AAPG Memoir 69, p. 177–200.

Robinson, A.G., and J.G. Gluyas, 1992, The duration of quartz cementation in sandstones, North Sea and Haltenbanken basins: Marine and Petroleum Geology 9, p. 324–327

Rose, P.R., 1987, Dealing with risk and uncertainty in exploration: how can we improve?: AAPG Bulletin, v. 71, p. 1–16.

Saller, A.H., D.A. Budd, and P.M. Harris, 1994, Unconformities and porosity development in carbonate strata: ideas from a Hedberg conference: AAPG Bulletin, v. 78, p. 857–872.

Sarg, J.F., 1988, Carbonate sequence stratigraphy, *in* C.K. Wilgus, B.S. Hastings, C.G.St.C. Kendall, H.W. Posamentier, C.A. Ross, and J.C. Van Wagoner, eds., Sea-level changes: An integrated approach: SEPM Special Publication 42, p. 155–181.

Scherer, M., 1987, Parameters influencing porosity in sandstones: a model for sandstone porosity prediction: AAPG Bulletin, v. 71, p. 485–491.

Schlager, W., 1992, Sedimentology and sequence stratigraphy of reefs and carbonate platforms: AAPG Continuing Education Course Note Series 34, 71 p.

Schmoker, J.W., 1984, Empirical relation between carbonate porosity and thermal maturity: an approach to regional porosity prediction: AAPG Bulletin, v. 68, p. 1697–1703.

Schmoker, J.W., and D.L. Gautier, 1988, Sandstone porosity as a function of thermal maturity: Geology, v. 16, p. 1007–1010.

Schmoker, J.W., and R.B. Halley, 1982, Carbonate porosity vs. depth: a predictable relation for South Florida: AAPG Bulletin, v. 66, p. 2561–2570.

Schmoker, J.W., and T. Hester, 1983, Porosity and thermal maturity of limestone bodies in Jurassic Swift Formation, Williston Basin, North Dakota: U.S. Geological Society Open-File Report 83-723, 7 p.

Schmoker, J.W., and T.C. Hester, 1990, Regional trends of sandstone porosity vs. vitrinite reflectance—a preliminary framework, *in* V.F. Nuccio and C.E. Barker, eds., Applications of thermal maturity studies to energy exploration: Rocky Mountain Section of SEPM, p. 53–60.

Schmoker, J.W., K.B. Krystinik, and R.B. Halley, 1985, Selected characteristics of limestone and dolomite reservoirs in the United States: AAPG Bulletin, v. 69, p. 733–741.

Schneidermann, N., and P.M. Harris, eds., 1985, Carbonate cements: SEPM Special Publication 36, 379 p.

Scholle, P.A., 1977, Chalk diagenesis and its relation to petroleum exploration: oil from chalks, a modern miracle?: AAPG Bulletin, v. 61, p. 982–1009.

Scholle, P.A., 1978, Porosity prediction in shallow versus deep water limestones—primary porosity preservation under burial conditions: SPE 7554.

Scholle, P.A., 1981, Porosity prediction in shallow vs. deepwater limestones: Journal of Petroleum Technology, p. 2236–2242.

Scholle, P.A., D.G. Bebout, and C.H. Moore, eds., 1983, Carbonate depositional environments: AAPG Memoir 33, 708 p.

Scholle, P.A., and R.B. Halley, 1985, Burial diagenesis: out of sight, out of mind!, *in* N. Schneidermann and P.M. Harris, eds., Carbonate cements: SEPM Special Publication 36, p. 309–334.

Schroeder, J.H., and B.H. Purser, eds., 1986, Reef diagenesis: New York, Springer-Verlag, 455 p.

Serna, M.J., 1984, Porosity prediction using amplitude mapping: case study of the Cretaceous M-2 Limestone Member of the Napo Formation, Ecuador: 4th Meeting of Petroleum Exploration in the Sub-Andean Basins, Bolivariano Symposium, Bogota, Colombia, Memoir V2, no. 29, 9 p.

Sluijk, D., and J.R. Parker, 1984, Comparison of predrilling predictions with postdrilling outcomes, using Shell's prospect appraisal system (abs.): AAPG Bulletin, v. 68, p. 528.

Sneider, R.M., 1990, Introduction: reservoir description of sandstones, *in* J.H. Barwis, J.G. McPherson, and J.R.J. Studlick, eds., Sandstone petroleum reservoirs: New York, Springer-Verlag, p. 1–3.

Sombra, C.L., and H.K. Chang, this volume, Burial history and porosity evolution of Brazilian Upper Jurassic to Tertiary sandstone reservoirs, *in* J.A. Kupecz, J. Gluyas, and S. Bloch, eds., Reservoir quality prediction in sandstones and carbonates: AAPG Memoir 69, p. 79–90.

Stoudt, E.L., and P.M. Harris, 1995, Hydrocarbon reservoir characterization: geologic framework and flow unit modeling: SEPM Short Course 34, 357 p.

Surdam, R.C., and L.J. Crossey, 1987, Integrated diagenetic modeling: a process-oriented approach for clastic systems: Annual Review of Earth and Planetary Sciences, v. 15, p. 141-170.

Taylor, T.R., and C.H. Soule, 1993, Reservoir characterization and diagenesis of the Oligocene 64-zone sandstone, North Belridge field, Kern County, California: AAPG Bulletin, v. 77, p. 1549–1566.

Thomas, A.R., W.M. Dahl, C.M. Hall, and D. York, 1993, ^{40}Ar/^{39}Ar analyses of authigenic muscovite, timing of stylolitization, and implications for pressure solution mechanisms: Jurassic Norphlet Formation, offshore Alabama: Clays and Clay Minerals, v. 41, p. 269–279.

Tinker, S.W., 1996, Building the 3-D jigsaw puzzle: applications of sequence stratigraphy to 3-D reservoir characterization, Permian Basin: AAPG Bulletin, v. 80, p. 460–485.

Tinker, S.W., and D.H. Mruk, 1995, Reservoir characterization of a Permian giant: Yates field, West Texas, *in* E.L. Stoudt and P.M. Harris, eds., Hydrocarbon reservoir characterization: SEPM Short Course 34, p. 51–128.

Tobin, R.C., this volume, Porosity prediction in frontier basins: a systematic approach to estimating subsurface reservoir quality from outcrop samples, *in* J.A. Kupecz, J. Gluyas, and S. Bloch, eds., Reservoir quality prediction in sandstones and carbonates: AAPG Memoir 69, p. 1–18.

Toomey, D.F., 1981, European fossil reef models: SEPM Special Publication 30, 546 p.

Tyler, N., W.E. Galloway, C.M. Garrett, and T.E. Ewing, 1984, Oil accumulation, production characteristics, and targets for additional oil recovery in major oil reservoirs of Texas: University of Texas Bureau of Economic Geology Circular 84-2, 31 p.

Vail P.R., R.M. Mitchum, and S. Thompson III, 1977, Seismic stratigraphy and global changes in sea level, part 3: relative changes of sea level from coastal onlap, *in* C.W. Payton, ed., Seismic stratigraphy applications to hydrocarbon exploration: AAPG Memoir 26, p. 63–97.

Van Wagoner J.C., R.M. Mitchum, K.L. Campion, and V.D. Rahmanian, 1990, Siliciclastic sequence stratigraphy in well logs, cores, and outcrops: AAPG Methods in Exploration Series 7, 55 p.

Van Wagoner, J.C. , H.W. Posamentier, R.M. Mitchum, Jr., P.R. Vail, J.F. Sarg, T.S. Loutit, and J. Hardenbol, 1988, An overview of the fundamentals of sequence stratigraphy and key definitions, *in* C.K. Wilgus, B.S. Hastings, C.G.St.C. Kendall, H.W. Posamentier, C.A. Ross, and J.C. Van Wagoner, eds., Sea-level changes: An integrated approach: SEPM Special Publication 42, p. 39–45.

Walderhaug, O., 1994, Precipitation rates for quartz cement in sandstones determined by fluid inclusion microthermometry and temperature-history modeling: Journal of Sedimentary Research, Section A, p. 324–333.

Walls, R.A., and G. Burrowes, 1985, The role of cementation in the diagenetic history of Devonian reefs, western Canada, *in* N. Schneidermann and P.M. Harris, eds., Carbonate cements: SEPM Special Publication 36, p. 185–220.

Waples, D.W., and H. Kamata, 1993, Modeling porosity reduction as a series of chemical and physical processes: *in* A.G. Doré et al., eds., Basin Modeling: Advances and Applications: Amsterdam, Elsevier, Norwegian Petroleum Society Special Publication 3, p. 303–320.

Waples, D.W., M. Suizu, and H. Kamata, 1992, The art of maturity modeling, part 2: alternative models and sensitivity analysis: AAPG Bulletin, v. 76, p. 47–66.

Weber, L.J., F.M. Wright, J.F. Sarg, E. Shaw, L.P. Harman, J.B. Vanderhill, and D.A. Best, 1995, Reservoir delineation and performance: applications of sequence stratigraphy and integration of petrophysics and engineering data, Aneth Field, southeast Utah, U.S.A., *in* E.L. Stoudt and P.M. Harris, eds., Hydrocarbon reservoir characterization: SEPM Short Course 34, p. 1–29.

Williams, L.B., R.L. Hervig, and K. Bjørlykke, 1997, New evidence for the origin of quartz cements in hydrocarbon reservoirs revealed by oxygen isotope microanalysis: Geochimica et Cosmochimica Acta, v. 61, p. 2529–2538.

Wilson, J.L., 1975, Carbonate facies in geologic history: New York, Springer-Verlag, 471 p.

Wilson, M.D., 1994, Case history — Jurassic sandstones, Viking Graben, North Sea, *in* M.D. Wilson, ed., Reservoir quality assessment and prediction in clastic rocks: SEPM Short Course 30, p. 367–384.

Wilson, M.D., and A.P. Byrnes, 1988, Porosity prediction in lithic sandstones (unpublished report), 234 p.

Wood, J.R., and A.P. Byrnes, 1994, Alternate and emerging methodologies in geochemical and empirical modeling, *in* M.D. Wilson, ed., Reservoir quality assessment and prediction in clastic rocks, SEPM Short Course 30, p. 395–399.

Zempolich, W.G., and L.A. Hardie, this volume, Geometry of dolomite bodies within deep-water resedimented oolite of the Middle Jurassic Vajont Limestone, Venetian Alps, Italy: analogs for hydrocarbon reservoirs created through burial dolomitization, *in* J.A. Kupecz, J. Gluyas, and S. Bloch, eds., Reservoir quality prediction in sandstones and carbonates: AAPG Memoir 69, p. 127–162.

Chapter 1

Porosity Prediction in Frontier Basins: A Systematic Approach to Estimating Subsurface Reservoir Quality from Outcrop Samples

R.C. Tobin
Amoco Corporation
Houston, Texas, U.S.A.

ABSTRACT

In frontier basins where subsurface data are limited, or absent altogether, the study of reservoir rocks exposed in surface outcrops may be the dominant (or only available) means of predicting subsurface reservoir quality. *This chapter provides a systematic, decision-tree–based procedure for using existing tools and techniques to evaluate potential subsurface reservoir quality when only surface outcrops are available.* This approach is applicable to both carbonate and terrigenous clastic reservoirs. With this system, outcrop samples are classified into one of ten lithofacies types whose reservoir properties are codependent on common diagenetic or burial processes. The classification subdivides outcrop samples into either "tight" or "porous" lithofacies, depending on the measured porosity relative to economic minimum. "Tight" rocks include six end-member lithofacies that were either cemented or compacted during burial, or were originally tight at the time of deposition. "Porous" rocks include four lithofacies types that are categorized by original depositional fabric and the degree of alteration by recent surface weathering. Risk assessment for each of the ten lithofacies types using existing geological tools and techniques is discussed, along with guidelines for estimating potential subsurface porosity and permeability. Case histories that illustrate the recommended process for assessing risk are described from China, Myanmar (Burma), and Turkey.

INTRODUCTION

Geological analyses of hydrocarbon systems often require surface outcrop studies, particularly in frontier basins where subsurface information is sparse. Outcrop exposures provide the explorationist with unique opportunities for observing surface structural features, lateral bedding and facies variations, and three-dimensional spatial configurations that are less directly observed in the subsurface. This information is especially beneficial in stratigraphic, sedimentologic, and structural modeling studies. The generally unlimited availability of rock material also favors studies in which large bulk samples must be used (e.g., source rock geochemistry, paleontology, geophysical rock properties).

Reservoir studies are also enhanced by outcrop investigations. *Reservoir facies* predictions benefit greatly from the three-dimensional characteristics observable on outcrop exposures. *Reservoir quality* predictions benefit from the unlimited sample size availability and from the potential for documenting the lateral variability in petrophysical rock properties within a given facies.

Despite these advantages, outcrop-based predictions of subsurface reservoir quality are less direct (and potentially less accurate) than those based on subsurface data alone because of the following limitations.

1. Reservoir facies exposed at the surface may have undergone a vastly different tectonic and burial history than their subsurface counterparts.
2. Diagenetic history and pore system evolution may be different than that of subsurface counterparts.
3. Recent outcrop diagenesis (leaching, cementation, sediment infill, etc.) may alter the composition and pore system characteristics of ancient reservoir facies.
4. Basin-margin reservoir facies exposed on outcrops are less likely to contain hydrocarbon shows than their subsurface counterparts, and any shows that are present may be weathered away or severely biodegraded.
5. Outcrop exposures may be dominated by basin-margin reservoir facies rather than basin-center facies.
6. Reservoir rock provenance may vary from the subsurface.

Therefore, interpretations of reservoir quality from outcrop data present a technical challenge to the explorationist.

SCOPE AND INTENT

Various approaches to reservoir quality prediction from subsurface data are commonly used in exploration and are widely published. These predictions may range from simple comparative analogs, where subsurface data are sparse, to more complex quantitative assessments of porosity and permeability from empirical calibrations (Bloch, 1991; Byrnes, 1994; Bloch and Helmold, 1995) or process-oriented geochemical models (Surdam and Crossey, 1987; Meshri, 1989; Meshri and Ortoleva,

1990; Wood, 1994). In contrast, surface outcrop studies of potential reservoir facies are less commonly used in exploration, and are not as frequently documented in the literature. With few exceptions (Goldstein, 1988; Scholle et al., 1991), most of the published outcrop studies to date are only marginally related to the prediction of subsurface reservoir quality. Additionally, there is no published account of a deliberate, systematic approach to predicting subsurface reservoir quality from outcrop samples that explorationists can use to guide their studies. As a result, outcrop evaluations are often plagued by inefficiency and incomplete technical investigation. For example, potentially porous reservoir facies are sometimes overlooked as viable exploration targets because of poor reservoir quality preservation at the surface, even though the same facies may be highly porous and permeable in the subsurface. Similarly, an unrealistically *low* degree of risk may be assigned to a porous facies exposed on outcrop whose original pore system has been greatly enhanced by recent weathering processes. The result of these problems may be poor risk assessment of reservoir quality in frontier areas that lack confirming subsurface well data.

To help alleviate these problems, this chapter is intended to provide a systematic approach for estimating the degree of risk associated with subsurface reservoir quality when only outcrop samples of prospective reservoir facies are available for study. This approach uses a decision-tree–based process flow diagram to evaluate the uncertainties associated with reservoir quality using existing well-tested exploration tools and technologies (Table 1). As a by-product of this evaluation process, surface rock samples are classified into one of ten logical groupings (Types 1–10) whose predictabilities are codependent on common burial or diagenetic phenomena (Figure 1). The systematic approach described in this chapter can be used as a guide by explorationists charged with making either qualitative, quantitative, or semi-quantitative predictions in both carbonate and terrigenous clastic strata in a time-effective manner. It is only intended to be a process-oriented approach to risk appraisal and is not intended as an all-inclusive solution to subsurface porosity prediction. However, the effective application of this approach requires that the explorationist consider the wide variety of porosity constructive and destructive processes that commonly affect subsurface reservoirs (Table 1). It also requires that the explorationist utilize information from a variety of related technologies in an integrated manner (Table 2).

RECOMMENDED APPROACH

The first recommended step in subsurface porosity prediction from surface data is to classify outcrop rock samples into one of ten logical categories whose present-day porosity and permeability values are the end products of common geologic and burial phenomena (Table 3). This classification is organized by a decision-tree flow diagram that leads the explorationist to discover the cause(s) for the present-day porosity and permeability of the rocks being studied (Figure 1). The classification then poses additional questions that require

Table 1. Geologic Factors Affecting Subsurface Reservoir Quality.

Geologic Factor	Effect on Porosity
Ancient destructive diagenesis (sediment infill, cementation, recrystallization, mechanical and chemical compaction)	Reduces porosity
Ancient constructive diagenesis (dissolution, fracturing)	Enhances porosity
Other ancient diagenesis (mineralogic replacement, authigenic clay growth, brecciation, tectonic deformation)	May reduce or enhance porosity
Framework composition/provenance	May control postdepositional diagenesis
Environment of deposition	Controls prediagenesis porosity
Paleoclimate	Affects EOD, weathering, karstification
Depth of burial	Indirectly related to porosity loss
Pressuring and overpressuring	Early overpressure may enhance porosity
Thermal maturity	Indirectly related to porosity loss
Erosional events/unconformities	May reduce or enhance porosity
Pore fluid migration (water)	Enhances cementation/dissolution
Pore fluid migration (oil)	Inhibits cementation/dissolution
Associated rock strata—seal	Affects pore fluid entrapment
Associated rock strata— source rock	Controls type of migrating pore fluids

a more detailed technical evaluation designed to assist the explorationist in predicting reservoir quality and assigning an appropriate degree of exploration risk. The following discussion will lead the reader through a number of key decision points illustrated in the decision tree shown in Figure 1.

First Decision: Porous or Tight?

The first decision point considered is whether the rock sample in question is currently "porous" or "tight" (Figure 1). This distinction does not rely on a universal porosity cutoff, but is dependent on the economic requirements of each play, and should include both the amount and type of porosity present. For example, a thin, deeply buried reservoir sandstone in a frontier area having small structural traps and lacking needed economic infrastructure (pipelines, transportation, etc.) might require 20–25% porosity and >500 md of permeability for the play to be economic (i.e., to ensure economically viable reserves and associated flow rates). In contrast, a thick, shallow, fractured carbonate play in an area having larger structural traps adjacent to an accessible pipeline might require only 5% porosity to be economic. In either case, if a large proportion of the pore system has ineffective microporosity, then the porosity cutoff used would have to be considerably higher (Figure 2). Therefore, when establishing local porosity cutoffs, it is important that the explorationist use only *effective* porosities. It is also critical to recognize that tight lithofacies exposed at the surface may have porous counterparts in the subsurface, and vice versa.

TIGHT ROCKS

Decision: Why Is the Sample Tight?

If the outcrop sample in question is deemed to be tight (effective porosity and permeability are below economic requirements), the next decision point to be considered is "Why is the sample tight?" Three possibilities exist. Either the original depositional fabric of the sample was tight, the original fabric has been so obscured by diagenesis that the cause of low porosity is uncertain, or the original sedimentary fabric was porous, but postdepositional diagenesis has reduced porosity to unacceptably low amounts. Each of these possibilities is described below.

Tight Depositional Facies (Rock Type 1)

If it is determined from petrographic observations that the original fabric was tight (i.e., a depositional facies lacking high initial macroporosity and permeability), the sample is considered to be a Type 1 lithology (Figure 1). For carbonate rocks, common examples of Type 1 lithofacies would be marl, lime mudstone or wackestone, sandy lime mudstone to micritic sandstone, or fine, dense crystalline dolomite. For clastic rocks, Type 1 lithofacies are usually shales or argillaceous siltstones, sandstones, and conglomerates (wackestones). Such lithofacies are usually associated with turbid, low-energy environments of deposition that are not conducive to the deposition of sediments with high initial macroporosity. Hence, the exploration risks associated with Type 1 lithofacies are normally considered to be quite high, unless paleogeographic

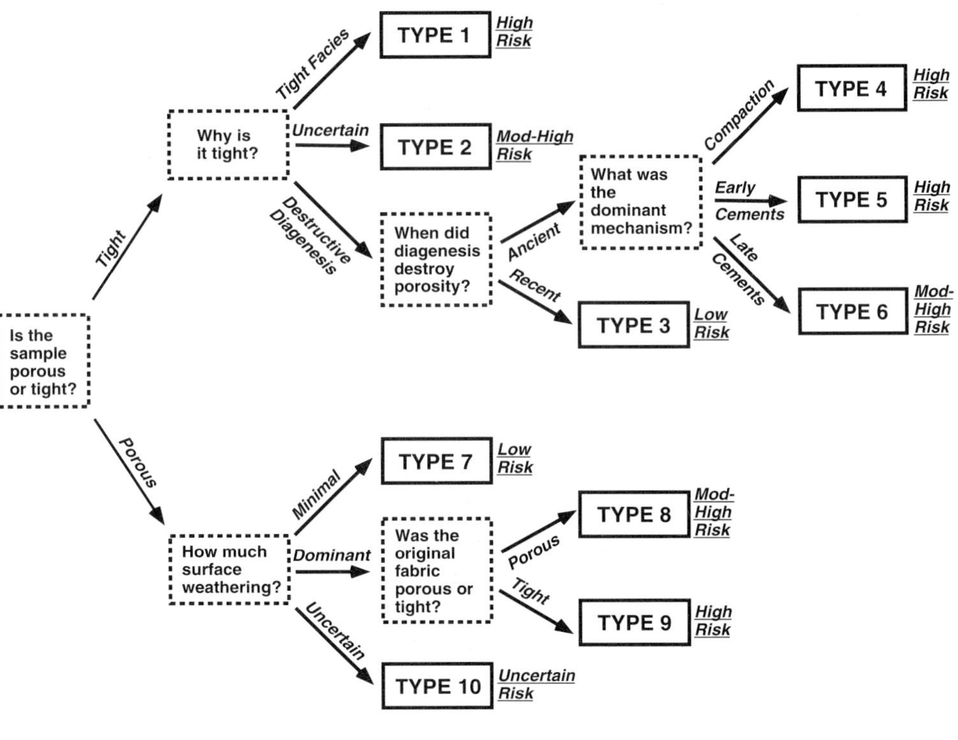

Figure 1. Decision-tree flow diagram used to evaluate the degree of exploration risk from the study of outcrop materials. Outcrop samples are classified into ten rock type categories based on their potential as subsurface reservoir facies. Complete descriptions of each rock type are given in the text and are summarized in Table 3.

reconstructions can be used to predict the occurrence of more porous depositional facies elsewhere.

Lower risks can also be predicted for Type 1 lithofacies under the following conditions: (1) Fracturing of brittle rock strata can create economic reservoirs out of nonporous or low-porosity facies, although the dominant effect is to significantly increase permeability rather than porosity. The likelihood of fracture development is related to structural position and various rock properties. In general, closely spaced fracture networks are favored by rocks with low matrix porosities that are fine grained, thinly bedded, and composed predominantly of brittle (nonductile) minerals such as quartz, feldspar, dolomite, and calcite (Nelson, 1985). (2) Karstification of tight carbonate facies can create economic porosity, but the prediction of porosity in karsted facies is severely limited by the lack of adequate diagenetic models (Saller et al., 1994). Total porosity in karstified reservoirs is generally low (3–6%), with associated low permeability except where fractured and leached. Karstic reservoirs may also be highly compartmentalized (Tobin, 1985; Kerans, 1988). (3) Dolomitization has been reported to create matrix porosity in originally tight, micritic facies (Weyl, 1960), although more recent evidence suggests that dolomitization alone may be insufficient to create a viable pore system. Mineralogically selective, post-dolomite dissolution, for example, may be necessary (Ottmann et al., 1976; Lucia and Major, 1994; Purser et al., 1994). (4) Subsurface (burial) dissolution may also create economic porosity, but only if there are dissolvable constituents within the rock and a sufficient plumbing system existed for subsurface dissolution to have been effective. For each of these four mechanisms, burial history reconstructions (time-temperature-depth profiles) may offer clues to past episodes in which porosity-constructive events could have taken place. Burial histories

and associated paragenetic sequences should always be included in risk assessment of Type 1 lithofacies before disregarding them as prospective reservoirs (Figure 3).

Uncertain Depositional Facies (Rock Type 2)

For some rock samples, the distinction between tight facies and tight diagenesis is uncertain because the original depositional fabric has been obscured by diagenesis (ancient or recent). These are referred to as Type 2 rocks (Figure 1). Examples of Type 2 rocks are recrystallized sparry limestones, some dolomites (particularly coarse crystalline, nonplanar dolomites of burial origin), and some quartzose sandstones whose original fabrics have been obscured by intense quartz cementation and associated pressure solution or incipient metamorphism. For these lithofacies, subsurface porosity prediction is uncertain, and exploration risk is considered to be quite high unless fracturing, karsting, dolomitization, or dissolution can be predicted elsewhere. For the explorationist faced with this type of facies, the recognition of original sedimentary fabric is of paramount importance. Tools that may be used to help identify these fabrics are "white-card" transmitted light microscopy (Folk, 1987), thin-section epifluorescence (Dravis and Yurewicz, 1985), and cathodoluminescence (Figure 4).

Destructive Diagenesis (Rock Types 3–6)

Some potential reservoir rocks exposed at the surface originally had porous depositional fabrics, but were subsequently altered by porosity-destructive diagenetic processes. Typical carbonate lithofacies in this group include grainstones, packstones, some planar dolomites, and some reefal facies. Typical clastic representatives are clean, matrix-poor sandstones and conglomerates (arenites). Risk assessment for this

Table 2. Related Technologies Used in Porosity Prediction.

Technology	Why Is It Important?
Sedimentology	Facies analysis, environment of deposition
Petrography	Microfacies, diagenesis, pore system description
Fluorescence	Depositional/diagenetic fabric recognition, pore geometry
Luminescence	Depositional/diagenetic fabric recognition
Geophysics	Facies analysis, unconformity recognition
Paleontology	Stratigraphy, environment of deposition, unconformity recognition
Core analysis	Porosity, permeability, pore geometry
Inorganic geochemistry	Diagenetic interpretations, unconformity recognition
Organic geochemistry	Source rock quality, migration timing
Fluid inclusion thermometry	Migration timing, diagenesis, thermal maturity
Thermal maturity analyses	Indirectly related to porosity, hydrocarbon phase preserved
Basin modeling	Timing of porosity creation/destruction events, depth of burial
Compaction simulation	Prediction of past burial depths and depth to porosity basement
Rock mechanics	Probability of fracturing

group requires that the timing and type of destructive diagenetic processes be identified, leading to the next decision points in the decision tree shown in Figure 1.

Decision: When Did Diagenesis Destroy Porosity?

For facies having initially porous depositional fabrics, the next question to consider is "When did the porosity-destructive diagenesis occur?" Two basic possibilities exist. Either the porosity was destroyed during recent outcrop exposure or it was destroyed during ancient diagenetic event(s). Both possibilities are discussed below.

Recent Pore Destruction (Rock Type 3)

Some outcrop samples show clear evidence of recent outcrop-related pore destruction, and are considered Type 3 lithofacies (Figure 5). Type 3 rocks originally contained economic amounts of primary, secondary, or dissolution-enhanced primary porosity that survived burial diagenesis prior to recent outcrop exposure. Upon exposure, the pore system of these rocks was subsequently destroyed by a variety of surface and near-surface diagenetic processes. These include recent sediment infill (e.g., terra rosa or other soil-forming processes), infill by weathering by-products such as iron oxide or clays, oil biodegradation resulting in pore-plugging bitumen, or surface to near-surface cementation.

Outcrop-related cements may be difficult to distinguish from ancient cements, although some petrographic criteria exist for their recognition. These include the presence of pendant or meniscus morphologies (particularly if they follow obvious by-products of burial diagenesis), isotopic and/or trace element compositions characteristic of meteoric origin and unrelated to prior diagenetic byproducts, and fluid inclusions suggestive of recent exposure (e.g., air inclusions and/or dominantly monophase aqueous inclusions, particularly if two-phase inclusions are present in earlier cement generations). The exploration risks associated with Type 3 (recent pore destruction) reservoir facies are considered to be relatively low because such facies clearly contained economic porosity prior to outcrop exposure. Therefore, porous counterparts probably exist somewhere in the subsurface, although they may not be ubiquitous. For the explorationist who wants to quantify the risk associated with Type 3 reservoirs, the following questions should be thoroughly investigated: (1) How much porosity was present in the sample prior to outcrop exposure? (2) Was the porosity of the sample well connected (permeable) prior to outcrop exposure? (3) How deep was this sample buried prior to outcrop exposure? (4) How much deeper could economic porosity survive in this sample beyond its estimated pre-outcrop depth of burial? (5) Are compaction-inhibiting processes (e.g., early overpressuring) or compaction-enhancing processes (e.g., pressure solution) likely to affect the porosity vs. depth estimates? (6) Is the pre-outcrop porosity primary or secondary? If it is primary, is there any potential for porosity enhancement by secondary dissolution elsewhere? These questions should lead the explorationist to a reasonable estimate of the range of porosity and permeability likely to be encountered at any given drilling depth. An example of this process is illustrated in Table 4.

Table 3. Summary of Outcrop Categories and Associated Reservoir Quality Risks.

Category	Name	Porosity	Typical Lithologies	Risk Assessment
Type 1	Tight depositional facies	Tight	Micritic limestone, marl and shale, sandy limestone, micritic dolomite, argillaceous siltstone, sandstone, or conglomerate	High risk unless fracturing, dolomitization, or porosity can be predicted
Type 2	Uncertain depositional facies	Tight	Recrystallized sparry limestone, some coarse, nonplanar dolomite, some quartz-cemented or metamorphosed quartz sandstones	High risk as above, unless original fabric can be determined
Type 3	Recent pore destruction	Tight	Originally porous sandstones and conglomerates, lime grainstones or packstones tightly cemented by recent weathering by-products	Very low risk for prospects shallower than pre-outcrop burial depth; variable risk for deeper prospects
Type 4	Dominantly compacted	Tight	Originally porous, but now tightly compacted sandstones, conglomerates, or nonmicritic carbonates	Very high risk for prospects that are as deep as pre-outcrop burial depth unless early overpressuring, rim cementation, or dissolution can be predicted
Type 5	Early near-surface cemented	Tight	Originally porous sandstones and conglomerates, lime grainstones, or packstones tightly cemented by ancient near-surface cements	High risk unless lateral cement pinchout or cement dissolution can be predicted
Type 6	Late burial cemented	Tight	Originally porous sandstones and conglomerates, lime grainstones, or packstones tightly cemented by ancient burial cements	Moderate to high risk unless lateral pinchout, dissolution, or diagenetic traps can be predicted
Type 7	Recent weathering minimal	Porous	Any porous lithology whose pore system is inherited from the subsurface (minimal recent weathering)	Very low risk for prospects shallower than pre-outcrop burial depth; variable risk for deeper prospects
Type 8	Weathered; depositional fabric porous	Porous	Originally porous depositional fabrics rendered tight by compaction or cementation, but leached by recent weathering	Moderate to high risk facies; risk assessment equivalent to Type 4, 5, or 6 as appropriate
Type 9	Weathered; depositional fabric tight	Porous	Originally tight depositional fabrics that have been leached by recent weathering processes	High risk facies; risk assessment equivalent to Type 1 lithofacies
Type 10	Recent weathering uncertain	Porous	Any reservoir lithology whose pore system contains appreciable amounts of secondary porosity of uncertain origin	Uncertain risk, but generally higher for increasing secondary porosity component

(A)

(B)

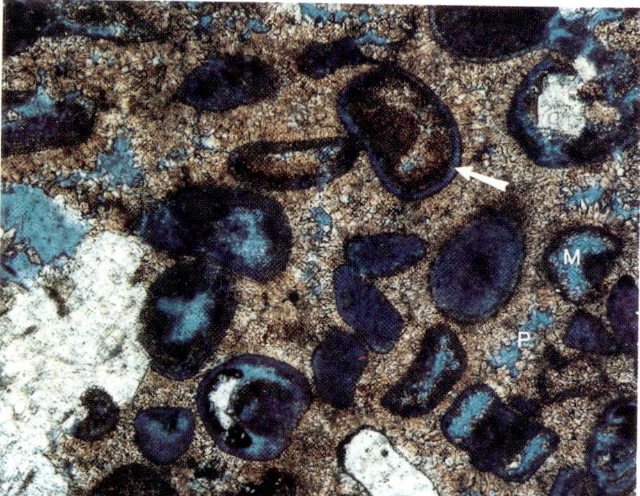

Figure 2. Porous reservoir facies containing appreciable amounts of ineffective microporosity. (A) Ooid-skeletal lime grainstone from the oil-producing Arab Zones in Dukhan field, Qatar, containing 20% total porosity, 40% of which is ineffective microporosity (purple color, note arrow) found within micritic grains. Total effective porosity for this sample is only 12%, and includes both primary (P) and secondary grain-moldic (M) pores (thin-section photomicrograph using plane-transmitted and ultraviolet fluorescent light; 80×). (B) Litharenite sandstone from western Siberia containing 17% total porosity, 35% of which is ineffective microporosity associated with authigenic clays, mostly pore-filling kaolinite (K). Total effective porosity for this sample is only 11% (SEM photomicrograph taken at 1400×).

Ancient Pore Destruction (Rock Types 4–6)

Most diagenetically altered outcrop samples show unmistakable evidence of ancient pore destruction resulting from some combination of compaction and cementation. Risk assessment for these rocks requires that the dominant diagenetic mechanism for pore destruction be identified, leading to the next decision point on the decision tree shown in Figure 1.

Decision: Ancient Pore Destruction— What Was the Dominant Mechanism?

Outcrop samples whose pore system was destroyed primarily by burial compaction are referred to as Type 4 (compacted) rocks. Cement-dominated samples are referred to as either Type 5 (early near-surface cemented) or Type 6 (late burial cemented) rocks depending on the timing of cement emplacement (Figure 1). In general, all three rock types are high-risk lithologies, although the specific degree of risk can be highly variable, depending on cement type, volumetric amount, timing, temperature, and presence of hydrocarbons. For this category, the explorationist must first determine the dominant mechanism of porosity loss before assessing exploration risk.

Compaction (Rock Type 4)

Type 4 lithofacies include nonargillaceous sandstones and conglomerates (arenites) and nonmicritic, grain-supported carbonates (lime grainstones, some packstones, some reef rocks, and some dolomites) whose pore systems have been destroyed by either mechanical compaction (grain rotation, slippage, rearrangement, repacking, plastic deformation, or grain breakage), pressure solution (intergranular or whole rock), or both (Figure 6). Type 4 rocks can contain minor amounts of cement, but the dominant mechanism for porosity loss is from compaction. Associated intergranular volumes are characteristically low. Although mechanical compaction can severely reduce porosity for sandstones of any composition, it is most effective in those containing an abundance of ductile lithic grains (Pittman and Larese, 1991). Similarly, mechanical compaction is most effective in grain-supported carbonate rocks containing ductile micritic grains (peloids, onkoids, some intraclasts) rather than hard, brittle grains like ooids or bioclasts (Moore, 1989, his figure 9.5). Compaction from pressure solution is most effective in sandstones containing an abundance of quartz and feldspar with minimal lithics (Pittman and Larese, 1991). Pressure solution is most likely to occur in grain-supported carbonate rocks that contain metastable (aragonitic) grain types (Wagner and Matthews, 1982) or insoluble components such as clays, quartz, and organics (Weyl, 1959). The presence of oil within pores appears to retard the effects of pressure solution (Dunnington, 1967; Feazel and Schatzinger, 1985).

Exploration risk for Type 4 reservoir facies is very high for prospects that are as deep or deeper than pre-outcrop burial depth. Risk can be considerably lower, however, under any of the following conditions: (1) early, shallow overpressuring can retard the rate of porosity loss from compaction (Scherer, 1987; Pittman and Larese, 1991). (2) Risk can also be lower if it can be demonstrated that early compaction-retarding rim cements are likely elsewhere. For example, early incipient quartz overgrowth cement in sandstones (Pittman and Larese, 1991) or early calcite rim cements or replacement dolomite in carbonate grainstones (Purser, 1978;

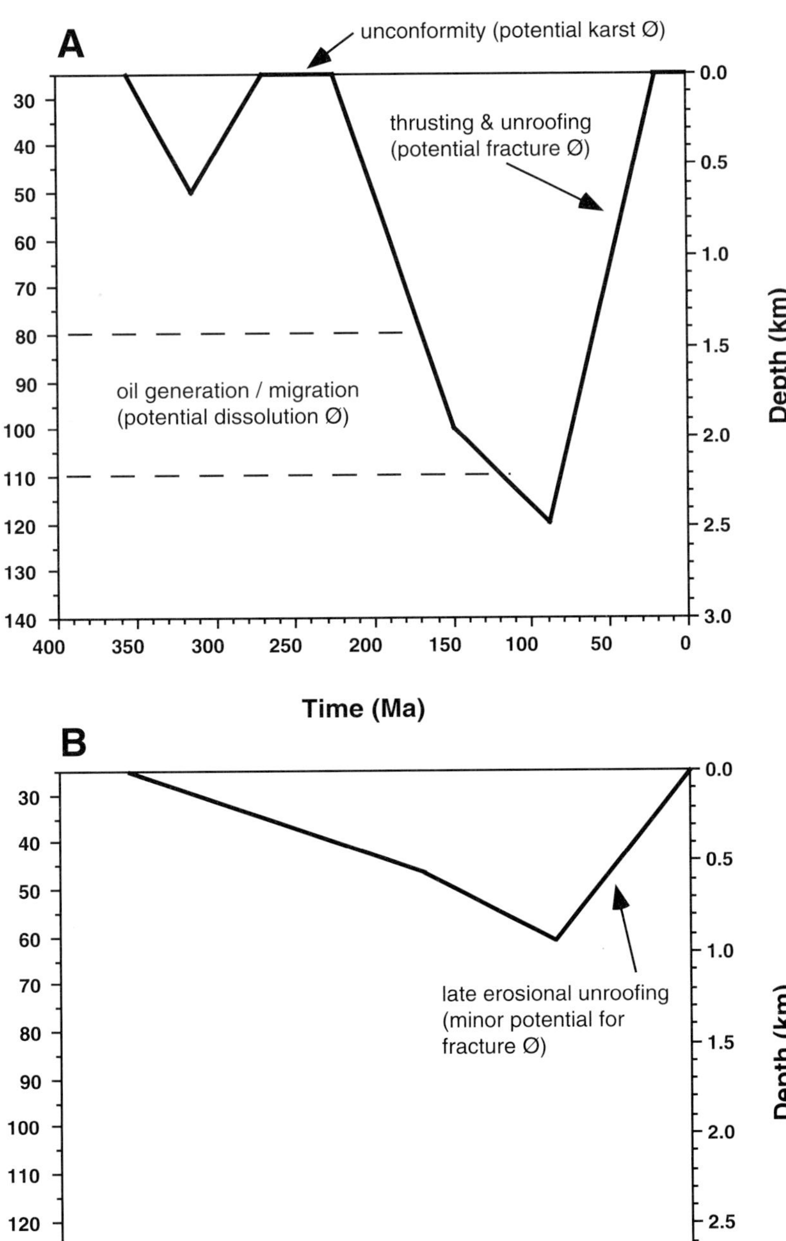

Figure 3. Two hypothetical burial history plots of a high risk, Type 1 reservoir facies (wrong depositional facies). In (A) two opportunities existed for secondary porosity development prior to late thrusting and possible fracture porosity enhancement. In (B) little or no opportunity for developing an economic porosity system existed. Therefore, a Type 1 reservoir with the burial history shown in (A) would be considered far less risky than the same facies with the burial history shown in (B). ϕ = porosity.

Moore and Druckman, 1981) can produce a rigid framework that is resistant to further compaction. Although small in volume, such cements take on the bulk of the overburden pressure, thereby inhibiting grain slippage/rotation, ductile grain deformation, and pressure solution. (3) Dissolution of various unstable rock components (cements and grains) can create secondary porosity that could yield an economic pore system, although the probability of significant porosity increase from the dissolution of a tightly compacted rock is fairly low because rocks generally have a very poor plumbing system for circulating undersaturated fluids.

Early Near-Surface Cementation (Rock Type 5)

Type 5 lithofacies are originally porous rocks that have been tightly cemented during early diagenesis by a variety of surface, near-surface, and shallow burial cements (Figure 7). Intergranular volumes found in Type 5 lithofacies are generally high because of the limited amount of compaction associated with shallow burial. For carbonate rocks, typical cements include calcite or aragonite of vadose, meteoric phreatic, marine or shallow burial origin, early dolomite, or evaporitic cements like anhydrite, gypsum, or halite. For sandstones, the most common early cements are

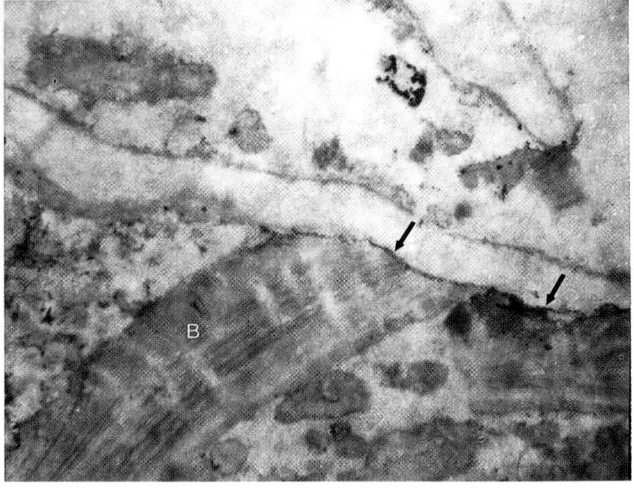

Figure 4. Photomicrographs illustrating the use of ultraviolet fluorescence to distinguish original depositional fabric in Type 2 lithofacies. (A) Plane-transmitted light view of a coarse crystalline, nonplanar dolomite lacking any obvious depositional texture (Devonian age, Canada). (B) Ultraviolet fluorescence view of the same area showing a well-defined, grain-supported skeletal fabric containing a variety of open-marine fossils such as the large punctate brachiopod shell in the lower left. Note also the pressure solution seam between the brachiopod and mollusk fragment (arrows). Both photographs were taken at 80×.

quartz, authigenic clays (especially chlorite), and calcite. Pore-filling iron oxides and other weathering by-products are also commonly associated with ancient subaerial exposure for both carbonates and clastics.

Exploration risks associated with Type 5 lithofacies are generally high because their pore systems are destroyed before hydrocarbon migration can occur. However, risks can be significantly lower depending on the timing, distribution, and chemical stability (solubility) of the cement phases in question. Therefore, risk assessment necessitates the prediction of two potential porosity-retaining scenarios, including (1) lateral pinchout of early cements elsewhere, and (2) ancient near-surface or burial dissolution of cements, or unstable framework grains engulfed within those cements.

Late Burial Cementation (Rock Type 6)

Type 6 lithofacies are originally porous rocks that have been tightly cemented during late diagenesis by a variety of deep burial cements (Figure 8). Intergranular volumes in Type 6 lithofacies are usually lower than those found in Type 5 lithofacies because of the delayed cementation that accompanies deep burial compaction. For carbonate rocks, the most common burial cements include coarse equant to poikilotopic calcite (both ferroan and nonferroan), anhydrite, nonplanar dolomite (baroque or saddle), ferroan dolomite, and ankerite. Some of these cements may be hydrothermal in origin, and may also include a variety of accessory minerals such as fluorite, galena, pyrite, and sphalerite. In addition to the same burial and hydrothermal cements found in carbonate rocks, sandstones may contain burial quartz, feldspar (usually albite), zeolites, and authigenic clays (kaolinite, illite, smectite).

In general, burial cements have a tendency to be more pervasive, laterally continuous, and chemically stable than near-surface cements. Therefore, the exploration risks associated with Type 6 lithofacies may be higher, particularly if the strata in question are thermally mature (dry-gas preservation window or above). At high maturity levels, no viable porosity-creating mechanisms exist to dissolve these cements (Tobin, 1991). However, risks can be significantly lower, depending on the timing and distribution of the cement phases in question. Some possible porosity-retaining scenarios include the following: (1) If the cements are hydrothermal in origin, they could be laterally restricted to fault zones, bedding contacts, or certain high-permeability carrier beds, and therefore could pinch out laterally into porous reservoir facies. (2) Ancient near-surface dissolution of burial cements may create secondary porosity elsewhere. (3) Ancient subsurface dissolution of burial cements could also create secondary porosity in Type 6 facies, but the probability of an effective dissolution mechanism is low, particularly if the strata are thermally mature. (4) Productive diagenetic traps (Rittenhouse, 1972; Wilson, 1977; Cant, 1986) could be predicted in the subsurface, especially if it can be demonstrated that oil migration has occurred prior to or during the initial stages of burial cementation, and the rock samples were taken from an area that lacked structural or stratigraphic closure at the time of migration. Burial cements that have very light $\delta^{13}C$ isotope signatures or contain fluorescing oil-filled fluid inclusions could represent past migration pathways that lead to productive diagenetic traps in higher structural positions in the subsurface.

Hybrid Lithofacies

Nearly all outcrop samples exhibit characteristics of two or more of the six rock types described, although one characteristic usually dominates. Some unusual hybrids can be found (e.g., a Type 1, tight depositional rock facies containing fractures that were cemented

Figure 5. Example of a Type 3 lithofacies (recent pore destruction) from the Ombilin Basin of Indonesia. This outcrop sample is a sublitharenite sandstone that has been pervasively cemented by hematite (black opaques) during recent outcrop weathering. Total porosity was reduced from 21% to 4% by recent cements (plane-transmitted light, 125×).

Figure 6. Example of a Type 4 lithofacies (tightly compacted) from the Chuxiong Basin, China. This sample is a very tightly compacted Triassic litharenite sandstone containing only trace amounts of microporosity. Intergranular volume for this sample is only 8%, indicating extreme mechanical and chemical compaction (pressure solution).

during recent outcrop exposure), but the most commonly recognized hybrids are the partially compacted and partially cemented reservoir rocks (Type 4 and Type 5/6 combination) and the early and late cement hybrids (Type 5 and 6). For hybrid lithofacies like these, petrographic point-count analyses are essential for determining the volumetric contribution of each mechanism toward total porosity destruction. Porosity potential in the subsurface must be predicted from a combination of the scenarios for each mechanism (compaction, near-surface cementation, burial cementation).

POROUS ROCKS

Decision: Degree of Recent Weathering?

All outcropping strata have been exposed to some degree of surface weathering. The duration of weathering may range from a just a few years to hundreds of thousands or even millions of years. Surface leaching processes can be minimal, or they can create significant amounts of secondary porosity that are not representative of the true subsurface conditions that existed prior to recent exposure. This possibility must be considered in risk assessment, particularly with regard to any petrophysical analyses of outcrop materials. The effects of surface dissolution can, of course, be minimized by avoiding heavily weathered exposures and by using a hammer (or a portable coring device) to obtain the freshest, least altered bedrock below the zone of intense weathering. However, if the efficacy of sample selection is uncertain, several factors should be considered in evaluating the probability of recent dissolution. These include the age of the outcrop (Is it a fresh roadcut? Or a mountain flank, fault escarpment, or stream cut exposed for the last 200,000 years?), the prevailing climatic conditions in the area (arid desert outcrops or tropical streamcuts?), outcrop proximity to human-induced weathering conditions (e.g., proximity to cultivated farmland with acidic groundwater runoff), and petrographic evidence of recent leaching (Table 5).

If the outcrop sample in question is deemed to be "porous" (i.e., effective porosity and permeability are above economic requirements), the next question to be asked is "How much surface weathering (and secondary porosity creation) has occurred?" (Figure 1). Three possibilities exist: (1) The outcrop has sustained minimal weathering, and most of the porosity found is inherited from the subsurface (Type 7 rocks); (2) outcrop weathering is substantial, and most of the porosity observed is the result of recent dissolution (Type 8 and 9 rocks); or (3) some recent weathering porosity is observed, but the amount is uncertain (Type 10 rocks). All three possibilities are discussed below.

Recent Weathering Minimal (Rock Type 7)

This group includes any reservoir rock whose pore system has survived intact throughout both burial and recent outcrop diagenesis (Figure 1). The pore system of Type 7 rocks represents indigenous porosity inherited from the subsurface, and may include not only primary inter- or intragranular pores but also secondary pores that were created by either near-surface or subsurface dissolution in the geologic past. It is therefore critical that inherited secondary porosity be distinguished from secondary porosity created during recent outcrop weathering. (Criteria for the recognition of recent weathering-related porosity are discussed in the section on rock Types 8 and 9.

This facies carries the least amount of exploration risk of any of the ten categories described in this chapter.

Table 4. Porosity Risk Assessment* for a Type 3 Reservoir Example.[†]

Question	Data Available	Answer
Pre-outcrop porosity?	Petrographic point count	14% macroporosity
Estimated permeability?	P vs. K crossplot from analog in adjacent basin	50–70 md
Pre-outcrop burial depth?	Best-analog compaction curve (from Pittman and Larese, 1991, their figure 20)	2 km
How much deeper to economic porosity basement?	Best-analog compaction curve (from Pittman and Larese, 1991, their figure 20); assumes a 10% economic porosity cutoff	Probable loss of porosity to 10% by 2.5 km
Overpressuring or pressure solution likely?	No incipient pressure solution noted; framework composition not conducive to pressure solution; no overpressures observed in adjacent basin	Low probability of either
Additional secondary porosity likely?	Petrographic description	Potential dissolution of unstable lithics would add another 8% porosity

*Risk assessment:
- economic porosity basement (10%) likely to be encountered at 2.5 km if no secondary dissolution
- economic porosity basement (10%) likely to be encountered deeper (>4.5 km) if secondary porosity is present
- minimum porosity likely at 2.5 km = 10%; maximum = 18%.

[†]Iron oxide cemented lithic sandstone.

Because of the limited weathering involved, samples from this facies are highly suitable for many types of routine and special core analyses. Questions that should be addressed as part of risk assessment are: (1) How much porosity and permeability are present? What other petrophysical properties can be determined from this facies? (2) What was the maximum pre-outcrop burial depth for this facies? (3) How much deeper could this facies have been buried before compaction would have destroyed economic porosity? (4) Is there any reason to suspect that early overpressuring could exist in the subsurface that could enhance porosity at depth? (5) Do petrographic observations detect any incipient destructive diagenesis that could be more intense at greater burial depths (or that might be laterally restricted at the same depth)? Is there any new diagenesis that can be predicted? (6) Is there any potential for further porosity enhancement from ancient near-surface or burial dissolution? What leaching mechanisms are likely? What would the pore system of this facies look like after dissolution? What would be the most likely porosity and permeability? (7) Based on available seismic or well data, how deep are potential traps (prospects) in the basin? Are they deeper than this sample has been buried prior to exposure? What is the probability of finding economic porosity at this depth (Table 6)?

Recent Surface Weathering Dominant (Rock Types 8 and 9)

This group includes any potential reservoir rock whose pore system is dominated by secondary porosity developed during recent outcrop exposure (Figure 1). Because of the intense weathering involved, such samples are not suitable for routine or special core analyses. Therefore, any estimates of subsurface porosity must be predicted by less direct means, as outlined below.

The distinction between Type 8 and Type 9 lithofacies is based on the original depositional fabric of the rock. Type 8 rocks have depositional fabrics that were originally porous, but have been subsequently destroyed by intense cementation or compaction prior to outcrop exposure and leaching. Therefore, these rocks should be considered as equivalents to either Type 4 (compacted) or Type 5/6 (cemented) lithofacies, depending on the predissolution rock fabric. Petrographic identification of secondary pore types and intergranular volume (IGV) can be used to distinguish between these two end members. Accordingly, risk assessment should follow the procedures outlined previously for Types 4, 5, and 6 lithofacies, with one exception: the amount of ancient secondary porosity creation, regardless of mechanism, could be similar to that created during recent outcrop exposure (assuming that the same rock components have been dissolved, and to the same extent). Thus, laboratory-measured porosity and permeability values from weathered outcrop samples could be representative of subsurface conditions that might exist if ancient dissolution has actually occurred.

Type 9 rocks have originally tight depositional fabrics that have remained tight throughout most of their burial history but have been subjected to surface leaching processes during recent outcrop exposure. Therefore, these rocks should be considered as equivalents to Type 1 rocks (tight depositional facies), and risk assessment should follow the precedures outlined for this facies.

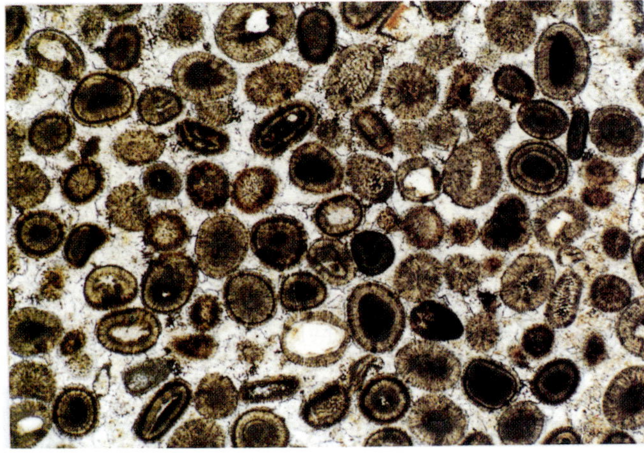

Figure 7. Example of a Type 5 lithofacies (early near-surface cemented). This Jurassic outcrop sample from Somalia is an ooid lime grainstone that was tightly cemented during early diagenesis by bladed and equant calcite (plane-transmitted light, 40×).

Degree of Recent Weathering Uncertain (Rock Type 10)

This category includes any reservoir lithology whose pore system contains appreciable amounts of secondary porosity of uncertain origin. Not surprisingly, most porous outcrop samples fall into this category, primarily because much of the physical evidence for recent dissolution is equivocal (Table 5), has been masked by a variety of diagenetic by-products, or is missing altogether. For these samples, the degree of risk associated with reservoir porosity remains uncertain.

CASE HISTORIES

Chuxiong Basin, Yunan Province, China

In this study, sandstone outcrops of Upper Triassic age were sampled and petrographically evaluated by Tobin and Nelis (1990) in an effort to characterize potential reservoir quality in two of Amoco's prospects. Both structures are interpreted as having been uplifted from a maximum burial depth of ~5 km to their present depth of ~3 km (based on available seismic data, sediment thickness estimates, and basin modeling). Given the thickness and areal extent of sandstone facies in this area, the average minimum porosity required for an economic gas play would be 12%.

Although a few examples of Type 1 (tight depositional facies) and Type 10 (abundant secondary porosity of uncertain origin) lithofacies are present, the vast majority of the outcrop samples collected are classified as Type 4 (nonporous, tightly compacted) and Type 8 (tightly compacted, but porous and weathered) rocks (Figure 1). Most of these samples are immature litharenites, feldspathic litharenites, or lithic arkoses that have suffered extreme primary porosity loss from intense mechanical and chemical compaction. Intergranular volume for these facies ranges from 8% to 29% (mostly 8%–12%), and intergranular cements are minimal, ranging from 3% to 5%.

Figure 8. Two examples of Type 6 lithofacies (burial cemented) from Trinidad. (A) A skeletal lime grainstone containing red algae (R), forams (F), and mollusks (M) is tightly cemented by poikilotopic ferroan calcite of burial origin. (B) A quartzarenite sandstone is tightly cemented by nonplanar ferroan dolomite of burial origin (plane-transmitted light, 80×).

The Type 8 lithofacies examined, although porous, exhibit unmistakable evidence for intense recent outcrop leaching, including abundant iron oxide staining, soil formation above outcrops, abundant iron oxide coatings in secondary pores, iron oxide rims and cleavage traces floating in secondary pores, and the ubiquitous occurrence of secondary pores engulfed within highly compacted rock fabrics. Type 8 rock samples are texturally and mineralogically equivalent to the Type 4 (tightly compacted) lithofacies from the same area, but are only exposed in outcrops that are downstream from cultivated farmland. It is believed that the higher acidic groundwater runoff associated with these types of exposures is responsible for the preferential dissolution observed. Therefore, the exploration risks associated with this facies are considered to be the same as that of Type 4 rocks.

Table 5. Petrologic Criteria for Distinguishing Recent Outcrop Dissolution.

Observational Scale	Level of Confidence	Description
Megascopic	Suggestive	Abundant iron oxide staining on outcrops
Megascopic	Suggestive	Soil or caliche formation on outcrops
Megascopic	Diagnostic	Recent karstic landforms and associated secondary porosity
Megascopic	Diagnostic	Soft, porous, weathered rims on outcrop surfaces with hard, tight rock beneath
Macroscopic	Diagnostic	Gradational dissolution rims on hand specimens
Microscopic	Suggestive	Abundant iron oxide coatings associated with secondary pores
Microscopic	Diagnostic	Iron oxide rims or cleavage or grain fracture traces "floating" in secondary pores
Microscopic	Suggestive	Late, postcompaction secondary pores in an otherwise tightly compacted rock
Microscopic	Suggestive	Secondary pores that postdate deep burial cements, fractures, or stylolites
Microscopic	Diagnostic	Secondary pores that postdate recent surface to near-surface cements
Microscopic	Diagnostic	Secondary pores that postdate oil entrapment by-products (e.g., bitumen)

By far, the most abundant rock type observed is the Type 4 variety. Porosities observed in this facies are considerably lower than the 12% required for an economic play (0–5%, mostly <3%). Therefore, this group is considered to be a very high-risk exploration target, unless a mechanism can be predicted for primary porosity preservation or secondary dissolution elsewhere.

Preservation of Primary Porosity

Early overpressuring can retard the rate of porosity loss from compaction, but shallow overpressures are not known to occur in this area. Similarly, early grain-coating rim cements can also retard compaction, but only minor amounts (0–6%, mostly <1%) of early quartz overgrowth cement are present in the outcrop samples examined. Furthermore, the high lithic content would probably limit the heterogeneous nucleation of quartz in these sandstones elsewhere, thereby reducing the porosity-preserving effectiveness of cementation. Alternatively, primary porosity could be more extensive in prospects that are shallower than pre-outcrop burial depth. However, vitrinite reflectance data indicate that pre-outcrop burial depth for the samples examined was ~4 km (based on a paleogeothermal gradient of 0.2% R_o/km and an assumed surface intercept of 0.2% R_o). Using an unrealistically optimistic linear compaction model, these samples are interpreted to have reached their economic porosity basement of 12% at a depth of 3 km, ~2 km less than the maximum burial depth (5 km) sustained by the two Amoco prospects (Figure 9). Experimental compaction studies, however, indicate that a linear compaction model is unrealistic; the effect of cementation and both mechanical and chemical compaction would be to reduce porosity to its economic basement at much shallower depths (Pittman and Larese, 1991). Thus, economically viable primary porosity preservation in this area is highly unlikely.

Creation of Secondary Porosity

Ancient dissolution of some of the unstable feldspars and lithic grains in these strata could yield economic porosity elsewhere. Two burial dissolution mechanisms are possible: dissolution by undersaturated water derived from shale compaction, or organic acid dissolution. The former is considered unlikely because the strata in question have been buried to a depth of ~4 km with no obvious signs of grain dissolution. If compaction water leaching had occurred, it would have taken place at considerably shallower depths. The latter mechanism (organic acids) is also considered unlikely for this area, because the undissolved sandstones examined are intercalated with organic-rich shales that have matured enough to have generated oil (~1.0% R_o), a level of maturity well past what is required for organic acids to form and migrate. The absence of organic acid dissolution may also be, in part, the result of two other factors: (1) the interbedded source rocks are dominated by gas-prone Type III kerogen, a potentially poor source of liquids (including organic acids); and (2) any acids or other types of undersaturated pore fluids that might have reached the sandstones in question would likely have been somewhat ineffective at creating secondary porosity because of the lack of an open, permeable pore system (destroyed during early burial by compaction). The only realistic mechanism for ancient dissolution would be meteoric (near-surface) leaching associated with paleoexposure surfaces. Ancient near-surface leaching is more likely to create significant secondary porosity because of the higher rock surface area exposed, exposure-related fracturing and pressure unloading, and the higher fluid flow rates involved. Therefore, porous sandstone reservoirs might exist below paleoexposure surfaces (unconformities). Accordingly, unconformity-related prospects may be less risky than the structural prospects previously identified.

Table 6. Porosity Risk Assessment* for a Type 7 Reservoir Example.

Question	Data Available	Answer
Measured porosity and permeability?	Petrographic point count + routine core analysis	14% macroporosity, 110 md permeability
Pre-outcrop burial depth?	Best-analog compaction curve (from Pittman and Larese, 1991, their figure 20)	1.5 km
How much deeper to economic porosity basement?	Best-analog compaction curve (from Pittman and Larese, their figure 20); assumes a 10% economic porosity cutoff	Probable loss of porosity to 10% by 2.5 km; 8% by 3 km
Overpressuring or pressure solution likely?	No incipient pressure solution noted; framework composition not conducive to pressure solution; no overpressures observed in adjacent basin	Low probability of either
Any destructive diagenesis likely?	Petrographic description	No incipient burial cements noted
Additional secondary porosity likely?	Petrographic description	Potential dissolution of unstable lithics would add another 6% porosity
How deep are prospects in the basin?	Seismic data only	Structural traps at 3 km

*Risk assessment:
- economic porosity basement (10%) likely to be encountered at 2.5 km if no secondary dissolution
- prospects at 3 km, likely primary porosity remaining = 8%
- potential for additional secondary porosity of 6%
- minimum porosity likely at 3 km = 8%; maximum = 14%.

Disposition of Prospect

Because of the high risks associated with reservoir quality in the subsurface, the two Amoco prospects under evaluation were not drilled. Subsequent wells drilled in this area by the Chinese have not penetrated the Triassic sandstone.

West-Central Myanmar

In an effort to assess the degree of reservoir risk prior to drilling, Murphy et al. (1991) described sandstone outcrop samples from the Paunggyi Formation (Paleocene to early Eocene in age) in the Chindwin Basin of Myanmar (Burma). The depth of the reservoir at the drilling prospect (the Yenan structure in Block B) was estimated to be ~6500 ft (2 km). Given the thickness and areal extent of sandstone facies in this area, the average porosity required for an economic oil play would have been ~15%.

Approximately two-thirds of the outcrop samples described in this study are immature litharenites and feldspathic litharenites containing minor amounts of porosity (mostly <3%). These samples include tightly compacted Type 4 lithofacies and Type 4/6 hybrids whose pore systems were destroyed by a combination of compaction and burial cementation (mostly calcite, dolomite and siderite, and minor quartz). The remaining samples are more mature quartzarenites, sublitharenites, and subarkoses. These are classified as Type 10 (recent weathering uncertain) rocks (for those containing more than 15% porosity), or as Type 5 (early near-surface cemented), Type 6 (late burial cemented), or Type 4/6 (compacted/burial cemented hybrid) rocks (for less porous examples). Intergranular volume is slightly lower for the immature sample group (mostly <25%) and higher for the more mature group (mostly >25%). Intergranular cements range from 2% to 32%.

Exploration risks are considered to be high for the immature, highly compacted sandstone facies (Type 4 and Type 4/6) in the Paunggyi Formation, unless primary porosity preservation or secondary dissolution can be predicted elsewhere. Early overpressuring is a potential mechanism for retarding the rate of porosity loss from compaction; overpressures are known to exist in the basin from previous drilling reports. However, the timing and both lateral and vertical extents of overpressure are uncertain. Similarly, early grain-coating rim cements can also retard compaction, but only minor amounts of early cement (quartz and some calcite) are present in the outcrop samples examined, and they do not appear to have significantly reduced the amount of porosity loss from compaction (based on IGV data). Alternatively, porosity could be higher for this facies if the drilling prospect is significantly shallower than pre-outcrop burial

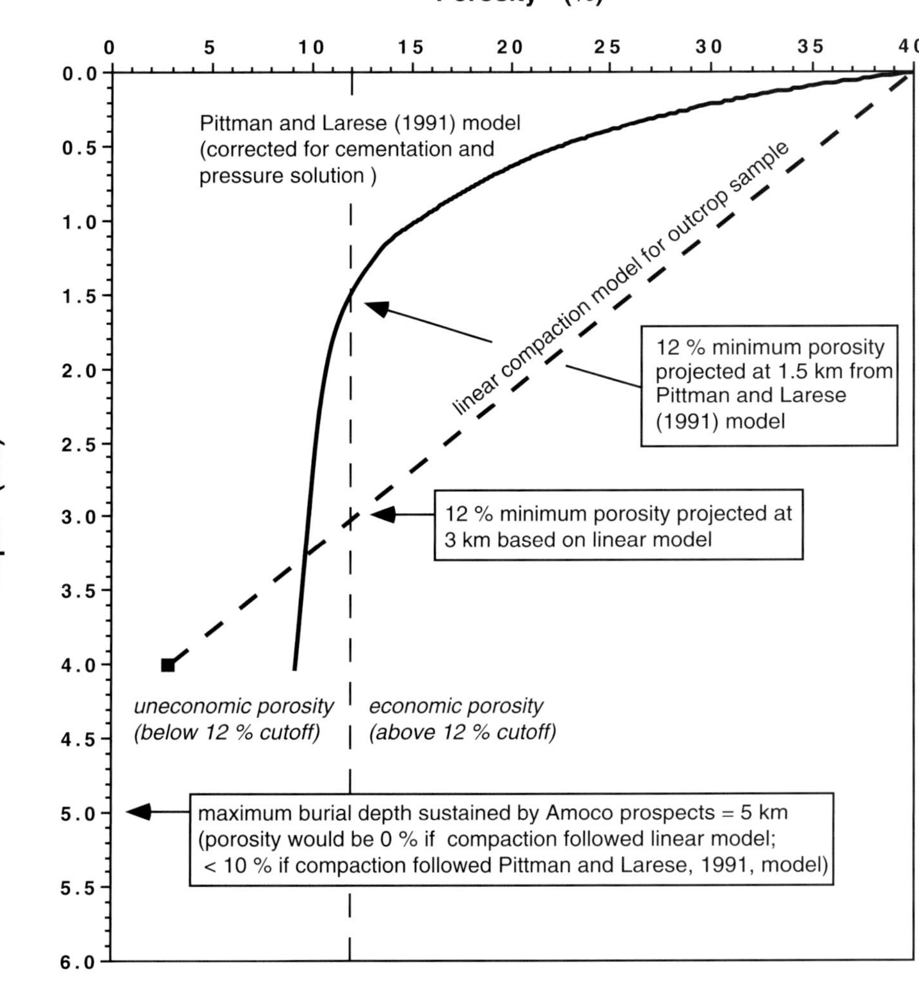

Figure 9. Prediction of pre-served primary porosity from compaction modeling in the Chuxiong Basin (China) example. The linear model (an unrealistically optimistic tool) predicts loss of porosity to the economic minimum (12%) at 3 km, which is the present-day depth of the drilling prospects but is 2 km less than maximum burial depth (5 km). The Pittman and Larese (1991) model (corrected for cementation and pressure solution) suggests that the economic porosity minimum would be encountered at much shallower depths (1.5 km). Therefore, economic porosity should not be expected in the drilling prospects.

depth. However, vitrinite reflectance data from this interval (0.5% R_o) indicate that pre-outcrop burial depth (7500 ft; 2.3 km) was only about 1000 ft (0.3 km) deeper than the drilling prospect (based on a paleo-geothermal gradient of 0.04% R_o/1000 ft and an assumed surface intercept of 0.2% R_o). Furthermore, experimental compaction studies by Pittman and Larese (1991, their figure 21) indicate that the 15% porosity basement for immature lithic sands like these would be considerably shallower than the drilling prospect [at ~2700 ft (0.8 km)] (Figure 10). This estimate does not include the additional risk associated with concomitant burial cementation observed in some of the outcrop samples. Dissolution of these cements (and/or unstable framework grains) is not considered to be a realistic mechanism for creating additional porosity because of the poor plumbing system that is characteristic of these highly compacted rocks.

Exploration risks are lower for the more mature sandstone facies (Types 5, 6, and 10). These sandstones have a less ductile framework composition, and consequently less compaction, but porosity values are still mostly below the 15% economic limit because of the effects of burial cementation. However, dissolution of cements and/or framework grains is more likely to be an effective mechanism for creating additional secondary porosity for these samples because of their more permeable plumbing system (for circulating undersaturated pore fluids). Indeed, some of the sandstones in this group (Type 10 facies) contain an appreciable amount of secondary porosity, but its origin (ancient or recent?) is uncertain. Either way, the combination of primary and secondary porosity for this facies could be in the 15%–20% range. Ancient subsurface dissolution and near-surface leaching associated with paleoexposure are viable mechanisms for creating secondary porosity for these facies.

In order to increase the odds of drilling success in this basin, Murphy et al. (1991) recommended that (1) a regional provenance study be conducted to map the localities of the more mature sandstone facies, and (2) seismic data be used to identify potential subaerial exposure surfaces. These criteria were intended to match the best sandstone compositions with the highest probability of ancient near-surface dissolution.

Disposition of Prospect

Because of the high risks associated with reservoir presence and quality in the subsurface, the Yenan prospect was not drilled. However, a second prospect

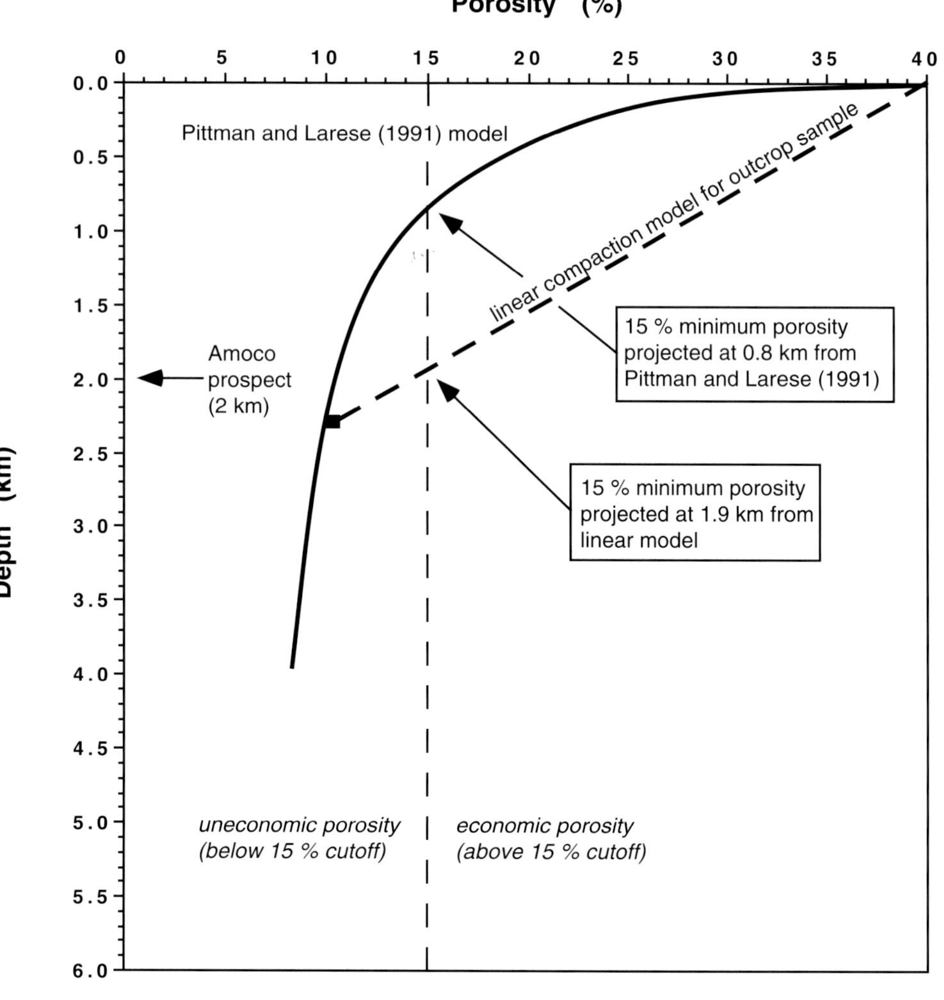

Figure 10. Prediction of preserved primary porosity from compaction modeling in the Chindwin Basin (Myanmar) example. The linear model (an unrealistically optimistic tool) predicts loss of porosity to the economic minimum (15%) at 1.9 km, slightly less than the present-day depth of the drilling prospect (2 km). The Pittman and Larese (1991) model suggests that the economic porosity minimum would be encountered at much shallower depths (0.8 km). Therefore, economic porosity should not be expected in the drilling prospect.

was drilled in the same block in early 1992 (Amoco #1 Uyu). This prospect encountered low-porosity lithic sandstone of slightly younger age (Miocene) whose framework composition is analogous to the Paunggyi sandstones described by Murphy et al. (1991). The pore system of this sand was greatly reduced below 2 km to <8% by a combination of mechanical compaction and cementation (mainly carbonates, zeolites, and authigenic clays). This mode of pore destruction closely matches predictions made by Murphy et al. (1991) in addition to three other independent Amoco studies of outcrop samples in Block B (Taylor et al., 1993).

Central Taurids, Turkey

In this study, carbonate outcrops of Jurassic age were petrographically studied by Tobin (1992) as part of an early risk evaluation of the central Taurids in southwestern Turkey. Two potential reservoirs were described: dolomites from the Jurassic Hendos Formation and limestones from the Jurassic Pisarcukuru Formation. The Hendos samples are medium to coarse crystalline, planar dolomites lacking any evidence of their original depositional fabric. Porosity for this group ranges from 2% to 6%, with permeabilities of <0.2 md. Therefore, these dolomites were initially considered

Type 2 rocks (uncertain depositional facies), with an associated high degree of exploration risk. Later petrographic data, however, suggested that this facies could have higher porosity and permeability values under the following conditions: (1) Incipient fracturing of this brittle lithology was observed in thin section. More intense fracturing elsewhere could increase the porosity by a few percentage points, and would greatly increase permeability. (2) Incipient paleokarst (vuggy and skelmoldic secondary porosity) was observed in all samples, suggesting the potential for higher porosities wherever karstic dissolution was more pervasive. (3) The rock matrix consists of planar dolomite crystals of near-surface to shallow burial origin and late planar to nonplanar dolomite of suspected burial origin. If the burial dolomite postdates hydrocarbon migration, this facies could contain up to 12% porosity (based on thin-section point-count data) in structures that existed at the time of migration because of the cement retarding effect of hydrocarbons (Wilson, 1977).

The Pisarcukuru Formation samples are medium-grained, well-sorted ooid-skeletal lime grainstones containing <1% porosity and 0.01 md of permeability. Porosity loss is the result of both compaction (63%; both mechanical and chemical) and early, near-surface cementation (37%; including isopachous micritic, isopachous bladed, syntaxial, and minor equant cements). Thus, these samples are considered Type 4 (compacted) and Type 5 (early near-surface cemented) hybrid lithofacies. The exploration risks associated with reservoir quality were initially regarded as high because of low porosity and permeability. Petrographic data, however, suggest that exploration risk is moderate for this lithofacies, because of the potential for the following: (1) lateral cement pinchout could result in up to 15% intergranular porosity, and (2) the dissolution of chemically unstable grains (including ooids, some foraminifera, and mollusk fragments) could have contributed additional porosity elsewhere.

Disposition of Play

Because of a variety of technical risk factors, this play was discontinued by Amoco, and no drilling prospects were recommended.

FUTURE RESEARCH

Predicting subsurface porosity and permeability from outcrop materials is risky business. To be successful, both ancient burial history and associated diagenesis as well as recent diagenesis and associated porosity modification must be accurately determined from petrographic or geochemical clues preserved in the rocks. These preserved signposts of complete diagenesis and porosity evolution are in part straightforward and in part extremely subtle. The weak links in this evaluation involve recognizing ancient and recent diagenesis for which petrographic criteria are limited or equivocal, such as petrographic criteria for recognizing and quantifying recent leached porosity, and criteria for recognizing recent pore-filling cements. Another weak link is the estimation of potential permeability in cemented reservoir rocks (Type 5 and Type 6 lithofacies) whose cements are predicted to be absent because of lateral pinchout or ancient dissolution. These areas of investigation are considered fertile ground for future research.

CONCLUSIONS

Outcrop observations can greatly assist in reservoir risk assessment, particularly in frontier basins where subsurface data are sparse. Outcrop exposures provide a three-dimensional view of sedimentary facies in addition to unlimited rock sample availability for laboratory analyses. The systematic, decision-tree methodology outlined in this chapter can greatly enhance the efficiency and completeness of outcrop-based reservoir prediction studies.

Every potential reservoir rock exposed in outcrop has a pore system that is the end product of its original depositional facies and subsequent diagenetic history, including both pre-exposure diagenesis and recent weathering effects. Therefore, reliable risk assessment must consider depositional and diagenetic history, including both ancient near-surface and subsurface diagenesis and recent weathering.

The discovery of tight reservoir facies in surface outcrop exposures does not necessarily mean that a high degree of risk should be assigned to subsurface porosity preservation. Tight facies may be assigned a low degree of risk whenever the following diagenetic conditions can be predicted: (1) recent destructive diagenesis of originally porous facies, (2) natural subsurface fracturing, (3) dolomitization of low-porosity limestone, (4) drilling prospect depths that are sufficiently shallower than the outcrop analog to preserve porosity, (5) early overpressuring or early thin rim cementation, (6) ancient secondary dissolution, (7) ancient karsting, (8) lateral or vertical cement pinchout in the subsurface, or (9) petroleum inhibition of cementation or other destructive diagenesis.

The discovery of porous reservoir facies at the surface does not necessarily guarantee that the same rocks will be porous in the subsurface. Recent surface leaching or fracturing can create secondary porosity that is not likely to exist in the same formation in the subsurface.

ACKNOWLEDGMENTS

The author thanks Dick Larese, Ron Nelson, and Paul Wagner for constructive reviews of an earlier version of this manuscript. Thanks are also extended to AAPG reviewers Andrew Leonard, Pascual Marquez, and Jon Gluyas. Many of the concepts presented in this paper are the by-products of fruitful conversations with fellow colleagues, in particular Dick Larese, Ione Taylor, Mary Nelis, Tim Murphy, and Christine Skirius. I am indebted to you all. Thanks also to Amoco for permission to publish this work.

REFERENCES CITED

Bloch, S., 1991, Empirical prediction of porosity and permeability in sandstones: AAPG Bulletin, v. 75, p. 1145–1160.

Bloch, S., and K.P. Helmold, 1995, Approaches to predicting reservoir quality in sandstones: AAPG Bulletin, v. 79, p. 97–115.

Byrnes, A.P., 1994, Empirical methods of reservoir quality prediction, *in* M.D. Wilson, ed., Reservoir quality assessment and prediction in clastic rocks: SEPM Short Course 30, p. 9–21.

Cant, D.J., 1986, Diagenetic traps in sandstones: AAPG Bulletin, v. 70, p. 155–160.

Dravis, J.J., and D.A. Yurewicz, 1985, Enhanced carbonate petrography using fluorescence microscopy: Journal of Sedimentary Petrology, v. 55, p. 795–804.

Dunnington, H.V., 1967, Aspects of diagenesis and shape change in stylolitic limestone reservoirs: Elsvier Publishing Co. Ltd. 7th World Petroleum Congress Proceedings, April 2–8, Mexico, v. 2, p. 339–352.

Feazel, C.T., and R.A. Schatzinger, 1985, Prevention of carbonate cementation in petroleum reservoirs, *in* N. Schneidermann and P.M. Harris, eds., Carbonate cements: SEPM Special Publication 36, p. 97–106.

Folk, R.L., 1987, Detection of organic matter in thin sections of carbonate rocks using a white card: Sedimentary Geology, v. 54, p. 193–200.

Goldstein, R.H., 1988, Cement stratigraphy of Pennsylvanian Holder Formation, Sacramento Mountains, New Mexico: AAPG Bulletin, v. 72, p. 425–438.

Kerans, C., 1988, Karst-controlled reservoir heterogeneity in Ellenburger Group carbonates of West Texas: AAPG Bulletin, v. 72, p. 1160–1183.

Lucia, F.J., and R.P. Major, 1994, Porosity evolution through hypersaline reflux dolomitization, in B.H. Purser, M.E. Tucker, and D.H. Zenger, eds., Dolomites: a volume in honor of Dolomieu: International Association of Sedimentologists Special Publication 21, p. 325–341.

Meshri, I.D., 1989, On prediction of reservoir quality through chemical modeling (abs.): AAPG Bulletin, v. 73, p. 390–391.

Meshri, I.D., and P.J. Ortoleva, 1990, Prediction of reservoir quality through chemical modeling: AAPG Memoir 49, 175 p.

Moore, C.H., 1989, Carbonate diagenesis and porosity: Amsterdam, Elsevier, Developments in Sedimentology 46, 338 p.

Moore, C.H., and Y. Druckman, 1981, Burial diagenesis and porosity evolution, Upper Jurassic Smackover, Arkansas and Louisiana: AAPG Bulletin, v. 65, p. 597–628.

Murphy, T.B., R.C. Tobin, and W.W. Dorsey, 1991, Reservoir risk assessment of Paleocene–Lower Eocene outcrop samples from west-central Myanmar: unpublished Amoco report, 51 p.

Nelson, R.A., 1985, Geological analysis of naturally fractured reservoirs: Houston, Gulf Publishing Company, 320 p.

Ottmann, R.D., P.L. Keyes, and M.A. Ziegler, 1976, Jay Field, Florida—a Jurassic stratigraphic trap, in J. Braunstein, ed., North American oil and gas fields: AAPG Memoir 24, p. 276–286.

Pittman, E.D., and R.E. Larese, 1991, Compaction of lithic sands: experimental results and applications: AAPG Bulletin, v. 75, p. 1279–1299.

Purser, B.H., 1978, Early diagenesis and the preservation of porosity in Jurassic limestones: Journal of Petroleum Geology, v. 1, p. 83–94.

Purser, B.H., A. Brown, and D.M. Aissaoui, 1994, Nature, origins and evolution of porosity in dolomites, in B.H. Purser, M.E. Tucker, and D.H. Zenger, eds., Dolomites: a volume in honor of Dolomieu: International Association of Sedimentologists Special Publication 21, p. 283–308.

Rittenhouse, G., 1972, Stratigraphic-trap classification, in R.E. King, ed., Stratigraphic oil and gas fields—

classification, exploration methods and case histories: AAPG Memoir 16, p. 14–28.

Saller, A.H., D.A. Budd, and P.M. Harris, 1994, Unconformities and porosity development in carbonate strata: ideas from a Hedberg Conference: AAPG Bulletin, v. 78, p. 857–872.

Scherer, M., 1987, Parameters influencing porosity in sandstones: a model for sandstone porosity prediction: AAPG Bulletin, v. 71, p. 485–491.

Scholle, P.A., L. Stemmerik, and D.S. Ulmer, 1991, Diagenetic history and hydrocarbon potential of Upper Permian carbonate buildups, Wegener Halvø area, Jameson Land Basin, East Greenland: AAPG Bulletin, v. 75, p. 701–725.

Surdam, R.C., and L.J. Crossey, 1987, Integrated diagenetic modeling: a process-oriented approach for clastic systems: Annual Review of Earth and Planetary Science, v. 15, p. 141–170.

Taylor, I.L., L.E. McRae, and G.O. Smith, 1993, Sedimentology and diagenesis of Tertiary sandstones from the Chindwin Basin, Myanmar (Burma): a case study for predicting reservoir quality from outcrop (abs.): AAPG Annual Convention Program, April 25–28, New Orleans, p. 188–189.

Tobin, R.C., 1985, Reservoir development in the Ellenburger Group of West Texas: a diagenetic Jambalaya (abs.): AAPG Bulletin, v. 69, p. 312.

Tobin, R.C., 1991, Pore system evolution vs. paleotemperature in carbonate rocks; a predictable relationship? (abs.): Organic Geochemistry, v. 17, p. 271.

Tobin, R.C., 1992, Petrography and core plug analyses of outcrop samples, central Taurids, southwestern Turkey: unpublished Amoco report, 26 p.

Tobin, R.C., and M.K. Nelis, 1990, Prediction of subsurface reservoir quality from outcrop samples collected in the Chuxiong Basin, Yunan Province, southern China: unpublished Amoco report, 34 p.

Wagner, P.D., and R.K. Matthews, 1982, Porosity preservation in the upper Smackover (Jurassic) carbonate grainstone, Walker Creek field, Arkansas: response of paleophreatic lenses to burial processes: Journal of Sedimentary Petrology, v. 52, p. 3–18.

Weyl, P.K., 1959, Pressure-solution and the force of crystallization: phenomenological theory: Journal of Geophysical Research, v. 64, p. 2001–2025.

Weyl, P.K., 1960, Porosity through dolomitization: conservation-of-mass requirements: Journal of Sedimentary Petrology, v. 30, p. 85–90.

Wilson, H.H., 1977, "Frozen-in" hydrocarbon accumulations or diagenetic traps—exploration targets: AAPG Bulletin, v. 61, p. 483–491.

Wood, J.R., 1994, Geochemical models, in M.D. Wilson, ed., Reservoir quality assessment and prediction in clastic rocks: SEPM Short Course 30, p. 23–40.

Chapter 2

Prediction of Porosity in Compacted Sands

Jon Gluyas[1]
BP Exploration de Venezuela SA
Caracas, Venezuela

Christopher A. Cade
BP Norge UA
Stavanger, Norway

ABSTRACT

We present a new porosity–depth relationship for clean, rigid grain (quartz, feldspar) sands under hydrostatic burial. This allows the prediction of porosity in uncemented sandstones to an accuracy of ±2.5 porosity units at 95% confidence levels. The relationship was derived using experimental data from laboratory compaction experiments and field data for buried uncemented sandstones from around the world. The equation is:

$$\phi = 50\exp\left(\frac{-10^{-3}z}{2.4+5\times10^{-4}z}\right)$$

where porosity (ϕ) is in percentages and depth (z) is in meters.

By scaling this relationship in terms of effective stress rather than depth, it can be used to provide an equally accurate prediction of porosity for uncemented sands in overpressured settings. This is done using the following equation:

$$z' = z - \left(\frac{u}{(\rho_r - \rho_w)g(1-\phi_\Sigma)}\right)$$

where z' = effective burial depth (in meters); z = burial depth (in meters); ρ_r = density of rock (in Kgm^{-3} [kilograms per cubic meter]) = typically 2650; ρ_w = density of water (Kgm^{-3}) = typically 1050; g = gravity (in ms^{-2} [meters per second squared]) = 9.8; ϕ_Σ= average porosity of overburden = typically 0.2; and u = overpressure (in MPa [megapascals]).

We propose that there is considerable value in a "compaction only" porosity–depth relationship. A compaction-only trend allows the accurate prediction of porosity in uncemented sandstones, and gives a maximum porosity baseline to which cement volumes, and resultant cemented sandstone porosities, can be compared. If both cemented and uncemented sandstone data are included to produce a "porosity loss–depth" relationship, the resultant scatter (typically ±5 porosity units for a given depth) in the relationship limits its usefulness.

[1] Present affiliation: Monument Oil and Gas, London, United Kingdom

Prior to drilling, the new relationships may be used either to predict the porosity of sands that are known to be uncemented or to place an upper limit on the porosity estimated for sandstones either known or suspected to be cemented.

INTRODUCTION

When exploring for oil and gas reservoirs, the prediction of porosity before drilling is an important part of the decision-making process. Higher porosity usually gives higher in-place and reserve volumes, and higher production flow rates (because permeability is typically proportional to porosity). The prediction of porosity can be made using several different methods:

- *Using global, regional, or local porosity–depth curves.* This is probably the most commonly used method. For sandstones covering an appropriate depth range, porosity is plotted against depth, and regression is used to establish a best-fit line or curve. This line or curve, or the equation that describes it, can then be used to predict porosity for the undrilled prospect. Examples of such curves take logarithmic (Athy, 1930; Weller, 1959), power function (Baldwin and Butler, 1985), or linear (Selley, 1978) forms, and they can give accurate porosity predictions, particularly when they describe porosity variation with depth for one sandstone type with a consistent burial and diagenetic character. When the curve has wider scope (for example, if it is based on data from sandstones of differing age, mineralogy, diagenetic history, or geographic location), the resultant scatter in the data will usually increase predictive uncertainty.
- *Using a wider group of porosity-controlling variables to predict porosity.* The porosity–depth curve uses depth as the only control on porosity. Scherer (1987) assembled a diverse group of sandstones from around the world for which there were data on a range of mineralogical and textural parameters, as well as age and depth. Using multiple regression, a predictive equation for porosity was established with a wide group of input variables. Porosity was principally correlated with burial depth, but there were other important variables that determined the degree to which porosity was reduced for each depth increment. For example, a sandstone with numerous ductile grains loses more porosity, for an equivalent depth of burial, than does a pure quartz sandstone, all other conditions being equal. The multiple-regression method of Scherer (1987) attempts to account for a wider group of controls, and produces well-constrained relationships between porosity and a range of rock properties. However, the method has serious limitations. For example, one of the primary porosity determinants in Scherer's data

set is sorting, and in many cases it will be difficult or impossible to predict sorting for an undrilled sandstone with any confidence.

In order to predict porosity from a specific rock property, prior to drilling, the particular rock property must be known or predictable. Depth to a prospect is usually well constrained, so a porosity–depth curve is easily applied. However, unless the porosity–depth curve is based on local data, the prediction uncertainty will commonly be too large to be useful.

We present an approach to porosity prediction based on the compaction process and parameters that are usually predictable: depth and pressure. The results of high-pressure laboratory compaction tests on quartzose sands are combined with porosity data from a varied data set of buried and uncemented sands to produce porosity–depth and porosity–effective stress relationships for the compaction process.

A third parameter, detrital mineralogy, may also be predictable in many cases, but in this chapter only quartzose (and quartzo-feldspathic) sands are considered. A similar approach applied to lithic sandstones is the subject of a separate paper in preparation.

This approach focuses on compaction only and is therefore applicable, on its own, to uncemented sands. Most published porosity prediction relationships (Athy, 1930; Selley, 1978; Baldwin and Butler, 1985), consider total porosity loss with burial, which includes both compaction and cementation; this in part accounts for the wide range of porosity, at any given depth, in their data sets. We propose that there is value in separating compaction and cementation effects, partly to narrow the range of predicted porosity at a given depth, but also because uncemented sandstones are a frequent exploration target.

POROSITY LOSS IN SANDSTONES

Recently deposited sands are usually highly porous, often >40% (Pettijohn, 1975). Buried sands and sandstones have lower porosities (Table 1). Porosity is reduced by two distinct and commonly independent processes: compaction and cementation. The difference between these two processes is most easily considered in terms of pore volume and bulk rock volume change. Compaction involves the reduction in pore space associated with shortening of the sand column under burial loading (reduction in both pore volume and bulk rock volume). Cementation, in contrast, involves a reduction in pore space without any reduction of bulk rock volume. Pore space is filled, partly or completely, by newly precipitated solid material.

In general, compaction is the dominant porosity-reduction process during early and shallow burial. Exceptions to this include cemented beach sandstones, calcrete paleosol sandstones, evaporitic sabkha sandstones, and sandstones that are exposed at a marine surface during a break in sedimentation. Such examples are not uncommon, but in terms of the sandstone component in the sediment column, they are rarely volumetrically significant.

As depth of burial and/or age increases, the relative importance of cementation tends to increase. There are, however, numerous exceptions to this pattern. For example, in the Gulf of Mexico (Table 1), uncemented sands are recorded at depths >3800 m. Also, there are many instances that demonstrate that the degree of cementation can vary widely even within the same formation, depth range, and geographical area. For example, in the Central Graben of the North Sea, both moderately and almost totally quartz-cemented sandstones of Upper Jurassic age occur at the same depth (~4000 m), but are areally separated by only a few hundred meters (Gluyas, this volume; Ramm et al., this volume). This variability produces much of the scatter on published porosity–depth plots. This scatter means that these relationships cannot give predictive accuracy to better than ±5 porosity units. We suggest that much of the spread in the published porosity–depth curves is derived from the mixing of compaction, cementation, and overpressure effects.

There are two reasons for considering compaction on its own. First, uncemented buried sandstones are not uncommon, and any predictive relationship for porosity in such lithologies should be based on data that exclude the impact of cementation. Second, where cemented sandstones are expected, a compaction-only trend gives a maximum porosity for a given depth. Deviations to values lower than this may be estimated from predictions of likely authigenic mineral volumes. Such predictions may come from diagenetic modeling, stratigraphic context, regional data, and other methods. Porosity loss due to cementation is considered elsewhere (Primmer et al., this volume; Gluyas and Coleman, 1992; Gluyas and Witton, this volume) and in other publications (Robinson and Gluyas, 1992; Gluyas et al., 1993a, b), but is outside the scope of this chapter.

DERIVATION OF A POROSITY–DEPTH RELATIONSHIP FOR COMPACTION

We have used two complementary approaches to derive a porosity–depth relationship for compacted sands. The first approach uses results from laboratory compaction experiments on quartzose sands, and involves conversion of the experimentally applied stress to burial effective stress and depth. The second approach uses field data from uncemented sands around the world to extend the porosity–depth trend. The close coincidence between the relationships derived from the two approaches gives considerable confidence in their use for porosity prediction.

Approach 1—Using Experimental Data

Most laboratory compaction experiments have a civil engineering or soil science application and are performed at much lower applied stresses than those involved in burial to depths in excess of a few hundred meters. There are, however, a few examples of triaxial compression experiments at higher stresses. Vesic and Clough (1968) published the results of tests at loads of ≤30 MPa (4350 psi) on medium-grained, uniform, slightly micaceous quartz sand. These stresses can be equated to hydrostatic burial to depths of ~1400 m (4500 ft). Vesic and Clough (1968) also offered a mathematical proof that under most deep-burial conditions (<100 MPa), the sand behaved as a linearly deformable solid, with a modulus of deformation proportional to the mean normal stress. In other words, the porosity–stress relationship is linear. In addition, Atkinson and Bransby (1978) state that at high stresses and under what they term normal consolidation conditions (no overpressure), sands will consolidate (compact) so that the relationship between incremental applied stress and volume change/porosity reduction is linear.

The conversion of laboratory pressures and resultant porosity values to effective stress or burial depth relationships raises two important issues. First, does laboratory compaction over necessarily short time periods (hours or days at most) involve the same processes as burial compaction over much longer geological time periods? Second, how can experimental stresses can be converted to burial stresses?

Burial Compaction Processes—Laboratory Replication

Sands consisting of quartz grains compact during burial by a combination of two processes. These processes are the mechanical response to stress (grain slippage, rotation, and fracturing) and the chemical process of pressure dissolution at grain contacts. Under near-surface conditions, all compaction is by mechanical processes. At a depth of ~1000 m (3000 ft), pressure and temperature become sufficiently elevated to permit pressure dissolution (Füchtbauer, 1967). From then on, compaction may include a combination of mechanical and chemical processes.

Sands that contain ductile grains in addition to quartz (or other rigid grains) commonly lose porosity more quickly than quartzose sands because the compactional process is different (Kurkjy, 1988). Examples of such grains are mudstone clasts, glauconite grains, phyllitic and schistose metamorphic grains, and micas. Not only do these grains deform more easily and rapidly than rigid grains, but their deformation will permit a greater degree of slippage and rotation of more-rigid grains such as quartz. As a result, a sand with such grains will lose porosity at a faster rate, particularly during early burial, than a quartzose sandstone under the same conditions. In addition, all compaction may be effected by mechanical processes, and pressure dissolution may be relatively unimportant. In this chapter, we present our findings for the compaction of quartzose (and similar) sands.

Table 1. Porosity, Depth, and Overpressure Data for Uncemented Sandstones.*

Well/Field (Interval)	Location	Depth (km)	Porosity (%)	Overpressure (MPa)	Age	Formation	Reference Comments
Frigg	NOCS	1.90	29	Hydrostatic	Eocene	Frigg	Abbots, 1991
Forties	UKCS	2.10	27	Hydrostatic	Paleocene	Forties	Abbots, 1991
Piper	UKCS	2.40	24	Hydrostatic	Jurassic	Piper	Abbots, 1991
Draugen	NOCS	1.60	30	Hydrostatic	Jurassic	Røgn	Bjørlykke et al., 1986
Abroath	UKCS	2.50	24	Hydrostatic	Paleocene	Forties	Abbots, 1991
Balmoral	UKCS	2.13	25	Hydrostatic	Paleocene	Andrew	Abbots, 1991
Maureen	UKCS	2.65	21.5	Hydrostatic	Paleocene	Maureen	Abbots, 1991
Montrose	UKCS	2.48	24	Hydrostatic	Paleocene	Forties	Abbots, 1991
Statfjord	NOCS	2.48	27	Hydrostatic	Jurassic	Brent Gp	Spencer et al., 1987
Statfjord	NOCS	2.55	23	Hydrostatic	Jurassic	Dunlin Gp	Spencer et al., 1987
Statfjord	NOCS	2.70	22	Hydrostatic	Triassic	Statfjord	Spencer et al., 1987
Odin	NOCS	2.00	29	Hydrostatic	Eocene	Frigg	Spencer et al., 1987
Nord Øst Frigg	NOCS	1.95	28	Hydrostatic	Eocene	Frigg	Spencer et al., 1987
Øst Frigg	NOCS	1.90	29	Hydrostatic	Eocene	Frigg	Spencer et al., 1987
Heimdal	NOCS	2.10	25	Hydrostatic	Paleocene	Heimdal	Spencer et al., 1987
Sleipner Øst	NOCS	2.33	26	Hydrostatic	Paleocene	Heimdal	Spencer et al., 1987
Cod	NOCS	2.95	20	Hydrostatic	Paleocene	Forties	Spencer et al., 1987
Gullfaks	NOCS	1.84	27	Hydrostatic	Jurassic	Brent Gp	Spencer et al., 1987
Gullfaks	NOCS	1.90	30	Hydrostatic	Jurassic	Dunlin Gp	Spencer et al., 1987
Gullfaks	NOCS	1.96	25.5	Hydrostatic	Triassic	Statfjord	Spencer et al., 1987
Troll	NOCS	1.40	30	Hydrostatic	Jurassic	Sognefjørd	Spencer et al., 1987
Gyda (field crest)	NOCS	3.87	24.5	14.0	Jurassic	Gyda	Gluyas, 1997
Ivanhoe	UKCS	2.39	24.6	Hydrostatic	Jurassic	Piper	Parker, 1991
Block 22/11	UKCS	2.40	25.3	Hydrostatic	Paleocene	Forties	unpublished
Well 22/6a-2	UKCS	2.27	27.5	Hydrostatic	Paleocene	Forties	unpublished
Well 22/17-1	UKCS	2.54	22	Hydrostatic	Paleocene	Forties	unpublished
Well 23/26a-2	UKCS	2.63	21.6	Hydrostatic	Paleocene	Forties	unpublished
Well 23/21-1	UKCS	2.57	23.2	3.0	Paleocene	Andrew	unpublished
Well 23/27-4	UKCS	2.75	22.5	3.7	Paleocene	Forties	unpublished

Well 30/6-2	UKCS	2.87	26	16.0	Paleocene	Andrew	unpublished
Gryphon	UKCS	1.74	33	Hydrostatic	Eocene	Balder	Newman et al., 1993
Melibur	Sumatra	0.75	37.5	Hydrostatic	Miocene	Sihapas	Gluyas and Oxtoby, 1995; 0.75 km = max. burial
Guapo	Trinidad	1.22	33	Hydrostatic	Pliocene	Cruse	Beard, 1985; max burial ≥ quoted depth
Fyzabad	Trinidad	1.37	33	Hydrostatic	Pliocene	Cruse & Forest	Low, 1985; max burial ≥ quoted depth
North Soldado	Trinidad	2.06	30	11.0	Pliocene	Manzanilla	Radowsky and Iqubal, 1985, as above
Caño Limon	Colombia	2.32	24.5	Hydrostatic	Eocene	Mirador	McCullough, 1990
Maui	New Zealand	2.74	25	Hydrostatic	Eocene	Kapuni	Abbot, 1990
War-Wink	U.S.A. (Texas)	4.60	38	40.0	Permian	–	Luo et al., 1994; sand composition not stated
East Breaks 165	U.S.A. (GOM)	3.84	30	20.3	Plio-Pleistocene	–	unpublished
East Breaks 165	U.S.A. (GOM)	3.32	30	16.1	Plio-Pleistocene	–	unpublished
Mississippi Canyon 20	U.S.A. (GOM)	3.00	31	15.8	Plio-Pleistocene	–	unpublished
Mississippi Canyon 20	U.S.A. (GOM)	1.84	31	5.8	Plio-Pleistocene	–	unpublished
Viosca Knoll 989	U.S.A. (GOM)	2.07	32	8.4	Plio-Pleistocene	–	unpublished
Mississippi Canyon 109	U.S.A. (GOM)	1.75	33	5.9	Plio-Pleistocene	–	unpublished

*NOCS = Norwegian Continental Shelf; UKCS = United Kingdom Continental Shelf; GOM = Gulf of Mexico.

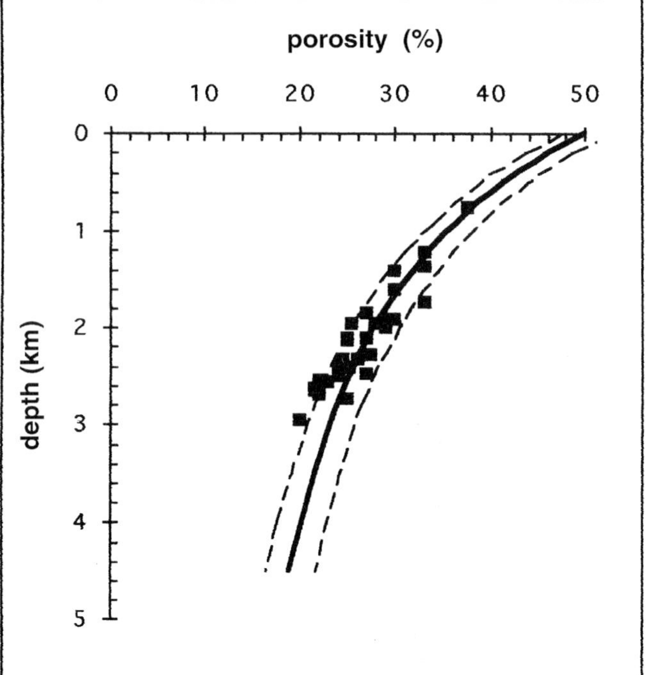

Figure 1. Squares = porosity to depth relationship for hydrostatically pressured (Table 1), uncemented, rigid-grain sandstones. Solid line = porosity to depth relationship for experimental (laboratory) compaction of natural sand under simulated hydrostatic conditions. Dashed lines = ±2.5% porosity variance from experimental curve of Atkinson and Bransby (1978).

The porosity losses sustained during laboratory compaction experiments occur through grain rearrangement, slippage, and fracture (Füchtbauer, 1967) rather than pressure dissolution and reprecipitation. As a result, the experiments may be expected to successfully reflect compaction in quartzose sands undergoing shallow burial and in lithic sands (with ductile grains) over most burial stress conditions. For quartz sands at greater depths, where chemical compaction processes (e.g., pressure dissolution) become important, the experiments cannot replicate burial compaction. Despite this, several observations allow the results of quartz sand compaction experiments to be used. The experiments of Vesic and Clough (1968) are restricted to relatively low stresses (burial to ~1400 m) and, over this depth range, compaction is predominantly by mechanical processes. In addition, porosity–depth data for buried sands show close agreement with the experimental results where the two relationships overlap; the linear consolidation trend described by Atkinson and Bransby (1978) matches observed data that show a similar linear trend at depths in excess of 1000 m (Figure 1).

Laboratory Stress—Conversion to Burial Depth

It is difficult, if not impossible, to exactly replicate the stress conditions of hydrostatic burial in a laboratory experiment. To achieve this requires a controlled, zero-stress confinement in all orientations with the exception of the vertical, and the application of a unidirectional vertical stress upon the sample. In addition, pore fluid pressure must be closely controlled to reflect hydrostatic fluid pressure increase during burial.

A triaxial test apparatus was used for the laboratory experiments of Vesic and Clough (1968). This gives the closest approximation to true burial. Confinement of the sample is controlled by the steel sample chamber and a membrane across which a pressure can be applied to an oil cushion. Moreover, pore fluid can be drained from the sample in such a way that pore fluid pressure can be controlled during the experiment. The simulation of burial loading is then applied, by a piston, to the top of the sample.

In these experiments, a constant cell pressure was applied to the oil cushion surrounding the sample, and pore fluid pressure within the sample was controlled so that no excess or overpressure (above hydrostatic) was allowed to develop. Although these triaxial tests do not mimic burial compaction exactly, they do, we believe, give a close approximation. In particular, they do not allow the buildup of overpressure in the pore fluid (which would occur if the pore fluid in the sample were to be completely confined), nor do they allow pore fluid to drain away freely during compaction, which would mean all vertical loading being applied at the grain contacts. To equate experimental stresses to burial stresses, the following conversion is used: 1 psi (experiment) = 1 ft of burial, or 0.02262 MPa = 1 m of burial.

For the purpose of making the conversion, it is assumed that fluid pressure in the experiments was maintained at the equivalent of hydrostatic for the applied stress. In a normally pressured (hydrostatic) sedimentary sequence, lithostatic load typically increases by 22.622 MPa/km. Using this conversion, the curve of Atkinson and Bransby (1978), using experimental data of Vesic and Clough (1968), may be redrawn as a porosity–depth curve for normally pressured clean, rigid-grain (quartz, feldspar) sand. Atkinson and Bransby (1978) do not quote an equation for their curve, so an empirical fit has been made to their experimentally derived (initially) loose-packed-sand porosity–stress curve.

The equation for this empirical fit is

$$\phi = 50\exp\left(\frac{-10^{-3}z}{2.4+5\times10^{-4}z}\right) \tag{1}$$

where porosity (ϕ) is in percentages and depth (z) is in meters.

This equation gives a close fit to the experimental data and is in close agreement with field data (Figure 1). In particular, it closely replicates the nonlinear deformation behavior of loose sands at low stress. Vesic and Clough (1968) performed a number of experiments with differently packed sand. Starting porosity ranged from the 50% used here to 40% for a well-packed sand. However, the initial differences in porosity reduced to ~1% at stresses equivalent to 1 km of burial.

Approach 2—Using Field Data for Uncemented Sands

Porosity data for a group of uncemented buried sandstones from a range of sedimentary basins have been collated (Table 1) and used to produce a predictive porosity–depth relationship. Figure 1 shows the experimentally derived curve (from equation 1) and porosity–depth data from this group of hydrostatically pressured clean sands. The sands in the group are currently at their maximum burial depth, have hydrostatic pore fluid pressure, and contain <5% ductile grains or dispersed argillaceous material. The close coincidence between the curve and the data in the depth range 1000–3000 m suggests that the conversion between experimental and burial stresses described in the preceding text is valid. With backup from the real buried sand data, the new compaction curve for uncemented sandstones provides a reliable means of predicting porosity for uncemented sands. Ninety-five percent of the buried sandstone data plotted in Figure 1 falls within 2.5 porosity units of the experimentally derived, and extrapolated, porosity–depth curve.

Prediction of Porosity in Clean, Overpressured Sands

In buried sands that have pore fluid pressures significantly higher than hydrostatic, anomalously high porosities can be preserved. The fluid overpressure (difference between actual pore pressure and hydrostatic pressure at the same depth) supports part of the burial loading, and thus reduces the effects of compaction. Overpressured sands commonly have higher porosity than hydrostatic sands have at the same depth. Compactional porosity reduction is the result of effective stress, which is the difference between lithostatic stress (the stress due to the weight of overlying sediments) and pore fluid pressure. An overpressured sand will have an effective stress that is equivalent to a hydrostatically pressured sand at a shallower depth (this shallower depth may be termed the "effective burial depth" of the overpressured sand); this depth difference is proportional to the magnitude of the overpressure. The concept of effective stress and effective burial depth can be used to correct our porosity prediction for an overpressured situation. The effective burial depth for an overpressured sand, or the depth under hydrostatic conditions at which the sand would have the same effective stress, is given by the equation

$$z' = z - \left(\frac{u}{(\rho_r - \rho_w)g(1 - \phi_\Sigma)} \right) \qquad (2)$$

where z = burial depth (in meters); z' = effective burial depth (in meters); ρ_r = density of overlying rock column (in Kgm^{-3}) = 2650 (suggested value); ρ_w = density of water (in Kgm^{-3}) = 1050 (suggested value); g = gravity (in ms^{-2}) = 9.8; ϕ_Σ = porosity (as a fraction of 1) = 0.2 (suggested value); and u = overpressure (in MPa).

The values given here for density are typical average values for sediments and formation brines. The

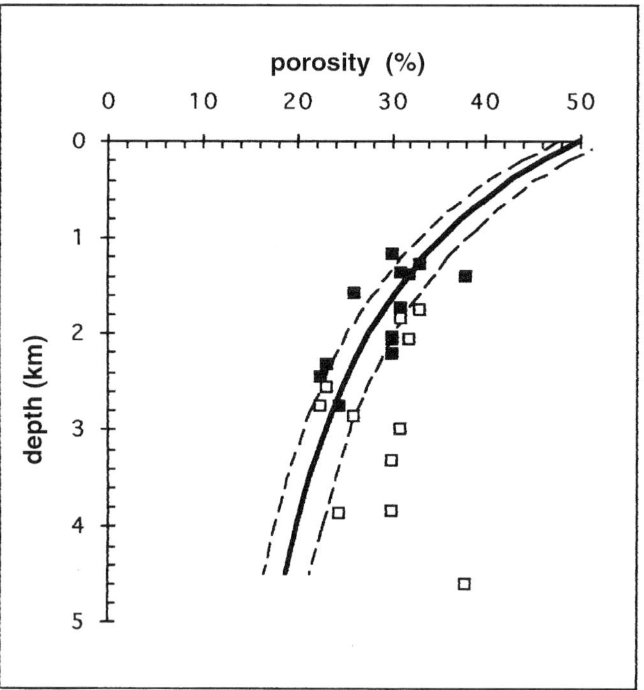

Figure 2. Open squares = porosity to depth relationship for overpressured, uncemented, rigid-grain sandstones. Solid squares = same sandstones as open squares but with depth recalculated as effective burial depth, where effective burial depth (in kilometers) = depth (in kilometers) − 0.08 × overpressure (in MPa). Solid line = porosity to depth relationship for experimental (laboratory) compaction of natural sand under simulated hydrostatic conditions. Dashed lines = ±2.5% porosity variance from experimental curve. For uncorrected data, 33% of predictions fall within the ±2.5% range on the mean. This improves to 50% after correction for overpressure. With a more generous distribution about the mean (±5%), prediction accuracy for overpressure-uncorrected data improves to 50%, while prediction accuracy for overpressure-corrected data improves to 92%.

suggested value for porosity is a typical figure for the average porosity (sands and muds) for a 3-km-thick column of 80% mud and 20% sand. With the suggested values, equation 2 works well for burial depths in the range of 2 to 4 km, and sand to mudstone ratios of 15:85 to 25:75. For shallow burial depths and/or unusual sand-to-shale ratios, the average porosity of the overburden can be calculated by integrating the area under simple empirically derived porosity–depth functions (Baldwin and Butler, 1985).

Equation 2 can be simplified to the following conversions in order to calculate the effect of 1 MPa (~140 psi) of overpressure in terms of effective depth differential. Using the above figures, 1 MPa of overpressure equates to ~80 m less burial. Thus,

$$z' = z - 80u \qquad (3)$$

can be used to derive effective burial depths for substitution into the porosity–depth equation.

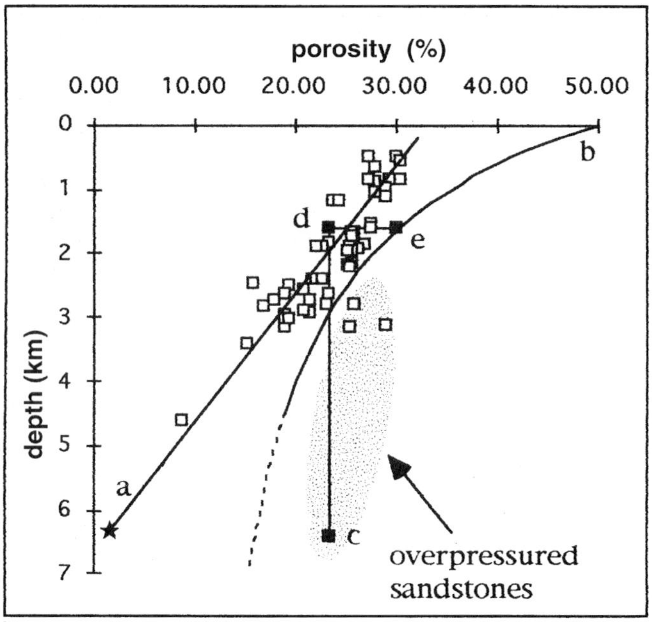

Figure 3. Porosity–depth plot for Cretaceous Tuscaloosa sandstones of Louisiana (open squares; Thomson, 1979), including those from Alma Plantation field (point c, solid squares). Simple regression of porosity on depth = a. Compaction curve from this chapter (point b). The Tuscaloosca sandstones at >6 km typically require mud weights of 16–17 lb/gal (1.59–1.69 kg/L) (Gill, 1980); equal to ~8500–9000 psi (~60 MPa) overpressure. From equation 2, this is 4.8 km less than the actual depth of 6.4 km (d). Using equation 1, we would predict a porosity of ~30% for the Tuscaloosa sandstones from Alma Plantation field compared with an actual average porosity for the sands of 23.5% (star). The fact that the Alma Plantation sandstones are partially cemented by chlorite has been ignored in this calculation. Thomson (1979) quotes 30% cement porosity (point e) for these sandstones; that is exactly as predicted from our compaction equation.

Figure 2 contains a plot of porosity against depth for a variety of overpressured sandstones. A plot of porosity against depth for these overpressured sandstones, but with their depths adjusted to effective burial depth using equation 2, is also shown in Figure 2. There is close agreement between the experimentally derived porosity–depth trend and the measured porosity/effective burial depth data for the overpressured sandstones.

DISCUSSION

We believe that the porosity–depth relationship presented in this chapter gives the explorer a valuable tool for the prediction of sandstone porosity ahead of drilling. The previously published global porosity–depth curves carry too much uncertainty for uses other than the prediction of average behavior of a sand under burial. The compaction equations presented here give an understanding of how porosity is lost as a function of effective stress. This allows the effect of overpressure as well as burial depth to be accounted for. Overpressured sands are generally more porous than their hydrostatically pressured counterparts. The porosity of overpressured sands is not accurately predicted by simple empirical porosity–depth relationships. The Cretaceous Tuscaloosa sandstone of the U.S. Gulf Coast area (Thomson, 1979) is a prime example of porous sandstone at depth that would have been predicted as having insufficient porosity for commercial flow rates, if a simple porosity–depth function from other Cretaceous sandstones in the area (Figure 3) had been used. However, had equations 1 and 3 been used along with an overpressure estimate, the maximum predicted porosity at 6.4 km would have been estimated to be 30%, rather than 2%, from the empirical relationship (compared with actual porosity of 23.5% minus cement porosity of 30%).

CONCLUSIONS

Porosity–depth functions are the most common method used for the prediction of sandstone porosity ahead of exploration drilling. Where they include sandstones with varying degrees of compaction, cementation, or overpressure, they will often carry a large range of uncertainty.

Experimental data on the relationship between sandstone porosity and confining stress provide the exploration geoscientist with an alternative method for predicting porosity at depth in exploration prospects. The equations presented in this chapter are derived from experimental data and have been tested against a diverse worldwide set of buried-sand data. They allow prediction of porosity to ±2.5 porosity units (at 95% confidence levels) for clean, normally pressured, uncemented sands. Moreover, through the link between porosity and effective stress, the equations deliver a methodology that allows prediction of anomalous porosity preservation due to the effects of overpressure.

ACKNOWLEDGMENTS

We thank BP Exploration for permission to publish this paper. We also thank Mike Bowman, David Epps, Shona Grant, Nick Milton, Steve Franks, and John Aggett for their thorough and constructive reviews.

REFERENCES CITED

Abbot, W.O., 1990, Maui field, *in* E.A. Beaumont and N.H. Foster, eds., Structural traps I: AAPG Treatise of Petroleum Geology, Atlas of Oil and Gas Fields, p. 1–25.

Abbots, I.L., 1991, United Kingdom oil and gas fields, 25 years commemorative volume: Geological Society of London Memoir 14, 573 p.

Athy, L.F., 1930, Density, porosity and compaction of sedimentary rocks: AAPG Bulletin, v. 14, p. 1–24.

Atkinson, J.H., and P.L. Bransby, 1978, The mechanics of soils: an introduction to critical state soil mechanics: London, McGraw Hill, 375 p.

Baldwin, B., and C.O. Butler, 1985, Compaction curves: AAPG Bulletin, v. 69, p. 622–626.

Beard, J.T., 1985, The geology of the Guapo field, *in* B. Carr-Brown and J.T. Christian, eds., Transactions of the 4th Latin American Geological Congress, Trinidad and Tobago 1979: Arima, Trinidad & Tobago Ltd, July 7–15, 1979, Port of Spain, p. 684–689.

Bjørlykke, K., P. Aagaard, D. Dypvik, D.S. Hastings, and A.S. Harper, 1986, Diagenesis and reservoir properties of the Jurassic sandstones from the Haltenbanken area, offshore mid-Norway, *in* Proceedings of the Norwegian Petroleum Society, Symposium on Habitat of Hydrocarbons—Norwegian Oil and Gas Finds: Stavanger, Norwegian Petroleum Society, p. 275–286.

Füchtbauer, H., 1967, Influence of different types of diagenesis on sandstone porosity, *in* W. Ruhl, ed., Proceedings of the 7th World Petroleum Congress, Mexico: Mexico City, vol. 2, p. 353–367.

Gill, J.A., 1980, Multiparameter log tracks, Tuscaloosa Woodbine pressures (abs.): Oil and Gas Journal, November, 3, p. 20–22.

Gluyas, J.G., this volume, Poroperm prediction for reserves growth exploration: Ula Trend, Norwegian North Sea, *in* J. Kupecz, J.G. Gluyas, and S. Bloch, eds., Reservoir quality prediction in sandstones and carbonates: AAPG Memoir 69, p. 201–210.

Gluyas, J.G., and M.L. Coleman, 1992, Material flux and porosity changes during diagenesis: Nature, v. 356, p. 52–53.

Gluyas, J.G., and N.H. Oxtoby, 1995, Diagenesis a short (2 million year) story—Miocene sandstones of central Sumatra, Indonesia: Journal of Sedimentary Research, v. A65, p. 513–521.

Gluyas, J.G., A.G. Robinson, D. Emery, S.M. Grant, and N.H. Oxtoby, 1993a, The link between petroleum emplacement and sandstone cementation, *in* J.R. Parker, ed., Petroleum geology of NW Europe: London Geological Society Publication, Proceedings of 4th Conference, p. 1395-1402.

Gluyas, J.G., A.G. Robinson, and S.M. Grant, 1993b, Geochemical evidence for a temporal control on sandstone cementation: AAPG Studies in Geology 36, p. 23–33.

Gluyas, J.G., and T. Witton, this volume, Porosity and permeability prediction for wildcat exploration drilling, Miocene Southern Red Sea, *in* J. Kupecz, J.G. Gluyas, and S. Bloch, eds., Reservoir quality prediction in sandstones and carbonates: AAPG Memoir 69, p. 163–176.

Kurkjy, K.A., 1988, Experimental compaction studies of lithic sands: M.S. thesis, Rosensteil School of Marine and Atmospheric Sciences, University of Miami, 101 p.

Low, B.M., 1985, The geology of the Fyzabad main field, *in* B. Carr-Brown and J.T. Christian, eds., Transactions of the 4th Latin American Geological Congress, Trinidad and Tobago:Arima, Trinidad & Tobago Ltd, July 7–15, 1979, Port of Spain, p. 714–719.

Luo, M., M.R. Baker, and D.V. LeMone, 1994, Distribution and generation of the overpressure system, eastern Delaware Basin, western Texas and southern New Mexico: AAPG Bulletin, v. 78, p. 1386–1405.

McCullough, C.N., 1990, Caño Limon field, Llanos Basin, Colombia, *in* E.A. Beaumont and N.H. Foster, eds., Structural traps II: AAPG Treatise of Petroleum Geology, Atlas of Oil and Gas Fields, p. 65–93.

Newman, M.St.J., M.L. Reeder, A.H.W. Woodruff, and I.R. Hatton, 1993, The geology of the Gryphon oil field, *in* J.R. Parker, ed., Petroleum geology of NW Europe: London Geological Society Publication, Proceedings of 4th Conference, p. 123–133.

Parker, R.H., 1991, The Ivanhoe and Rob Roy fields, Block 15/21a-b, UK North Sea, *in* I.L. Abbots, ed., United Kingdom oil and gas fields, 25 years commemorative volume: Geological Society of London Memoir 14, p. 331–338.

Pettijohn, F.J., 1975, Sedimentary rocks (3d ed.): New York, Springer-Verlag, 628 p.

Primmer, T.P., C.A. Cade, I.J. Evans, J.G. Gluyas, M.S. Hopkins, N.H. Oxtoby, P.C. Smalley, E.A. Warren, and R.H. Worden, this volume, Global patterns in sandstone diagenesis: their application to reservoir quality prediction for petroleum exploration, *in* J. Kupecz, J.G. Gluyas, and S. Bloch, eds., Reservoir quality prediction in sandstones and carbonates: AAPG Memoir 69, p. 61–78.

Radowsky, B., and J. Iqubal, 1985, Geology of the North Soldado field, *in* B. Carr-Brown and J.T. Christian, eds., Transactions of the 4th Latin American Geological Congress, Trinidad and Tobago 1979; Arima, Trinidad & Tobago Ltd, July 7–15, 1979, Port of Spain, p. 759-769.

Ramm, M., A.W. Forsberg, and J. Jahren, this volume, Porosity depth trends in deeply buried Upper Jurassic reservoirs in the Norwegian Central Graben: an example of porosity preservation beneath the normal economic basement by grain-coating micro-quartz, *in* J. Kupecz, J.G. Gluyas, and S. Bloch, eds., Reservoir quality prediction in sandstones and carbonates: AAPG Memoir 69, p. 177–200.

Robinson, A.G., and J.G. Gluyas, 1992, Model calculations of sandstone porosity loss due to compaction and quartz cementation: Marine and Petroleum Geology, v. 9, p. 319–323.

Scherer, M., 1987, Parameters influencing porosity in sandstones: a model for sandstone porosity prediction: AAPG Bulletin, v. 75, p. 485–491.

Selley, R.C., 1978, Porosity gradients in North Sea oilbearing sandstones: Journal of the Geological Society of London, v. 135, p. 119–132.

Spencer, A.M., et al., 1987, Geology of the Norwegian oil and gas fields: Stavanger, Graham and Trotman, 443 p.

Thomson, A., 1979, Preservation of porosity in the deep Woodbine-Tuscaloosa trend, Louisiana: Gulf Coast Association of Geological Society Transactions, v. 30, p. 396–403.

Vesic, A.S., and G.W. Clough, 1968, Behaviour of granular material under high stresses: Journal of Soil Mechanics Foundation Division, v. 94, p. 661–688.

Weller, J. M., 1959, Compaction of sediments: AAPG Bulletin, v. 43, p. 273–310.

Chapter 3

Porosity Variation in Carbonates as a Function of Depth: Mississippian Madison Group, Williston Basin

Alton Brown
ARCO Exploration and Production Technology Company
Plano, Texas, U.S.A.

ABSTRACT

Log-determined porosities of argillaceous limestone, limestone, dolomitic limestone, and dolomite of the Mississippian Madison Group in the Williston Basin were analyzed to determine the influence of carbonate mineralogy, shale content, and fabric on porosity loss with depth of burial. Carbonate mineralogy and shale content strongly influence the rate of porosity loss. Argillaceous carbonates lose porosity at the greatest rate with burial, followed by clean limestone, dolomitic limestone, and dolomite. Average porosity of grain-supported limestone is not systematically higher than average porosity of mud-supported limestone in the same depth range, but there is a significant difference in the respective porosity range. Moderately to deeply buried (1.5–3 km) limestones with a grain-supported texture have a small percentage of high-porosity samples, whereas porosity distributions in matrix-supported limestones at equal burial depth cluster around the mean porosity and lack a tail of high-porosity samples. This effectively limits economic porosity in moderately to deeply buried Madison limestones to grain-supported rocks (packstones and grainstones).

Results of this study reveal characteristics of basin-scale porosity loss mechanisms. Secondary porosity formed during burial is not evident in the porosity–depth profiles. Porosity loss is strongly influenced by mineralogy; clay content greatly accelerates the rate of porosity loss in limestones. In these rocks, dolomite porosity higher than limestone porosity at a given maximum burial depth is due primarily to selective preservation of dolomite porosity. Porosity decreases with increasing temperature in rocks with otherwise similar burial (effective stress) history. The observed porosity–depth relationships roughly follow an exponential trend; this may indicate that there is some sort of feedback between porosity and the porosity reduction mechanism.

Data generated in this study can be used to predict porosity distribution at a given depth in the Mississippian strata of the Williston Basin if no other information is available. Average limestone porosity at moderate to deep burial is significantly less than the porosity required for economic development of unfractured petroleum accumulations, so average porosity cannot be used as an estimate of economic porosity in a prospect. However, the distribution of porosity in a depth range can be used to estimate the risk associated with encountering sufficient thickness of economic porosity. The presence or absence of potentially economic porosity is best evaluated as a risk statement. For this reason, the porosity cumulative frequency distribution in a given depth range is a particularly useful tool because it can be interpreted in terms of expected thickness of porosity higher than a given threshold value. If information about vertical spatial correlation of porosity is available, the distribution can be interpreted in terms of risk of finding a minimum net thickness of carbonate exceeding a threshold porosity level. These methods can be used in other wildcat exploration settings where proper calibration data have been collected. The results of this study can be used as a guide to understanding porosity distribution with depth in other Paleozoic carbonates, and perhaps be directly applied to other late Paleozoic carbonates in cratonic settings.

INTRODUCTION

Porosity in sediments is strongly influenced by depositional and early diagenetic environments. Upon burial, porosity is usually lost as pressure, temperature, and time of exposure to diagenetic environments increase. The path of the porosity loss is of major economic importance because the path determines the distribution of porosity in buried rocks, and directly influences the likelihood of economic hydrocarbon accumulations.

Porosity in chalks and deep-water limestones has long been recognized to be predominantly influenced by depth of burial (Scholle, 1978). However, early diagenesis has long been believed to have a much greater effect on porosity in shallow-water carbonate rocks than burial diagenesis (Choquette and Pray, 1970). Subsequent work has demonstrated that burial is a major control on average porosity distribution in shallow-water limestones in some basins (Schmoker and Halley, 1982; Schmoker, 1984).

The observed correlations between depth of burial and porosity raise many questions. What are the possible effects of sampling strategy on porosity–depth trends? What is the effect of lithology types on porosity loss with burial? What are the effects of depositional texture on porosity loss? How can one evaluate the tremendous scatter characteristic of shelf limestone porosity vs. depth data in order to make predictions about the likelihood of encountering economic porosity?

In an attempt to answer some of these questions, porosity data from wells penetrating the Madison Group (Mississippian) carbonates of the Williston Basin were analyzed for correlations to lithology, depositional fabric, depth, and temperature gradient. Results presented here document the strong effect of carbonate mineralogy and relatively weak effect of depositional texture on average porosity trends with depth of burial. The major effect of depositional texture is preservation of a broader distribution of porosity in grain-supported limestone textures than in mud-supported limestone textures in a given depth range.

PREVIOUS WORK

Previous studies of porosity–depth relationships in shallow-water carbonates evaluated two types of data: measurements of reservoir porosity of petroleum accumulations (Schmoker et al., 1985), and measurements of porosity of the carbonate on a basinal scale (Schmoker and Halley, 1982).

Reservoir porosity studies such as those done by Schmoker et al. (1985) provide a basis for evaluating the properties of discovered and undiscovered economic reservoirs. These types of data provide estimates of the expected porosity and other reservoir characteristics of fields once they are discovered. Reservoir properties may or may not reflect the properties of the basin-scale carbonate, because for a reservoir to be economic, it must have some minimum reservoir quality. The substantial fraction of carbonates that have low porosity are not represented in the field databases. This means that the porosity distribution of reservoirs may not indicate the likelihood of encountering economic porosity during wildcat exploration, but it may indicate the porosity likely to be found in economic discoveries.

Basin-scale studies of shallow-water carbonate porosity–depth relationships initiated with the now-classic study by Schmoker and Halley (1982). Because all carbonates were sampled in the approach used by those authors, conclusions based on their data apply to basin-scale trends of porosity development. They clearly demonstrated the systematic decrease of porosity with burial depth, evident not only in intermediate to deep samples (Schmoker and Halley, 1982), but also in shallow samples (Halley and Schmoker, 1983). These data also document the different porosity-loss pathways of dolomite and limestone, and the approximately exponential shape to the trend of porosity loss (Schmoker and Halley, 1982). A subsequent study by Schmoker (1984) evaluated a number of carbonate porosity–depth trends and found that, in general, they followed a log-linear relationship to time-temperature index (TTI), a measure of thermal exposure.

Two problems crop up in the previous studies: (1) using reservoir data to characterize basinal porosity trends and (2) possible bias in selection of basin-scale data. As long as the differences between the uses of reservoir data and basinal data are recognized, no confusion results. Just as reservoir data cannot be used to characterize the basin-scale changes in limestone reservoir quality in an unbiased manner, the basin-scale data analyses include data that have uneconomic porosity, so averages of these data do not reflect the porosity of expected discoveries. This distinction has not always been clear in previous studies. Schmoker (1984) used reservoir data in order to characterize basin-scale carbonate properties in addition to other basin-scale data. Although this does not invalidate his results, some of the scatter in the trends of porosity to TTI may be explained by the use of reservoir data sets to characterize a basin-scale process.

Some previous basin-scale studies used data collection techniques that can introduce a bias of unknown magnitude. For example, the sampling approach of Schmoker and Halley (1982) introduced a bias to their data. They measured average porosity of intervals with relatively constant porosity. This approach does not measure porosity on a volumetric basis, because a short interval of low porosity carries as much weighting as a much longer interval of high porosity. As average porosity decreases with depth, the high-porosity intervals generally become shorter and the low-porosity intervals become longer. This means that the shallow intervals may be systematically biased toward low-porosity, whereas the deeper intervals may be biased toward high porosity. Although interval length varies by a factor of 7 in their data, the variable interval length has not introduced enough error to invalidate the conclusions of Schmoker and Halley (1982). Quantitative use of this data set for testing models of porosity loss may be affected by this bias, however.

This brief review indicates why this study was undertaken in the manner it was. The main goal of the study was to evaluate basin-scale trends of porosity evolution. This requires careful consideration of the sampling strategy in order to collect an unbiased, basin-scale database. The secondary goal is developing methods of quantitative prediction of wildcat risk for reservoir quality. It is believed that the only successful strategy for predicting economic porosity is to consider the distribution of all porosity within the interval of interest: uneconomic porosity as well as potentially economic porosity levels.

STUDY AREA AND METHODS

Setting

The Madison Group of the Williston Basin was chosen for study for the following reasons. (1) The same general stratigraphic interval could be sampled at various depths of burial. By sampling rocks of a narrow age range, time effects on porosity loss can be minimized. (2) A large number of well logs with modern porosity logging packages are available over a large geographic area. This removes possible geographic bias. (3) Modern depths of burial in the study area in eastern and central Williston Basin are close to maximum burial experienced by the basin, although there has been minor Cenozoic erosion around the margin of the basin. Thus, present subsurface temperatures are probably close to the maximum temperatures to which the carbonates were exposed. Williston Basin subsidence is somewhat episodic, but samples from different burial depths have very similar relative subsidence curves (Figure 1). This means that differences in burial history are not likely to affect porosity evolution. (4) A variety of carbonate mineralogies and textures is present in the Madison Group. Mineralogies can be identified from log analysis due to the relatively simple mineralogical composition of the carbonates. This allows for accurate porosity determination for a range of carbonate lithologies. Generalized carbonate fabric data are available from cuttings descriptions.

The Madison Group is a Kinderhookian to Merimecean, argillaceous carbonate, carbonate, and evaporite unit that in fills the Williston Basin by progradation from east and south North Dakota (Peterson and MacCary, 1987). The group shoals upward from argillaceous limestone and shale near the base (Lodgepole Limestone) through interbedded carbonates and anhydrite (Mission Canyon Formation) to salina salts (Charles Formation) at the top of the group. Facies tracts generally prograde to the west and northwest, resulting in distribution of nearly all major lithofacies over essentially all parts of central and western North Dakota, although the facies tracts are not exactly contemporaneous.

Data Collection and Analysis

Data were collected from the Lodgepole and Mission Canyon formations of the Madison Group. Thirty-one wells were selected to sample the Madison Group at a range of depths and geographic areas in western and central North Dakota (Figure 2, Table 1). Porosities and mineralogies were determined from wireline logs. Digitized wireline well logs were not available, so lithology and porosity were determined by manual cross-plotting techniques of data from paper copies. To create a bias-free data set, porosity and lithology were collected at exact 3.3-m (10-ft) depth marks on wireline well logs, starting from the base of the Charles Salt through the base of the Lodgepole limestone. Only carbonate lithologies were analyzed; shale units in the Lodgepole Formation and the evaporite beds were not evaluated.

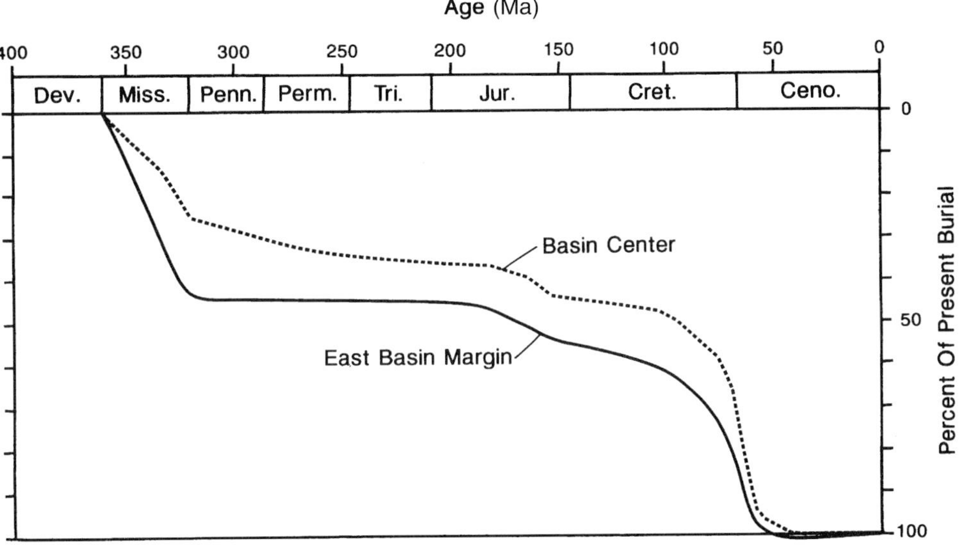

Figure 1. Comparison of burial history between a basin center well (#7 in Table 1) and a basin margin well (#32 in Table 1). The burial depth is scaled in percentage of present-day burial. The major difference in relative burial history is the preservation of the late Paleozoic–early Mesozoic age strata in the basin center and its absence in the basin margin. The basin margin well has also been exhumed somewhat more than the basin center well, but the exhumation in both cases probably does not exceed 300 m. The similarity of burial histories indicates that the porosity changes correlate to relative burial depths, not differences in burial history. Figured burial curves are constructed from undecompacted formation thicknesses.

If the log readings were unreliable at the 10-ft depth mark due to borehole conditions or bed edge effects, the interpretations were made at a depth of 0.6 m (2 ft) above the 10-ft mark. Because of the equal and arbitrarily spaced sampling interval, the sample set provides an unbiased estimate of the different lithologies and porosity in the rocks.

Necessary borehole corrections were made before mineralogical and porosity evaluation from wireline log readings. Argillaceous carbonates were identified by high gamma-ray (GR) response (>30° API units after mud weight and caliper correction) combined with elevated neutron log porosity and depressed sonic log response. Porosity of argillaceous limestones was interpreted from compensated density logs, using a grain density of 2.71 g/cm^3. Density porosity estimates are relatively insensitive to changes in matrix mineralogy in this setting because the matrix density of limestone is similar to that of the silicate minerals. In these rocks, the predominant silicate mineral is illite, which has a density of 2.77 g/cm^3 (Ellis et al., 1988), somewhat higher than 2.71 g/cm^3 assumed in the porosity model. Also, small quantities of pyrite (grain density of 5.0 g/cm^3) are routinely reported in descriptions of cuttings of the argillaceous limestones in the studied wells. These compositional differences can lead to an actual matrix density slightly higher than the assumed 2.71 g/cm^3, resulting in a small systematic bias for argillaceous samples toward low porosity. This bias is thought to be <3% in the worst case. Reported negative porosity probably represents in-situ porosity <1% combined with a matrix density >2.71 g/cm^3.

Limestones, dolomitic limestones, and dolomites were distinguished by compensated density log–compensated neutron log crossplots. The plots also provided porosity estimates. Because the wells had slightly different porosity logging tools, different charts were used on the wells as appropriate. Over intervals with questionable mineralogy (such as halite- or anhydrite-cemented limestone), cuttings description, compensated sonic tool response, and resistivity tool responses were used to confirm mineralogy. Halite- and anhydrite-cemented limestones were not included in the data set.

In addition to the porosity and lithology information, texture, temperature, and effective stress were estimated for all of the depth intervals for which porosity was measured. Depositional fabrics were estimated from commercial sample logs provided by the AMSTRAT (American Stratigraphic) Company. Most wells had AMSTRAT cuttings logs; those that did not had AMSTRAT cuttings logs available within a few miles of the analyzed wells (Figure 2). These descriptions, in most cases, could be readily correlated to the study wells. Some wells lack textural data due to lack of cuttings descriptions. Dolomites had poor description of depositional texture. Argillaceous limestones were invariably described as mudstones or wackestones. For these reasons, the effect of texture on dolomite, dolomitic limestone, and argillaceous limestone porosity was not investigated. AMSTRAT cuttings descriptions were used to group limestones by texture into four textural classes (Dunham, 1962): mudstones, wackestones, packstones, and grainstones. Boundstones were not observed in the Madison Group. Because so few grainstones were described, packstones and grainstones were combined into the single category of grain-supported rocks for some analyses.

Thermal gradient varies significantly over the Williston Basin and is apparently not directly related

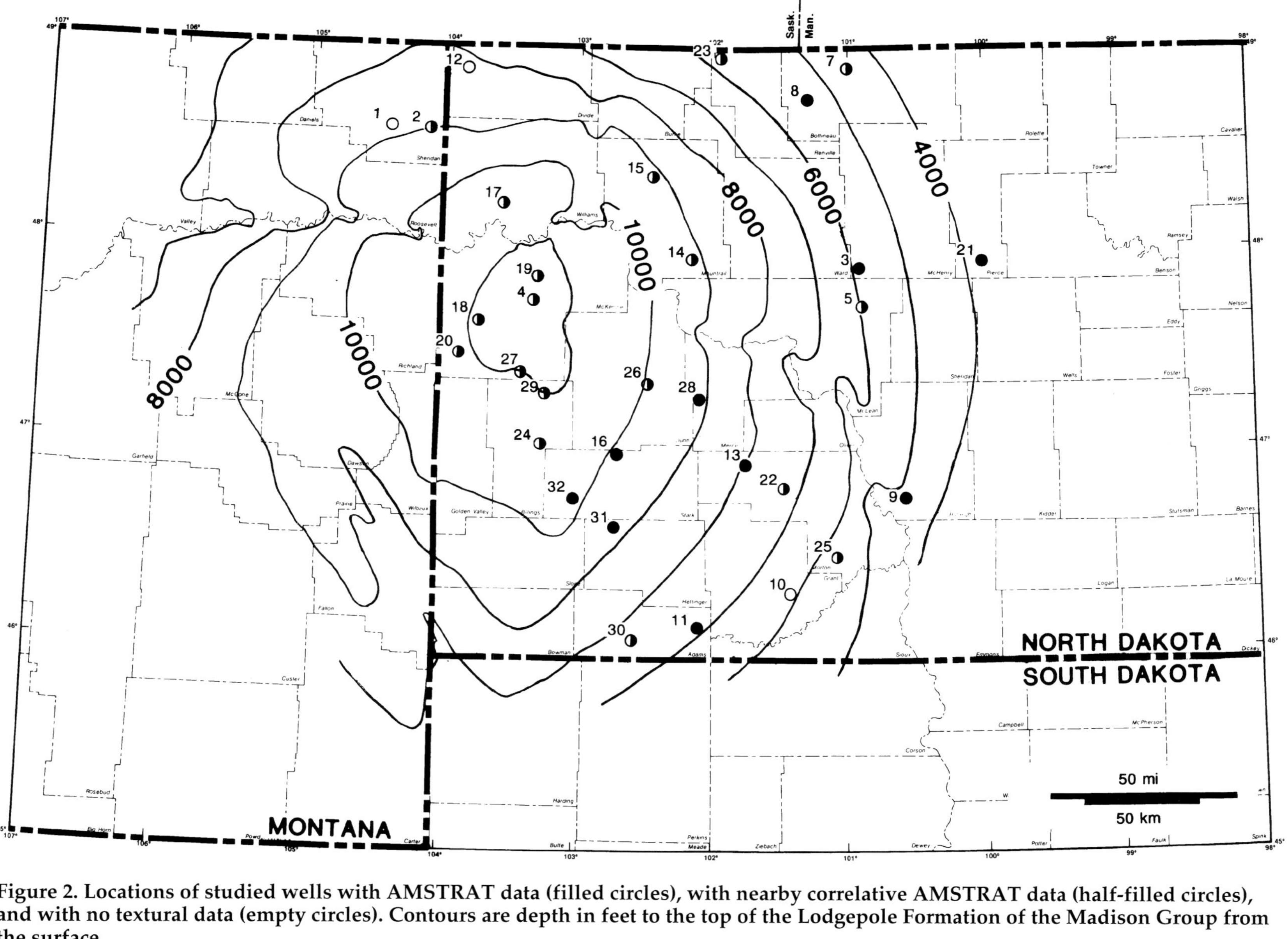

Figure 2. Locations of studied wells with AMSTRAT data (filled circles), with nearby correlative AMSTRAT data (half-filled circles), and with no textural data (empty circles). Contours are depth in feet to the top of the Lodgepole Formation of the Madison Group from the surface.

Table 1. Studied Wells.

No.	Well	Location
1	FUCE Jayhawk-Nelson # 43x-30	sec. 30 T33N/R56E (MT)*
2	Gulf Lee Mae #1-33-1a	sec. 33 T33N/R58E (MT)
3	ARCO Wunderlich #1	sec. 22 T151N/R80W
4	Amoco Sondrol #1	sec. 10 T149N/R99W
5	ARCO Klain #1	sec. 26 T149NR80W
7	Placid Rosenthal # 36-5	sec. 36 T163N/R80W
8	Cities Service Rice #1	sec. 27 T161N/R82W
9	Asmera Welch #1	sec. 31 T138N/R78W
10	Gas Prod. Enterprise BN #1	sec. 27 T132N/R86W
11	Energetics, Inc. Soelberg #23-7	sec. 7 T130N/R91W
12	Tenneco #1-1 Reistad	sec. 1 T162N/102W
13	Amoco Richter #1	sec. 26 T140 N/R88W
14	Mitchell - Elberg # 1-35	sec. 35 T152N/R90W
15	Gulf Juma #1-1-1D	sec. 1 T156N/R92W
16	Hunt Barta #1	sec. 5 T140N/R95W
17	Hunt - Treffry #1	sec. 30 T155N/R100W
18	Shell USA #22-24	sec. 24 T148N/R103W
19	Texaco Luttin #1	sec. 27 T151N/R99W
20	Shell Quinell #31-14	sec. 14 T146N/R104W
21	Getty Vetter #1	sec. 34 T152N/R73W
22	AMOCO Karch #1	sec. 6 T138N/R85W
23	Shell Lindbad #41-16	sec. 16 T163N/R87W
24	Adobe Oil 23-31X Luptak	sec. 31 T141N/R99W
25	Pennzoil Railroad Bend #2-24	sec. 2 T134N/R83W
26	Terra Resources Borth #1-35	sec. 35 T145N/R93W
27	Gulf Rough Rider Federal #1-21-30	sec. 21 T145N/R100W
28	Conoco Entze #29-1	sec. 29 T144N/R90W
29	Amoco Thompson #8-1a	sec. 29 T144N/R99W
30	Amoco J. Christman #1	sec. 28 T130N/R95W
31	Amoco Kenny # 1	sec. 14 T136N/R96W
32	Supron Privatsky #1	sec. 26 T138N/R98W

*MT = Montana.

to the amount of basin subsidence (DeFord et al., 1976). The differences in thermal gradient can be used to determine the effects of temperature differences on porosity loss with depth. As the carbonate in each well is buried, temperature increases, so that the present temperature is only the most recent part of a rock's thermal history. For this reason, thermal gradients rather than present temperatures are used to distinguish temperature populations. The wells analyzed were divided into three groups by the thermal gradient present within the well: high gradient (>31°C/km; >1.7°F/100 ft), moderate gradient (25.5°–31°C/km; 1.4–1.7°F/100 ft), and low gradient (<25.5°C/km; <1.4°F/100 ft). Because the surface temperatures of these wells are almost identical, the ranking by thermal gradient also ranks the wells by temperature at any depth. Temperature differences between the wells in high thermal gradients and low thermal gradients are about 22°C at 3 km (40°F at 10,000 ft burial).

The pore pressure of analyzed Mississippian carbonates was estimated from the potentiometric surface map of Miller and Strauz (1980). The present-day vertical effective stress was calculated by subtracting the pore pressure from the geostatic load. It is assumed that the vertical weight of overlying strata is the maximum stress on the rock, so the vertical effective stress is the maximum effective stress.

The porosity, texture, temperature, effective stress, and depth data were analyzed with the SAS (Statistical Analysis Services, Inc.) mainframe statistical package. The porosity distribution was evaluated in several ways. First, samples were grouped into arbitrary depth intervals of 152 or 305 m (500 or 1100 ft), depending upon sample density. Porosities of different mineralogies and textures were then averaged over the depth interval. The porosity averaged by lithology or texture was then plotted against depth. The advantage of this approach over a depth-regression model is that depth intervals with high sample size do not influence the porosity estimated for depth intervals with smaller sample size. Variation of porosity with depth is also not constrained to a particular functional form (such as linear and exponential).

Porosity was also evaluated by plotting the depth averages of porosity from wells with the thermal gradient ranges discussed above. The significance of temperature and other variables in an overall regression model was also evaluated by SAS GLM, a general linear

Table 2. Porosity Correlations to Gamma-Ray Intensity.

Well #	Depth Investigated (ft)	Regression Slope*	Zero G-R Intercept *	R^2
7	3800–4100	−0.00188	0.142	0.58
8	4540–4880	−0.0024	0.128	0.49
3	5300–5700	−0.00178	0.07	0.66
4	10,180–10400	−0.00012	−0.020	0.026

* Least-squares model between fractional porosity and gamma-ray (G-R) intensity in API units.

model similar to ANOVA. This was used to determine the statistical significance of class (discontinuous) variables such as the well number and texture, as well as the continuous variables of depth and effective stress, on the average porosity.

The effect of shale content on limestone porosity was of particular interest, because clay content is proposed to enhance pressure dissolution (Weyl, 1959) and, therefore, to decrease porosity by cementation in the burial environment. The interval chosen for this study is the Lodgepole Formation, which has argillaceous carbonate mudstones and wackestones interbedded with shale-free carbonate mudstones and wackestones. The relationship between porosity and shale content is best tested by comparing the density porosity to a shaliness indicator within single boreholes. Based on cuttings description, G-R intensity increases with increasing shale content and is therefore a suitable shaliness indicator for these rocks. Four of the study wells (listed in Table 2) were digitized, the G-R intensity corrected for borehole effects, and the corrected density porosity plotted as a function of G-R intensity in each well. Effects of shale content were then evaluated by linear regression for each of the four wells (Table 2).

Average porosity data for Madison Group petroleum accumulations were also compiled along with the dominant mineralogy of the reservoir rock. As noted by Lindsay (1985), reservoirs north and east of the basin center are predominantly limestone, whereas those to the south and west are predominantly dolomitized or partially dolomitized reservoirs. Reservoir lithology and porosity data were compiled from Tonnesen (1985), Tyler (1962), and field papers by Kupecz (1984), Lindsay and Kendall (1985), LeFever and LeFever (1991), Beach and Griffin (1992), and DeMis (1992). The data set is limited in two ways. First, average porosity in fractured limestone traps (such as Mondak field) is not reported in field papers, so the average reservoir porosity compiled here is believed to represent matrix porosity averages. Second, some of the field descriptions in the compilation volumes (Tyler, 1962; Tonnesen, 1985) report a limestone porosity, whereas examination of well logs indicates a dolomitic limestone reservoir. For this reason, not all reservoirs reported as limestone in this compilation may actually be limestone.

RESULTS

Effect of Carbonate Lithology

Mean porosity decreases as a function of depth for all lithologies investigated (Figure 3). The porosity–depth trends decrease in a manner quite similar to exponential porosity loss with depth, as can be seen where the porosity is plotted on a logarithmic scale (Figure 4). Figure 5 shows mean, standard deviation, and maximum porosity for each depth interval and lithology type.

Clay-free limestones (those with <30° API units corrected G-R response) show a systematic, gradual decrease of porosity with depth (Figure 5a). The smoothness of the porosity decrease probably reflects the large sample size for clean limestones compared with other carbonate lithologies sampled. Dolomitic limestones (Figure 5c) and dolostones (Figure 5d) have a more erratic porosity decrease with increasing depth. This probably reflects the small sample size at each depth interval and the more complex diagenetic history of dolomitization. At least two types of dolomite were included in the samples of this study. Dolomite in some wells is the high G-R marker bed dolomites, whereas in others, the dolomite is the product of more pervasive dolomitization. Kupecz (1984) also reports that high-porosity dolomite is associated with petrographic evidence of anhydrite secondary dissolution. Porosity of the dolomitic limestone samples is affected by both secondary anhydrite dissolution and by variable fraction of dolomitization. The rate of porosity loss of dolomite with increasing depth is only slightly less than that of dolomitic limestone. The major difference between the two trends is that dolomite has an average porosity higher than dolomitic limestone at the shallowest sample depths.

Argillaceous limestones have the lowest porosity at any depth (Figure 3). Porosity decreases rapidly to 3% at 1.5–2.6 km (5000–8500 ft), and then decreases slowly with increasing depth (Figure 5b). The samples from depths to 2.6–3.0 km (8500–10,000 ft) have an average density porosity that is negative. The negative density porosity of the argillaceous limestones probably represents porosity <1% combined with a matrix density greater than the assumed 2.71 g/cm^3, as discussed above.

Effect of Clay Content

In the three wells with samples of shallow to intermediate depth (<2 km; <6000 ft), density porosity systematically decreases with increasing G-R intensity (Figure 6a; Table 2). The observed decrease in density porosity with increasing shale content cannot be fully explained by any reasonable change in matrix density; therefore, the systematic porosity decrease is interpreted to be caused primarily by increasing clay content.

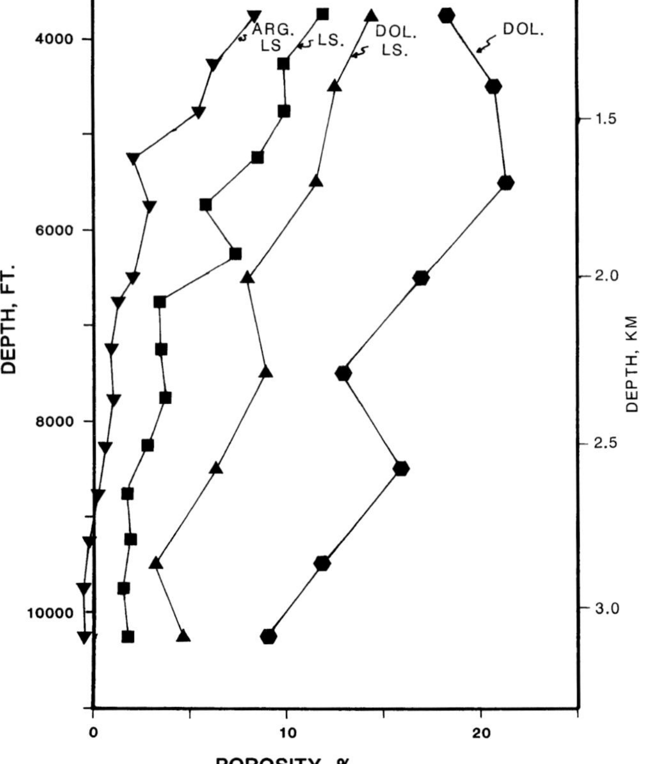

Figure 3. Mean porosity of carbonate lithologies as a function of present-day burial depth. Symbols represent average depth and average porosity for that depth and lithology. ARG. LS. = argillaceous limestone; LS. = clean limestone; DOL. LS. = dolomitic limestone; DOL. = dolomite.

The well with deeply buried (~3.3 km; 11,000 ft) samples shows no significant correlation between density porosity and GR intensity (Table 2; Figure 6b). The lack of correlation is believed to be caused by the relatively small range of porosity in the evaluated depth interval. This is consistent with the regional pattern shown in Figure 3; namely, clean limestones have very low porosity at this depth, so any further decrease in porosity with increasing clay content would not be significant. Much of the variation in log-determined porosity in Figure 6b is probably caused by variations in matrix density; this is the reason that many of the porosity measurements have negative values.

The intercept of the regression equations in Table 2 theoretically represents the porosity of a limestone with a GR intensity of zero. The intercept porosities for the first three wells in Table 2 are significantly higher than porosities shown for clean limestones at comparable depths in Figures 3 and 5. The difference is caused by the fact that almost all limestones have a GR response >10° API units. Clean limestones can have ≤30° API units of corrected GR. Porosity calculated from the regression equations using the actual values of GR intensity is within the range of average porosity shown on Figure 3.

Effect of Texture

The general linear regression model demonstrates that fabric has a significant effect on limestone porosity; this effect is much less significant than the effect of depth and that of effective stress (Table 3). Specific influence of texture was investigated by examining the mean porosity and porosity frequency distribution of different textures at different depth ranges.

Figure 7a plots the means of porosities of different limestone textures as a function of depth. At shallow and deep depths, average porosity of the different textures is quite similar. At intermediate depths (1.5–2.5 km; 5500–8500 ft), different limestone textures at the same depth range have different mean porosities. Grainstones appear to have consistently higher mean porosity than packstones, wackestones, and mudstones over this interval. These porosity differences between packstones, wackestones, and mudstones are not consistent from depth interval to depth interval, so no systematic pattern is evident from the data. This inconsistency probably results from the distinctly non-normal distribution of porosity within each texture type (Figure 8).

The fraction of high-porosity samples (porosity >8%) in different limestone textures has systematic differences that are not evident from the analysis of mean porosity (Figure 7b). At depths <1.5 km (5000 ft), all limestone textures have high percentages of samples with high porosity. From 1.5 to 2.6 km (5000–8500 ft), the grainstones and packstones have a greater percentage of high-porosity rocks than do the mudstones and wackestones. Below 2.6 km (8500 ft), all limestone textures have a low percentage of samples with porosity >8%. The systematic differences in economic porosity over the intermediate depth range are especially significant because no systematic trend in mean porosity of the different limestone textures was evident over the same depth interval.

The preservation of higher porosities in the grainstone and packstone textures relative to the mud-supported textures, without significant differences in the mean porosity, indicates that the distributions of porosities within the various fabric types are strikingly different. Grain-supported rocks have a tail of high-porosity values that is missing or much smaller in the mud-supported rocks in samples >2.1 km (7000 ft) (Figure 8).

Effect of Temperature and Effective Stress

The average porosity in clean limestones appears to decrease with increasing thermal gradients at almost all depths (Figure 9). In the general linear model, temperature is of marginal statistical significance for the model of limestone porosity as a function of depth, fabric, well, and temperature (Table 3). Although the overall effect of temperature is small, in the depth range between 2.6 and 3.0 km (8500–10,000 ft), the temperature effect is significant. The effect is believed to be most significant in this range because at shallower depths, temperature differences are very small for the different thermal gradient ranges; deeper in the basin, the average porosity in all wells is low.

Pore pressure variations are not significant compared to the geostatic load, so effective stress and depth are highly correlated. Linear regression between

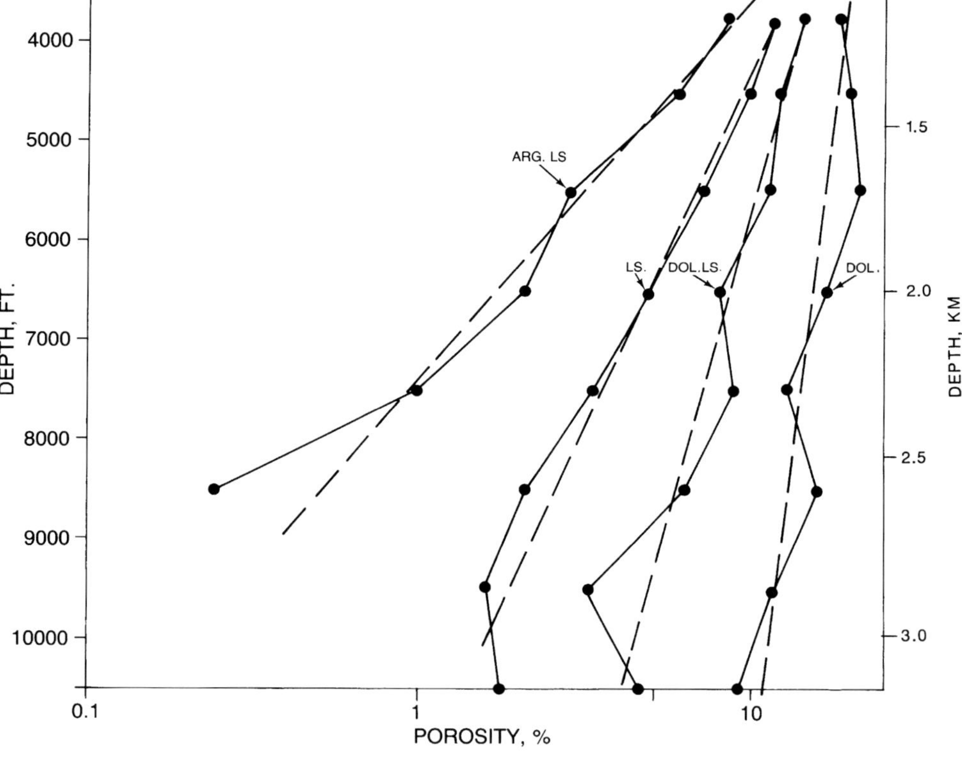

Figure 4. Porosity trends on a semilogarithmic plot. The porosity loss plots close to a straight line in all of the lithologies, indicating that the porosity loss functions are approximately exponential. Low-porosity averages have a large scatter on semilogarithmic plots due to the inaccuracy of log-determined porosity at low values. ARG. LS. = argillacious limestone; LS. = clean limestone; DOL. LS. = dolomitic limestone; DOL. = dolomite.

clean limestone porosity and the three depth-related variables (depth, effective stress, and temperature) yields very similar R^2 values. Given the strong covariance between these variables, it cannot be demonstrated unambiguously that effective stress is the primary control on mean porosity rather than some other factor correlated with present-day burial, such as temperature or stress history.

DISCUSSION

Controls on Porosity

The data generated in this study demonstrate that Mississippian carbonates of the Williston Basin exhibit systematic porosity loss with increasing depth of burial. Although secondary porosity may increase porosity locally, or other secondary processes may result in increased permeability, which increases the economic potential of a carbonate, the overall trend is one toward decreasing average porosity with increasing depth of burial for all carbonate rock types. Effective stress, temperature, stress history, or some other variable correlated with depth actually causes the reduction in porosity, not the depth itself. Because of the strong covariance of these variables in this data set, the actual relative influence of the different variables cannot be ascertained.

The difference between the basin-scale porosity trend and the reservoir porosity trend is quite striking. The porosity reduction with depth in economic limestone petroleum accumulations in the Madison Group is nowhere near as great as the average porosity reduction in all Madison Group limestones (Figure 10). Instead, average porosity of petroleum accumulations

in limestone has a modest systematic decrease in average reservoir porosity with increasing depth. Dolomitic limestones and dolomites have average reservoir porosity higher than that of limestones in the deeper part of the basin.

The major control on rate of average porosity loss with depth is carbonate mineralogy. Dolomite and dolomitic limestones lose porosity at a slower rate than do limestones (Figure 4). Because of the location of control wells, the sample size of dolomite lithology is too small for quantitative evaluation; however, the decreasing average porosity from dolomite through dolomitic limestone to limestone indicates that the dolomite average porosity estimates are consistent with porosity trends in other data. The progressive decrease in porosity with decreasing dolomite content over a narrow depth range was also noted by Kupecz (1984, her figure 45) in Billings anticline fields.

Argillaceous limestones have a lower overall porosity and a faster rate of porosity loss than do the clean carbonates at similar depths (Figure 3). Porosity becomes lower as the clay content of the limestone increases (Figure 6; Table 2).

Porosity differences between rock types appears to be predominantly a burial feature, not inherited from initial (zero-depth) porosity. Extrapolation of the exponential porosity trends to zero burial depth gives an estimate of zero-depth porosity. Zero-depth porosities of argillaceous limestones are highest, whereas dolomites have the lowest zero-depth porosity. This is the opposite porosity ranking than that seen over the entire depth range of study, but this ranking is consistent with some observations of the Cenozoic average porosity trends.

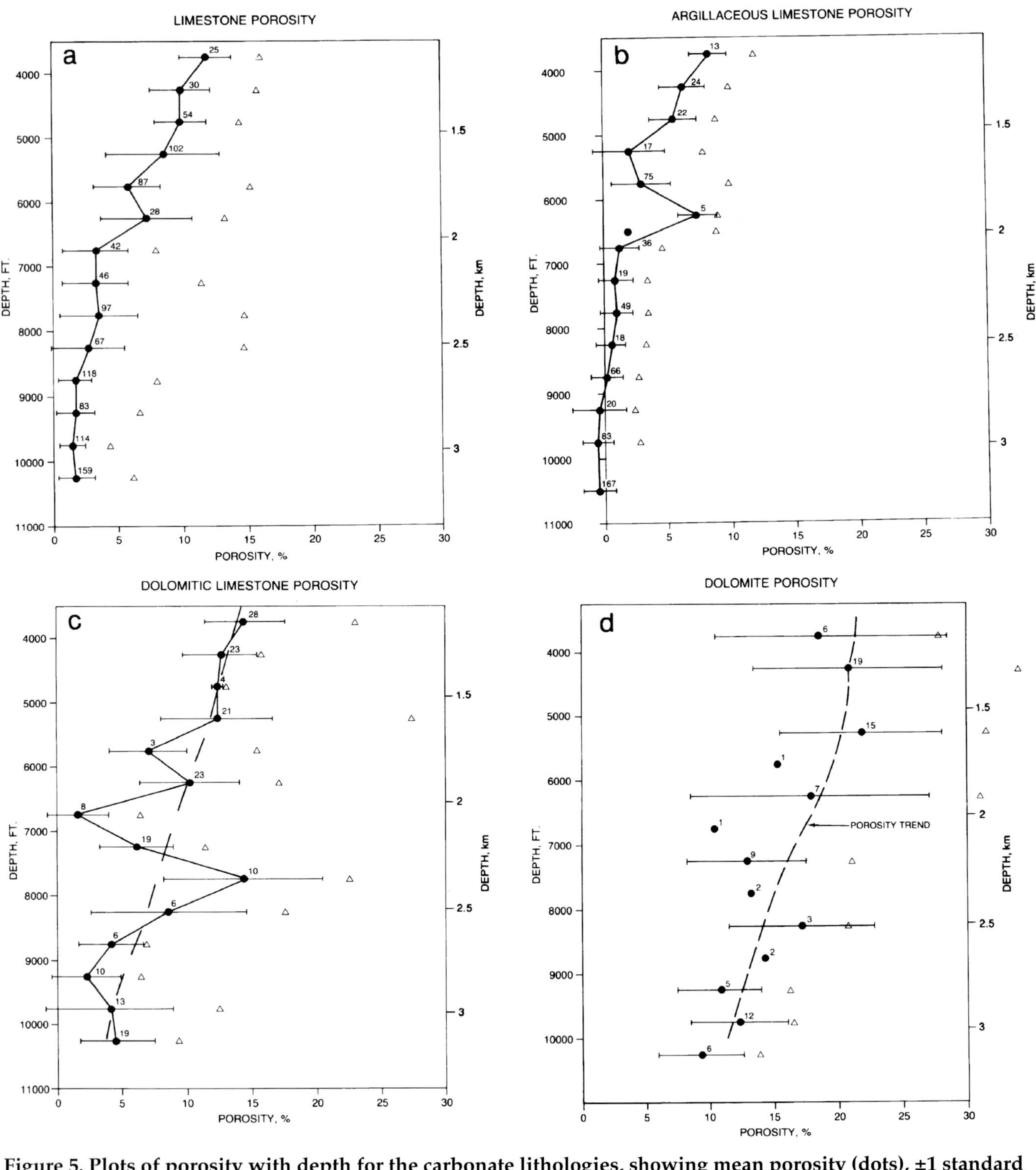

Figure 5. Plots of porosity with depth for the carbonate lithologies, showing mean porosity (dots), ±1 standard deviation (horizontal bars), and maximum porosity (triangles). The number at each depth refers to the number of measurements of that lithology in that depth interval. (a) Limestone, (b) argillaceous limestone, (c) dolomitic limestone, and (d) dolomite. Dashed lines are porosity trends of the dolomite and dolomitic limestone data.

For example, average shallow-buried dolomites have a lower porosity than coexisting limestones (Halley and Schmoker, 1983). If the zero-depth porosity is an indicator of preburial porosity, then the preburial porosity of the different carbonate lithologies is not responsible for their relative porosity in the subsurface.

In contrast to the mineralogy effect, texture has little systematic effect on average limestone porosity. At shallow depths of burial, all limestone textures have similar average porosity (Figure 7a). At intermediate depths, texture has an effect on average porosity, but it is not systematic. This has two implications: (1) differences in

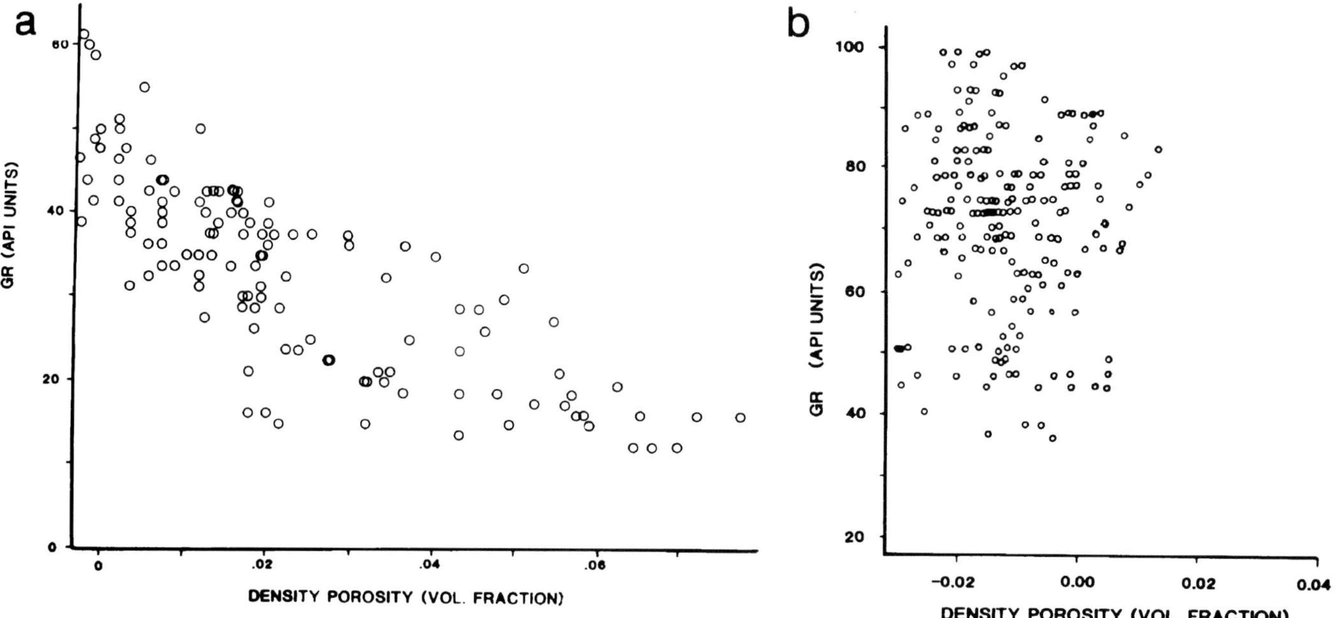

Figure 6. Plot of density porosity against shaliness as indicated by gamma-ray (GR) intensity. (a) Shallow carbonates (ARCO Wunderlich #1) show porosity decreasing with increasing shaliness as indicated by GR intensity. (b) Porosity in the deeply buried Amoco Sondrol #1 well has no systematic relationship to shaliness as indicated by the GR intensity. Lack of correlation is due to the small overall porosity variation in this argillaceous carbonate.

average porosity in different limestone textures develop during burial diagenesis; they are not exclusively features inherited from initial porosity or early diagenesis; and (2) the erratic effect of texture on porosity may indicate that the textural features do control porosity loss, but the subdivisions used here are too broad to identify the specific depositional fabrics and facies that selectively preserve porosity.

However, the fraction of rock with high porosity is clearly a function of texture (Figure 7b). Moderately to deeply buried grain-supported rocks are much more likely to have high porosity than are mud-supported rocks buried to the same depth; that is, a deeply buried lime packstone is likely to have an average porosity quite similar to that of a lime wackestone or lime mudstones buried to the same depth (Figure 7a). However, some beds in the lime packstones will have high porosity, whereas none of the beds in the lime wackestones and lime mudstones will.

One of the major results of this study is to document that average porosity in grain-supported and mud-supported limestones is about the same at equivalent burial depths. Only the fraction of grain-supported limestones that retain substantially higher-than-average porosity can become economic limestone reservoirs without fracturing. Porosity of some grain-supported rocks is not occluded as rapidly as the porosity of other grain-supported rocks or mud-supported rocks. As a result, the near-normal distribution of porosity characteristic of shallow-buried, grain-supported limestones evolves into a negatively skewed distribution with burial (Figure 8). This process is treated statistically here, but deterministic, physicochemical processes such as

timing of mineralogical stabilization, early cementation, early diagenetic fabric alteration, or pore size or geometry actually control which grain-supported rock may preserve its porosity with burial.

Mechanisms of Porosity Loss

The porosity trends indicate that porosity-destructive processes dominate over burial secondary porosity creation. As in other shelf limestones, the Madison Group carbonates clearly lose porosity by cementation rather than mechanical compaction. This is evident from thin-section photomicrographs of numerous diagenetic and field studies in the basin. The most likely process generating the cement is pressure dissolution near the site of cementation. This interpretation is supported by the correlation of porosity loss with clay mineral content. Clay minerals have been postulated to increase the effectiveness of pressure dissolution (Weyl, 1959).

Allochthonous carbonate cementation (i.e., carbonate cements derived from parts of the basin removed from the site of cementation by a distance of kilometers or hundreds of meters of burial) cannot be ruled out, but it is judged unlikely for two reasons. First, porosity loss is correlated with depth and not geographic position. If allochthonous cements were precipitated from moving water, there should be an asymmetry of cement distribution related to position of recharge and discharge of the water, and patterns should be less dependent on depth. Second, each lithology follows its own porosity-reduction pathway. If reduced by allochthonous cement, all lithologies should follow a similar porosity-reduction path, so

Table 3. Effects of Variables on Limestone Porosity.

Variable	Sum of Squares	Significance*
Model 1: Limestone Porosity = ϕ (Depth, Fabric, Effective Stress, Temperature)**		
Mode	8747.9	0.0001 (R^2 = 0.5)
Error	8744.6	---
Depth (ft)	5.3 3	0.426
Fabric (class variable)	98.03[†]	0.02
Effective Stress (psi)	148.16[†]	0.0001
Temperature (°F)	10.7[†]	0.2702
Model: Limestone Porosity = ϕ (Well, Fabric, Effective Stress, Depth, Temperature)**		
Mode	11,142.8	0.0001 (R^2 = 0.637)
Error	6349.6	---
Well (class variable)	10,380.9 (2395[†])	0.0001

*Probability of the regression coefficient to equal zero as determined from the t-test; the lower the number, the greater the significance of the variable.
**Regression coefficients cannot be shown here because GLM model is a type of ANOVA. The purpose of table is to show signficance of variables.
[†]Sum of squares if added last to the model in stepwise fashion.

that the main influence is the initial porosity or pore-size distribution, not the mineralogy of the host rock.

Other results of this study provide insights into the properties of the porosity-reducing process, whatever the process turns out to be. The fact that average porosity loss follows an approximately exponential decrease is evidence that the process of porosity loss has some sort of feedback. It is possible that the rate of porosity loss is proportional to the porosity of the rock. It is also possible that increasing depth or temperature actually decreases the rate of porosity loss in limestone by increasing rock ductility (decreasing stress differences, which may lead to dissolution) or by locking of stylolite surfaces and decreasing the area at which cement is generated.

Dolomites are generally found to lose porosity with depth at a slower rate than limestones of the same age in this basin (Figure 3). This clearly indicates that dolomite porosity is being selectively preserved with respect to limestone. Dolomite porosity in Paleozoic basins higher than porosity of surrounding limestone has been interpreted as evidence that dolomitization creates porosity (Weyl, 1960). Whether dolomitization actually creates porosity in geological settings remains controversial, but the porosity vs. depth trends for this basin substantiate the selective preservation mechanism as the dominant reason for the difference in limestone and dolomite porosity, at least in this Paleozoic formation. It is postulated that the lower rate of porosity loss in dolomite is related to its higher bulk modulus (Ellis et al., 1988), or related to slower diffusion or precipitation kinetics. A higher bulk modulus results in decreased chemical potential change with stress during pressure dissolution (Paterson, 1973).

Porosity Prediction from Average Porosity and Porosity Distribution

The type of data presented here can be used to predict the average porosity and the proportion of high (i.e., reservoir) porosity at a given depth. Such predictions can be used for two purposes: (1) estimation of velocities and densities as a function of depth for synthetic seismic reflection data sections and (2) estimation of both porosity risk and reserves for prospects with a given depth, lithology, and fabric.

Average porosity can be used to estimate average rock properties (density, velocities) where no site-specific data (sonic logs or vertical seismic reflection data profiling) are available. The porosity range can be used to estimate the range of acoustic impedance possible within a single lithology at a depth. Ranges of porosity at a given depth can also be used to model an expected range of acoustic impedance at boundaries between two carbonate lithologies.

Although average porosity is a qualitative indicator of the likelihood of finding economic porosity, the average porosity is not likely to be the average of the petroleum-charged porosity. Petroleum selectively charges the fraction of the rock with the best-quality rock properties within the trap. Also, those sections of low reservoir quality that are charged with petroleum may not be economically recoverable. For these reasons, economic porosity in many settings (especially Paleozoic carbonates) is usually substantially greater than the average porosity.

The fraction of the total rock likely to exceed an economic threshold porosity is a better predictor for the likelihood and amount of economic porosity, because this is the fraction of rock that is likely to be charged by petroleum if trapping conditions are favorable. If a distribution type (such as normal, log normal, or Cauchy) is assumed, it can be fitted to the frequency diagrams for porosity at different depths. The expected fraction of the carbonate section exceeding an arbitrarily chosen porosity threshold can then be determined for the distribution (Figure 11).

A simple, empirical method of estimating the fraction of economic porosity in an interval with a given fabric or lithology is proposed as an alternative procedure that requires no assumption of a distribution. Porosity data are plotted on a cumulative frequency plot, with frequency plotted on a probability scale, and

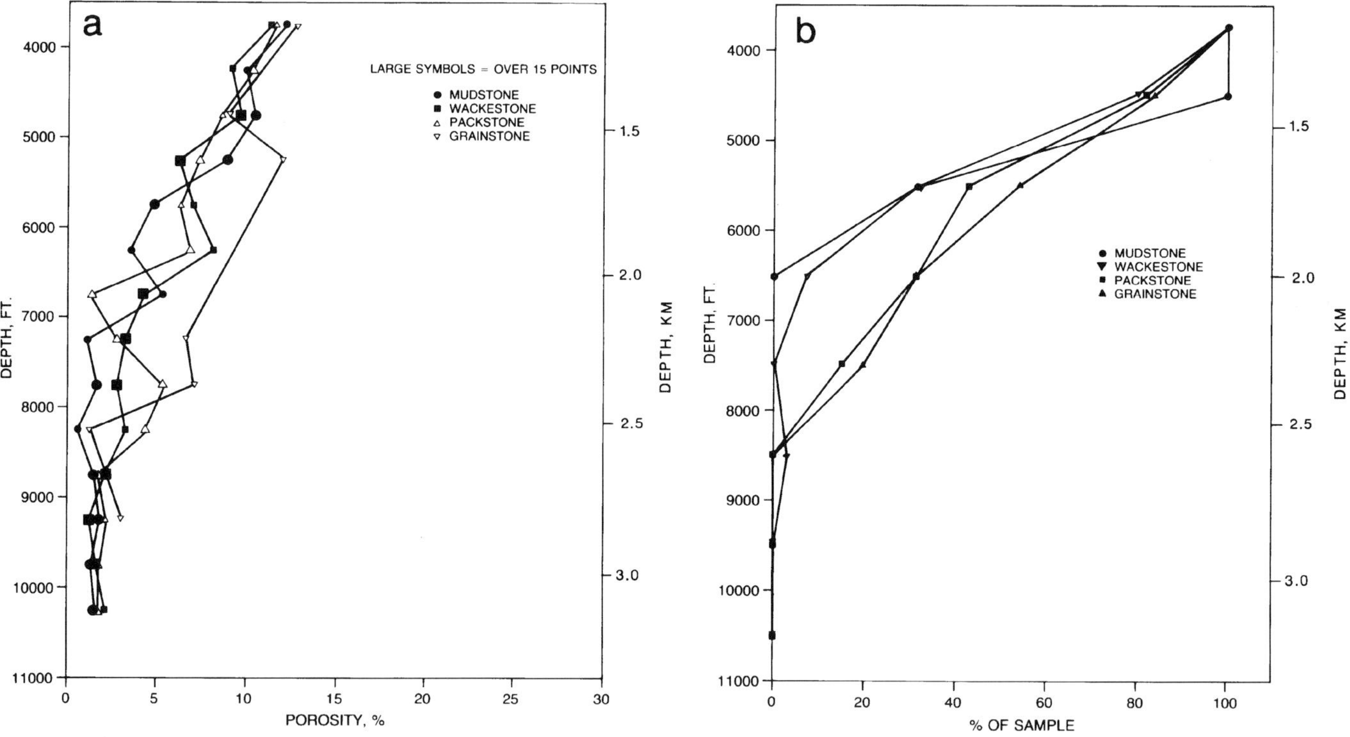

Figure 7. Porosity of limestone fabrics. (a) Mean porosity of fabrics in depth intervals. There is no consistent ranking of the mudstone, wackestone, or packstone porosities from depth interval to depth interval. The grainstone samples are too small to be statistically distinguished from the packstones at the same depth given the large standard deviations of the porosity means. (b) Percentage of high-porosity samples (ϕ >8%) in carbonate fabrics from different depth intervals. Between 7.5 and 2.5 km, (500–8500 ft), the framework-supported carbonates have a significantly higher percentage of samples with porosities >8%. Below 2.5 km (8500 ft), all carbonates have low percentages of high-porosity samples

the porosity plotted on either a linear or a logarithmic scale. The economic threshold porosity is chosen on the horizontal axis, based on drilling experience in the area. The cumulative porosity greater than the threshold is determined from the porosity cumulative frequency distribution by reading the cumulative probability on the vertical axis (Figure 11).

As an example, the porosity cumulative frequency distributions for mudstones and packstones at different average depths are plotted on a linear porosity scale (Figure 12a). Due to the small sample size of porosities of given depths and lithologies, considerable scatter occurs near the tails of the distributions, so smoothed curves were fitted to the tails of the data by sight (Figure 12b). If desired, more quantitative fits could be made with an assumed distribution. For our example, let us assume that a minimum of 8% porosity is needed in an oil-saturated reservoir at 3 km (10,000 ft) to be considered pay. Figure 12b indicates that the fraction of mudstones exceeding this porosity is off the scale, and estimated to be about 0.01%. In contrast, packstones should have about 3% of the section with porosity exceeding 8%. If 30 m (100 ft) of each lithology were penetrated, 3 mm (~0.1 in.) of the mudstone interval would have porosity greater than 8%, whereas 1 m (~3 ft) of the packstone would exceed this porosity. If a minimum of 3 m (10 ft) of pay were required for economic production, the chances of encountering adequate thickness and

porosity in the mudstone would be remote. Although the risk for encountering the same thickness of porosity in the packstone may be high, it may be acceptable if economic factors are favorable, or if an especially thick section of packstone were expected from facies models.

Several points should be made about the previous example and this approach to reservoir quality prediction. First, it is essential that the porosity distribution be developed from unbiased data. If, for example, porosity data were collected only from intervals with some minimum threshold porosity, then the fraction of economic porosity is relative to the total thickness of the porosity with the minimum porosity level, not relative to the thickness of the carbonate body as a whole. This gives the prediction much greater uncertainty, because the fraction of the total carbonate thickness with the threshold minimum porosity level must also be estimated. Likewise, if only maximum porosity data are collected, porosity prediction has less power, because much production from economic reservoirs comes from intervals below the maximum porosity.

Second, porosity in carbonates is typically spatially correlated in vertical sections; that is, high-porosity samples tend to lie near other high-porosity intervals in a vertical section. The problem is that the thickness of the correlated interval (referred to as "bed" here) is not known. In the previous example (Figure 12), 1 m of economic porosity was predicted in the packstone interval.

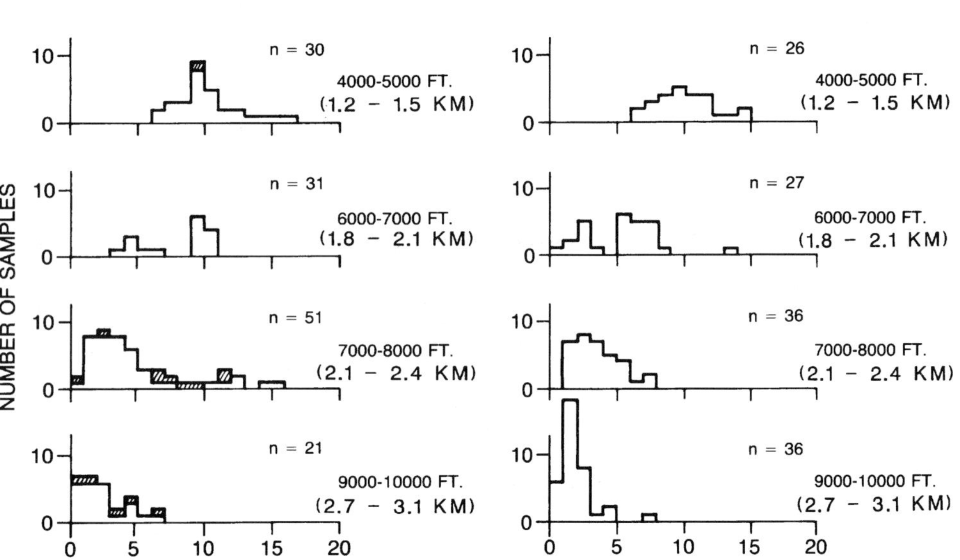

Figure 8. Porosity distribution histograms of samples from representative depths. The histograms indicate the number of samples (vertical scale) with the given porosity (horizontal scale, in percentages). The grainstone samples are indicated by cross hatch in the packstone and grainstone histograms.

It is possible that this occurs as five beds each 20 cm thick, as a 1-m-thick bed, or as some other distribution. The analysis presented here will not distinguish between the alternatives. Statistical analyses of vertical sections in both carbonate and siliciclastic intervals indicate that as the interval between samples decreases, the porosities of the samples are more likely to be correlated (Kittridge et al., 1990). Results of these types of studies indicate that vertical spatial correlation is on the order of feet to a few tens of feet (1–10 m) (Kittridge et al., 1990).

Vertical spatial correlation of porosity varies from location to location, and can best be calibrated by local data. If an average spatial correlation is assumed from local calibration data, the risk of finding a thickness of economic porosity greater than the expected economic porosity thickness can be calculated by the binomial sampling theory. The number of independent trials is the gross thickness of the unit divided by the thickness of independent units as estimated from semivariograms, and the probability of success of each trial is determined from the porosity cumulative frequency distribution, as discussed above.

Williston Basin Porosity Prediction

The data set collected here can strictly be used only for prediction of porosity in Mississippian carbonates of the Williston Basin. The preferred method for wildcat exploration is described in the previous section. From the calculations presented here, the depth limit for economic (>8%) porosity in grain-supported Mississippian limestones of the Williston Basin is ~2.5 km (8000 ft; 20% probability) to 3 km (10,000 ft; 2% probability), depending on which probabilities are at acceptable risk for random drilling.

Although specific compositional or textural subdivisions have been able to distinguish depositional facies with significantly higher porosity in field studies (e.g., pisolitic facies at Glenburn field; Gerhard, 1985), the gross textural subdivisions used here could

not. In many exploration settings, these sorts of gross textural subdivisions are likely to be the only available information. This means that one rarely has the higher quality depositional facies information necessary to predict the presence of a depositional facies shown to have higher average porosity.

Williston Basin Madison Group fields with dolomite reservoir rocks are concentrated in the southern and western part of the basin. For this reason, the basinwide sampling pattern resulted in too few dolomite samples to apply the cumulative frequency analysis approach used for the limestones. The higher variability of the dolomite average porosity vs. depth is interpreted to be caused by small sample size and a more variable diagenetic history than the limestone. Porosity does not seem to be a problem with Madison Group dolomites down to the maximum depth examined as part of this study.

Although strictly applicable to the Williston Basin Mississippian rocks, the trends developed from these data can also be used as a guide to porosity prediction in other Paleozoic cratonic basins. Specifically, a high risk for limestone porosity is expected in Late Paleozoic reservoirs buried much deeper than 3 km (10,000 ft). Late Paleozoic dolomite reservoirs are not expected to have much of a reservoir quality problem due to burial cementation down to 3 km. However, the significance of evaporitic cementation on porosity was not evaluated in this study, and it is likely that anhydrite or halite cementation could significantly reduce porosity for those dolomites associated with evaporitic sections. Of course, dolomites buried with low initial porosity are not likely to develop substantial porosity with burial, so these results can only be applied to dolomites with high initial porosity.

Comparison with Other Porosity Trends

Comparison of different chalk and limestone porosity–depth trends and porosity–TTI trends indicates that porosity data from one basin cannot be directly used to

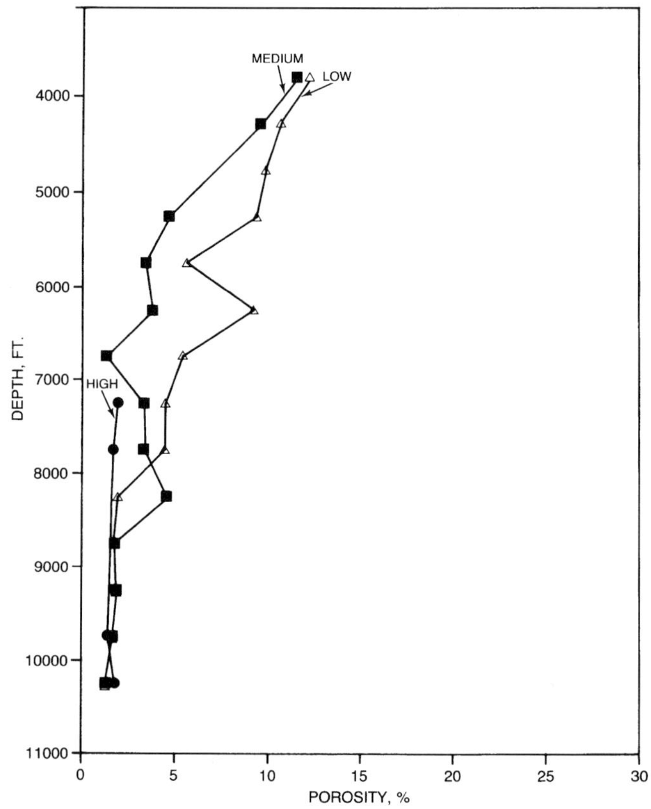

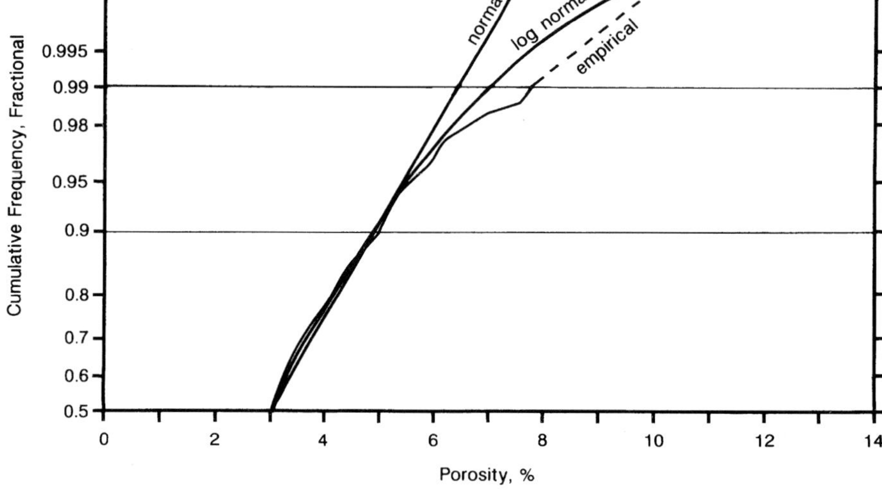

Figure 9. Temperature gradient effect on limestone porosity. Gradients are divided into high (circles), medium (squares), and low (triangles), as discussed in the text.

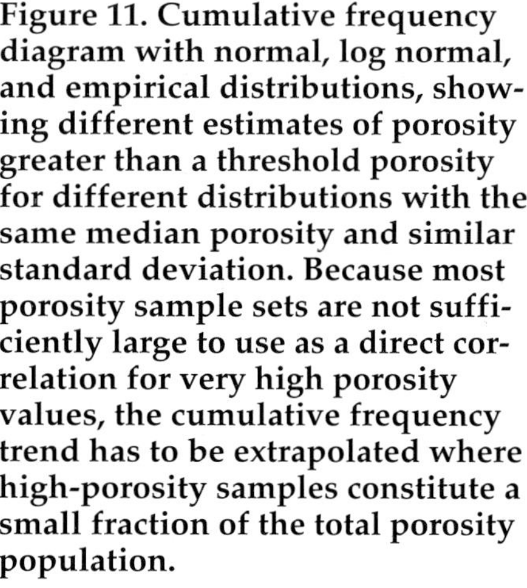

Figure 10. Comparison of average limestone porosity trend developed here (solid line) with reservoir porosity of Madison Group fields of different reservoir mineralogy. Some fields with limestone may be dolomitic limestone or dolomite.

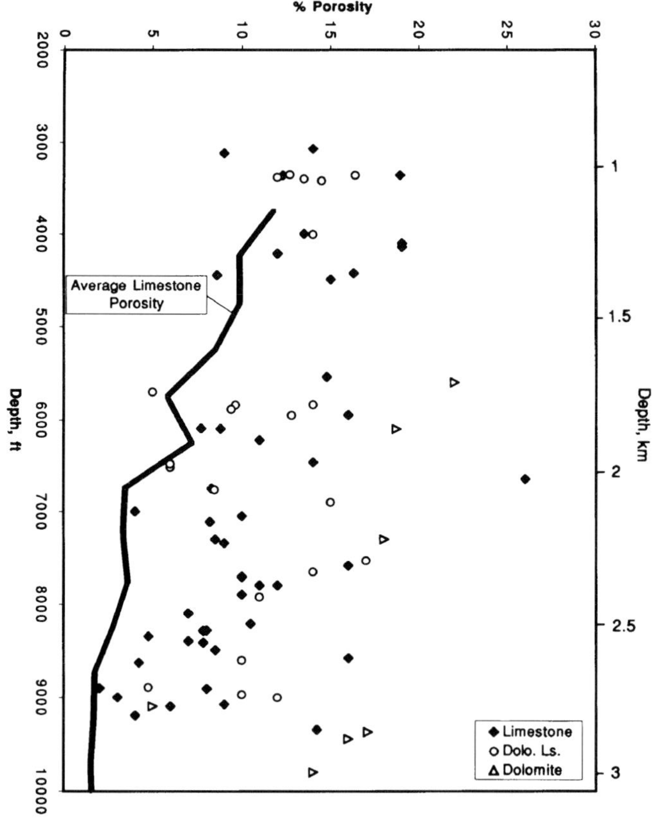

Figure 11. Cumulative frequency diagram with normal, log normal, and empirical distributions, showing different estimates of porosity greater than a threshold porosity for different distributions with the same median porosity and similar standard deviation. Because most porosity sample sets are not sufficiently large to use as a direct correlation for very high porosity values, the cumulative frequency trend has to be extrapolated where high-porosity samples constitute a small fraction of the total porosity population.

estimate carbonate porosity in another basin (Schmoker, 1984). However, where basin-scale, shallow-water limestone data are compared between basins, older limestones have an average porosity that is lower than that of younger limestones at the same maximum burial depth (Figure 13). This indicates that time is important for porosity reduction, in addition to effective stress and

temperature, as postulated by Schmoker (1984). The effect appears to be somewhat systematic, and provides hope that a generic fundamental relationship between limestone porosity and burial can be developed.

Although the relative magnitudes of the effects on effective stress, time, and temperature on porosity loss cannot be ascertained from this study due to its

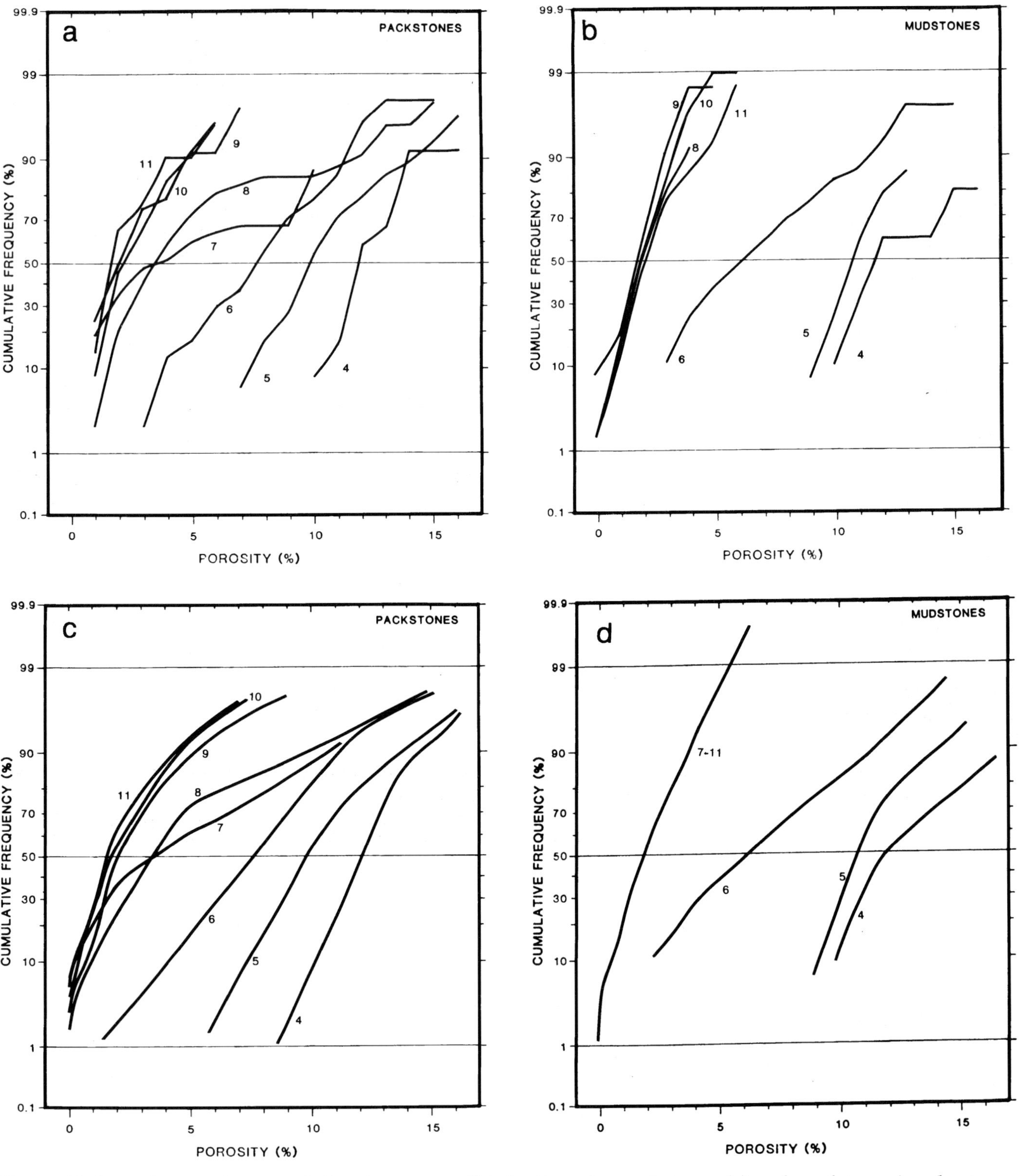

Figure 12. Sample porosity cumulative frequency distributions for packstones (a) and mudstones/wackestones (b) plotted on a probability scale. Smoothed porosity cumulative frequency distributions for packstones (c) and mudstones/wackestones (d). Numbers along the cumulative frequency curves correspond to depth range: 4 = 900–1200 m (3000–4000 ft); 5 = 1200–1500 m (4000–5000 ft); 6 = 1500–1800 m (5000–6000 ft); 7 = 1800–2100 m (6000–7000 ft); 8 = 2100–2400 m (7000–8000 ft); 9 = 2400–2700 m (8000–9000 ft); 10 = 2700–3000 m (9000–10,000 ft); and 11 = 3000–3300 m (10,000–11,000 ft).

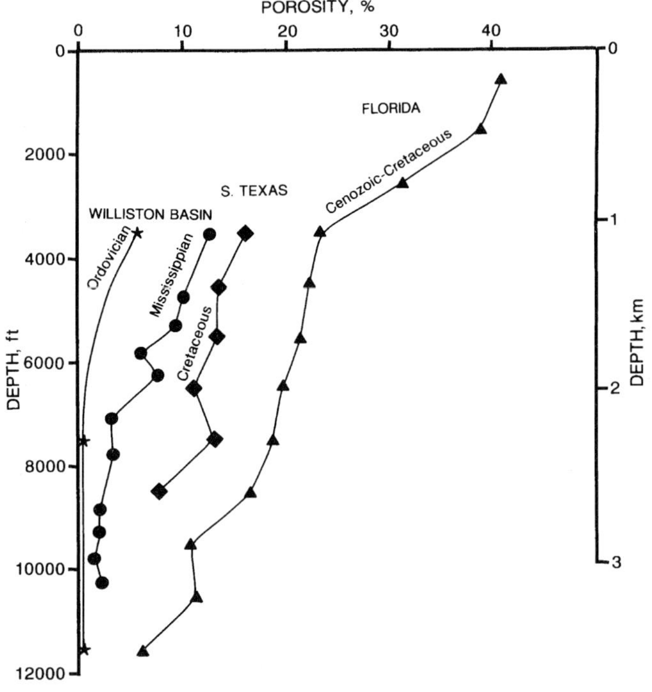

Figure 13. Limestone porosity trend compared to other quantitative porosity trends for shelf limestones. Florida data from Schmoker (1984). Texas Cretaceous data are unpublished core analysis trends collected by R.G. Loucks (1985, personal communication). Ordovician data are average wireline-log limestone porosity from three wells penetrating the Red River Formation, collected as part of this study.

design, it has been demonstrated that temperature does have an effect on porosity loss independent of time and effective stress. Temperature and time have been postulated to be the major controls (Schmoker, 1984), but this cannot be verified in this study, and regression equations seem to indicate that effective stress is still dominant over temperature.

CONCLUSIONS

Average porosity decreases as a function of depth in Mississippian carbonates from the Williston Basin. This porosity decrease is approximately exponential for all carbonate lithologies. The greatest control on rate of porosity loss with depth is the lithology of the carbonate. Argillaceous limestones lose porosity at the greatest rate and have the lowest porosity at all depths analyzed. Clay-free limestone porosity decreases faster with depth than does dolomitic limestone porosity, and dolomite porosity decreases the least with burial depth. The effect of limestone fabric on average porosity is quite small, but fabric has a strong influence on range of porosity at a given depth, and thus on the presence of high (economic) porosity. In Williston Basin Mississippian limestones, the selective occurrence of economic porosity in grain-supported rocks is due to the selective preservation of porosity in a small fraction of the grain-supported rocks, while porosity in most grain-supported rocks and all mud-supported rocks is systematically destroyed. The exact geological mechanism for selective preservation of porosity cannot be determined from this type of study. Increased thermal gradient enhances porosity loss in limestone.

The expected net thickness of economic porosity can be estimated from cumulative frequency distributions of porosity samples. The distributions are skewed significantly in moderately to deeply buried samples, so a normal distribution cannot be assumed for prediction of abundance of high porosity. If the average thickness of beds with similar porosity levels can be estimated, these estimates can be converted into quantitative risk factors using standard binomial sampling theory.

Because the Williston Basin is a well-drilled petroleum province, the application of this study to the Williston Basin is limited. In most parts of the basin, porosity can be mapped and the drilling location chosen to enhance the likelihood of encountering adequate porosity. The drilling is not random, so the odds of encountering porosity are significantly greater than those presented here, assuming reservoir quality of nearby wells is carefully assessed. However, this study demonstrates the method by which porosity loss in carbonates in other, less well drilled settings can be evaluated and the method by which scatter of porosity data can be used to predict the risk for encountering porosity exceeding a threshold value.

The results seen here confirm the general trends observed elsewhere. (1) Average carbonate porosity does decrease with depth (Schmoker and Halley, 1982). (2) Limestones lose porosity with depth at a faster rate than do dolomites with equivalent burial histories (Schmoker and Halley, 1982). (3) Average porosity of a limestone at a given depth decreases with increasing age (Schmoker, 1984). These generalizations can be used as a guide to evaluate new deep plays for which little empirical data are available.

ACKNOWLEDGMENTS

The author thanks Bob Loucks, Jim Hickey, Julie Kupecz, James Schmoker, Jerry Lucia, and Andrew Horbury for reviews. I also thank ARCO Exploration and Production Technology Co. for permission to release this study, which was completed as an internal study in 1984. Gulf Coast Cretaceous porosity vs. depth data were provided by Bob Loucks.

REFERENCES CITED

Beach, D.K., and J.W. Griffin, 1992, Stanley field—U.S.A. (Williston Basin, North Dakota), *in* N.H. Foster and E.A. Beaumont, compilers, Stratigraphic traps III: Tulsa, Oklahoma, AAPG Treatise of Petroleum Geology, Atlas of Oil and Gas Fields , p. 389–420.

Choquette, P.W., and L. Pray, 1970, Geologic nomenclature and classification of porosity in sedimentary carbonates: AAPG Bulletin, v. 54, p. 207–250.

DeFord, R.K., et al., eds., 1976, Geothermal gradient map of North America: Tulsa, Oklahoma, AAPG, scale 1:5,000,000, 2 sheets.

DeMis, W.D., 1992, Elkhorn Ranch field—U.S.A. (Williston Basin, North Dakota), *in* N. H. Foster and E. A. Beaumont, compilers, Stratigraphic traps III: Tulsa, Oklahoma, AAPG Treatise of Petroleum Geology, Atlas of Oil and Gas Fields, p. 369–388.

Dunham, R.J., 1962, Classification of carbonate rocks according to depositional texture, *in* W. Ham, ed., Classification of carbonate rocks: AAPG Memoir 1, p. 108–121.

Ellis, D., J. Howard, C. Flaum, D. McKeon, H. Scott, O. Serra, and G. Simmons, 1988, Mineral logging parameters: nuclear and acoustic: Technical Review, v. 36, p. 38–52.

Gerhard, L.C., 1985, Porosity development in the Mississippian pisolitic limestones of the Mission Canyon Formation, Glenburn field, Williston Basin, North Dakota, *in* P.O. Roehl and P.W. Choquette, eds., Carbonate petroleum reservoirs: New York, Springer Verlag, p. 192–205.

Halley, R.B., and J.W. Schmoker, 1983, High-porosity Cenozoic carbonate rocks of south Florida: progressive loss of porosity with depth: AAPG Bulletin, v. 67, p. 191–200.

Kittridge, M.G., L.W. Lake, F.J. Lucia, and G.E. Fogg, 1990, Outcrop/subsurface comparisons of heterogeneity in the San Andres Formation: SPE Formation Evaluation, September 1990, p. 233–240.

Kupecz, J., 1984, Depositional environments, diagenetic history, and petroleum entrapment in the Mississippian Frobisher-Alida interval, Billings anticline, North Dakota: Colorado School of Mines Quarterly, v. 79, no. 3, 62 p.

LeFever, R.D., and J.A. LeFever, 1991, Newburg and South Westhope fields—U.S.A. (Williston Basin, North Dakota), *in* N.H. Foster and E.A. Beaumont, compilers, Stratigraphic traps II: Tulsa, Oklahoma, AAPG Treatise of Petroleum Geology, Atlas of Oil and Gas Fields, p. 161–187.

Lindsay, R.F., 1985, Madison Group (Mississippian) reservoir facies of Williston Basin, North Dakota: AAPG Bulletin, v. 69, p. 279–280.

Lindsay, R.F., and C.G.St.C. Kendall, 1985, Depositional facies, diagenesis and reservoir character of Mississippian cyclic carbonates in the Mission Canyon Formation, Little Knife field, Williston Basin, North Dakota, *in* P.O. Roehl and P.W. Choquette, eds., Carbonate petroleum reservoirs: New York, Springer-Verlag, p. 177–190.

Miller, W.R., and S.A. Strauz, 1980, Preliminary map showing freshwater heads for the Mission Canyon and Lodgepole Limestones and equivalent rocks of Mississippian age in the Northern Great Plains of Montana, North Dakota, South Dakota, and Wyoming: U.S. Geological Survey Open File Report 80–729, map, 1 sheet.

Paterson, M.S., 1973, Nonhydrostatic thermodynamics and its geologic applications: Reviews of Geophysics and Space Physics, v. 11, p. 355–389.

Peterson, J.A., and L.M. MacCary, 1987, Regional stratigraphy and general petroleum geology of the U.S. portion of the Williston Basin and adjacent areas, in Williston Basin, *in* M.W. Longman, ed., Anatomy of a cratonic oil province: Denver, Colorado, Rocky Mountain Association of Geologists, p. 9–44.

Schmoker, J.W., 1984, Empirical relation between carbonate porosity and thermal maturity: an approach to regional porosity prediction: AAPG Bulletin, v. 68, p. 1697–1703.

Schmoker, J.W., and R.B. Halley, 1982, Carbonate porosity vs. depth: a predictable relation for South Florida: AAPG Bulletin, v. 66, p. 2561–2570.

Schmoker, J.W., K. Krystinik, and R. Halley, 1985, Selected characteristics of limestone and dolomite reservoirs in the United States: AAPG Bulletin, v. 69, p. 733–741.

Scholle, P.A., 1978, Porosity prediction in shallow vs. deep water limestones: 53d Annual Fall Technical Conference of the Society of Petroleum Engineers, Houston, Texas, October 1978, SPE Preprint SPE 7554, 6 p.

Tonnesen, J.J., 1985, ed., Montana oil and gas fields: proceedings (2 volumes): Billings, Montana, Montana Geological Society, 1217 p.

Tyler, C.D., ed., 1962, Oil and gas fields, North Dakota Symposium: Bismarck, North Dakota, North Dakota Geological Society, 220 p.

Weyl, P.K., 1959, Pressure solution and the force of crystallization—a phenomenological theory: Journal of Geophysical Research, v. 64, p. 2001–2025.

Weyl, P.K., 1960, Porosity through dolomitization: conservation of mass requirements: Journal of Sedimentary Petrology, v. 30, p. 85–90.

Chapter 4

Predicting Reservoir Quality Using Linear Regression Models and Neural Networks

K.M. Love
Exxon Production Research Co.
Houston, Texas, U.S.A.

C. Strohmenger
BEB Erdgas und Erdöl GmbH
Hannover, Germany

A. Woronow
Exxon Production Research Co.
Houston, Texas, U.S.A.

K. Rockenbauch
BEB Erdgas und Erdöl GmbH
Hannover, Germany

ABSTRACT

A method for predicting the three-dimensional distribution of reservoir attributes has been developed by integrating geological and statistical models. The general method, applicable to carbonate and siliciclastic reservoirs, has been demonstrated by predicting the distribution of dolomite, calcitized dolomite, porosity, and permeability from regional to field scales in the Permian Zechstein 2 Carbonate of northern Germany.

The first step in the prediction process consists of identifying factors potentially responsible for reservoir quality distribution. For the Zechstein 2 Carbonate, the resulting geologic model suggested that paleofaults and related fracture systems controlled the distribution of nonporous calcite (calcitized dolomite) by acting as conduits for calcitizing fluids originating from anhydrites underlying the carbonates.

The next step in the prediction process involves determining if the geologic model provides variables that can be used to predict the variable of interest given the predrill data available. If not, then other predictor variables, not necessarily cause-and-effect variables but ones whose values are known predrill, are required. Although a geologic model for Zechstein diagenesis elucidated the probable cause-and-effect relationship regarding the distribution of mineral types, it provided no means for predicting the geographic distribution of mineral types, because data on the distribution of paleofault and paleofracture systems cannot be obtained. For pragmatic purposes, models must both predict the desired parameter at the necessary scale and use predictor variables whose values are known prior to drilling. For the Zechstein 2 Carbonate, linear regression models using facies and location (x-y coordinates and depth) accomplished practical predictions of mineral distribution. The fact that location provides significant predictions indicates that calcite and dolomite occur in a spatially organized manner, reflecting the geologic processes that caused the calcitization of the dolomite. Because paleostructure presumably controlled calcite distribution, separate models were developed for structurally distinct subareas. The use of structural subdivisions provided a way to account for different types of calcite distribution caused by different types of fault and fracture systems.

Although mineralogy is a dominant control on reservoir quality in the Zechstein 2 Carbonate, the porosity and permeability distributions reflect additional factors. Like the mineralogy distribution, however, the porosity and permeability distributions have a dominant nonrandom spatial component, and therefore can be predicted reliably using location information. Because the spatial distribution of porosity and permeability in the Zechstein 2 Carbonate is highly complex, a nonparametric predictive technique (an artificial neural network) was implemented. It produced models that surpassed those of linear regression.

Although cast here in terms of a particular application, the methodology is general, and such predictive models can be used to generate maps and cross sections of predicted parameters within any reservoir. In addition, sets of point values generated by the models can be loaded into visualization software to provide three-dimensional representations of the predicted parameters.

INTRODUCTION

Geologic studies commonly provide a means to link reservoir quality to one or more controlling factors. If these factors subsequently predict predrill reservoir quality at the necessary scale, reservoir risk can be reduced. In some cases, however, the controlling factors cannot be identified, or, more commonly, knowledge of the controlling factors does not permit prediction at a pragmatic scale. For such cases, quantitative models based on linear or nonparametric methods that rely, at least in part, on location variables (x-y coordinates and depth) may provide a useful means for predicting the three-dimensional distribution of nonrandomly distributed parameters.

This chapter gives a general approach to the resolution of such pragmatic prediction issues, using a case study for illustrative purposes. The Upper Permian Zechstein 2 Carbonate of northern Germany (Figure 1) provides an example of a reservoir-quality problem where cause-and-effect models failed to generate practical predictions. Carbonates of the second Zechstein cycle (the Ca2 or Stassfurt Carbonate) constitute northern Germany's most prolific carbonate gas play; consequently, many efforts focus upon characterizing reservoir quality. Prediction of depositional facies provides one key to reservoir-quality prediction (Strohmenger et al., 1996), but the Zechstein 2 Carbonate underwent an extensive calcitization of dolomite ("dedolomitization") (Figure 2) that generally destroyed porosity and permeability and was not depositional-facies-specific (Strohmenger et al., 1993). Calcitization generally increases basinward, but its lateral distribution has been difficult to predict within individual slope facies. Thus, predicting the distribution of this nonporous diagenetic calcite vs. porous dolomite was identified as a crucial first step toward predicting reservoir quality, especially within thick slope deposits. As a result, a geologic model was developed to explain the mechanism of calcitization in the hope of using the model to predict calcite vs. dolomite. The resulting model indicated, however,

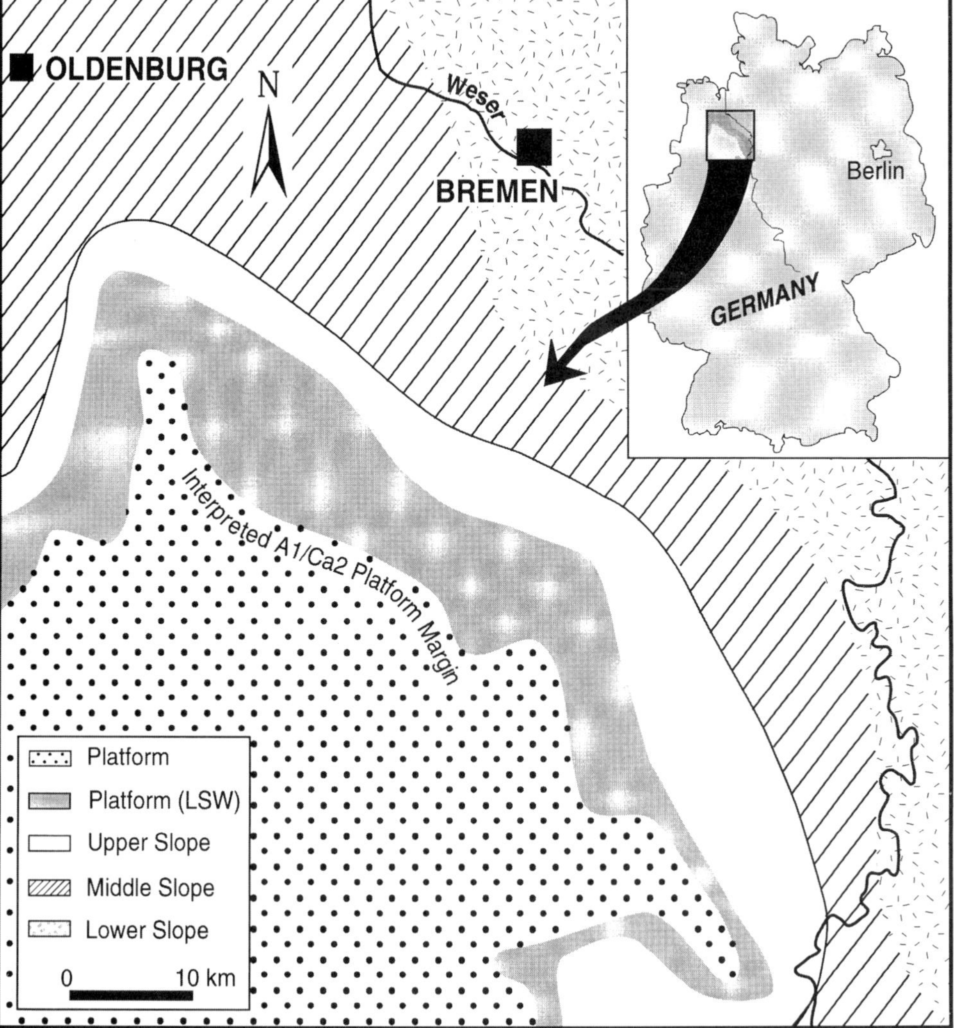

Figure 1. Location of study area (outlined in black) within the Upper Permian Zechstein 2 Carbonate (Ca2) of the Southern Zechstein Basin in northern Germany. LSW = lowstand wedge.

that paleofaulting and paleofracturing were responsible for the calcite distribution, thus providing no variables that could be used directly to predict calcite distribution at the desired scale prior to drilling. As a result, statistical models were sought that used values of variables accessible before drilling as proxies for the unattainable values of the cause-and-effect variables. For the purpose of predicting calcite, linear regression models were used.

Although prediction of calcite improved reservoir-quality prediction in the case of the Zechstein 2 Carbonate, other factors influence porosity and permeability distribution. Thus, models were developed to predict porosity and permeability distribution directly. Because of the functional complexity of the porosity and permeability distributions, artificial neural network models (a form of artificial intelligence) and linear regression models were used. The objective of the models was to predict porosity and permeability in as much detail as possible ahead of the drill. Although the models are not capable of replicating the high-frequency variations of porosity and permeability that occur within a facies, trends within facies can be predicted.

DATA

A statistical study of factors useful for predicting reservoir-quality distribution requires a database containing variables likely to be either directly or indirectly related to reservoir quality. For the Zechstein 2 Carbonate study, an existing database at BEB Erdgas und Erdöl GmbH was expanded to include data for hypothesized reservoir-quality controls. Core data included mineralogy, facies, subfacies, porosity, and permeability from 287 wells. The cores provided good coverage of the facies present in a given area, and data from core plugs generally were available every 15 cm throughout a core. Although each core did not necessarily cover the entire Ca2 interval, enough data from surrounding wells were available to adequately represent all facies present in a given area. Well log, structural, geochemical, thickness, and location data also were available. Because of the large amount of core available, all porosity and permeability values used for model development came from measurements on core plugs, rather than from well logs. For reasons discussed later, the data were divided into ten subsets, ranging in number of wells from 7 to 81, and in number of samples from 616 to 6990.

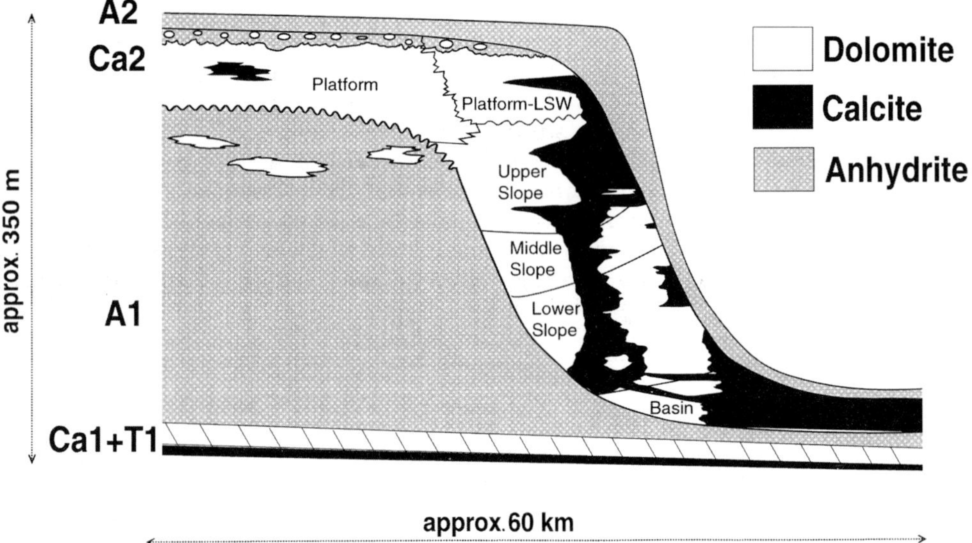

Figure 2. Schematic cross section through the Zechstein 2 Carbonate showing distribution of depositional facies [platform, platform-LSW (low-stand wedge), upper slope, middle slope, lower slope, and basin] and mineralogy (dolomite vs. calcite). The distribution of calcite within the slope is not facies dependent.

Data Smoothing

An important problem in predicting porosity and permeability using core-plug data is the large variation of the values over small distances within a core (high-frequency, high-excursion data). For example, porosity values commonly differ by an order of magnitude (e.g., from 2% to 20%) among several core plugs separated by <1 m. These differences may be due to geologic factors (subfacies changes or diagenesis), sampling bias (choosing the unusual specimen for analysis), or erroneous measurements.

No existing method has the ability to predict such abrupt, centimeter-scale excursions, but prediction at this level is generally not necessary. However, because models cannot predict the abrupt change, the residual values (the observed porosity/permeability value minus the model-predicted value) from these models will be large. To make the variance of the calibration data set more commensurate with the predicted data set, data can be "smoothed." Kacewicz (1994) found that smoothing data improved the performance of a neural network. A simple way to accomplish data smoothing involves using an average value of porosity or permeability for each well or for each major subdivision within a well such as facies; however, models typically can predict a higher frequency of variation than this, so information would be lost through such "oversmoothing."

To avoid oversmoothing, a smaller window of observations for averaging can be used. In addition, a moving average can be calculated. Although this diminishes the abruptness of the changes in porosity-permeability values, a single high-excursion value still has a large influence. Thus, we used a tapered moving average. This weights the values closer to the middle of the window more heavily than those at the ends. For the Zechstein 2 Carbonate, a weighted moving average of five measurements (a window typically <1 m) was calculated so that the middle value was weighted most heavily (0.4), then the two adjacent values less (0.2 each), and the top and bottom values least (0.1 each).

The same procedure was used with the five depths associated with the five porosity and permeability values to obtain an average depth value. The averaging procedure did not cut across facies boundaries.

METHODS

Prediction Techniques

Although statistical procedures can use different variables to *predict* the distribution of a parameter, the ability to predict does not imply cause and effect. Establishing cause and effect is not necessary to accomplish the goal of prediction. All variables that significantly predict parameters of interest may aid in understanding cause and effect, but their utility in prediction models depends on the ability to estimate their values away from well control. If, for example, the percentage of anhydrite cement in the Zechstein 2 Carbonate significantly predicts carbonate mineralogy, this might give clues about the calcitization process; however, cement content would be difficult to estimate in undrilled localities, so its input value in an equation to predict calcite vs. dolomite is uncertain, and the variable is not useful in practical estimation. In fact, a poor estimate of the input value may adversely affect predictive capabilities.

Regression

Regression analysis was used extensively to identify which variables significantly predict mineralogy, porosity, and permeability in the Zechstein 2 Carbonate. In addition, predictive models were developed using forward stepwise regression with backward elimination, which is a method to select a few predictor variables from a large number of potential predictor variables. Using this procedure, individual variables enter into and exit from an evolving model (Draper and Smith, 1981; Bowerman and O'Connell, 1990). For the Zechstein modeling, a significance level of 0.15 was chosen for a variable to enter into and to remain in the model.

Although stepwise regression helps to reduce the number of predictor variables, it does not necessarily provide the best regression model, nor can the remaining predictor variables be considered the most important.

The regression analyses made extensive use of indicator variables [that is, variables that assume discrete values (e.g., 0 and 1)] to identify different categories of a variable (Kleinbaum and Kupper, 1978). In this study, for example, an indicator variable was created for mineralogy, where mineralogy = 0 if the sample (a core plug) is dolomite, but mineralogy = 1 if the sample is calcite. If regression models are constructed such that a 0,1 variable is the dependent (predicted) variable, and hypotheses about the regression will be tested, it may be advisable to use a logistic function. However, *if the purpose of the regressions is only to predict,* as is the case in this study, such transforms are not mandatory.

Indicator variables allow the inclusion of such qualitative data as mineral type and facies in quantitative models. The number of (0,1) indicator variables required to represent one type of information (e.g., facies) is $n - 1$, where n is the number of different categories for the information (e.g., upper slope, platform, etc.) (Kleinbaum and Kupper, 1978). If five different facies occur, which is the case for the Zechstein 2 Carbonate, then four indicator variables are required to represent all the facies, as the "missing" variable is represented when the other four variables equal 0. For the Zechstein data, the lower slope facies was not used as a facies variable; thus, when the platform, platform-lowstand-wedge, upper slope, and middle slope facies variables were all 0 for a particular sample, that sample represented the lower slope facies.

If the relationship between the predicted and the predictor variables is complex, continuous variables, such as depth or thickness, can be transformed to new variables to possibly improve the regression model. Common transformations include logs, squares, square roots, and reciprocals of the original variables. Because of the complex spatial distribution of porosity and permeability in the Zechstein 2 carbonate, several different transformations of the location variables were made in attempts to accommodate nonplanar variations. For example, logarithms, squares, and reciprocals of the spatial data were offered to the prediction models, and commonly provided improved predictive ability (e.g., $\ln(\text{depth})$, X^2, $1/Y$).

Artificial Neural Networks

Although prediction of calcite in the Zechstein 2 Carbonate was relatively straightforward using linear models, prediction of porosity and permeability using linear regression was commonly improved by the addition of terms higher than second order. This suggested a high level of complexity (nonlinearity?); for this reason, an adaptive nonparametric prediction method was sought that might better predict the porosity/permeability distribution. One such method is an artificial neural network. A network consists of interconnected computing cells; weights are assigned to the connections between cells. These weights are used in conjunction with input data to predict some outcome. The network learns to predict a desired outcome by iteratively modifying the weights and comparing the predicted result with the actual result (Haykin, 1994).

This study used BPNET, a back-propagation neural network developed by author A. Woronow. The program randomly splits an input data file into a training data set (from which the network learns how to predict porosity or permeability) and a test data set that is not used during training (to assess the prediction effectiveness of the network as learned from the training data set). The test data consisted of ~10% of a data set. As with regression, the withheld data constitute a critical part of the evaluation of the predictive capabilities; they provide the only means to evaluate how well the predictive tool will work when presented with new data. The program was run on a 486-33 PC, and the neural network used one hidden layer, eight nodes, and a sigmoidal logistic function (Haykin, 1994).

For each data set, the network was allowed to learn until no further effective improvement occurred. The time required to reach this state ranged from ~30 min to 3 hr, depending on the size and complexity of the data set, although the program was allowed to run beyond the cessation of improvement to ensure that a later "breakthrough" in learning did not occur. In two cases, the program was allowed to run for ~14 hr to further check for this possibility; no additional learning occurred. The normal training times corresponded to between 500 and 3000 learning cycles, where a cycle is one pass through each case in the training data set

Unlike for the regression models, formed variables were not important for the neural network models, because a network can effectively develop its own linear and nonlinear transformations of variables to provide better prediction. If, however, a known relationship occurs (e.g., it is known that one variable is related to another by a particular function), the introduction of that transformed variable to the neural network could help the network learn faster. However, such functional relationships are not known in this case.

RESULTS AND IMPLICATIONS FOR GENERAL RESERVOIR-QUALITY PREDICTION

Mineral Distribution and Structural Influences

The mineral distribution in the Zechstein 2 Carbonate reflects a general division between dolomites in the platform deposits, dolomites plus calcites in the slope deposits, and calcites in the basinal deposits (Figure 2). However, departures from this broad pattern provided clues for development of a geologic model for calcitization. In particular, the platform deposits contain several relatively small areas with distinctively high percentages of calcite, closely corresponding to the positions of "bald highs"—tectonically high areas on the platform, bounded by deep-rooted fault systems, where the Zechstein 2 Carbonate is absent due to removal during Cretaceous tectonic activity. This relationship provided the

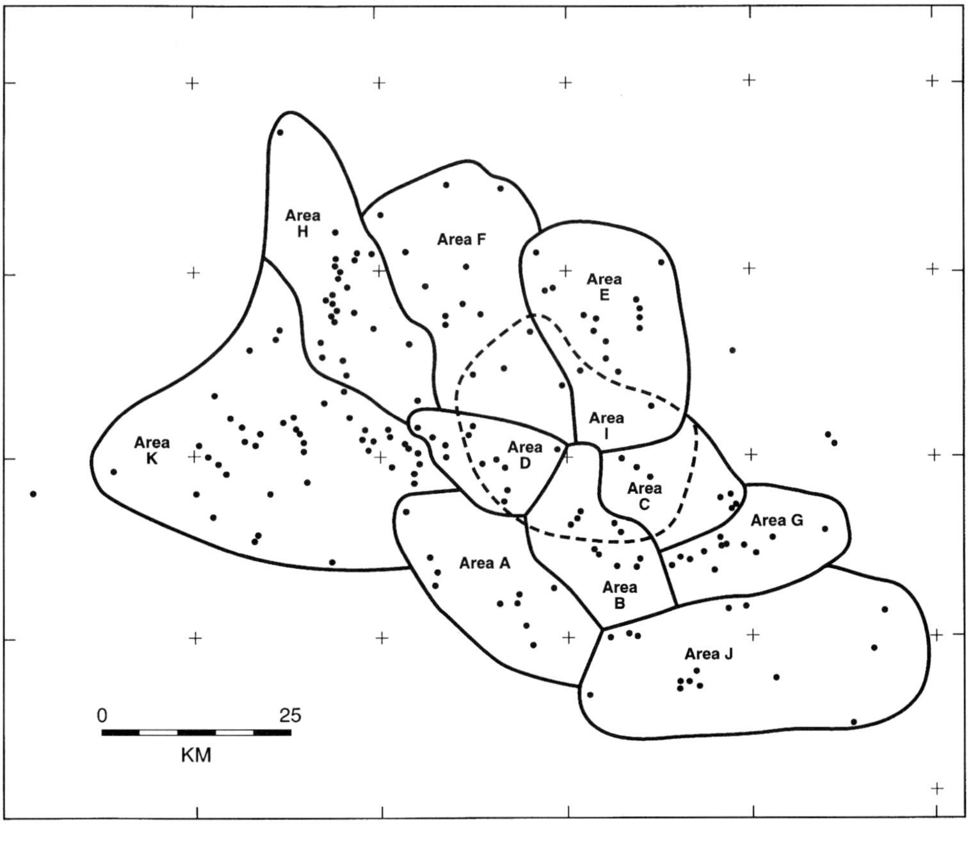

Figure 3. Map showing subdivision of area into "structurally homogeneous" subareas. Black dots represent well locations.

working hypothesis that calcitization of dolomite was related to fault and fracture systems. The large volumes of the calcium-sulfate–saturated waters required for the calcitization of dolomite can move from surrounding anhydrite formations through dolomitic units via fractures and other permeable zones (Clark, 1980; Warren, 1991), providing a mechanism for calcitization of the Zechstein 2 Carbonate. Calcitization of dolomite has been linked to fractures and faults in several formations, whether associated with evaporites or not, including the Mississippian Madison Limestone (Budai et al., 1984) and the Cretaceous Edwards Limestone (Abbott, 1974). Although this geologic model highlights a plausible cause-and-effect relationship regarding the distribution of calcite, it is ineffectual in predicting the geographic distribution of calcite at a pragmatic (fine) scale. Such predictions would require detailed paleofault and paleofracture data that are not available. Consequently, other predictor variables were sought, keeping in mind that structure was an important factor that might somehow be incorporated into predictive models.

In the Zechstein 2 Carbonate, the same facies have been subjected to different degrees of calcitization in different parts of the study area, likely reflecting differences in the access of calcitizing fluids to the areas. For this reason, formulating one equation to predict calcite for the entire area produced inadequate results. The formation thus was divided into subareas (Figure 3) that were delimited based on "structural homogeneity"; that is, because structure presumably played a substantial role in influencing the movements of diagenetic fluids, subareas enclose deposits that experienced similar

structural histories (e.g., a spur-and-graben subarea, and a structurally complex, folded and faulted subarea). For each of these structural subareas, a separate mineralogy-predicting equation was generated using only data from wells located within that subarea. Table 1 indicates the number of wells and samples for each of these subareas. The existence of borderlines between two subareas creates the possibility that predictions for one location can be made using two equations (one for each of the subareas) and, furthermore, that those two predictions might be meaningfully different. This was checked by comparing values generated for one location by the two different equations; in most instances, the values matched closely. After initially defining ten subareas, predictions were requested for a location at the juncture of several subareas. To check the predictions (which were extrapolations, because the location was outside the well control for each subarea), a new subarea was defined with the desired prediction location near the center (dashed outline in Figure 3, Area I). The predictions generated from this model were very similar to those generated by the existing models, although the new model (which does not represent a structurally defined area) did a poorer job of predicting withheld data than did the models for surrounding areas.

Porosity and permeability regression equations and neural network models were generated for each of the subareas defined during development of the mineralogy-predicting equations. The same subareas were deemed useful because calcitization is the dominant control on porosity and permeability; thus,

Table 1. Porosity and Permeability Prediction Results of Regression and Neural Network Models for the Zechstein 2 Carbonate for Withheld Test Data Sets.*

Subarea[†]	Number of Wells in Subarea	Number of Samples in Subarea	Porosity Prediction Accuracy		Permeability Prediction Accuracy	
			% Within ±2 porosity %	% Within ±4 porosity %	% Within ±1 ln(k) md	% Within ±1 ln(k) md
A	7	616	**38**	**75**	**52**	**87**
			40	60	46	83
B	20	1434	**34**	**62**	**38**	**79**
			15	29	38	68
C	10	875	**60**	**76**	**42**	**77**
			49	68	35	64
D	23	1807	**42**	**76**	**55**	**85**
			34	61	36	73
E	14	908	**70**	**88**	**64**	**88**
			57	84	48	73
F	32	5271	**42**	**67**	**42**	**75**
			33	56	34	67
G	23	4195	**49**	**75**	**47**	**80**
			38	57	44	76
H	32	6990	**60**	**79**	**49**	**80**
			45	68	42	73
I	30	3336	**42**	**73**	**44**	**76**
			27	53	37	63
J	15	1320	**58**	**87**	**68**	**91**
			44	78	52	78
K	81	6925	**36**	**63**	**41**	**68**
			26	49	34	60

*Neural network results are in boldface type.
[†]Subareas are outlined in Figure 3.

subareas defined by calcitization controls would be similar to those defined by porosity and permeability controls.

Predictor Variables for Calcite, Porosity, and Permeability

Predicting the distribution of reservoir quality requires variables whose values are known prior to drilling. For the Zechstein 2 Carbonate, depositional facies and location variables meet this requirement; thus, equations to predict calcite, porosity, and permeability were developed using only facies and location variables and their transformations (e.g., logarithms). These predictor variables were used for both the regression and neural network models. Although not available or not as useful for the Zechstein 2 Carbonate, formation thicknesses and seismic attributes may be good predrill predictor variables.

Depositional Facies

For the Zechstein 2 Carbonate, depositional facies provide important information about mineralogy, porosity, and permeability due to both depositional and diagenetic differences among facies. The depositional facies characterized in this formation are the platform, platform-lowstand-wedge, upper slope, middle slope, lower slope, and basin. A detailed depositional

framework and sequence stratigraphic model (Strohmenger et al., 1993) had been constructed prior to this reservoir-quality work, so that the succession of facies could be predicted ahead of drilling a well.

Location

Location variables (x-y coordinates and depth) play a crucial role in predicting three-dimensional distributions of parameters. For the Zechstein 2 Carbonate, location variables significantly predict mineralogy, porosity, and permeability. In addition, the predrill values for x and y are known, and a reasonable estimate can be made for depth by combining well log and seismic data. Depth units are in meters, and, for convenient use in the models, the x-y coordinates (e.g., 3452555.0, 5810250.5) were divided by 10^6. In addition, as described in the Methods section, transformations of location variables were used extensively for the regression models.

As with facies, the mathematical functions of the location variables that best predict mineralogy, porosity, and permeability differ in different subareas. When making a prediction for a facies in one subarea, location provides information about smaller-scale areal differences in reservoir-quality distribution; for example, calcite is not distributed uniformly throughout the upper slope within one subarea, but rather may be concentrated in the upper, eastward portion of that facies.

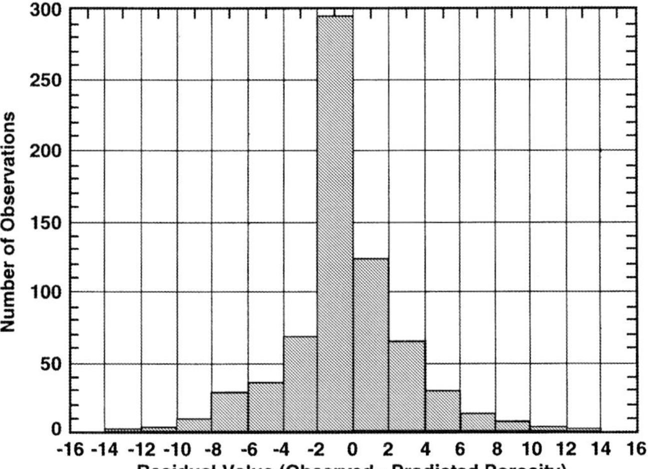

Regression Model (Withheld Data)

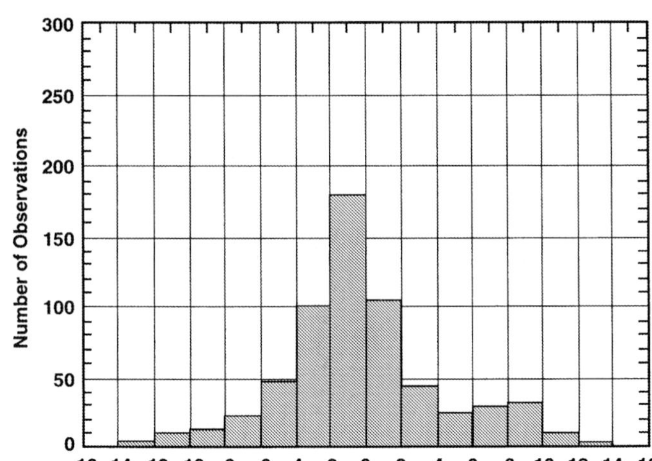

Neural Network Model (Withheld Data)

Figure 4. Histograms of residual values (observed value minus predicted value) from porosity prediction models for one structural subarea (H) within the Zechstein 2 Carbonate. The regression model predicted 45% of the values within ±2 porosity %, and 68% within ±4 porosity %; the neural network predicted 60% of the values within ±2 porosity %, and 79% within ±4 porosity %.

Calcite Prediction Models

When not used to extrapolate, the regression models used to predict Zechstein 2 Carbonate mineralogy almost always generate values between 0 and 1, interpreted as "likelihoods" of calcite (vs. dolomite). Values from 0 to 1 result because the input datum for each core plug was either a 0 (for dolomite) or a 1 (for calcite). This binary input was appropriate because, at a core-plug scale, the carbonate samples were almost always >95% calcite or dolomite. One of the conundrums of this method lies in interpreting the continuous values generated by the discrete models. Because

a dolomite sample is represented by the value 0 and a calcite sample by the value 1, these are the only values that have an unambiguous interpretation. However, when an equation is used to predict calcite or dolomite in an area, a number between 0 and 1 may result. If the predicted value for mineralogy is close to 0 or 1 (a reasonable definition of close in this case would be within ~0.3), then the mineralogy reasonably can be assigned to essentially calcite or essentially dolomite. For example, a value of 0.75 can be interpreted as representing mostly calcite, whereas a value of 0.25 indicates mostly dolomite. The greater ambiguity arises as to the meaning of an intermediate value such as 0.5; this could be interpreted as meaning half calcite and half dolomite *or* as an indeterminate mineralogy. Because of this ambiguity, cases were examined with values near 0.5 to establish a "ground truth" for interpretation. Predicted values of 0.5 generally corresponded to wells having both calcite and dolomite within the targeted interval. In practice, the probability of calcite in a given facies corresponded closely to the actual percentage of calcite in that facies.

Prediction Equations

Equation 1 is an example of a regression equation used to predict calcite for one structural subarea within the Zechstein 2 Carbonate. Values for the facies variables are 0 or 1, depending on which facies is being used for the prediction. If, for example, one desires a prediction for the upper slope facies, that variable would be set to 1, and all other facies variables would be set to 0.

$$\text{Likelihood of Calcite} = -125.73 + 3.65(X) + 3.51(Y^2) - 0.86[(\ln(\text{depth})] + 0.19(\text{lowstand}-\text{wedge platform facies}) + 0.23(\text{upper slope facies}) + 0.38(\text{middle slope facies}) \tag{1}$$

As discussed in the Methods section, all variables remaining in the predictive equations at the completion of the regression procedure are significant at the 0.15 level. The magnitude of the coefficients in the regression equation depends on the magnitude of the variable associated with that coefficient. Although using location variables and their transforms introduces collinearity into regression equations, collinearity is not a problem when the equations are used only to predict, as they are in this study. However, problems are caused by collinearity if one wants to interpret the regression coefficients (e.g., if one wants to know which variables are most important in each equation). Again, the equations developed for this study were used only to predict, not to test hypotheses regarding the equations.

Porosity and Permeability Prediction Models

Unlike the mineralogy models, the porosity and permeability predictive models generate estimates of porosity and permeability values rather than likelihoods. Equation 2 is an example of a regression equation used to predict porosity. Values for the

Regression Model (Withheld Data)

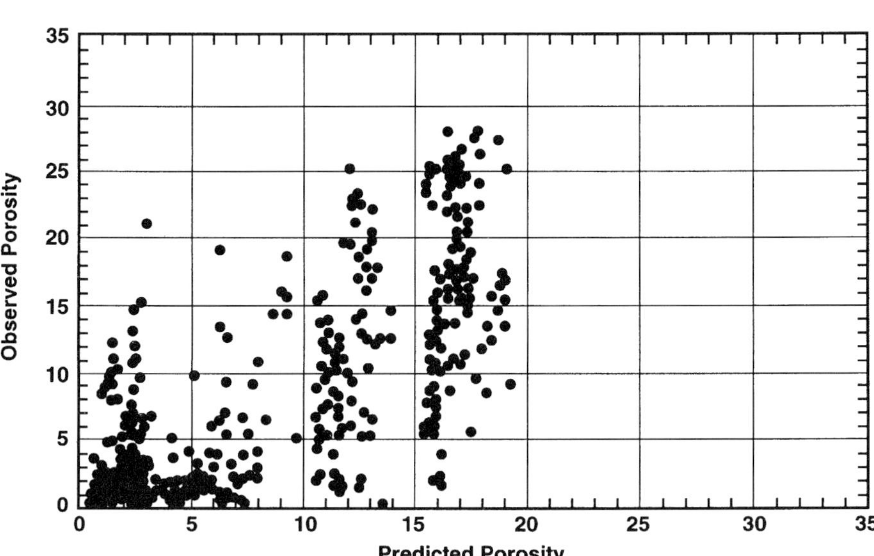

Figure 5. Plot of predicted vs. observed values for regression and neural network model for one sub-area (H). The vertical groupings of points from the regression analysis relate to facies, as the contributions from facies must enter the equation additively. The neural network plot does not show these groupings; the predictions are free from the linear constraint.

Neural Network Model (Withheld Data)

facies variables are again 0 or 1, depending on which facies is being used for the prediction. As mentioned, the neural network program produces matrices of weights by which values of the predictor variables are multiplied to obtain a porosity or permeability prediction.

Because depth is present in the equation, the prediction is being made for one point (one x,y,z location). To construct a set of horizontal and/or vertical predictions (as along a borehole), the equation is solved numerous times, changing x,y,z and facies as appropriate.

$$\text{Porosity} = -2537.7 + 24.4(X^2) + 403.1(Y) - 13.5[(\ln(\text{depth})] + 10.9(\text{platform facies}) + 3.9(\text{lowstand–wedge platform facies}) - 4.5(\text{upper slope facies}) - 4.4(\text{middle slope facies}) \quad (2)$$

Comparison of Neural Network and Regression Results

Two generalities for porosity and permeability prediction for the Ca2 emerged during the analyses: porosity is more easily predicted than permeability, and neural networks predict porosity and permeability better than multiple regression does. The best way to compare the predictive capabilities of various models is to compare the residuals (i.e., the actual porosity or permeability values minus those predicted by the model) for a set of data not used to develop the model. Regardless of the goodness-of-fit of any model to the data from which it was generated, the model must be able to predict values for new data within a reasonable tolerance.

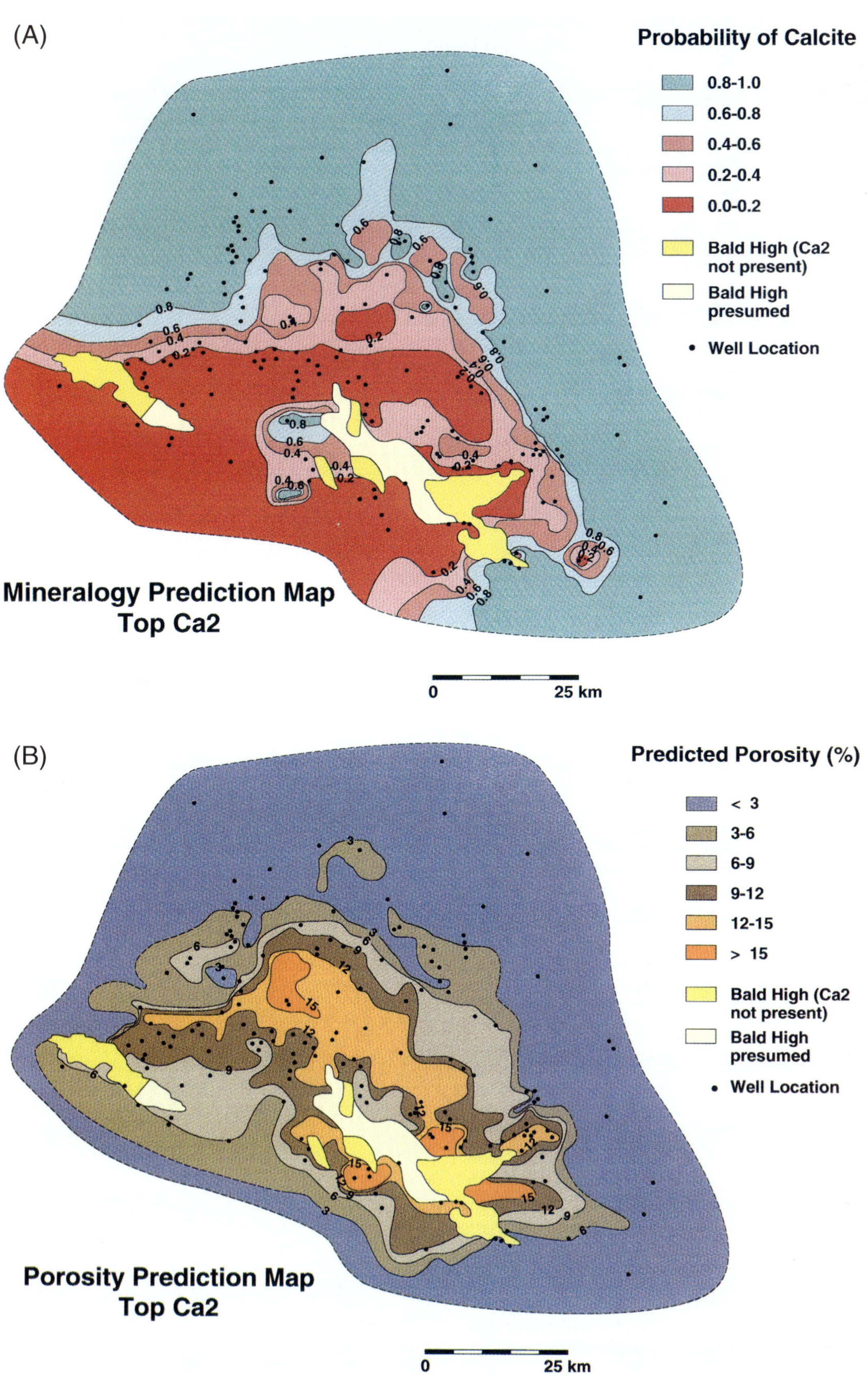

Figure 6. Prediction maps for the top of the Zechstein 2 Carbonate. The maps were made by superimposing a grid over the area with individual cells sized 2.5×2.5 km, and generating predictions at every grid x-y intersection. The yellow areas labeled "Bald High" in the legend indicate areas where the Zechstein 2 Carbonate is absent over tectonic features. (A) Mineralogy prediction map. (B) Porosity prediction map.

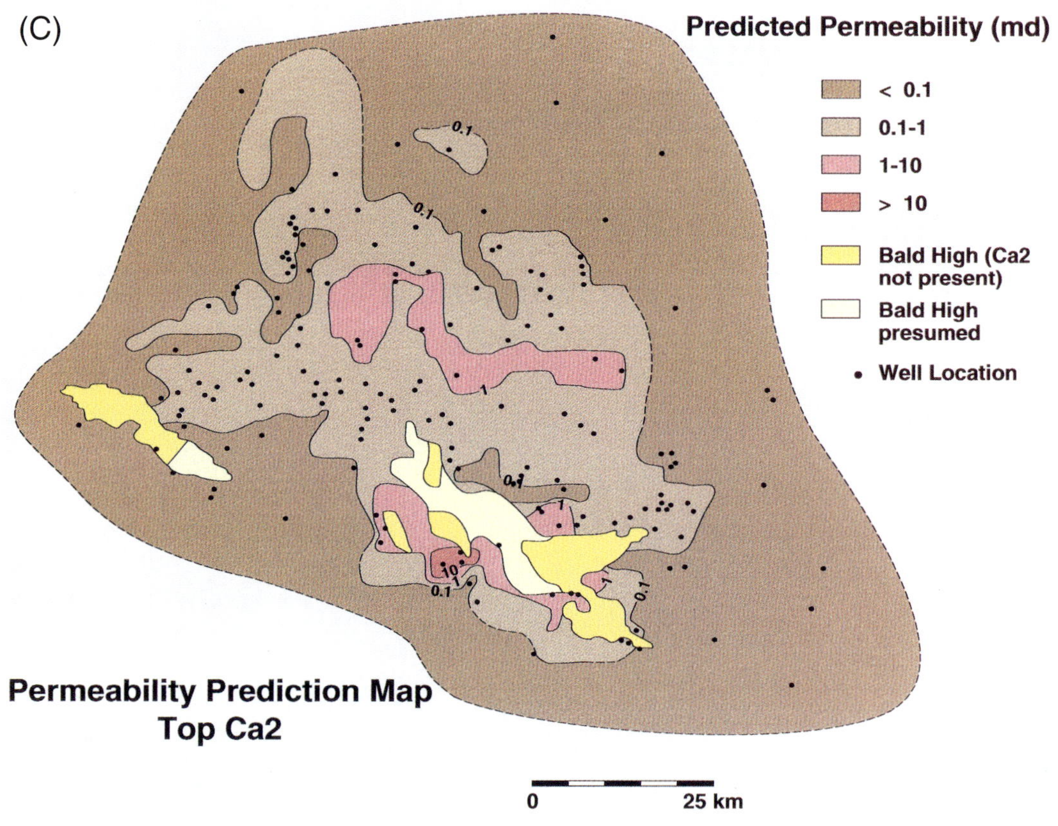

Figure 6. (C) Permeability prediction map.

Table 1 shows the results for both neural network and regression model prediction of porosity and permeability for each of the structural subareas into which the Zechstein 2 Carbonate was divided. The columns in Table 1 indicate tolerances (e.g., ± 2 porosity %) about a prediction. The values in Table 1 show the percentages of withheld porosity and permeability values that were predicted within a given tolerance. Within each subarea, the same data were withheld for both the regression and neural network models (the training and calibration data sets randomly selected by the neural network model were saved and used for the regression model). The data in Figure 4 show the residual values from the regression and neural network models for one of the structural subareas.

Differences exist among the structural subareas with respect to the ability to predict porosity and permeability. Subarea B (Figure 3) is one of the most difficult areas in which to make predictions; this may reflect a more complex distribution. In contrast, subarea E models predict quite well, reflecting the simpler porosity and permeability distribution. In general, in subareas for which porosity prediction is difficult, permeability prediction is also a problem.

Table 1 shows that, in general, the neural network models predicted a higher percentage of values within the stated tolerance than did the regression models. However, the regression models perform about as well as the neural network models in predicting average porosity and permeability values for facies or even in predicting general trends within a facies (such as decreasing or increasing porosity). Two pieces of information are available to help interpret average porosity and permeability values in the case of the Zechstein 2 Carbonate: mineralogy predictions and an understanding of the geologic variability. If, for example, the mineralogy predictions indicate a very high probability of calcite in the upper slope facies, but the predicted average porosity is high, one would predict that high-porosity dolomite layers exist, because almost all of the calcite in the upper slope is less than 3% porosity.

If one must predict more detail than that provided by the average, neural network models are recommended, as they are when there is a large difference between the neural network and regression results or where all predictions are poor. Plots of predicted vs. observed values (Figure 5) provide a useful demonstration of the differences in predictive ability between the regression and neural network models for a typical subarea. One of the most important differences illustrated in Figure 5 is that the regression model predicts in "groupings" (visible in Figure 5 as vertical clusters of points), whereas the neural network predictions are continuous. The regression clusters relate to the facies; because the regression model is a linear model, the effects of different facies must enter the model additively—hence the jump in porosity from one facies to the next. Second, the regression model is constrained in

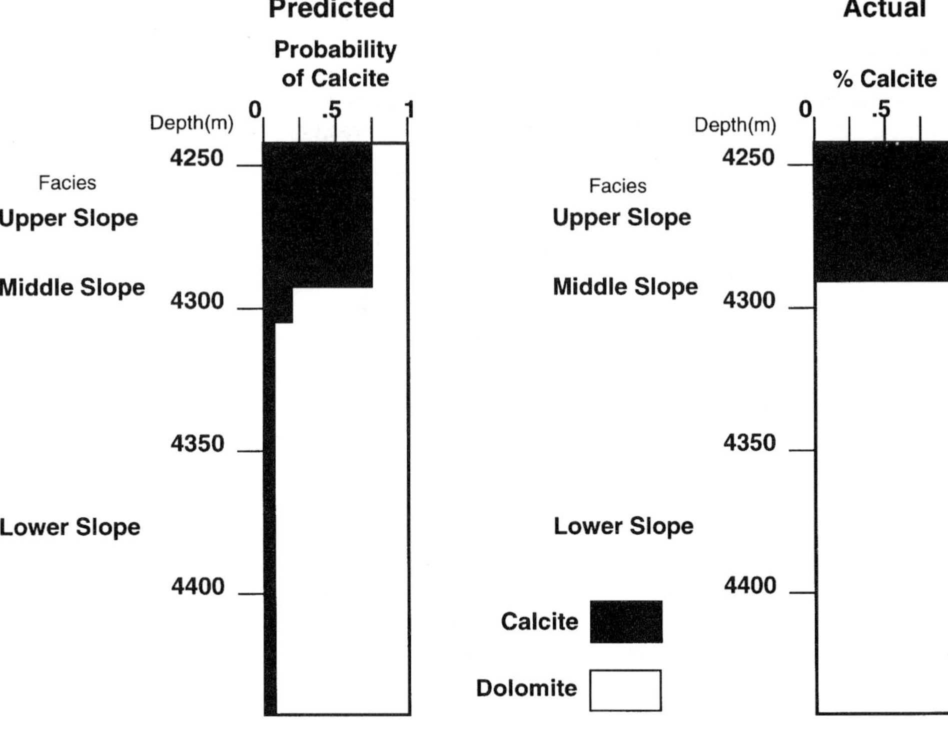

Figure 7. Example of mineralogy prediction for an "undrilled" well. The data from the well were withheld during model development, and then a prediction was made for the well location and facies succession. The column on the left shows the predicted probability of calcite for each facies; for example, the upper slope facies has a 0.75 probability of calcite. The column on the right shows the actual mineralogy—the well had a calcite "cap" at the top, and the remainder was dolomite, as predicted by the low probabilities of calcite shown on the left (<0.1).

this case to predicting porosity values no higher than ~19%, even though values in the calibration set were as high as ~32%. This is again a result of the linear constraint, which the neural network does not have. As shown in Figure 5, the neural network model predicted values up to ~28%. For some cases, however, the regression models suffice, and have the advantage of relative ease of use, as they can be solved using a calculator.

APPLICATIONS AND TROUBLESHOOTING

When x-y coordinates and depth are used in predictive equations, the equations can generate three-dimensional predictions. Using these equations, three-dimensional representations of the predictions can be created, and two-dimensional maps or cross sections can be constructed. For the Zechstein 2 Carbonate, maps were made showing the distribution of predicted mineralogy, porosity, and permeability for different horizons. For example, Figure 6 shows maps of the mineralogy, porosity, and permeability predicted at the top of the formation. Such maps provide an overview of the predicted distributions in a given area, and constitute a prelude to more detailed prediction work, as illustrated in Figure 7 and Table 2.

Because the predictive models for permeability were generated from core-plug permeabilities, they predict matrix, rather than fracture, permeability. Thus, in areas known to produce from fractures, permeabilities predicted by the models must be used with caution. This limitation can be avoided if reliable well-log–derived permeabilities are available and if

other desired predictor variables such as facies can be ascertained from well logs where no core exists. Another potential problem arises when input values (location, facies) for the equations are not within the range of values used to develop the model (i.e., when extrapolating rather than interpolating). In such cases, an inconsistent prediction may result, such as a negative porosity. Caution must be exercised when relying upon extrapolations, even if the answer appears to be "reasonable."

Although data smoothing led to better predictions, all the models still produced poorly predicted values (e.g., where the observed and predicted values differed by >10 porosity %) (Figure 4). Given this, the question arises: What, if anything, do the poorly predicted values have in common? For any one model, are these points all from one well or one facies, indicating that the model predicts poorly for an entire well location or facies? Or do the poorly predicted points have nothing in common, simply reflecting a "random" scattering across the different wells and facies? If the points reflect random scattering, there is a negligible effect on the ability to predict values representative for a proposed well location. If, on the other hand, the poorly predicted points originate from one area, a separate predictive model, perhaps using additional predictor variables, could be tried. This should be checked before applying predictions. After checking many of the predictions for the Zechstein 2 Carbonate, the random scattering scenario was accepted; for example, few wells were found in which the majority of predictions within a facies had residual values greater than about 4 porosity units.

Table 2. Comparison of Predicted and Actual Porosity for Two Wells.*

Well	Dep. Facies	Probability of Calcite (%)	Interpreted Mineralogy	Observed Mineralogy	Predicted Porosity (Average) (%)	Actual Porosity (Average) (%)
1	Plat-LSW	50	Dolomite and Calcite	Mostly Calcite	16	15
	Upper Slope	30	Mostly Dolomite	Mostly Dolomite	11	12
	Middle Slope	<10	Dolomite	Dolomite	7	7
2	Upper Slope	70	Mostly Calcite	Calcite	2	2
	Middle Slope	40	Mostly Dolomite	Mostly Dolomite	8	10
	Lower Slope	60	Mostly Calcite	Mostly Calcite	2	2

*Data from these wells were withheld during development of the linear regression equation for this structural subarea. The boldface columns indicate input data (well location and facies). The equation generated a probability of calcite, from which an interpretation of mineralogy was made and compared to the observed mineralogy observed in the well. In addition, porosity predictions were made.

Data Requirements for Spatial Predictions

The Zechstein data set used to illustrate the procedure for predicting reservoir quality is fairly large, consisting of data from 287 wells. However, such a large data set is not required for making spatial predictions. In fact, because the Zechstein data were divided into subareas, some of the predictive models were developed using fewer than 10 wells (Table 1). Although a general rule for the necessary density of data cannot be devised, several factors should be considered: (1) number of wells and number of measurements of the parameter of interest, (2) spacing of wells, and (3) complexity of the spatial distribution of the parameter. As the complexity of the distribution increases, more wells at a closer spacing are required to achieve useful predictions.

CONCLUSIONS

Optimally, predictive models use reservoir-quality controls, rather than surrogates, as model inputs. However, the three-dimensional distribution of parameters such as porosity, permeability, and mineralogy can be predicted at a pragmatic scale even where cause-and-effect models are not available or do not provide predictions at the required scale. For prediction purposes, location variables, along with any other significant variables whose values are known prior to drilling, may provide predictive capabilities in statistical models. Prediction of complexly distributed parameters commonly improves by using neural networks. Overall, predictive models should be judged by their success or failure, not only by their use of geologic variables thought to be related to the predicted parameters through cause and effect. Such models, however, may not be reliable for extrapolation purposes.

ACKNOWLEDGMENTS

The authors thank BEB Erdgas und Erdöl GmbH, the W.E.G. publication committee, and Exxon Production Research Co. for their permission to publish this paper. Reviews by Tom Jones, Dave Pevear, Alton Brown, and Marek Kacewicz are appreciated. An anonymous reviewer also contributed comments.

REFERENCES CITED

Abbott, P.L., 1974, Calcitization of Edwards Group dolomites in the Balcones fault zone aquifer, south-central Texas: Geology, v. 1, p. 359–362.

Bowerman, B.L., and R.T. O'Connell, 1990, Linear statistical models: an applied approach (2d ed.): Boston, PWS-Kent Publishing Co., 1024 p.

Budai, J.M., K.C. Lohmann, and R.M. Owen, 1984, Burial dedolomite in the Mississippian Madison Limestone, Wyoming and Utah thrust belt: Journal of Sedimentary Petrology, v. 54, p. 276–288.

Clark, D.N., 1980, The diagenesis of Zechstein carbonate sediments; in H. Fuechtbauer and T. Peryt, eds., The Zechstein Basin with emphasis on carbonate sequences: Contributions to Sedimentology, no. 9, p. 167–203.

Draper, N., and H. Smith, 1981, Applied regression analysis (2d ed.): New York, Wiley & Sons Inc., 709 p.

Haykin, S., 1994, Neural networks: New York, Macmillan College Publishing Co., 696 p.

Kacewicz, M., 1994, Model-free estimation of fracture aperture with neural networks: Mathematical Geology, v. 26, p. 985–994.

Kleinbaum, D.G., and L.L. Kupper, 1978, Applied regression analysis and other multivariable methods: Boston, Duxbury Press, 556 p.

Strohmenger, C., M. Antonini, G. Jäger, K. Rockenbauch, and C. Strauss, 1996, Zechstein 2 Carbonate reservoir facies distribution in relation to Zechstein sequence stratigraphy (Upper Permian, northwest Germany): an integrated approach: Bull. Centre Rech. Explor. Prod. Elf Aquitaine, v. 20, p. 1–35.

Strohmenger, C., K.M. Love, J.C. Mitchell, and K. Rockenbauch, 1993, Sedimentology and diagenesis of the Zechstein Ca^2 Carbonate, Late Permian, Northwest Germany (abs.): AAPG Bulletin, v. 77, p. 1668.

Warren, J.K., 1991, Sulfate dominated sea-marginal and platform evaporative settings, *in* J.L. Melvin, ed., Evaporites, petroleum and mineral resources: Developments in Sedimentology 50, p. 69–187.

Chapter 5

Primmer, T.J., C.A. Cade, J. Evans, J.G. Gluyas, M.S. Hopkins, N.H. Oxtoby, P.C. Smalley, E.A. Warren, and R.H. Worden, 1997, Global patterns in sandstone diagenesis: their application to reservoir quality prediction for petroleum exploration, *in* J.A. Kupecz, J. Gluyas, and S. Bloch, eds., Reservoir quality prediction in sandstones and carbonates: AAPG Memoir 69, p. 61–77.

Global Patterns in Sandstone Diagenesis: Their Application to Reservoir Quality Prediction for Petroleum Exploration

Tim J. Primmer
BP Exploration
Dyce, Aberdeen, Scotland, United Kingdom

Chris A. Cade
BP Norway Ltd.
Forus, Norway

Jonathan Evans
BP Exploration
Poole, Dorset, England, United Kingdom

Jon G. Gluyas
BP Exploration de Venezuela SA
El Rosal, Caracas, Venezuela

Mark S. Hopkins
BP Exploration
Sunbury on Thames, England, United Kingdom

Norman. H. Oxtoby
University of London, Department of Geology
Egham Hill, England, United Kingdom

P. Craig Smalley
Edward A. Warren
BP Exploration
Sunbury on Thames, England, United Kingdom

Richard H. Worden
Department of Geology, Queen's University
Belfast, Northern Ireland, United Kingdom

ABSTRACT

Sandstones that share common detrital mineralogies, depositional environments, and burial histories also share common diagenetic histories. A survey of the diagenetic history of 100 sandstones from around the world has recognized five common, repetitive, and predictable styles of diagenesis in which similar diagenetic mineral assemblages have been observed.

The five diagenetic styles are: (1) *quartz,* commonly with lesser quantities of neoformed clays (e.g., kaolinite and/or illite) and late-diagenetic, ferron carbonate; (2) *clay minerals* (illite or kaolinite) with lesser quantities of quartz or zeolite and late-diagenetic carbonate; (3) *early diagenetic (low-temperature) grain-coating clay mineral cements* such as chlorite, which may inhibit quartz cementation during later burial; (4) *early diagenetic carbonate or evaporite cement,* often localized, which severely reduces porosity and net pay at very shallow burial depths; and (5) *zeolites,* which occur over a wide range in burial temperature, often in association with abundant clay (usually smectite or chlorite) and late-diagenetic, nonferroan carbonates.

The quartz diagenetic style is the most common and accounts for 40% of the sample set. It is also most likely to occur in mineralogically mature sandstones, while early diagenetic carbonates and zeolites dominate in mineralogically immature sandstones. Presence or absence of clay appears to be independent of both initial sand mineralogy and depositional environment. However, when clay is present, the type appears to vary as a function of initial sand mineralogy and depositional environment.

Large quantities of quartz are unusual cements in sequences that have never been hotter than ~75°C, while illite precipitation at temperatures below ~100°C is rare. Zeolite composition changes systematically from clinoptilolite at ~25°C to laumonite at temperatures >100°C.

The repetitive nature and simplicity of these five styles can help predict modifications in reservoir quality due to burial. An accurate prediction of the reservoir quality in sandstones forms the basis of an accurate porosity and permeability prediction ahead of drilling wells in petroleum exploration, development, or production.

INTRODUCTION

The quality of a petroleum reservoir is a function of both its porosity and permeability. In sandstones, these parameters are controlled by initial sediment composition and its subsequent modification during burial and lithification. Overburden stress acts during burial to compact the sand, reducing porosity and constricting pore throats. Mineral precipitation and dissolution also affect pore space and may completely change the pore structure. Extreme, locally intense cementation can also lead to reservoir compartmentalization, as in the Troll and Murchison fields of the North Sea (Gibbons et al., 1993; Prosser et al., 1993).

Accurate assessment of reservoir potential ahead of drilling is a critical factor throughout the petroleum exploration and production cycle. In the early stages of exploration, the limits of economic basement is a key datum to define; the depth at which sediments are insufficiently permeable to sustain economic production is difficult to define. In a more mature exploration area, accurate prediction of reservoir quality anomalies may become more important, especially where sandstones that are more porous and permeable than would be expected are found, given their burial depth. Prediction of reservoir quality during production tends to concentrate on understanding the spatial architecture or localized heterogeneity in porosity and permeability needed in the course of reservoir management (petroleum production and fluid injection).

This chapter presents a pragmatic method for reservoir quality prediction, one that has been applied successfully to many oil and gas fields. Given detrital mineral composition, burial depth, and overpressure, the premise of this method is that porosity and permeability can be predicted for uncemented sands and cemented sandstones.

A key concept discussed below is "cementation style," the relationship between detrital composition, burial depth, temperature, and cement type. Summarized data from published literature and BP's in-house reports are used to illustrate various styles of diagenesis, each of which has a specific impact on reservoir quality. Examples of porosity and permeability prediction are presented in this volume (Evans et al.; Gluyas; Gluyas and Witton) and elsewhere (Cade et al., 1994; Evans et al., 1994).

PREDICTION OF DIAGENETIC STYLES

Sandstones with similar geological histories (e.g., sediment composition, depositional environment, facies associations, and burial history) would be expected to develop similar styles of diagenesis. Although diagenetic cement reduces porosity on a simple volume-for-volume basis, it is important to establish a style containing distinct cements with different habits and distibutions at the pore scale (e.g., thin pore lining or blocky pore blocking). Similar volumes of different cements can have dramatically different effects on permeability. For example, Pallatt et al. (1984) describe the disproportionate effect that small quantities of authigenic illite can have on permeability. Clearly, establishing a "style" of diagenesis can help provide a framework in which modification of reservoir quality by postdepositional processes can be predicted more quantitatively. The global database review discussed below shows how sediment composition, depositional environment, and burial temperature combine to establish particular styles of diagenesis. The improved understanding based on this analysis can be used to predict likely changes in the porosity and permeability of sandstones during burial.

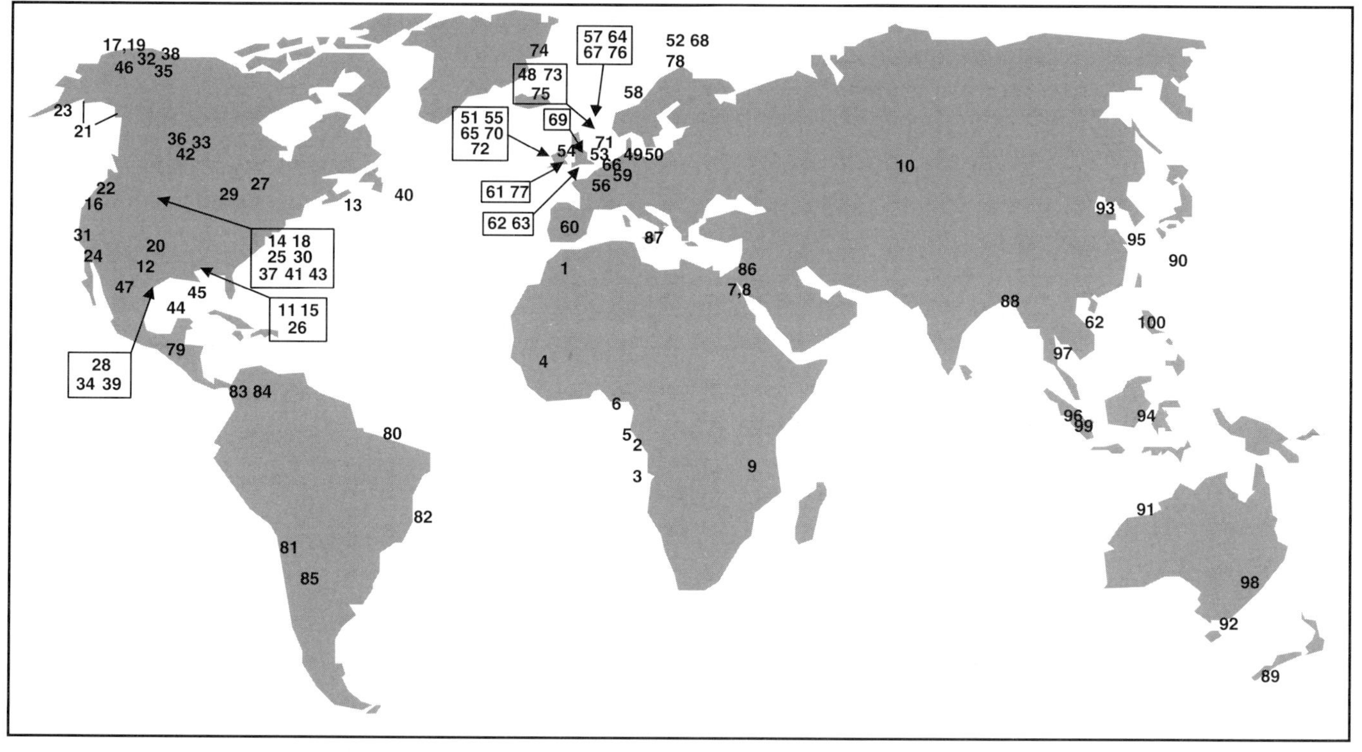

Figure 1. Location of reviewed studies.

The Database

The database comprises studies of 100 stratigraphically discrete sandstone units (Figure 1; Table 1). This is not intended to be an exhaustive review of all available literature around the world, but rather a selection of examples that are representative of reservoir quality variation observed in the regions around the world. Many of the studies reviewed are from North America (mainly the Gulf Coast, the western United States, and Canada). Publications from that part of the world constitute a third of the data reviewed. BP's in-house studies (primarily from the U.K. continental shelf and Porcupine Basin) supplement the more-limited open literature available for NW Europe, and form another third of the data set. The remaining third are studies from other parts of the world (mainly Africa, South America, and SE Asia). The fact that >80% of the units studied are Mesozoic or younger reflects the bias of past work to those reservoirs that have an economic importance in oil and gas exploration.

The database encompasses a wide range of relevant geological attributes (depositional environment, sandstone composition, and maximum burial temperature). The data are dominated by fluvial, deltaic, and shallow-marine sandstones (Figure 2). The relatively small aeolian data set probably reflects the poor preservation potential of this depositional environment. The small number of good deep-marine examples shows how poorly represented this depositional setting is in some of the well-studied parts of N. America (Wyoming and the Gulf Coast) and the North Sea. There is also a significant lack of good descriptions of diagenesis in lacustrine environments. The few examples considered here, either saline lake or temperate lake deposits, are grouped with eolian sands or fluviodeltaic depositional environments, respectively.

The compositional maturity of various depositonal environments is also shown in Figure 2. Typically, eolian and shallow-marine sands are more mature than fluvial or deltaic sands, reflecting the degree of reworking usually encountered in these sorts of depositional environments. In contrast, the compositional immaturity of the deep-marine examples may reflect sampling bias, because a significant number of these studies are from active volcanic margins (e.g., the West Coast of the United States), rather than passive margins or failed rifts sourced by cratonic basement.

Although the selected data are drawn from the most comprehensive studies available, data on maximum burial temperature are sparse and often poorly constrained. Estimates of maximum burial temperature were available for just over 60% of the cases studied and range from 25° to 300°C. In an attempt to estimate the effect of burial temperature (when good field data were not available), fluid inclusion, stable isotope, and organic maturation data have been interpreted where appropriate.

RESULTS

Five Styles of Diagenesis in Sandstones

Five common diagenetic styles have been identified (Figure 3). Each has a distinctive diagenetic mineral assemblage. Their characteristics are:

1. *Quartz dominated*, which often occurs in association with smaller quantities of neoformed clays

Table 1. List of Studies Reviewed.

	Region/Country	Formation	Age	Reference
1	Atlas Mountains, Morocco	Adrar N'Dguoe	Ordovician	Evans, 1990
2	S. Gabon Rift, offshore W Africa	"Presalt"	E. Cretaceous	Giroir et al., 1989
3	Angola Margin, offshore W Africa	"Presalt"	E. Cretaceous	Girard et al., 1989
4	W Mali	Souroukoto	L. Proterozoic	Girard and Deynoux, 1991
5	W Gabon, offshore W Africa	(Various)	L. Cretaceous	Pittman and King, 1986
6	Niger Delta, Nigeria	Agbada	Tertiary	Lambert and Shaw, 1982
7	Gulf of Suez, Egypt	Rudeis	Miocene	Evans, 1990
8	Ras Budran, Egypt	Nubian	Paleozoic/E. Cretaceous	BP in-house
9	Ruhuhu Basin, Tanzania	Karoo	Triassic	Wopfner et al., 1990
10	W. Siberia, Russia	Vartorsk	E. Cretaceous	BP in-house
11	Alabama Gulf Coast, USA	Norphlet	L. Jurassic	Dixon et al., 1989
12	N Texas, USA	Gray	L. Carboniferous	Land and Dutton, 1978
13	Scotian Basin, offshore E Canada	(Various)	L. Jurassic/E. Cretaceous	Jansa and Urrea, 1990
14	Wyoming, USA	Frontier	L. Cretaceous	Tillman and Almon, 1979
15	Louisiana Gulf Coast, USA	Woodbine/ Tuscaloosa	L. Cretaceous	Thomson, 1979
16	SW Oregon, USA	Umpqua	Paleocene/Eocene	Burns and Etheridge, 1979
17	N Alaska, USA	Sag River	L. Triassic/E. Jurassic	Mozley and Hoernle, 1990
18	Wyoming, USA	Shannon	L. Cretaceous	Rangathan and Tye, 1986
19	N Alaska, USA	Nanushuk/Colville	E.-L. Cretaceous	Smosna, 1988
20	N Texas, USA	Mobeetie	L. Carboniferous	Dutton and Land, 1985
21	NE Pacific Coast, USA	(Various)	Tertiary	Galloway, 1979
22	W Oregon, USA	(Various)	Eocene	Chan, 1985
23	S Alaska, USA	(Various)	Jurassic-Paleogene	Bolm et al., 1983
24	California, USA	Santa Ynez	Paleogene	Helmold and Van de Kamp, 1984
25	Wyoming, USA	U. Minnesula	E. Permian	Market and Al-Shaieb, 1984
26	Mississippi/Alabama, USA	Norphlet	L. Jurassic	McBride et al., 1987
27	S Ontario, Canada	Cataract	Silurian	O'Shea and Frape, 1988
28	Texas Gulf Coast, USA	Travis Peak	E. Cretaceous	Dutton and Diggs, 1990
29	Michigan, USA	St. Peter	Ordovician	Barnes et al., 1991
30	Wyoming, USA	Upper Almond	L. Cretaceous	Meshri and Walker, 1990
31	California, USA	Stevens	Miocene	Boles, 1984
32	N Alaska, USA	Kuparak	E. Cretaceous	Eggert, 1987
33	Alberta, W Canada	Clearwater	E. Cretaceous	Hutcheon et al., 1989
34	Texas Gulf Coast, USA	Frio	Oligocene	Milliken et al., 1981
35	N Alaska, USA	Ivishak	Permo-Triassic	Melvin and Knight, 1984
36	Alberta, W. Canada	Belly River	L. Cretaceous	Ayalon and Longstaffe, 1988
37	Wyoming, USA	Tensleep	Carboniferous	Manckiewicz and Steidtmann, 1979
38	N Alaska, USA	Kekiktuk	L. Carboniferous	Bloch et al., 1990
39	Texas Gulf Coast, USA	Wilcox	Eocene	Land and Fisher, 1987
40	Grand Banks, offshore E Canada	Hibernia	E. Cretaceous	Brown et al., 1990
41	Wyoming, USA	Upper Muddy	L. Cretaceous	Almon and Davies, 1979
42	Alberta, W Canada	Viking	E. Cretaceous	Reinson and Foscolos, 1986
43	Wyoming	Lower Muddy	L. Cretaceous	Almon and Davies, 1979
44	Offshore Gulf of Mexico	(Unspecified)	Miocene-Holocene	Whynot, 1986
45	Louisiana Gulf Coast, USA	(Unspecified)	Plio-Pleistocene	Milliken, 1985
46	N Alaska, USA	Kuparak	E. Cretaceous	BP in-house
47	N Mexico	Baucarit	Miocene	Cocheme et al., 1988
48	C North Sea, UKCS	Marnock	Triassic	Smith et al., 1993
49	Schleswig-Holstein, N Germany	Dogger	M. Jurassic	Horn, 1965
50	Bornholm, Denmark	Hasle	E. Jurassic	Larsen and Friis, 1991
51	Porcupine Basin, offshore W Ireland	(Unspecified)	Permo-Triassic	BP in-house
52	Barents Shlef, Svalbard	Helvetiafjellet	E. Cretaceous	Edwards, 1979
53	S North Sea, UKCS	Rotliegend	E. Permian	Gluyas and Leonard, 1995
54	Irish Sea, UKCS	Sherwood	E. Triassic	Macchi et al., 1990

Table 1. (continued.)

	Region/Country	Formation	Age	Reference
55	Porcupine Basin, offshore W Ireland	(Unspecified)	E.-M. Jurassic	BP in-house
56	Paris Basin, France	Chaunoy	Triassic	Worden, 1995
57	N North Sea, UKCS	Brae	L. Jurassic	Gluyas and Coleman, 1992
58	Haltenbanken, NOCS	Garn	M. Jurassic	Ehrenberg, 1990
59	Rhine Graben, Germany	Bundsandstein	Triassic	Evans, 1990
60	Iberian Range, E Spain	(Various)	Permo-Triassic	Morad et al., 1990
61	Celtic Sea, offshore SE Ireland	Wealden	E. Cretaceous	BP in-house
62	Dorset, S England	Bridport	E. Jurassic	Morris and Shepperd, 1982
63	Dorser, S England	Sherwood	E. Triassic	Strong and Milodowski, 1987
64	N North Sea, UKCS	Magnus	L. Jurassic	Emery et al., 1993
65	Porcupine Basin, offshore W Ireland	(Unspecified)	E. Cretaceous	Britoil in-house
66	S North Sea, offshore Holland/Germany	"J1"–"J4"	M.-L. Jurassic	BP in-house
67	N North Sea, UKCS	Brent	M. Jurassic	Glasmann et al., 1989
68	Barents Shelf, Svalbard	Helvetiafjellet	E. Cretaceous	Edwards, 1979
69	E. Midlands, England	Crawshaw	L. Carboniferous	Warren, 1987
70	Porcupine Basin, offshore W Ireland	(Unspecified)	L. Carboniferous	BP in-house
71	S North Sea, UKCS	(Unspecified)	L. Carboniferous	Cowan, 1989
72	Porcupine Basin, offshore W Ireland	(Unspecified)	M.-L. Jurassic	BP in-house
73	Inner Moray Firth, UKCS	Beatrice	E. Jurassic	Haszeldine et al., 1984
74	E Greenland	Vardekloft	M. Jurassic	BP in-house
75	C North Sea, NOCS	Ula	L. Jurassic	Oxtoby et al., 1995
76	N North Sea, UKCS	Statfjord	E. Jurassic	BP in-house
77	Celtic Sea, offshore SE Ireland	Greensand	L. Cretaceous	BP in-house
78	Barents Sea, NOCS	Stø	E. Jurassic	Riches et al., 1986
79	S Guatemala	(Unspecified)	Neogene-Holocene	Davies et al., 1979
80	Potiguar Basin, NE Brazil	Pendencia	E. Cretaceous	Moraes, 1991
81	N Chile	Puilactis	L. Cretaceous/Paleocene	Hartley et al., 1991
82	Campos Basin, offshore SE Brazil	Campos	L. Cretaceous-Eocene	Moraes, 1989
83	Llanos Basin, Colombia	Mirador	Eocene	Cazier et al., 1995
84	Llanos Basin, Colombia	Guadalup	L. Cretaceous	Cazier et al., 1995
85	Huaco, W. Argentina	Huaco	Neogene	Damanti and Jordan, 1989
86	S Israel	Helez	E. Cretaceous	Shenhav, 1971
87	Calabria/Sicily, S Italy	Stilo-Capo-d'Orlando	Miocene	Cavazza and Dahl, 1990
88	Bengal Basin, Bangladesh	Bengal	Neogene	Imam and Shaw, 1985
89	Southland Syncline, New Zealand	Murihiku	Triassic/Jurassic	Boles and Coombs, 1977
90	Daito Ridge and Basin, offshore NW Pacific	(Unspecified)	Eocene	Lee, 1988
91	NW Shelf, offshore Australia	Mungaroo	M. Jurassic	BP in-house
92	Gippsland Basin, offshore SE Australia	Latrobe	L. Cretaceous/Paleogene	Surdam et al., 1989
93	Gulf of Bohai, N China	Shahejie	Eocene/Oligocene	BP in-house
94	E Borneo, Indonesia	Mahakan	Teriary	Rinckenbach, 1988
95	Yellow Sea, offshore China	Fourth and Fifth	Paleogene	BP in-house
96	C Sumatra, Malaysia	Sihapas	Miocene	Gluyas and Oxtoby, 1995
97	Pattani Basin, Gulf of Thailand	(Unspecified)	Miocene	Trevena and Clark, 1986
98	Queensland, E. Australia	Surat	Cretaceous	Hawlader, 1990
99	S Sumatra, Malaysia	Air Benakat	Miocene	BP in-house
100	N Luzon, Philippines	Cagayan	Plio-Pleistocene	Mathisen, 1984

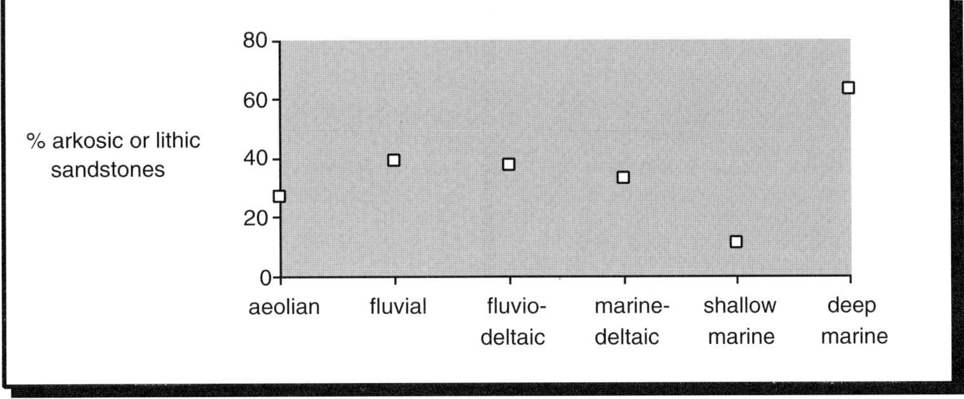

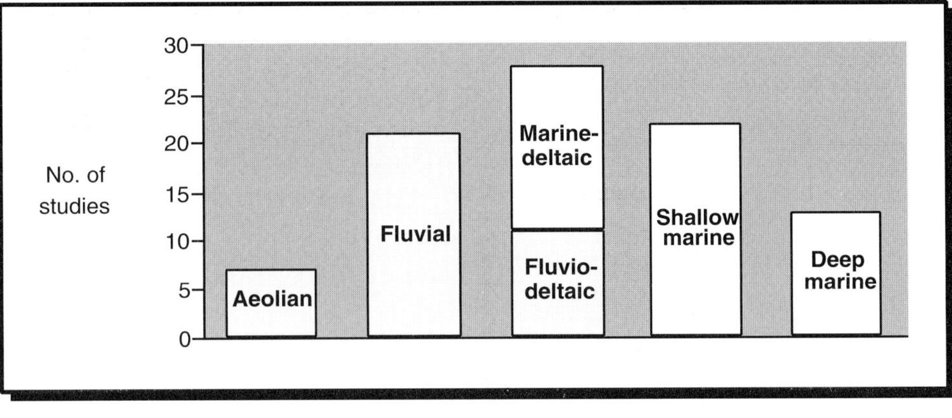

Figure 2. Division of selected studies by gross depositional environment (top) and compositional maturity (bottom). All lacustrine examples have been grouped in with either eolian or fluviodeltaic environments (see text). Compositional maturity of each gross depositional environment is shown in terms of the proportion of arkosic or lithic sands in each depositional environment.

(e.g., kaolinite and/or illite) and late-stage, high-temperature, ferroan carbonate.

2. *Clay minerals dominated*, such illite or kaolinite with smaller quantities of quartz or zeolite and late-diagenetic carbonate.

3. *Early diagenetic (low-temperature) grain-coating clay mineral cements,* such as chlorite. These may inhibit or restrict subsequent quartz cementation during burial to higher temperatures. This can, with help from overpressuring, maintain higher porosity than might be expected when buried to considerable depths (>3.5 km).

4. *Early diagenetic carbonate or evaporite cement dominated*, often localized, which severely reduces porosity and net pay from very shallow burial depths.

5. *Zeolite dominated*, which occur over a wide range in burial temperatures, often in association with abundant clay (usually smectite or chlorite) and late-diagenetic nonferroan carbonates.

It is apparent that quartz-dominated diagenesis (representing 40% of the total) is the most common diagenetic style seen in the selected studies (Figure 4). It is also notable that the specific association of early diagenetic grain-coating clay with inhibition of later quartz cement is more common than diagenesis dominated by clay minerals alone. In ~10% of cases, early or late diagenetic carbonates were the predominant cements, and a similar number of cases contained significant quantities of zeolite. However, evaporite minerals were significant cements in <1% of the examples investigated.

Figure 5 shows the frequency of occurrence of particular clay minerals in the clay-dominated and early grain-coating clay with quartz styles of diagenesis. Chlorite is the most commonly occurring clay mineral, mainly as a result of its frequent occurrence as an early grain-coating clay. Both kaolinite and illite are less important; they most commonly occur in subordinate quantities with quartz in quartz-dominated diagenesis. Where carbonates occur as significant cements, Fe-calcite and Fe-dolomite are the more common late-diagenetic cements, whereas siderite and dolomite are more common as early cements (Figure 6). Calcite is common both as an early cement and as a late stage cement.

Effect of Sediment Composition and Depositional Environment on Diagenetic Style

The controls exerted on diagenetic style by differences in primary sediment composition and depositional environment are shown in Figure 7. Quartz cements are more common in sands deposited in environments where sediment reworking has produced compositionally more mature (quartzose) sands (e.g., eolian, deltaic, and shallow-marine environments), whereas sandstones that are mineralogically immature are likely to be cemented by carbonates and zeolites, regardless of whether sands are arkosic or lithic. On the other hand, there appears to be no apparent correlation between sand composition and clay-dominated styles of diagenesis. Zeolites seem to be most common in deep-marine sands, but this may reflect the subsample investigated; the majority were from active volcanogenic

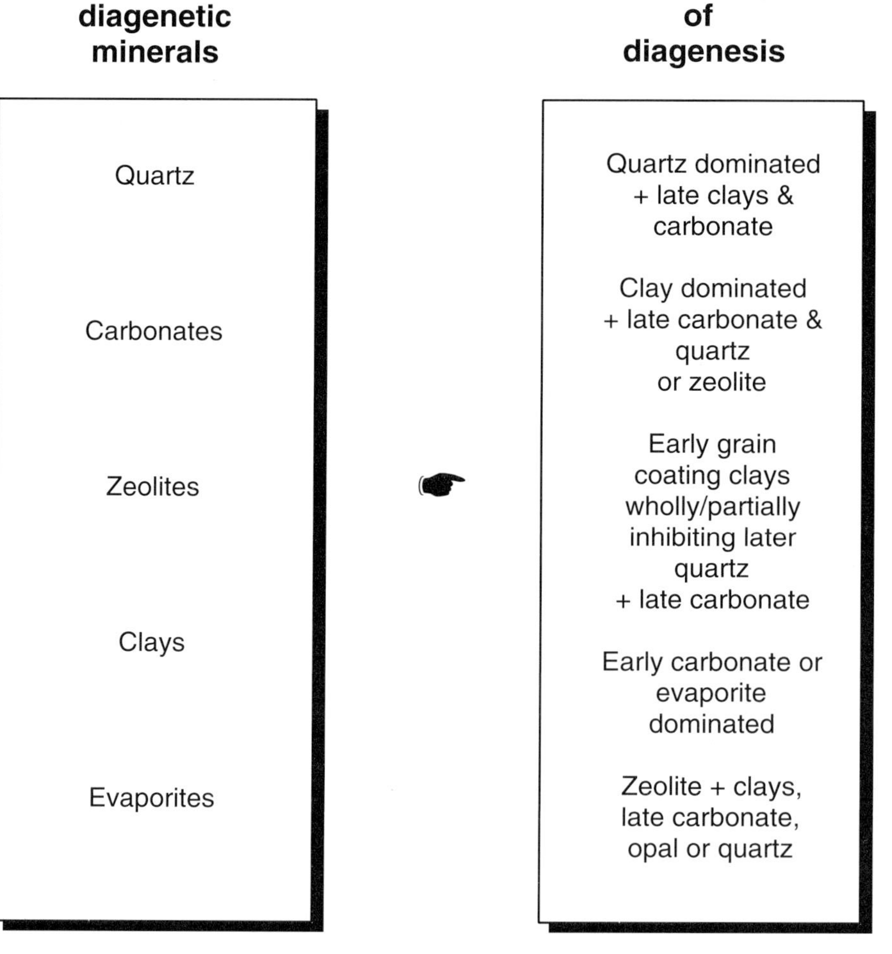

Figure 3. Schematic illustration of the relationship between commonly occurring diagenetic cements (shown in order of decreasing abundance) and their associated styles.

margins rather than passive margins sourced from cratonic basement. Late carbonate cements are more prevalent (by a factor of 2 or more), but there appears to be no correlation between depositional environment and early carbonate or clay-dominated styles of diagenesis.

Although no correlation appeared to exist between the occurrence of clay-dominated styles of diagenesis and either sediment composition or depositional environment, both of these factors influence the type of diagenetic clay present in other styles of diagenesis. Kaolinite occurs in more mature (quartzose, subarkosic/lithic) sands and is the most common clay in all depositional environments except eolian settings. Illite is more common in quartzose/subarkosic sands deposited in eolian or fluvial sands, and chlorite is most common in deltaic and shallow-marine sands. Chlorite is also the most abundant diagenetic clay in immature sands, whereas smectite occurs only in immature sands and is most common in deep-marine depositional environments.

Sediment composition and depositional environment control the occurrence of some types of early and late carbonate cements. Specifically, early siderite cements are most common in relatively mature sands deposited in fluvial and marginal-marine settings. This is in contrast to early diagenetic dolomite and early diagenetic calcite cements, which do not seem to be correlated with either sediment composition or depositional environment. Late-diagenetic ferroan dolomite cements are most frequently encountered in subarkosic or sublithic sands, whereas late calcite cements are far more abundant in less mature arkosic or lithic sands.

Effect of Maximum Burial Temperature on Diagenetic Style

Estimates of maximum burial temperature indicate that quartz cements precipitate over a wide range of burial temperatures, although fluid inclusion studies suggest that minimum temperatures of 75°C are usually required for precipitation (Figure 8). Whereas authigenic clay minerals such as kaolinite and chlorite appear to form over a wide range of temperatures, many studies indicate that illite seems to require significantly elevated burial temperatures, usually >100°C (Trevena and Clark, 1986; McBride et al., 1987; Cowan, 1989; Girard et al., 1989; Glasmann et al., 1989; Ehrenberg, 1990; Barnes et al., 1992; Emery et al., 1993; Robinson et al., 1993). As noted elsewhere, different zeolites are stable over relatively narrow temperature ranges in different sedimentary and tectonic environments (Iijima, 1988, and references quoted therein). Temperature estimates for the most commonly

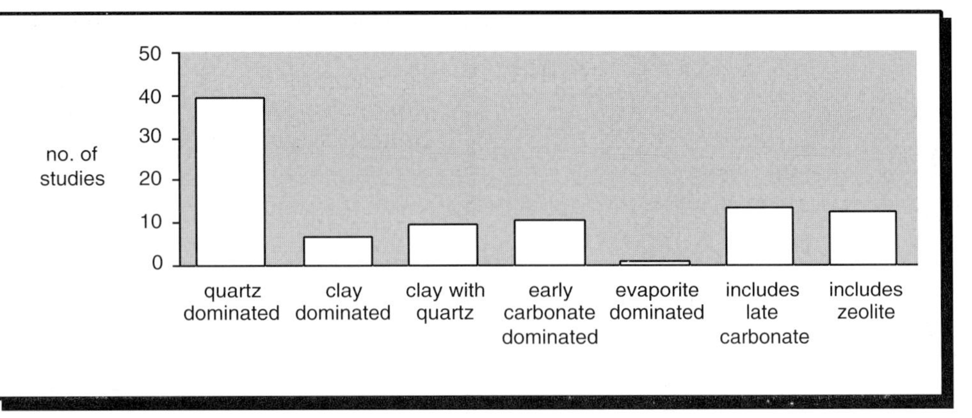

Figure 4. Occurrence of different styles of diagenesis in total data set.

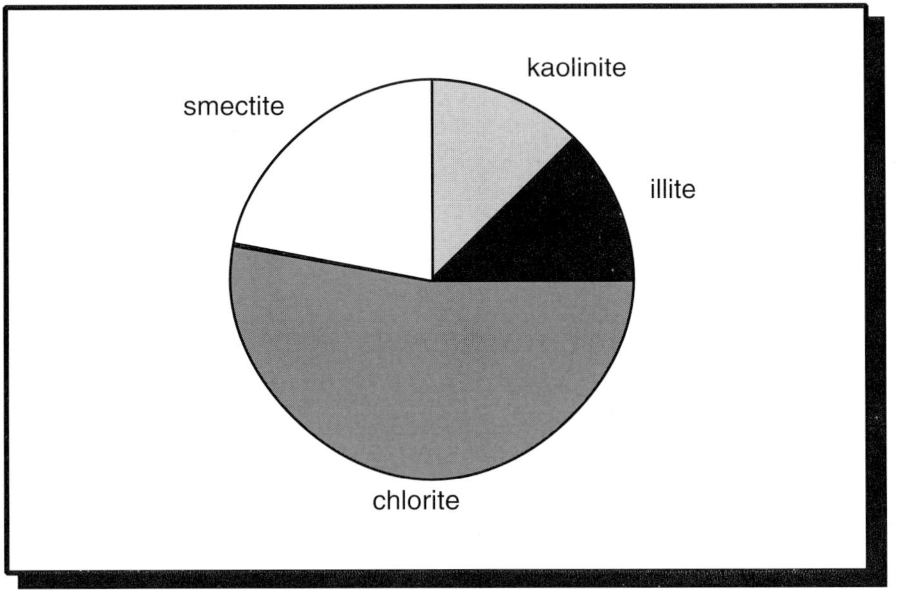

Figure 5. Occurrence of different clay minerals in clay-dominated or clay with quartz styles of diagenesis.

observed zeolites in this review are 15°–85°C for clinoptilolite, 85°–120°C for heulandite, and in >120°C for laumonite.

Effect of Cement Import on Diagenetic Style

There is a continuing debate about mass balance (i.e., the extent to which material is supplied or removed from a sediment during diagenesis) and the different possible sources of cement in particular (Hayes, 1979; Bjørlykke, 1984; Houseknecht, 1988; Gluyas and Coleman, 1992). Answers to the question of whether sandstones act as open or closed systems during burial depend on the size of the system envisaged. Obviously, on a basin scale, the system is largely closed to outside influences, but on the scale of the individual sandstone pore, the system is open. Between these extremes, at the scale of each stratigraphically distinct sandstone unit, diagenesis appears to be a largely isochemical process, hence the noted close relationship between sediment composition and diagenetic style. However, in cases where there is no close relationship between sediment composition and diagenetic style, some external control such as the import of cementing components from surrounding sediments must be invoked. In these cases, the gross depositional environment and sediments from facies associated with the sandstones under scrutiny become a more significant influence. For example, in some clay-dominated styles of diagenesis involving illite in compositionally mature eolian sands, import of potassium (among other components) from associated evaporites appears necessary (McBride et al., 1987; Gluyas and Leonard, 1995). Although reliable data on absolute mineral abundances in this review are relatively sparse, it has been reported elsewhere (Curtis, 1978; Boles, 1981; Gluyas, 1985) that cement can be imported to sandstones from a number of sources. These are summarized in Table 2.

Amount of Cement

So far this chapter has considered diagenesis in terms of the relative abundance of constituent

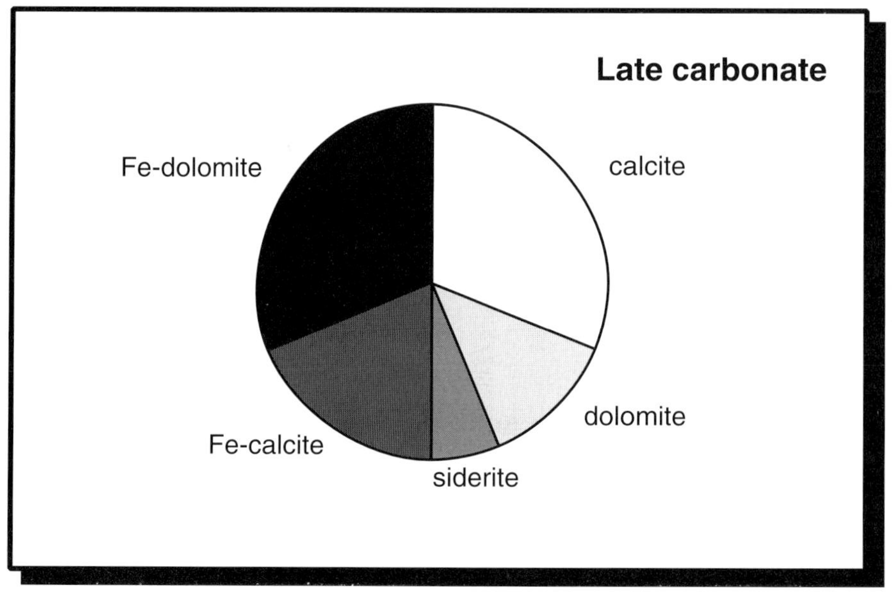

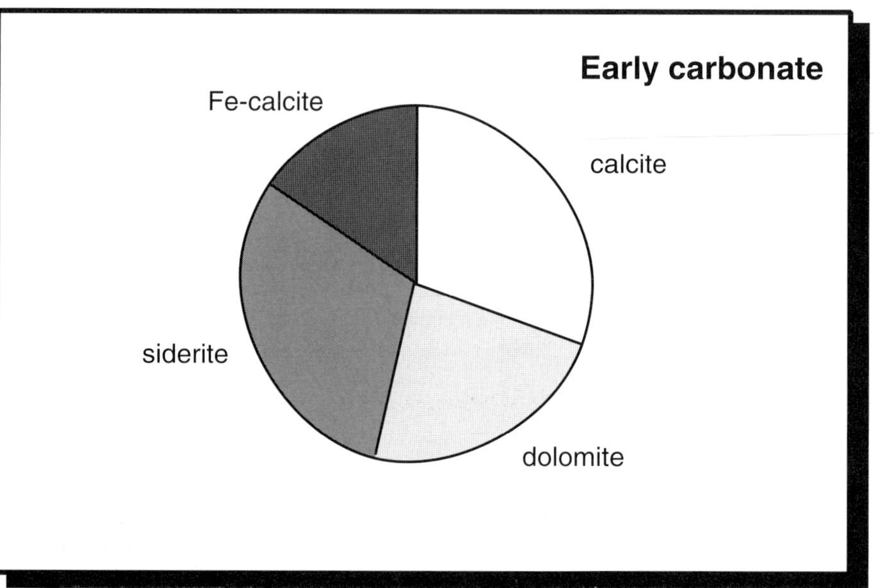

Figure 6. Occurrence of different carbonates in situations where early carbonate cement is dominant (top) or where significant late carbonate occurs (bottom).

cements. Although quantitative modal analysis (point count) data on mineral abundance exist in the data reviewed, they are of variable quality, vintage, and reliability. This makes consistent comparisons of one study with another difficult. However, based on available data, some tentative volumetric ranges can be assigned to each of the styles of diagenesis established in Table 3.

To tackle the problem of predicting porosity and permeability, the diagenetic history of a sandstone needs to be reconstructed, and cement volumes need to be estimated. One approach to this problem is to try to link the cement abundance range to another variable (in the case of quartz cement, an increase in depth/temperature of burial often corresponds to an increase in cement volume). Armed with the ranges in Table 3, a pragmatic approach is to use a "most likely" value within the range tabulated with the ranges themselves to generate a "most likely" estimate of range in porosity and permeability. This approach is discussed briefly below and is given in more detail by Gluyas and Witton (this volume).

CONCLUSIONS

The contributions of each of the principal factors controlling diagenesis (e.g., sediment composition, depositional environment, burial temperature, and mass import into sandstones) are shown in Figure 9. The present study is not an exhaustive treatment of clastic diagenesis, but aims to describe the main factors controlling five important styles of diagenesis.

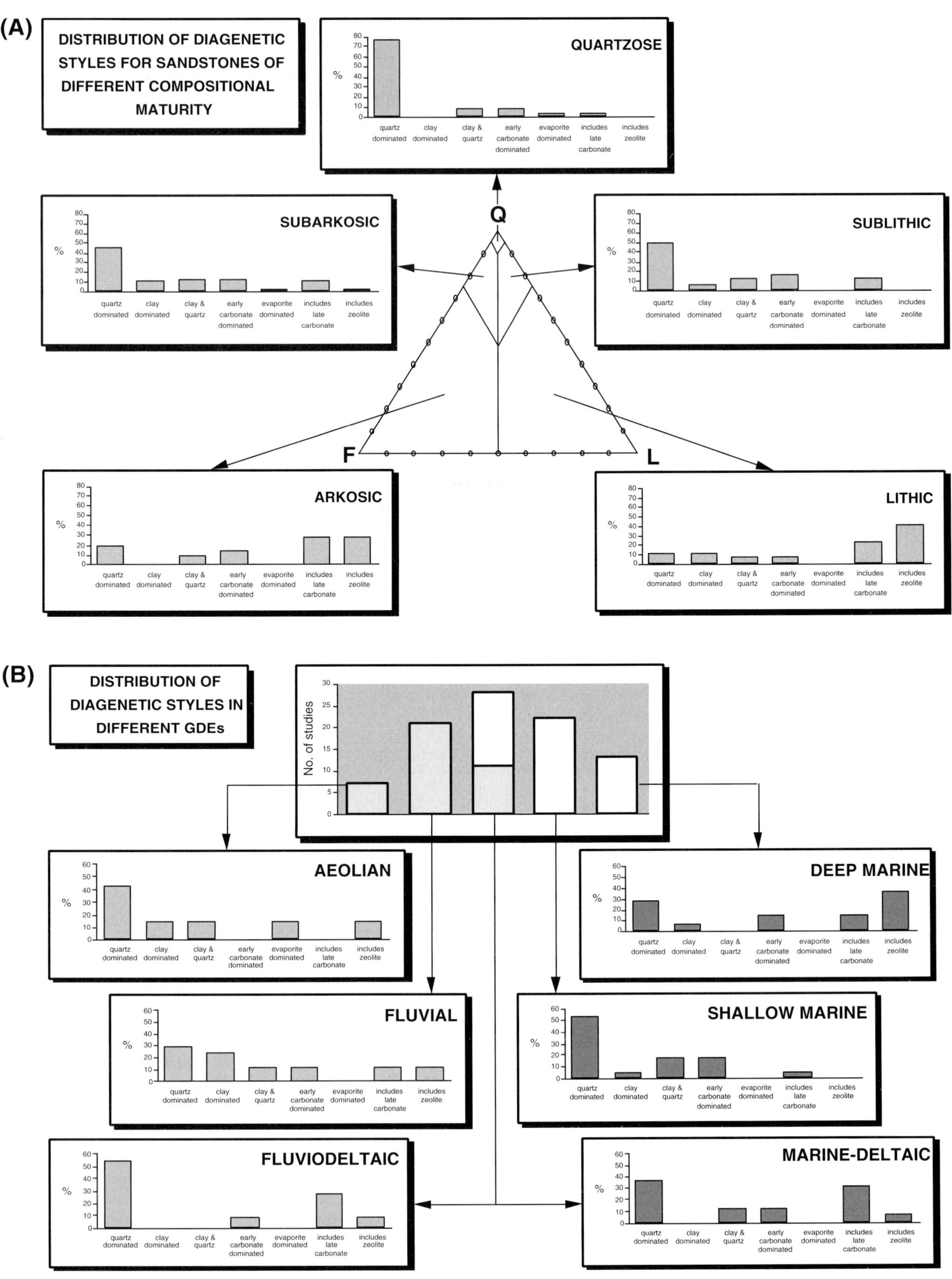

Figure 7. (A) The influence of different sand compositions on diagenetic style (sediment composition is expressed in terms of compositional maturity using the scheme of Dott, 1964). (B) The influence of gross depositional environment (GDE) on diagenetic style.

Table 2. External Sources of Cements in Sandstones.

Source	At Low Temperatures Supply	At Higher Temperatures Supply
Mudrocks	Fe^{2+} for chlorite, carbon for early carbonates	SiO_2 for quartz, carbon for late carbonates
Carbonates	Ca^{2+} and carbon for early carbonates	Ca^{2+} and carbon for late carbonates
Evaporites	Ca^{2+} for early carbonates	K^+ for illite, $CaSO_4$ in remobilized evaporites (e.g., anhydrite)

The basic framework of Figure 9 illustrates the different silicate cements that are likely to result from different starting materials at different temperatures in different depositional environments. Additional parameters are included to show the conditions at which carbonate cements are developed, together with some of the more frequently observed products from material influx into the sandstone.

This chapter has integrated the results of 100 studies of diagenesis in sandstones worldwide and established a series of regionally consistent patterns of diagenesis. Given a certain minimal amount of information regarding sediment composition, depositional environment, and burial depth and temperature, it seems possible to predict the likely diagenetic history of any sandstone. Although variations in detail from area to area or sandstone to sandstone will exist, and exceptions to the patterns shown in Figure 9 will arise, we expect the findings outlined in this review will generally hold true.

THE IMPACT

Prediction of Porosity and Permeability

Besides authigenic cements, the main factor that influences porosity and permeability in sedimentary rocks is compaction. Compaction curves determined from laboratory experiments enable porosity to be estimated as a function of burial depth, overpressure, and ductile grain/clay content (Kurkjy, 1988; Gluyas and Cade, this volume). These estimates can be further refined by taking into account the most likely diagenetic cement predicted at the given depth/temperature of burial for a particular style of diagenesis in the formation of interest.

Simulations from sphere-pack models (Bryant et al., 1993) have indicated that permeability can be calculated directly as a function of porosity, grain size, sorting, and the type of cement present (Cade et al., 1994; Evans et al., 1994). With porosity-depth trends established, analysis of the effect of different styles

Table 3. Approximate Ranges in Cement Volumes for Different Styles of Diagenesis.

Style of Diagenesis	Range in Volume of Principal Cement	Range in Volume of Ancillary Cements
Quartz dominated	5–15% (increases with temperature of burial)	3–5% clay, <5% late carbonate*
Clay dominated	10–20% (only illite dominated increases with temperature of burial)	<5% quartz, <5% late carbonate*
Early clay/late quartz	5–10% clay, <5% quartz	<5% late carbonate*
Early carbonate/ evaporite dominated	≤20–30% (increases in proximity to evaporites/saline lake deposits)	
Zeolite	5–20% (increases with increasing lithic content)	≤10% clay, ≤10% late carbonate*

*Can be locally ≤20–30%, completely occluding porosity.

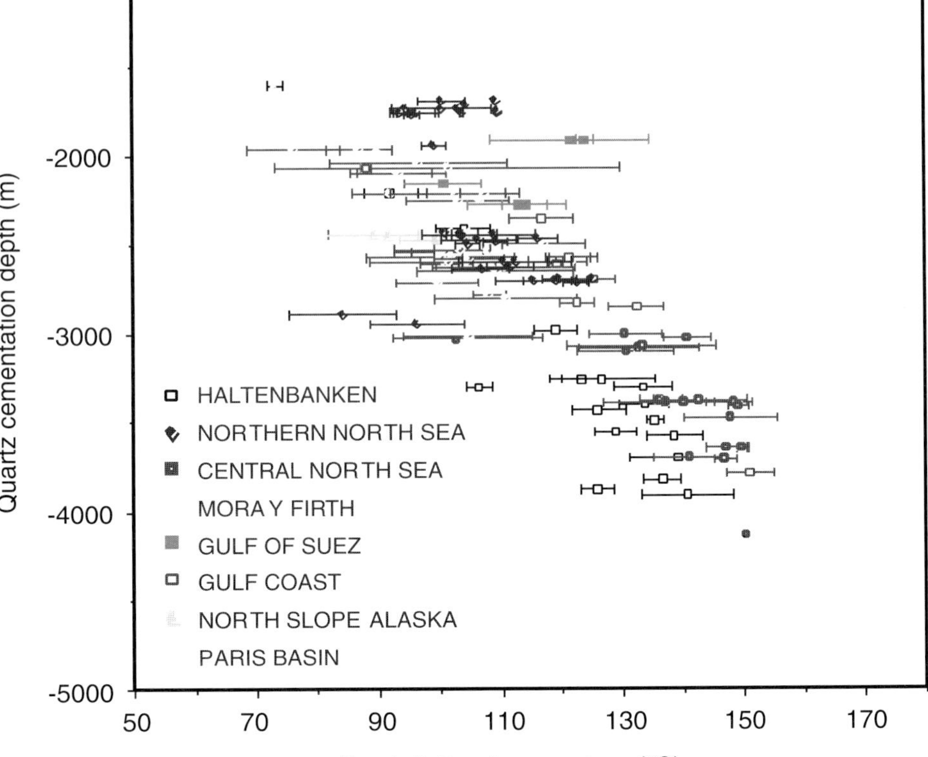

Figure 8. Mean homogenization temperatures (±1 σ) of fluid inclusions trapped within quartz cements from sandstones (from Gluyas et al., 1993). These data are taken to indicate that quartz cementation can occur at any temperature (but over a restricted temperature range) above a minimum threshold of 75°C.

of diagenesis on permeability can be made in terms of a series of characteristic porosity-permeability curves for a given sand grain size and sorting. Two hypothetical examples are shown in Figure 10, in which a clean, compositionally mature quartz-cemented sandstone buried to 3000 m is compared to a less mature, subarkosic quartz-kaolinite-illite–cemented sand of similar grain size buried to the same depth/temperatures of burial. Permeability in the subarkosic quartz-kaolinite-illite–cemented sand is consistently lower than the compositionally more mature quartz-cemented sand at any given porosity.

Graphical representations such as in Figure 10 can be used as "maps" during exploration to predict modifications in porosity and permeability relationships as a result of a specific style of the cementation. Predictions of overall reservoir effectiveness at any given depth of burial can be made, allowing a porosity threshold (and, hence, a depth threshold) to be estimated from the poroperm relationship established. Specific examples and additional details are given elsewhere in this volume (Gluyas, this volume; Gluyas and Witton, this volume).

ACKNOWLEDGMENTS

We thank BP Exploration for permission to publish this paper. We also thank other former members of the Reservoir Quality Prediction Team: Kourosh Amiri, Andrew Brayshaw, Steve Bryant, Dominic Emery, Shona Grant, Andrew Hogg, Clive Maile, Fiona Neall, Joyce Neilson, Hugh Nicholson, and Andrew Robinson. Thanks also to Mike Bowman, Malcolm McClure, Mark Osborne, Dick Swarbrick, and Sal Bloch for their constructive reviews.

REFERENCES CITED

Almon, W.R., and S.H. Davies, 1979, Regional diagenetic trends in the Lower Cretaceous muddy sandstone, Powder River Basin, *in* P.A. Scholle and P.R. Schluger, eds., Aspects of diagenesis: SEPM Special Publication 26, p. 379–400.

Ayalon, A., and F.J. Longstaffe, 1988, Oxygen isotope studies of diagenesis and pore water evolution in the Western Canada sedimentary basin: evidence from the Upper Cretaceous Belly River Sandstone, Alberta: Journal of Sedimentary Petrology, v. 58, p. 489–504.

Barnes, D.A., C.E. Lundgren, and M.W. Longman, 1992, Sedimentology and diagenesis of the St. Peter Sandstone, Central Michigan Basin, United States: AAPG Bulletin, v. 76, p. 1507–1532.

Bjørlykke, K., 1984, Secondary porosity: how important is it?, *in* D.A. McDonald and R.C. Surdam, eds., Clastic diagenesis: AAPG Memoir 37, p. 277–286.

Bloch, S., J.H. McGowen, J.R. Duncan, and D.W. Brizzolara, 1990, Porosity prediction, prior to drilling, in sandstones of the Kekituk Formation (Mississippian),

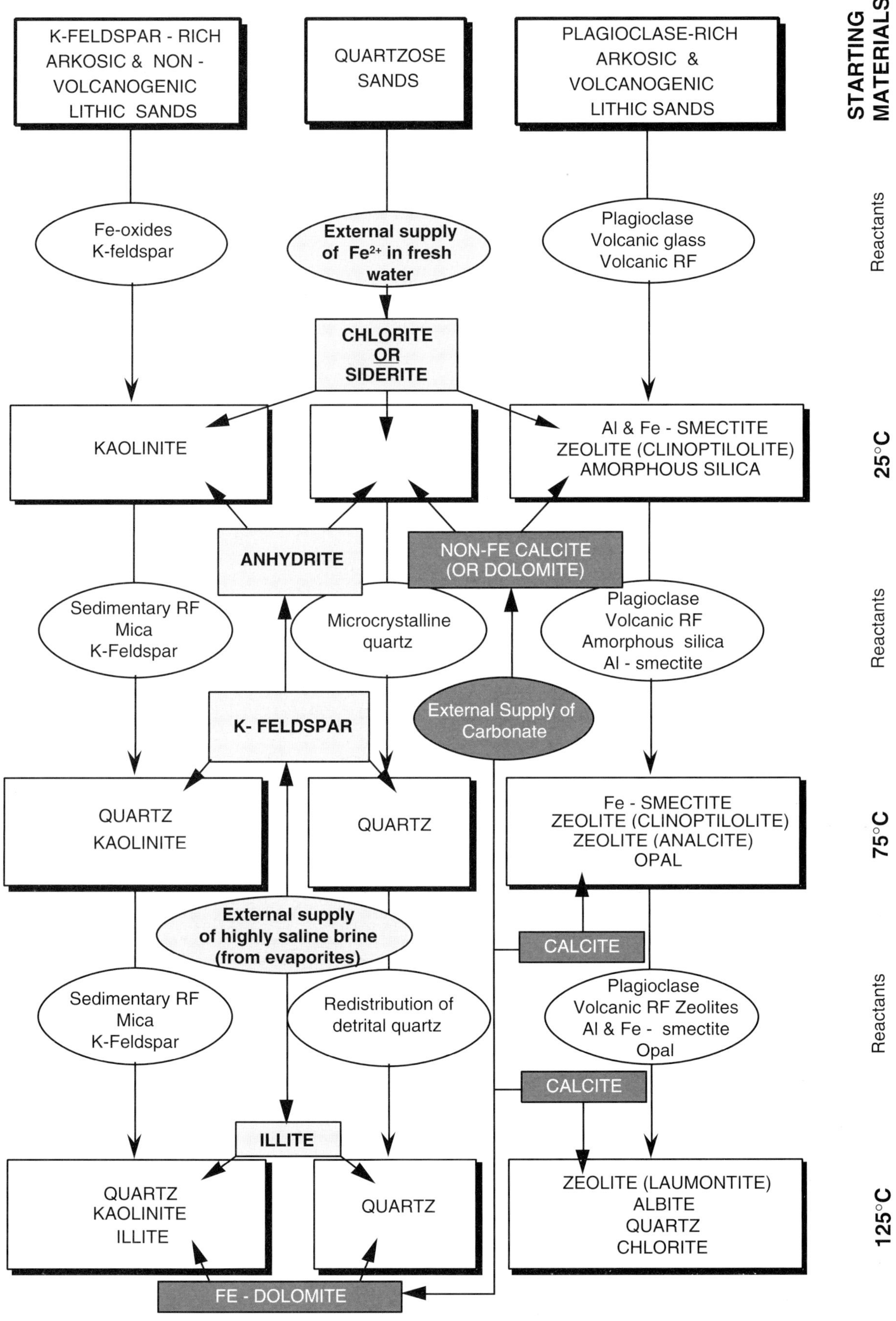

Figure 9. Flow chart illustrating the combined control of sediment composition, depositional environment, and burial temperature on diagenetic cements in sandstones.

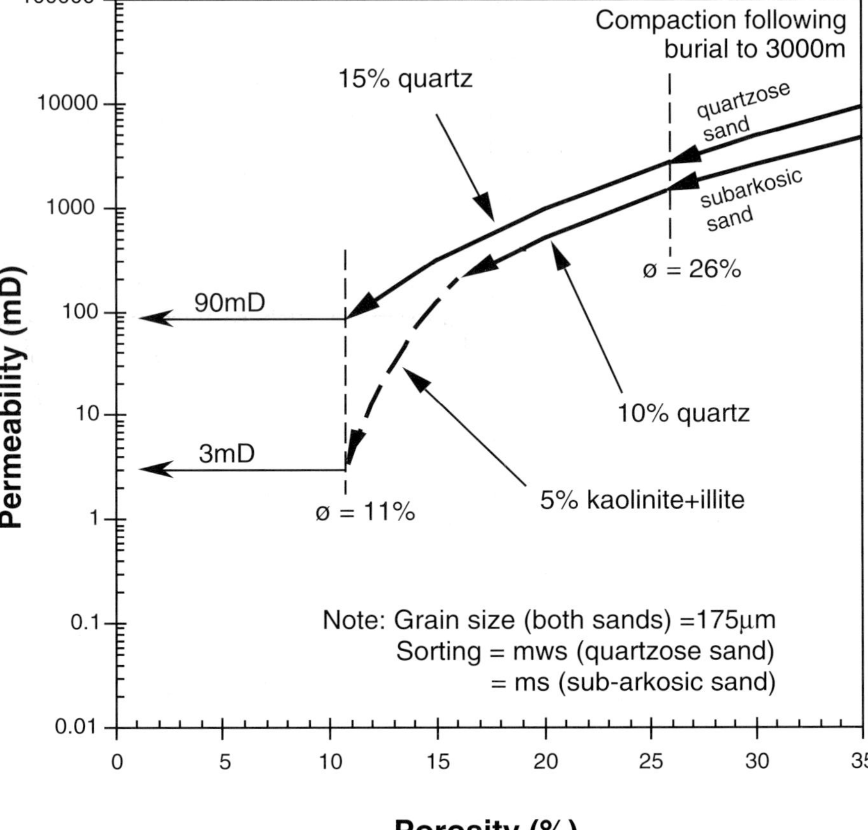

Figure 10. Example porosity (φ)-permeability trends constructed using the principles outlined by Cade et al. (1994) for two sands of similar grain size but different detrital composition that have experienced similar burial histories. Slightly different sorting character-reflecting differences in compositional maturity are also assumed. Note that even with the same degree of compaction- and cementation-related porosity loss, the predicted permeability of the quartzose sands is 30 times larger than the subarkosic sand.

North Slope of Alaska: AAPG Bulletin, v. 74, p. 1371–1385.

Boles, J.R., 1981, Clay diagenesis and the effects on sandstone cementation, *in* F.J. Longstaffe, ed.: Mineralogical Association of Canada Short Course in Clays and the Resource Geologist, p. 148–168.

Boles, J.R., 1984, Secondary porosity reactions in the Stevens Sandstone, San Joaquin Valley, California, *in* D.A. McDonald and R.C. Surdam, eds., Clastic diagenesis: AAPG Memoir 37, p. 217–224.

Boles, J.R., and D.S. Coombs, 1977, Zeolite facies alteration of sandstones in the Southland Syncline, New Zealand: American Journal of Science, v. 277, p. 982–1012.

Bolm, J.G, T.H. McCulloh, and R.J. Stewart, 1983, Diagenesis of sandstones in the Lower Cook Inlet, Alaska, and its implications for petroleum plays: Journal of the Alaska Geological Society, v. 3, p. 25–31.

Brown, D.M., K.D. McAlpine, and R.W. Yole, 1990, Sedimentology and sandstone diagenesis of the Hibernia Formation in Hibernia oil field, Grand Banks of Newfoundland: AAPG Bulletin, v. 73, p. 557–575.

Bryant, S.L., C.A. Cade, and D.W. Mellor, 1993, Permeability prediction from geologic models: AAPG Bulletin, v. 77, p. 1338–1350.

Burns, I.K., and F.G. Etheridge, 1979, Petrology and diagenetic effects of lithic sandstones: Paleocene and Eocene Umpqua Formation, Southwest Oregon, *in* P.A. Scholle and P.R. Schluger, eds., Aspects of diagenesis: SEPM Special Publication 26, p. 307–317.

Cade, C.A., I.J. Evans, and S.L. Bryant, 1994, Analysis of permeability controls—a new approach: Clay Minerals, v. 29, p. 491–501.

Cavazza, W., and J. Dahl, 1990, Notes on the temporal relationships between sandstone compaction and precipitation of authigenic minerals: Sedimentary Geology, v. 69, p. 37–44.

Cazier, E.C., A.B. Hayward, G. Espinosa, J. Velandia, J.H. Mugnoit, and W.G. Leel, Jr., 1995, Petroleum geology of the Cusiana Field, Llanos Basin Foothills, Colombia: AAPG Bulletin, v. 79, p. 1444–1463.

Chan, M., 1985, Correlations of diagenesis with sedimentary facies in Eocene sandstones, western Oregon: Journal of Sedimentary Petrology, v. 55, p. 322–333.

Cocheme, J.-J., A. Demant, L. Aguirre, and D. Hermitte, 1988, Heulandite in the sedimentary filling of the "basin and range" (Baucarit Formation) of the northern Sierra Madre Occidental, Mexico (abridged English version): Compte Rendu Academie de Sciences Paris, v. 307, Serie II, p. 643–649.

Cowan, G., 1989, Diagenesis of Upper Carboniferous sandstones: southern North Sea Basin, *in* M.K.G. Whateley and K.T. Pickering, eds., Deltas: sites and traps for fossil fuels: Geological Society of London Special Publication 41, p. 57–73.

Curtis, C.D., 1978, Possible links between sandstone diagenesis and depth-related geochemical reactions in enclosing mudstones: Journal of the Geological Society of London, v. 135, p. 107–118.

Damanti, J.F., and T.E. Jordan, 1989, Cementation and compaction history of synorogenic foreland basin sedimentary rocks from Huaco, Argentina: AAPG Bulletin, v. 73, p. 858–873.

Davies, D.K., W.R. Almon, S.B. Bonis, and B.E. Hunter, 1979, Deposition and diagenesis of Tertiary–Holocene volcaniclastics, Guatemala, *in* P.A. Scholle and P.R. Schluger, eds., Aspects of diagenesis: SEPM Special Publication 26, p. 281–306.

Dixon, S.A., D.M. Summers, and R.C. Surdam, 1989, Diagenesis and preservation of porosity in Norphlet Formation (Upper Jurassic), southern Alabama: AAPG Bulletin, v. 73, p. 707–728.

Dott, R.H., Jr., 1964, Wacke, graywacke and matrix—what approach to immature sandstone classification?: Journal of Sedimentary Petrology, v. 34, p. 625–632.

Dutton, S.P., and T.N. Diggs, 1990, History of quartz cementation in the Lower Cretaceous Travis Peak Formation, East Texas: Journal of Sedimentary Petrology, v. 60, p. 191–202.

Dutton, S.P., and L.S. Land, 1985, Meteoric burial diagenesis of Pennsylvanian arkosic sandstones, southwestern Anadarko Basin, Texas: AAPG Bulletin, v. 69, p. 22–38.

Edwards, M.B., 1979, Sandstone in Lower Cretaceous Helvetiafjellet Formation, Svalbard: bearing on reservoir potential of Barents Shelf: AAPG Bulletin, v. 63, p. 2193–2203.

Eggert, J.T., 1987, Sandstone petrology, diagenesis and reservoir quality, Lower Cretaceous Kuparuk River Formation, Kuparuk River field, North Slope Alaska (abs.), *in* I. Taileur and P. Weimer, eds., Alaska North Slope geology, volume 1: Pacific Section of the SEPM and the Alaska Geological Society, p. 108.

Ehrenberg, S.N., 1990, Relationship between diagenesis and reservoir quality in sandstones of the Garn Formation, Haltenbanken, mid-Norwegian continental shelf: AAPG Bulletin, v. 74, p. 1538–1558.

Emery, D., P.C. Smalley, and N.H. Oxtoby, 1993, Synchronous oil migration and cementation in sandstone reservoirs demonstrated by quantitative description of diagenesis: Philosophical Transactions of the Royal Society of London, v. A344, p. 115–125.

Evans, A.L., 1990, Miocene sandstone provenance relations in the Gulf of Suez: insights into synrift unroofing and uplift history: AAPG Bulletin, v. 74, p. 1386–1400.

Evans, I.J., 1989, Geochemical fluxes during shale diagenesis: an example from the Ordovician of the Moroccan High Atlas: Ph.D. thesis, Reading University, 262 p.

Evans, I.J., 1990, Quartz dissolution during shale diagenesis: implications for quartz cementation in sandstones: Chemical Geology, v. 84, p. 239–240.

Evans, I.J., S.L. Bryant, and C.A. Cade, 1993, Modelling the effect of diagenetic cements on sandstone permeability, *in* J. Parnell, A.H. Ruffell, and N.R. Moles, eds.: Geofluids '93, p. 212–214.

Evans, J., C. Cade, and S. Bryant, this volume, A geological approach to permeability prediction in clastic reservoirs, *in* J.A. Kupecz, J. Gluyas, and S. Bloch, eds., Reservoir quality prediction in sandstones and carbonates: AAPG Memoir 69, p. 91–101.

Galloway, W.E., 1979, Diagenetic control of reservoir quality in arc-derived sandstones: implications for petroleum exploration, *in* P.A. Scholle and P.R. Schluger, eds., Aspects of diagenesis: SEPM Special Publication 26, p. 251–262.

Gibbons, K., T. Hellen, A. Kjemperud, S.D. Nio, and K. Vebbenstad, 1993, Sequence architecture, facies development and carbonate-cemented horizons in the Troll Field reservoir, offshore Norway, *in* M. Ashton, ed., Advances in reservoir geology: Geological Society of London Special Publication 69, p. 1–31.

Girard, J.P., and M. Deynoux, 1991, Oxygen isotope study of diagenetic quartz overgrowths from Upper Proterozoic quartzites of western Mali, Taoudeni Basin: implications for quartz cementation: Journal of Sedimentary Petrology, v. 61, p. 406–418.

Girard, J.P., S.M. Savin, and J.L. Aronson, 1989, Diagenesis of the Lower Cretaceous arkoses of the Angola Margin: petrologic, K/Ar dating and $^{18}O/^{16}O$ evidence: Journal of Sedimentary Petrology, v. 59, p. 519–538.

Giroir, G., E. Merino, and D. Nahon, 1989, Diagenesis of Cretaceous sandstone reservoirs of the South Gabon Rift Basin, West Africa: mineralogy, mass transfer and thermal diffusion: Journal of Sedimentary Petrology, v. 47, p. 482–493.

Glasmann, J.R., P.D. Lundegard, R.A. Clark, B.K. Penny, and I.D. Collins, 1989, Geochemical evidence for the history of diagenesis and fluid migration: Brent Sandstone, Heather field, North Sea: Clay Minerals, v. 24, p. 255–284.

Gluyas, J.G., 1985, Reduction and prediction of sandstone reservoir potential, Jurassic North Sea: Philosophical Transactions of the Royal Society of London, v. A315, p. 187–202.

Gluyas, J.G., this volume, Poroperm prediction for reserves growth exploration: Ula Trend Norwegian North Sea, *in* J.A. Kupecz, J. Gluyas, and S. Bloch, eds., Reservoir quality prediction in sandstones and carbonates: AAPG Memoir 69, p. 201–212.

Gluyas, J., and C.A. Cade, this volume, Prediction of porosity in compacted sands, *in* J.A. Kupecz, J. Gluyas, and S. Bloch, eds., Reservoir quality prediction in sandstones and carbonates: AAPG Memoir 69, p. 19–28.

Gluyas, J.G., and M.L. Coleman, 1992, Material flux and porosity changes during sediment diagenesis: Nature, v. 356, p. 52–53.

Gluyas, J.G., and A.J. Leonard, 1995, Diagenesis: of the Rotliegend Sandstone: the answer ain't blowin' in the wind: Marine and Petroleum Geology, v. 12, p. 491–497.

Gluyas, J.G., and N.H. Oxtoby, 1995, Diagenesis a short (2 million year) story—Miocene sandstones of Central Sumatra, Indonesia: Journal of Sedimentary Research, v. A65, p. 513–521.

Gluyas, J.G., A.G. Robinson, D. Emery, S.M. Grant, and N.H. Oxtoby, 1993, The link between petroleum emplacement and sandstone cementation, *in* J.R. Parker, ed., Petroleum Geology of Northwest Europe: Proceedings of the 4th Conference, p. 1395–1402.

Gluyas, J.G., and T. Witton, this volume, Poroperm prediction for wildcat exploration prospects: Miocene Epoch, Southern Red Sea, *in* J.A. Kupecz, J. Gluyas, and S. Bloch, eds., Reservoir quality prediction in sandstones and carbonates: AAPG Memoir 69, p. 165–178.

Hartley, A., S. Flint, and P. Turner, 1991, Analcime: a characteristic authigenic phase of Andean alluvium, northern Chile: Geological Journal, v. 26, p. 189–202.

Haszeldine, R.S., I.N. Samson, and C. Cornford, 1984, Quartz diagenesis and convective fluid movement: Beatrice oilfield, UK North Sea: Clay Minerals, v. 19, p. 391–402.

Hawlader, H.M., 1990, Diagenesis and reservoir potential of volcanogenic sandstones—Cretaceous of the Surat Basin, Australia: Sedimentary Geology, v. 66, p. 181–195.

Hayes, J.B., 1979, Sandstone diagenesis—the hole truth, *in* P.A. Scholle and P.R. Schluger, eds., Aspects of diagenesis: SEPM Special Publication 26, p. 127–139.

Helmold, K.P., and P.C. Van de Kamp, 1984, Diagenetic mineralogy and controls on albitization and laumonite formation in Paleocene arkoses, Santa Ynez Mountains, California, *in* D.A. McDonald and R.C. Surdam, eds., Clastic diagenesis: AAPG Memoir 37, p. 239–276.

Horn, V.D., 1965, Diagenese und porositat des Dogger-beta-Hauptsandsteines in den Olfeldern Plon-Ost und Preetz: Erdol Und Kohle-Erdgas-Petrochemie, v. 17, p. 249–258.

Houseknecht, D., 1988, Intergranular pressure-solution in four quartz sandstones: Journal of Sedimentary Petrology, v. 58, p. 228–246.

Hutcheon, I., H.J. Abercrombie, P. Putnam, R. Gardner, and H.R. Krouse, 1989, Diagenesis and sedimentology of the Clearwater Lake Formation at Tucker Lake: Bulletin of Canadian Petroleum Geology, v. 37, p. 83–97.

Iijima, A., 1988, Diagenetic transformations of minerals as exemplified by zeolites and silica minerals—a Japanese view. Part I: zeolitic diagenesis, *in* G.V. Chillingarian and K.H. Wolf, eds., Diagenesis II: Developments in Sedimentology, v. 43, p. 147–189.

Imam, M.B., and H.J. Shaw, 1985, The diagenesis of neogene clastic sediments from the Bengal Basin, Bangladesh: Journal of Sedimentary Petrology, v. 55, p. 665–671.

Jansa, L.F., and V.H.N. Urrea, 1990, Geology and diagenetic history of overpressured sandstone reservoirs, Venture gas field, offshore Nova Scotia, Canada: AAPG Bulletin, v. 74, p. 1640–1658.

Kurkjy, K.A., 1988, Experimental compaction studies of lithic sands: Master's thesis, University of Miami, Miami, Florida.

Lambert-Aikhionbare, D.O., and H.F. Shaw, 1982, Significance of clays in the petroleum geology of the Niger Delta: Clay Minerals, v. 17, p. 91–103.

Land, L.S., and S.P. Dutton, 1978, Cementation of a Pennsylvanian deltaic sandstone: isotope data: Journal of Sedimentary Petrology, v. 48, p. 1167–1176.

Land, L.S., and R.S. Fisher, 1987, Wilcox Sandstone diagenesis, Texas Gulf Coast: a regional isotopic comparison with the Frio Formation, *in* J. Marshall, ed., Diagenesis of sedimentary sequences: Geological Society of London Special Publication 36, p. 219–235.

Larsen, O.H., and H. Friis, 1991, Petrography, diagenesis and pore-water evolution of a shallow marine sandstone Hasle Formation, Lower Jurassic, Bornholm, Denmark: Sedimentary Geology, v. 72, p. 269–284.

Lee, Y.I., 1988, Chemistry and origin of zeolites in sandstones at DSDP sites 445 and 446, Daito Ridge and Basin Province, Northwest Pacific: Chemical Geology, v. 67, p. 261–273.

Macchi, L., C.D. Curtis, A. Levison, K. Woodward, and C.R. Hughes, 1990, Chemistry, morphology and distribution of illites from Morcambe gas field, Irish Sea, offshore United Kingdom: AAPG Bulletin, v. 74, p. 296–308.

MacDonald, H., P.M. Allan, and J.P.B. Lovell, 1987, Geology of oil accumulation in Block 26/28, Porcupine Basin, offshore Ireland, *in* J. Brooks and K. Glennie, eds.: Petroleum Geology of NW Europe, p. 643–651.

Manckiewicz, D., and J.R. Steidtmann, 1979, Depositional environments and diagenesis of the Tensleep Sandstone, *in* P.A. Scholle and P.R. Schluger, eds., Aspects of diagenesis: SEPM Special Publication 26, p. 319–336.

Markert, J.C., and Z. Al-Shaieb, 1984, Diagenesis and evolution of secondary porosity in Upper Minnelusa sandstones, Powder River Basin, Wyoming, *in* D.A. McDonald and R.C. Surdam, eds., Clastic diagenesis: AAPG Memoir 37, p. 367–389.

Mathisen, M.E., 1984, Diagenesis of Plio-Pleistocene nonmarine sandstones, Cagayan Basin, Philippines: early development of secondary porosity in volcanic sandstones, *in* D.A. McDonald and R.C. Surdam, eds., Clastic diagenesis: AAPG Memoir 37, p. 177–193.

McBride, E.F., L.S. Land, and L.E. Mack, 1987, Diagenesis of eolian and fluvial feldspathic sandstones, Norphlet Formation, Upper Jurassic, Rankin County, Mississippi, and Mobile County, Alabama: AAPG Bulletin, v. 71, p. 1019–1034.

Melvin, J., and A.S. Knight, 1984, Lithofacies, diagenesis and porosity of the Ivishak Formation, Prudhoe Bay Area, Alaska, *in* D.A. McDonald and R.C. Surdam, eds., Clastic diagenesis: AAPG Memoir 37, p. 347–366.

Meshri, I.D., and J.M. Walker, 1990, A study of rockwater interaction and simulation of diagenesis in the Upper Almond Sandstones of the Red Desert and Washakie Basins, Wyoming, *in* I.D. Meshri and P.J. Ortoleva, eds., Prediction of reservoir quality through chemical modeling: AAPG Memoir 49, p. 55–83.

Milliken, K.L., 1985, Petrology and burial diagenesis of Plio-Pleistocene sediments, northern Gulf of Mexico: Ph.D. thesis, University of Texas at Austin, Austin, Texas, 112 p.

Milliken, K.L., L.S. Land, and R.G. Loucks, 1981, History of burial diagenesis determined from isotopic geochemistry, Frio Formation, Brazoria County,

Texas: AAPG Bulletin, v. 65, p. 1397–1413.

Morad, S., I.S. Al-Aasm, K. Ramseyer, R. Marfil, and A.A. Aldahan, 1990, Diagenesis of carbonate cements in Permo-Triassic sandstones from the Iberian Range, Spain: evidence from chemical composition and stable isotopes: Sedimentary Geology, v. 67, p. 281–295.

Moraes, M.A.S., 1989, Diagenetic evolution of Cretaceous–Tertiary turbiditic reservoirs: AAPG Bulletin, v. 73, p. 598–612.

Moraes, M.A.S., 1991, Diagenesis and microscopic heterogeneity of lacustrine deltaic and turbiditic sandstone reservoirs, Lower Cretaceous, Potiguar Basin, Brazil: AAPG Bulletin, v. 75, p. 1758–1771.

Morris, K.A., and C.A. Shepperd, 1982, The role of clay minerals in influencing porosity and permeability characteristics in the Bridport Sands of Wytch Farm, Dorset: Clay Minerals, v. 17, p. 41–54.

Mozley P.S., and K. Hoernle, 1990, Geochemistry of carbonate cements in the Sag River and Shublik formations, Triassic/Jurassic, North Slope, Alaska: implications for the geochemical evolution of formation waters: Sedimentology, v. 37, p. 817–836.

O'Shea, K.J., and S.K. Frape, 1988, Authigenic illite in the Lower Silurian Cataract Group sandstones of southern Ontario: Bulletin of Canadian Petroleum Geology, v. 36, p. 158–167.

Oxtoby, N.H., A.W. Mitchell, and J.G. Gluyas, 1995, The filling and emptying of the Ula oil field, Norwegian North Sea, *in* J.M. Cubitt and W.A. England, eds., The geochemistry of reservoirs: Geological Society of London Special Publication 86, p. 141–158.

Pallatt, N., M.J. Wilson, and W.J. McHardy, 1984, The relationship between permeability and the morphology of diagenetic illite in reservoir rocks: Journal of Petroleum Technology, v. 36, p. 2225–2227.

Pittman, E.D., and G.E. King, 1986, Petrology and formation damage control, Upper Cretaceous sandstone, offshore Gabon: Clay Minerals, v. 21, p. 781–790.

Prosser, D.J., J.A. Dawes, A.E. Fallick, and B.P.J. Williams, 1993, Geochemistry and diagenesis of stratabound calcite cement layers within the Rannoch Formation of the Brent Group, Murchison Field, North Viking Graben (Northern North Sea): Sedimentary Geology, v. 87, p. 139–164.

Rangathan,V., and R.S. Tye, 1986, Petrography, diagenesis and facies control on porosity in Shannon Sandstone, Hartzog Draw Field, Wyoming: AAPG Bulletin, v. 70, p. 56–69.

Reinson, G.E., and A.E. Foscolos, 1986, Trends in sandstone diagenesis with depth of burial, Viking Formation, southern Alberta: Bulletin of Canadian Petroleum Geology, v. 34, p. 126–152.

Riches, P., I. Traub-Sobott, W. Zimmerie, and U. Zinkernagel, 1986, Diagenetic peculiarities of potential Lower Jurassic reservoir sandstones, Troms 1 area, off northern Norway, and their tectonic significance: Clay Minerals, v. 21, p. 565–584.

Rinckenbach, T., 1988, Diagenese minerale des sediments petroliferes du delta fossile de la Mahakam: Ph.D. thesis, L'Universite Louis Pasteur, Strasbourg.

Robinson, A.G., M.L. Coleman, and J.G. Gluyas, 1993, The age of illite cement growth, Village Fields area, southern North Sea: evidence from K-Ar ages and $^{18}O/^{16}O$ ratios: AAPG Bulletin, v. 77, p. 68–80.

Shenhav, H., 1971, Lower Cretaceous sandstone reservoirs, Israel: petrography, porosity, permeability: AAPG Bulletin, v. 55, p. 2194–2224.

Smith, R.I., N. Hodgson, and M. Fulton, 1993, Salt control on Triassic reservoir distribution, UKCS Central North Sea, *in* J.R. Parker, ed., Petroleum geology of Northwest Europe: Proceedings of the 4th Conference, p. 547–558.

Smosna, R., 1988, Low-temperature, low-pressure diagenesis of Cretaceous sandstones, Alaskan North Slope: Journal of Sedimentary Petrology, v. 58, p. 644–655.

Strong, G.E., and A.E. Milodowski, 1987, Aspects of the diagenesis of the Sherwood Sandstones of the Wessex Basin and their influence on reservoir characteristics, *in* J. Marshall, ed., Diagenesis of sedimentary sequences: Geological Society of London Special Publication 36, p. 325–337.

Surdam, R.C, L.J. Crossey, E.S. Hagen, and H.P. Heasler, 1989, Organic-inorganic interactions and sandstone diagenesis: AAPG Bulletin, v. 73, p. 1–23.

Thomson, A., 1979, Preservation of porosity in the deep Woodbine-Tuscaloosa trend, Louisiana: Gulf Coast Association of Geological Society Transactions, v. 30, p. 396–403.

Tillman, R.W., and W.R. Almon, 1979, Diagenesis of the Frontier Formation offshore bar sandstones, Spearhead Ranch field, Wyoming, *in* P.A. Scholle and P.R. Schluger, eds., Aspects of diagenesis: SEPM Special Publication 26, p. 337–378.

Trevena, A.S., and R.A. Clark, 1986, Diagenesis of sandstone reservoirs of Pattani Basin, Gulf of Thailand: AAPG Bulletin, v. 70, p. 299–308.

Warren, E.A., 1987, The application of a solution-mineral equilibrium model to the diagenesis of Carboniferous sandstones, Bothamsall oil field, East Midlands, England, *in* J. Marshall, ed., Diagenesis of sedimentary sequences: Geological Society of London Special Publication 36, p. 55–69.

Whynot, J.D., 1986, Mineralogy and early diagenesis of deep Gulf of Mexico Basin sediments: Ph.D. thesis, Texas A&M University, 112 p.

Wopfner, H., S. Markwort, and P.M. Semkiwa, 1990, Early diagenetic laumontite in the Lower Triassic Manda beds of the Ruhuhu Basin, southern Tanzania: Journal of Sedimentary Petrology, v. 61, p. 65–72.

Worden, R.H., 1996, Carbonate cements in the Triassic sandstones of the Paris Basin, France: origin and effects.

Chapter 6

Burial History and Porosity Evolution of Brazilian Upper Jurassic to Tertiary Sandstone Reservoirs

Cristiano Leite Sombra
PETROBRÁS, Centro de Pesquisa e Desenvolvimento
Rio de Janeiro, RJ, Brazil

Chang, Hung Kiang
Universidade Estadual Paulista, Instituto de Geociências e Ciências Exatas
São Paulo, Saõ Paulo, Brazil

ABSTRACT

The parameter time-depth index (TDI) is applied in this study to quantify empirically the influence of burial history on sandstone porosity evolution. The TDI, expressed in kilometers per million years of age, is defined as the area in the burial history diagram enclosed by the burial curve of the reservoir and the axes of the diagram. In practice, reservoir depths during burial history are integrated at regular time intervals of 1 m.y. The calculations exclude present-day bathymetry or paleobathymetry.

Sandstone reservoirs from several sedimentary basins along the Brazilian continental margin (Santos, Campos, Espírito Santo, Cumuruxatiba, Recôncavo, Sergipe, Alagoas, and Potiguar) were analyzed to investigate the evolution of porosity against TDI. These Upper Jurassic to Tertiary sandstones lie in depths of 700 to 4900 m, and are hydrocarbon charged (oil or gas). Average porosities of most of these reservoirs were obtained from core analysis, and a few porosity data were taken from well log interpretations. Detrital constituents of the sandstones are mainly quartz, feldspar, and granitic/gneissic rock fragments. Sandstones were grouped into three main reservoir types, based on composition (detrital quartz content) and grain sorting: Type I (average quartz content <50%) are very coarse grained to conglomeratic, poorly to very poorly sorted lithic arkoses. Rock fragments are mainly granitic/gneissic and coarse grained. Type II (average quartz content ranging from 50% to 70%) are fine- to coarse-grained (pebbles absent or occurring in small percentages), moderately sorted arkoses. Type III (average quartz content >80%) are fine to coarse, moderately to poorly sorted quartz arenites or subarkoses.

Plots of average porosity against depth show great dispersion in porosity values; such dispersion is mostly due to differences in the reservoir burial histories. However, plotting porosity values against the TDI for individual reservoir types produces well-defined trends. The decrease in porosity is less marked in Type III reservoirs, intermediate in Type II, and faster in Type I. Such plots suggest that it is possible to make relatively accurate porosity predictions based on reservoir TDI, texture, and composition, within the constraints of reservoir depth/age and basin tectonics analyzed in this study.

INTRODUCTION

The initial (depositional) porosity of sandstones depends mainly on their grain sorting (Beard and Weil, 1973). In the first stages of burial, porosity is mainly reduced by mechanical compaction. At intermediate to advanced burial stages, porosity changes are mainly governed by chemical reactions (pressure solution, cementation, and dissolution). The bulk effect of these mechanical and chemical events results in general trends of decreasing porosity with increasing depth. Perturbations in such general trends may be introduced by many different parameters, such as framework composition, early and late cementation, clay coatings, dissolution, pore fluid composition, pressure (Nagtegaal, 1980), geothermal gradient (Galloway, 1974), time-temperature exposure (Schmoker and Gautier, 1988), and duration of burial (Scherer, 1987; Bruhn et al., 1988).

The importance of time during the evolution of reservoir quality points to a kinetic control on the evolution of porosity. This has been observed both in laboratory experiments and in subsurface data sets. De Boer (1976) concluded, after simulating porosity reduction in quartz-rich sandstone in the laboratory as a function of pressure, temperature, time and pore fluid (Figure 1), that: (1) porosity decreases with increasing pressure, temperature, and time; (2) if the pore fluid is oil, the porosity reduction is slightly smaller than if the pore fluid is water; and (3) time, alone, can account for porosity reduction even if temperature and pressure are kept constant. Siever (1983) suggested that relationships among burial histories, thermal regimes, and rates of diagenetic reactions could be compared with petrologic information to deduce at what stage in its history a sediment would have accumulated sufficient time and thermal energy to accomplish a given extent of reaction. Franks and Forester (1984) proposed that the occurrence of CO_2 in dissolved gases in the Gulf Coast was kinetically controlled. Dutta (1986) estimated the kinetic parameters for the smectite-illite transformation based on

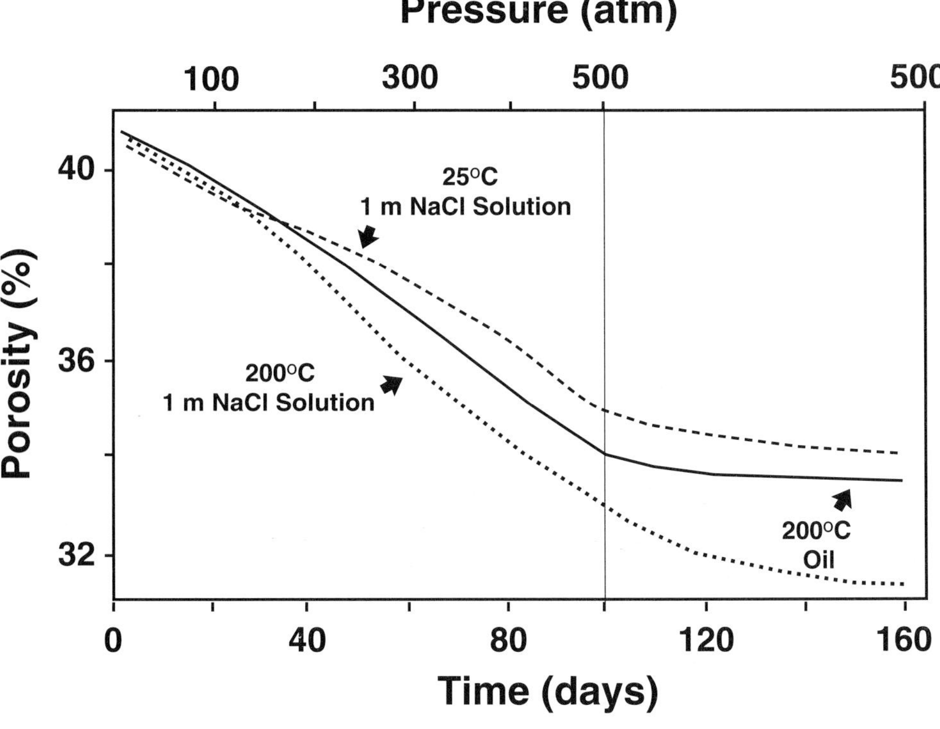

Figure 1. Experimental simulations of porosity variations as a function of time, temperature, pressure, and pore fluid (de Boer, 1976).

subsurface data. Other investigators have studied the important by-product reactions of the smectite-illite transformation, such as the generation of organic acids (Crossey et al., 1986) and cementation (Boles, 1978). Schmoker and Gautier (1988) suggested that sandstone porosity decreases in the subsurface as a power function of thermal maturity. Bruhn et al. (1988), who analyzed porosity–depth trends in sediments of the rift phase of Brazilian basins, observed that offshore reservoirs were more porous than onshore ones, and suggested that these differences were related to differences in the burial histories. Statistical analysis of the influence of 13 distinct parameters on compaction in basins of average geothermal gradients led to the conclusion that the first-order parameters are age (time of burial), detrital quartz content, maximum depth of burial, and sorting (Scherer, 1987). Bloch (1991) pointed out that the most important parameters for empirical prediction of porosity and permeability were grain size/sorting, detrital composition, and temperature history or pressure history, or both. Dixon et al. (1989) interpreted the diagenetic evolution of the deep Norphlet Formation in a time-temperature framework.

Many attempts have been made to simulate diagenetic processes based on a time-temperature scenario. Angevine and Turcotte (1983) simulated pressure solution. Leder and Park (1986) simulated quartz cementation. Surdam et al. (1989) constructed a diagenetic model based mainly on time-temperature–controlled generation or destruction of organic acids. Waples and Kamata (1993) modeled porosity reduction as a series of chemical and physical processes, but they did not recommend the use of their model at its current stage of development to predict production characteristics of specific reservoirs. Different points of view are also arising from recent research, bringing new interpretations to the cementation and dissolution events. In the diagenetic model of Smith and Ehrenberg (1989), temperature-controlled equilibria among feldspar, clay, and carbonate minerals control dissolution/precipitation of carbonate phases; time does not play an important role. Sombra et al. (1990b) did not detect kinetic control on CO_2 occurrence in natural gases of Brazilian sedimentary basins. Gluyas and Coleman (1992) argue that any successful model of cementation by silica must consider source, transportation, and precipitation mechanisms. Numerical models could become possible, but they are hard to test because duration of cementation, and thus flux, is almost impossible to constrain. Subsequently, Gluyas et al. (1993) examined data from the Garn Formation, Haltenbanken, that argue against a direct depth control on quartz cementation, suggesting that cementation took place within a restricted time period, associated with rapid subsidence and heating. In this case, origin of cementation would be associated with a particular time, not a particular temperature or pressure. Either way, it seems that diagenetic models are not well known to the point that they can be properly quantified.

The uncertainties in the quantification of diagenetic processes make empirical models still valuable and operational tools for porosity prediction. Sombra (1990) defined a new parameter, the time-depth index (TDI), which reflects time-temperature-pressure exposure and can be easily obtained from burial history diagrams. In this chapter, the TDI is used to estimate the influence of burial history on sandstone porosity evolution, quantitatively and empirically. The validity of the relationship between porosity and TDI was tested in a data set composed of 38 Late Jurassic to Tertiary age sandstone reservoirs of 7 sedimentary basins along the Brazilian continental margin (onshore and offshore). Three main compositional/textural sandstone types were considered in this study, based on detrital quartz content and grain sorting.

SANDSTONE RESERVOIRS FROM THE BRAZILIAN CONTINENTAL MARGIN

The main characteristics and the origin of the sandstone reservoirs from the Brazilian continental margin included in this study are described in this section.

General Aspects

Porosity data are mainly from core analysis, when available. In some conglomeratic or unconsolidated sandstone reservoirs, porosity data are from well log analysis. Average porosity calculations excluded calcite concretions, which typically represent <20% of the net pay. In oil/gas fields where the reservoirs have been cored in several wells, a single well was chosen to represent that field. Reservoirs rich in early diagenetic clays introduced by mechanical infiltration, such as the prerift Late Jurassic fluvial sandstones of the Sergi Formation, were not included. Strongly bioturbated or thin-bedded sandstone/shale sequences were not included in the data set either. All of the studied reservoirs are hydrocarbon saturated, either oil or gas bearing. Reservoir ages range from Late Jurassic to Tertiary. Depths range from 700 to 4900 m (2300–16,000 ft). Temperatures range from 50° to 150°C (122°–302°F), with all basins having normal geothermal gradients. Pressures are either within or close to the normal pressure gradient.

Compositional/Textural Reservoir Types

Sandstone reservoirs were grouped in three main types (I, II, and III), based on framework composition and texture. Sandstones vary from fine grained to conglomeratic, very poorly sorted to very well sorted; quartz, feldspar, and granitic/gneissic rock fragments are the main constituents. Quartz content ranges between 40% and 100%. Petrographic analyses used in this study were taken from previous works (Figure 2): Sombra et al. (1990a) studied Cretaceous marine turbidites of the Santos Basin; Moraes (1989) reported on Cretaceous and Tertiary marine turbiditic sandstones of the Campos Basin; Chang et al. (1983) analyzed Cretaceous and Tertiary marine turbiditic sandstones of the Cumuruxatiba and Espírito Santo basins; Bruhn (1985) studied Cretaceous lacustrine turbiditic sandstones of the Candeias Formation, Recôncavo Basin. Barroso (1987) and Lanzarini and Terra (1989) analyzed Upper Jurassic fluvio-eolian prerift sandstones of the Recôncavo Basin;

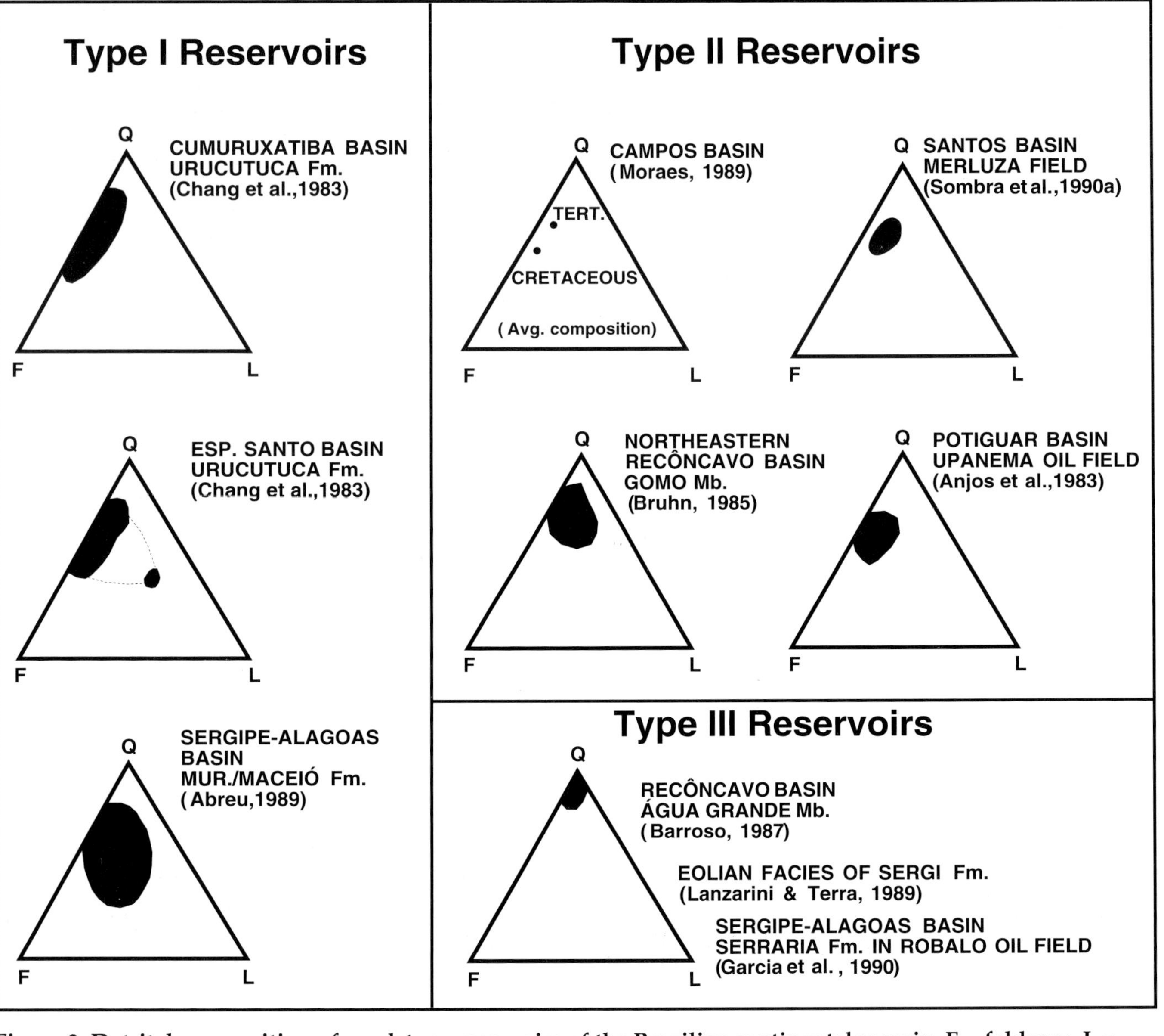

Figure 2. Detrital composition of sandstone reservoirs of the Brazilian continental margin. F = feldspar, L = lithic fragments, Q = quartz.

Abreu (1989) studied Cretaceous lacustrine and marine turbiditic sandstones of the Maceió Formation, Sergipe-Alagoas Basin; Anjos et al. (1990) and Souza (1990) reported on Cretaceous lacustrine fan-deltaic, fluvial, deltaic, and turbiditic sandstones of the Pendencia Formation, Potiguar Basin; Garcia et al. (1990) studied Upper Jurassic diagenetic quartz arenites of the Serraria Formation, Sergipe-Alagoas Basin. Detrital quartz content (measured in percentages), including mono- and polycrystalline grains, was the parameter determined to be the main framework composition indicator. These three main reservoir types, called Types I, II, and III, are described as follows:

Type I reservoirs (low quartz content, <50%) are lithic, conglomeratic sandstones that are poorly to very poorly sorted. Rock fragments are granitic/gneissic. Quartz content of these reservoirs is difficult to ascertain based on previous petrological works, because samples for thin sections were biased toward the sandy fractions. Visual estimations of the pebble content from rock fragments were made in order to estimate the quartz content. Also included in this reservoir type are feldspar-rich, medium-grained, moderately sorted sandstones with quartz content <50%. Type I reservoirs include mostly apron and fan-deltaic deposits.

Type II reservoirs (intermediate quartz content, 50%–70%) are fine- to coarse-grained arkoses that are moderately to poorly sorted. Pebbles are absent or occur in low content. Type II reservoirs represent 58% of the data set in this study and include mostly massive slope/basin turbidites in addition to fluvial, deltaic-lacustrine, and fan-deltaic deposits.

Type III reservoirs (quartz content >80%) are fine- to coarse-grained subarkoses or quartz arenites that are moderately to well sorted. This type includes eolian sands or diagenetic quartz arenites.

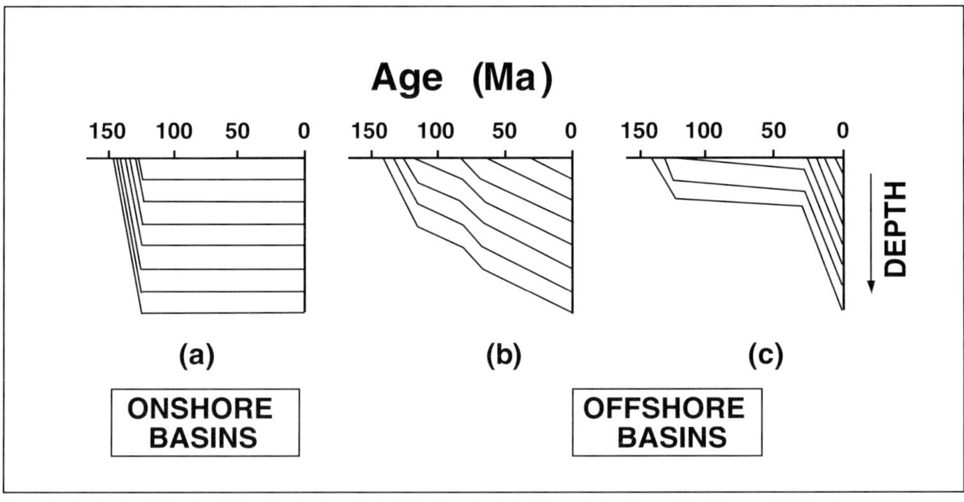

Figure 3. Typical schematic burial history diagrams of onshore (a) and offshore (b and c) reservoirs from the Brazilian continental margin. If we compare onshore and offshore reservoirs that lie today at similar depths, onshore reservoirs were buried first.

Burial History

The Brazilian offshore sedimentary basins are considered to have originated during regional extensional tectonics, with the breakup of Gondwana resulting in the separation of South America and Africa (Ponte and Asmus, 1978; Chang et al., 1988). Chang et al. (1988) showed how the stratigraphic evolution of Brazilian offshore basins fit into the model of basin development of McKenzie (1978). This uniform extension model has two stages of development, which can be summarized as: (1) crustal thinning as a consequence of stretching of the lithosphere, followed by a passive upwelling of hot asthenosphere, which is responsible for the initial rift subsidence; and (2) subsequent cooling of the lithosphere, which will further amplify the initial subsidence, producing thermal or postrift subsidence. This model explains the burial histories at the offshore Brazilian continental margin basins that display two main sedimentation phases: Lower Cretaceous continental sedimentation associated with the rifts, followed by a evaporitic and transitional Aptian deposition, which underlies the Cretaceous to Recent open-marine sediments of the postrift stage. Sedimentation rate in the rift phase was controlled mainly by the degree and rate of extension of the lithosphere. Further subsidence, in addition to thermal subsidence, was influenced by climate, sea level fluctuations, and sedimentary supply, and resulted in local variations that can be found from basin to basin, or even within one basin.

On the onshore portions of the marginal basins, initial subsidence associated with the rift stage predominates, with insignificant thermal subsidence. The sedimentary record is composed almost entirely of Lower Cretaceous continental deposits (lacustrine and fluvial deposits). In those areas, the crustal extension/thinning occurred mainly in the crust, such as predicted by the nonuniform extension models of Royden and Keen (1980) and Wernicke (1985). Because of the differences in burial histories, if we compare onshore and offshore reservoirs that lay today at similar depths, the onshore ones were buried first (Figure 3).

Porosity vs. Depth Relationship

There is no clear relationship between average porosity and depth in the reservoirs that compose the data set in this study (Figure 4). Even after analyzing specific reservoir Types I, II, and III, regression analysis of porosity vs. depth reveal very low correlation coefficients (Table 1). The plot of porosity vs. depth for Type II reservoirs (Figures 5, 6), which represent 58% of the data set, shows the absence of any clear relationship between these variables. However, reservoirs in offshore wells have systematically higher porosities than the ones in onshore wells (Figure 5), and younger reservoirs have systematically higher porosities than the older ones (Figure 6). Bruhn et al. (1988), who studied rift deposits along the Brazilian continental margin, observed that reservoirs were more porous offshore than onshore, arguing that differences in burial histories were responsible.

TIME-DEPTH INDEX

The TDI of a reservoir, as defined by Sombra (1990), can be calculated in the burial history diagram of any well. The index corresponds to the area in the diagram enclosed by the burial curve of the reservoir and the axes of the diagram (shown in Figure 7). The TDI is expressed in kilometers times million years of age. In practice, reservoir depths during burial history are integrated at regular time intervals of 1 m.y.

It seems that the first attempt to analyze porosity evolution against an integration of burial depth was that from Block et al. (1986). They integrated the burial depth, for six wells in the Haltenbanken area, within the pressure solution domain, i.e., the product of burial time and depth between 1525 m (5000 ft) and present-day depth. For more details, also see Bloch (1994).

The calculation of the TDI ignores the present-day bathymetry and paleobathymetry. Water depth does affect compaction in most situations. The effective stress along grain-to-grain contacts can be defined as the vertical stress (total weight of the overburden; i.e., sediment/fluid plus water column above the reservoir)

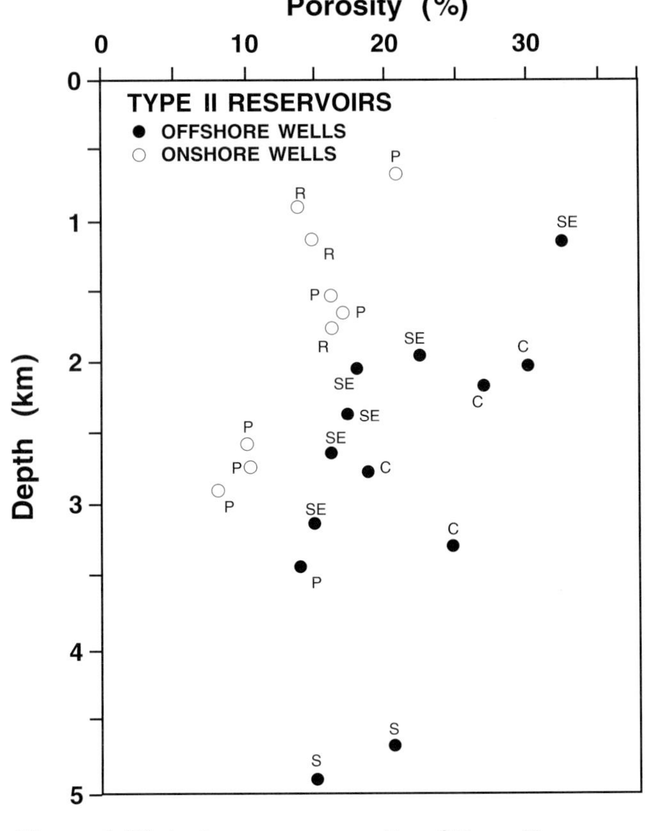

Figure 4. Plot of average porosity of reservoirs studied against depth. AL = Alagoas Basin; C = Campos Basin; ES = Espírito Santo/Cumuruxatiba basins; P = Potiguar Basin; R = Recôncavo Basin; S = Santos Basin; SE = Sergipe Basin.

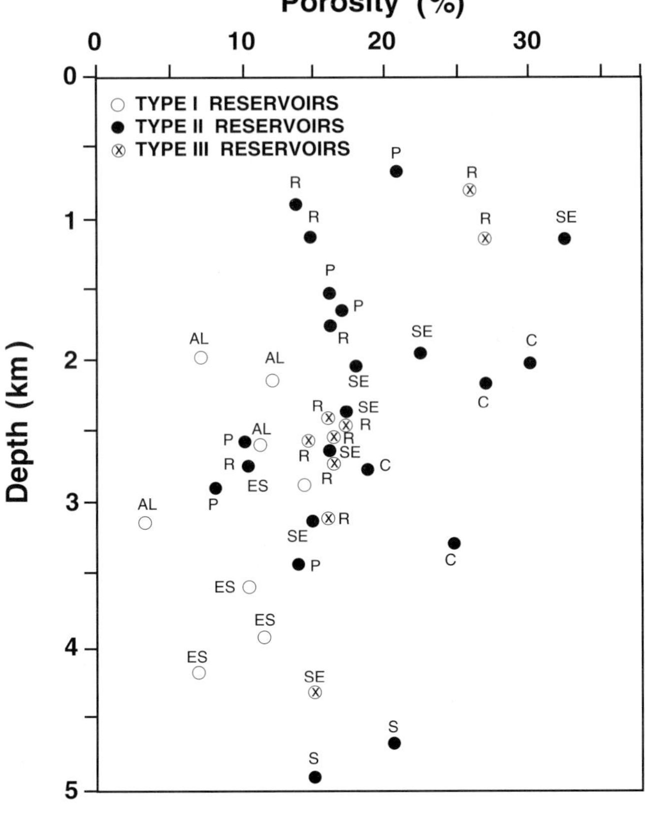

Figure 5. Plot of average porosity of Type II reservoirs against depth, for onshore and offshore wells. Reservoirs located offshore are more porous. C = Campos Basin; P = Potiguar Basin; R = Recôncavo Basin; S = Santos Basin; SE = Sergipe Basin.

minus fluid pressure (Terzaghi and Peck, 1967). An increase in effective stress due to water column will be counterbalanced by an equivalent increase of reservoir fluid pressure in reservoirs with approximately normal pressure gradients, or with hydrostatic communication.

The TDI represents a simplification of the procedure proposed by Bloch et al. (1986) and Bloch (1994). Pressure solution is probably a kinetically controlled process and, in many areas, the pressure-solution domain may not be well known. The TDI also represents a simplification of that proposed by Schmoker and Gautier (1988) to predict porosity evolution. The thermal parameter, which is important because it affects the mechanical strength of the grains and the

susceptibility for chemical transformations, such as pressure solution, has not been incorporated into the calibration. The omission stems from the desire to maintain the relationship as simple and as operational as possible by eliminating parameters that introduce uncertainties (because the reservoirs in the study have been deposited in the same tectonic context). The thermal parameter is intrinsically present in the TDI parameter. The TDI reflects the evolution of reservoir depth during its burial, so it is a number that contains data related to the evolution of effective pressure and temperature. Vitrinite reflectance is kinetically controlled, and there is good correlation between Brazilian reservoirs' TDI and vitrinite reflectance in the associated shales (Figure 8).

The TDI was calculated for all the reservoirs included in this study. The plots of average porosity vs. TDI for the three reservoir types defined (I, II, and III) show very clear trends of decreasing porosity with increasing TDI (Figures 9–11). Regression analysis of porosity on TDI (exponential model) obtained very good correlation coefficients (Table 2).

Type I reservoirs (lithic, conglomeratic sandstones), which are texturally and compositionally the most immature and presented lower initial porosities, also display a very rapid porosity decay with increasing

Table 1. Simple Linear Regression of Porosity on Depth for Reservoir Types I, II, and III.*

Reservoir Type	b	a	n	$r^2(\%)$
I	11.3	–4.33E–4	8	1.0
II	20.2	–8.30E–4	21	1.3
III	28.1	–3.88E–3	8	73.3

*Porosity (%) = b + a; Depth (km); n = number of points in data set.

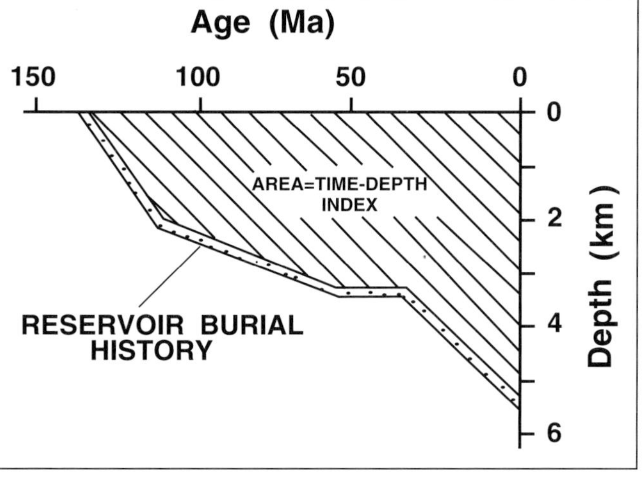

Figure 6. Plot of average porosity of Type II reservoirs against depth, by age. Younger reservoirs are more porous. C = Campos Basin; P = Potiguar Basin; R = Recôncavo Basin; S = Santos Basin; SE = Sergipe Basin.

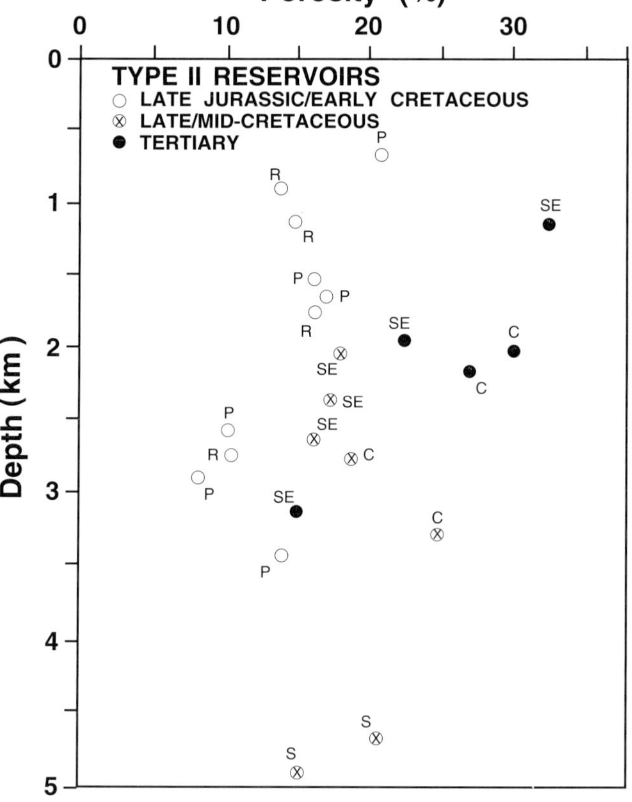

Figure 7. Burial history diagram of a hypothetical reservoir, illustrating the significance of the time-depth index (i.e., area enclosed by the axes and the reservoir burial curve).

diagenetic processes; quartz-rich sandstones were deposited in eolian settings. Diagenetic quartz arenites resulted from the leaching of feldspars and rock fragments close to regional unconformities, as in the Serraria Formation, Sergipe Basin (Garcia et al., 1990). Porosity destruction due to intense silica cementation tends to be an important diagenetic event in quartz-rich sandstones at elevated temperatures such as 100°–150°C (212°–302°F) (Bjørlykke et al., 1989). In such reservoirs, the presence of hydrocarbons is essential for porosity preservation because they retard diagenesis. In a well from Sergipe Basin, at 4300 m (14,100 ft) the average porosity of the Serraria Formation is 15% above the oil-water contact and near zero below this contact (Garcia et al., 1990), with quartz cementation responsible for porosity destruction.

DISCUSSION

Diagenetic processes related to porosity destruction are grouped into either compaction or cementation. Compaction consists of two modes, mechanical and chemical. In the former, porosity reduction is caused by grain rearrangement in response to applied stress, usually associated with incremental overburden. Chemical compaction results from rearrangement of framework grains that underwent chemical dissolution, particularly along the regions of major contacts or stress concentrations. This remobilization produces an additional reduction of volume compared with pure mechanical compaction. Diagenetic and mass balance studies performed in siliciclastic sequences of three Brazilian basins of the equatorial margin (Chang, 1983) led to the conclusion that compaction (mechanical and chemical) is the main diagenetic factor controlling porosity decay during burial. Lundegard (1992) analyzed a large database of point count from diverse sandstones and concluded that compaction (mechanical and chemical), although being generally underappreciated, is probably the dominant mechanism of porosity loss in sandstones.

TDI (Figure 9). The data set for this reservoir type is very limited, and only cautious conclusions can be made for TDI values >200 km × Ma based on observed trends. However, we expect to find very low porosities (<10%) for TDI values >200 km × Ma.

Type II reservoirs (feldspar rich) represent the best documented reservoir type (58% of the data set). A good trend of porosity decline with increasing TDI can be seen for this reservoir type (Figure 10). A shift to increased porosity with depth is observed with the comparison of Type I and Type II reservoirs. Type II is invariably more porous. Two outlier points were documented as a typical case of porosity preservation at great depth due to early chlorite coatings. Sombra et al. (1990a) concluded that the porosity preserved due to early chlorite coatings was 9% and 4% in these two wells, after comparing chlorite-coated and chlorite-free sandstones.

Type III (quartz-rich) are the most porous reservoirs, displaying a good trend of porosity decline vs. TDI (Figure 11). However, the data for this reservoir type are limited, and this trend must be viewed cautiously. The chemical and mechanical stability of quartz is probably responsible for the highest porosity preservation in this reservoir type. Quartz enrichment in these reservoirs was related to either depositional or

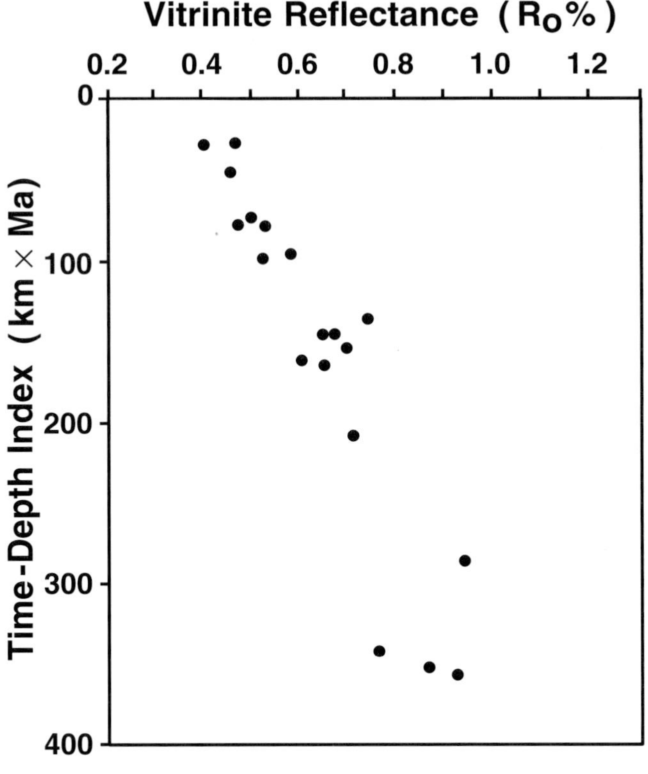

Figure 8. Plot of time-depth index of some reservoirs of the Brazilian continental margin against vitrinite reflectance in the associated shales.

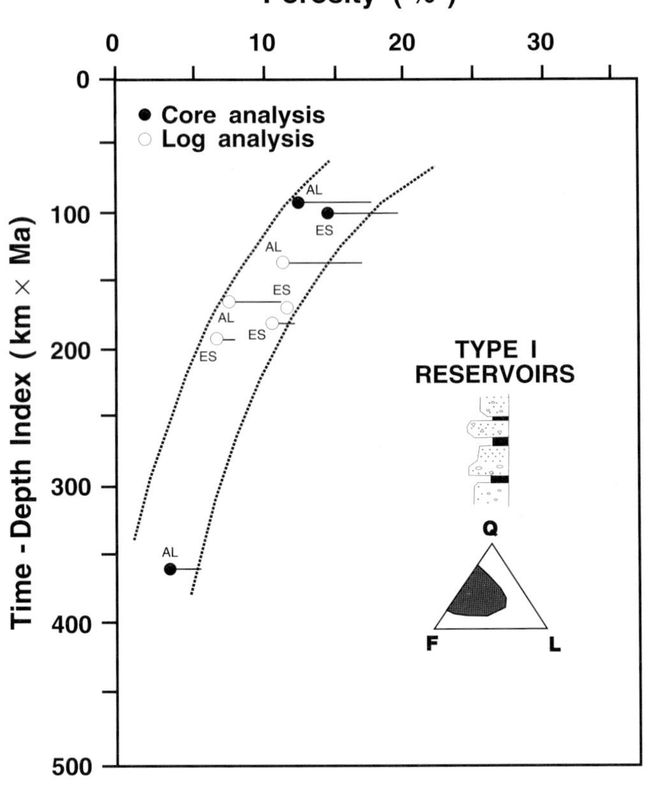

Figure 9. Plot of porosity (average and maximum) against time-depth index for Type I reservoirs. Broken lines enclose average porosity values. AL = Alagoas Basin; ES = Espírito Santo/Cumuruxatiba basins; F = feldspar; L = lithic fragments; Q = quartz.

The role of cementation is that of porosity reduction by filling the void space with authigenic mineral precipitation. As reasoned by Bjørlykke et al. (1989), this mode of porosity decrease occurs without loss in the bulk volume, in contrast to the bulk volume reduction that results from mechanical compaction. Therefore, a porosity trend produced solely by cementation or dissolution should not significantly affect the commonly observed trend of porosity reduction with depth. This statement will hold more strongly if diagenetic transformations occurring in the sandstones are relatively isochemical, as it has been suggested by several authors (Chang, 1983; Bjørlykke et al., 1988; Giles and de Boer, 1990).

Compaction is essentially bulk volume reduction that involves the removal of a fluid phase from a porous solid. The physics involved in the compaction of two-phase porous media (solid framework and fluid) has been the subject of many reports (Sharp and Domenico, 1976; Bethke, 1985; Nakayama and Lerche, 1987; Mello, 1994). Mello (1994) presents an excellent overview of the rheology of compacting porous sediments on a geological time scale. Sediment rheology must be accounted for to properly model porosity reduction. At room temperature, most consolidated sedimentary rocks are brittle, which means that they behave elastically until they fail. At higher temperature and pressure, buried sediments behave like ductile material. Both brittle and ductile deformations are permanent and irreversible. Observations of sediments at geological time scale indicate that sediments exhibit three rheological components: elastic, plastic, and viscous. The predominance of each behavior is dependent on composition, temperature, state of stress, degree of lithification, and length of time. For instance, sediment deformation on a short time scale responds elastically, because there is not enough time for the fluids to be removed. A complete rheological model that is geologically realistic is an elasto-viscoplastic model (Mello, 1994).

The porosity trend exhibited for the three types of reservoirs is at least qualitatively consistent with the

Table 2. Exponential Regressions of Porosity (%) on Time-Depth Index (km × Ma) for Reservoir Types I, II, and III.*

Reservoir Type	b	a	n	r^2(%)
I	3.07	−4.79E−3	8	84.7
II	3.34	−3.08E−3	21	82.2
III	3.46	−1.92E−3	8	82.9

*Porosity (%) = exp (b + a TDI); n = number of points in data set.

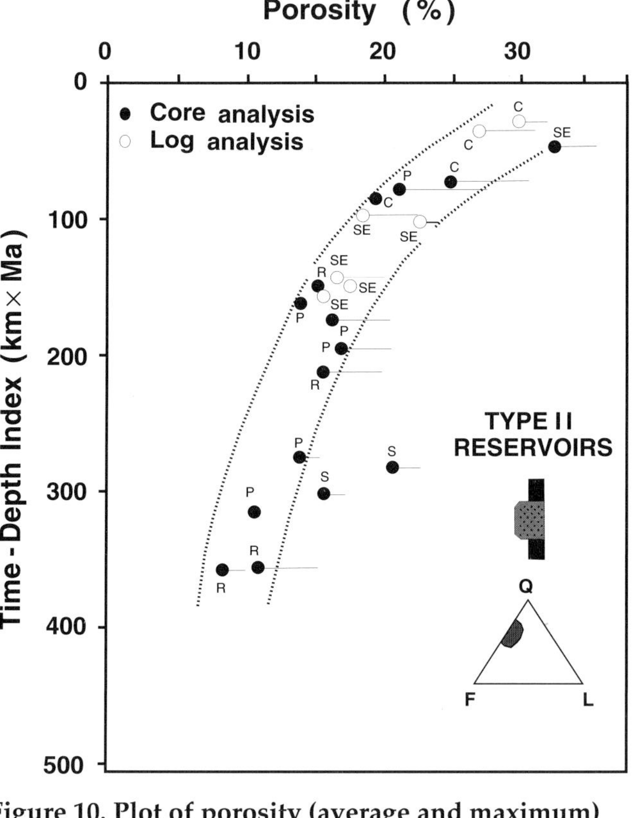

Figure 10. Plot of porosity (average and maximum) against time-depth index for Type II reservoirs. Broken lines enclose average porosity values. Outliers had porosity preserved by early chlorite coatings (Sombra et al., 1990a). C = Campos Basin; P = Potiguar Basin; F = feldspar; L = lithic fragments; Q = quartz; R = Recôncavo Basin; S = Santos Basin; SE = Sergipe Basin.

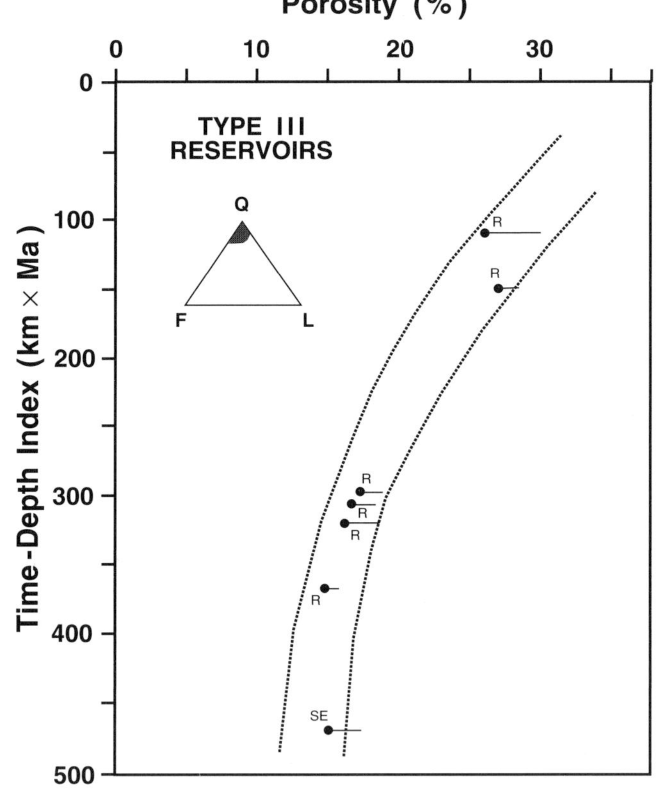

Figure 11. Plot of porosity (average and maximum) against time-depth index for Type III reservoirs. Broken lines enclose average porosity values. F = feldspar; L = lithic fragments; Q = quartz; R = Recôncavo Basin; SE = Sergipe Basin.

variables listed above and influential on the rheological behavior. For instance, Type II or Type III reservoirs have a higher quartz content, and subsequently a low content of lithic (ductile) rock fragments. As a result, the compaction trend shows less porosity reduction. This compositional control dominates because the thermal regimes and burial histories are very similar.

CONCLUSIONS

Burial history plays a very important role in the evolution of sandstone porosity along the Brazilian continental margin, in addition to detrital composition and texture. Reservoirs that have resided at maximum burial longer tend to be less porous than the ones that achieved maximum burial late in their burial history, indicating time is a factor in porosity destruction. The more mineralogically and texturally mature sandstones lose porosity at a slower rate than the immature ones during progressive burial. The decay in porosity is poorly related to the present-day depth of the sandstone reservoirs, but it is closely related to the evolution of depth during burial. Good relationships were obtained between porosity and TDI, a parameter that reflects the evolution of reservoir depth over geologic time.

Porosity prediction of a sandstone reservoir along the Brazilian continental margin is possible with information about its mineralogy, texture, and burial history. Exceptions observed were sandstones that contained early chlorite coatings, which preserved porosity. The ranges of detrital composition, texture, age, depth, temperature, and burial history of these sandstones should be considered when making porosity predictions.

ACKNOWLEDGMENTS

We thank AAPG reviewers M. Emery and J. Schmoker. We thank J. Gluyas, S. Bloch, and Carlos H. L. Bruhn for suggestions and discussions. We also thank Sylvia Anjos, Luis F. De Ros, and Rogério Schiffer de Souza for exchanging ideas. We thank PETROBRÁS for granting permission to publish this paper.

REFERENCES CITED

Abreu, C.J., 1989, Predicting reservoir quality in the Cretaceous Maceió Member of the Sergipe-Alagoas Basin, northeast Brazil: Master's thesis, University of Cincinnati, Cincinnati, Ohio, 106 p.

Angevine, C.L., and D.L. Turcotte, 1983, Porosity reduction by pressure solution: a theoretical model for quartz arenites: Geological Society of America Bulletin, v. 94, p. 1129–1134.

Anjos, S.M.C., C.L. Sombra, R.S. Souza, and R.N. Waick, 1990, Potencial de reservatórios profundos na Formação Pendência, Bacia Potiguar Emersa: Boletim de Geociências da Petrobrás, v. 4, p. 509–530.

Barroso, A.S., 1987, Diagênese e eficiência de recuperação dos reservatórios do Campo de Araças, Bacia do Recôncavo, Brasil: Master's thesis, Universidade Federal de Ouro Preto, Ouro Preto, Brasil, 160 p.

Beard, D.C., and P.K. Weil, 1973, Influence of texture on porosity and permeability of unconsolidated sands: AAPG Bulletin, v. 57, p. 349–369.

Bethke, C.M., 1985, A numerical model of compaction-driven groundwater flow and heat transfer and its application to the paleohydrology of intracratonic sedimentary basins: Journal of Geophysical Research, v. 90, p. 6817–6828.

Bjørlykke, K., A. Mo, and E. Palm, 1988, Modelling of thermal convection in sedimentary basins and its relevance to diagenetic reactions: Marine and Petroleum Geology, v. 5, p. 338–351.

Bjørlykke, K., M. Ram, and G.C. Saigal, 1989, Sandstone diagenesis and porosity modification during basin evolution: Geologisches Rundschau, v. 78, p. 243–268.

Bloch, S., 1991, Empirical prediction of porosity and permeability in sandstones: AAPG Bulletin, v. 75, p. 1145–1160.

Bloch, S., 1994, Case histories—offshore Mid-Norway/Taranaki Basin, New Zeland/San Emigdio area, California, *in* M. D. Wilson, ed., Reservoir quality assessment and prediction in clastic rocks: SEPM Short Course 30, p. 357–365.

Bloch, S., R.K. Sucheki, J.R. Duncan, and K. Bjørlykke, 1986, Porosity prediction in quartz-rich sandstones: Middle Jurassic, Haltanbanken area, offshore central Norway (abs.): AAPG Bulletin, v. 70, p. 567.

Boles, J. R., 1978, Active ankerite cementation in the subsurface Eocene of Southwest Texas: Contributions to Mineralogical Petrology, v. 68, p. 13–22.

Bruhn, C.H.L., 1985, Sedimentação e evolução diagenética dos turbiditos eocretácicos do Membro Gomo, Formação Candeias, no compartimento nordeste da Bacia do Recôncavo, Bahia: Master's thesis, Universidade Federal de Ouro Preto, Ouro Preto, Brasil, 203 p.

Bruhn, C.H.L., C. Cainelli, and R.M.D. Matos, 1988, Habitat do petróleo e fronteiras exploratórias nos rifts Brasileiros: Boletim de Geociências da Petrobrás, v. 2, p. 217–254.

Chang, H.K., 1983, Diagenesis and mass transfer in Cretaceous sandstone-shale sequences, offshore Brazil: Ph.D. thesis, Northwestern University, Evanston, Illinois, 339 p.

Chang, H.K., S.M.C. Anjos, and C.R. Drug, 1983, Características dos reservatórios e evolução diagenética da sequência turbidítica do Cretáceo Superior e Terciário Inferior das Bacias do Espírito Santo e Cumuruxatiba: Rio de Janeiro, Petrobrás internal report.

Chang, H.K., R.O. Kowsmann, and M.F. Figueiredo, 1988, New concepts on the development of east

Brazilian marginal basins: Episodes, v. 11, p. 194–202.

Crossey, L.J., R.C. Surdam, and R.W. Lahann, 1986, Application of organic/inorganic diagenesis to porosity prediction, *in* D. L. Gautier, ed., Roles of organic matter in sediment diagenesis: SEPM Special Publication 38, p. 147–156.

de Boer, R.B., 1976, Thermodynamical and experimental aspects of pressure solution, *in* J. Cadek and T. Paces, eds., Proceedings of the International Symposium on Water-Rock Interactions: Geological Survey, Prague, 1974, p. 381–387.

Dixon, S.A., D.M. Summers, and R.C. Surdam, 1989, Diagenesis and preservation of porosity in Norphlet Formation (Upper Jurassic), southern Alabama: AAPG Bulletin, v. 73, p. 707–728.

Dutta, N.C., 1986, Shale compaction, burial diagenesis and geopressures: a dynamical model, solution and some results, *in* J. Burrus, ed., Thermal modeling in sedimentary basins: Paris, Editions Technip, Collection Colloques et Seminaires 44, p. 149–172.

Franks, S., and R. Forester, 1984, Relationships among secondary porosity, pore fluid chemistry and carbon dioxide, Texas Gulf Coast, *in* W.S. McDonald and R.C. Surdam, eds., Clastic diagenesis: AAPG Memoir 37, p. 63–80.

Galloway, W.E., 1974, Deposition and diagenetic alteration of sandstone in a Northeast Pacific arc-related basin: implications for graywacke genesis: Geological Society of America Bulletin, v. 85, p. 379–390.

Garcia, A.J.V., L.F. de Ros, R.S. Souza, and C.H.L. Bruhn, 1990, Potencial de reservatórios profundos na Formação Serraria, Bacia de Sergipe-Alagoas: Boletim de Geociências da Petrobrás, v. 4, p. 467–488.

Giles, M.R., and R.B. de Boer, 1990, Origin and significance of redistributional secondary porosity: Marine and Petroleum Geology, v. 7, p. 378–397.

Gluyas, J., and M. Coleman, 1992, Material flux and porosity changes during sediment diagenesis: Nature, v. 356, p. 52–54.

Gluyas, J.G., S.M. Grant, and A.G. Robinson, 1993, Geochemical evidence for a temporal control on sandstone cementation, *in* A.D. Horbury and A.G. Robinson, eds., Diagenesis and basin development: AAPG Studies in Geology 36, p. 23–33.

Lanzarini, W.L., and G.J.S. Terra, 1989, Fácies sedimentares, evolução da porosidade e qualidade de reservatório da Formação Sergi, Campo de Fazenda Boa Esperança, Bacia do Recôncavo: Boletim de Geociências da Petrobrás, v. 3, p. 365–375.

Leder, F., and W.C. Park, 1986, Porosity reduction in sandstone by quartz overgrowth: AAPG Bulletin, v. 70, p. 1713–1728.

Lundegard, P.D., 1992, Sandstone porosity loss—a "big picture" view of the importance of compaction: Journal of Sedimentary Petrology, v. 62, p. 250–260.

McKenzie, D., 1978, Some remarks on the development of sedimentary basins: Earth and Planetary Science Letters, v. 40, p. 25–32.

Mello, U.T., 1994, Thermal and mechanical history of sediments in extensional basins: Ph.D. thesis, Columbia University, New York, 395 p.

Moraes, M.S., 1989, Diagenetic evolution of Cretaceous–

Tertiary turbidite reservoirs, Campos Basin, Brazil: AAPG Bulletin, v. 73, p. 598–612.

Nagtegaal, P.J.C., 1980, Diagenetic models for predicting clastic reservoir quality: Barcelona, Revista del Instituto de Investigaciones Geologicas, v. 34, p. 5–19.

Nakayama, K., and I. Lerche, 1987, Two-dimensional basin analysis, *in* B. Doliges, ed., Migration of hydrocarbons in sedimentary basins: Paris, Editions Technip, p. 597–611.

Ponte, F.C., and H.E. Asmus, 1978, Geological framework of the Brazilian continental margin: Geologisches Rundschau, v. 67, p. 201–235.

Royden, L., and C.E. Keen, 1980, Rifting processes and thermal evolution of the continental margin of eastern Canada determined from subsidence curves: Earth and Planetary Science Letters, v. 51, p. 343–361.

Scherer, M., 1987, Parameters influencing porosity in sandstones: a model for sandstone porosity prediction: AAPG Bulletin, v. 71, p. 485–491.

Schmoker, J., and D.L. Gautier, 1988, Sandstone porosity as a function of thermal maturity: Geology, v. 16, p. 1007–1010.

Sharp, J.M., Jr., and P.A. Domenico, 1976, Energy transport in thick sequences of compacting sediments: Geological Society of America Bulletin, v. 87, p. 390–400.

Siever, R., 1983, Burial history and diagenetic reaction kinetics: AAPG Bulletin, v. 67, p. 684–691.

Smith, J.T., and S.N. Ehrenberg, 1989, Correlation of carbon dioxide abundance with temperature in clastic hydrocarbon reservoirs—relationship to inorganic chemical equilibrium: Marine and Petroleum Geology, v. 6, p. 129–135.

Sombra, C.S., 1990, O papel da história de soterramento na evolução da porosidade de arenitos (bacias marginais Brasileiras): Boletim de Geociências da Petrobrás, v. 4, p. 413–428.

Sombra, C.L., L.M. Arienti, M.J. Pereira, and J.M. Macedo, 1990a, Parâmetros controladores da porosidade e da permeabilidade nos reservatórios clásticos profundos do Campo de Merluza, Bacia de Santos, Brasil: Boletim de Geociências da Petrobrás, v. 4, p. 451–466.

Sombra, C.L., T. Takaki, G.I. Henz, and A.S. Barroso, 1990b, CO_2 in natural gases of Brazilian sedimentary basins (abs.): AAPG Bulletin, v. 4, p. 768.

Souza, R.S., 1990, Controle deposicional e diagenético dos reservatórios profundos do Campo de Pescada, Bacia Potiguar: Boletim de Geociências da Petrobrás, v. 4, p. 531–553.

Surdam, R.C., T.L. Dunn, H.P. Heasler, and D.B. MacGowan, 1989, Porosity evolution in sandstone/shale systems, *in* I.E. Hutcheon, ed., Burial diagenesis: Mineralogical Association of Canada, Short Course Handbook, v. 15, p. 61–134.

Terzaghi, K., and R.B. Peck, 1967, Soil mechanics in engineering practice (2d ed.): New York, J. Wiley & Sons, Inc., 729 p.

Waples, D.W., and H. Kamata, 1993, Modelling porosity reduction as a series of chemical and physical processes, *in* A.G. Doré et al., eds., Basin modelling: advances and applications: Norwegian Petroleum Society (NPF) Special Publication 3, p. 303–320.

Wernicke, B., 1985, Uniform-sense simple shear of the continental lithosphere: Canadian Journal of Science, v. 22, p. 108–125.

Chapter 7

A Geological Approach to Permeability Prediction in Clastic Reservoirs

Jonathan Evans
BP Exploration
Poole, Dorset, England, United Kingdom

Chris Cade
BP Norge
Stavanger, Norway

Steven Bryant
ENIRICERCHE
Milan, Italy[1]

ABSTRACT

Permeability is a key parameter in determining the economic value of a hydrocarbon accumulation; however, our ability to predict the magnitude and range of permeability in undrilled areas is poor. Traditional methods of permeability prediction are empirical and rely on developing relationships between permeability and other parameters that may be predicted with greater confidence, such as porosity or lithology. These empirical methods may work well in areas where there is sufficient calibration data, but extrapolation away from well data is prone to large errors (often by orders of magnitude).

An alternative approach to permeability prediction is to model the effect of geological processes such as burial and cementation on the pore structure of the rock and, hence, calculate the change in permeability. Through understanding the effect of various geological processes on permeability, it is then possible to predict permeability from geological models. This approach has applications in both data-rich and undrilled areas.

The quantitative insight into which factors affect the permeability has been provided by computer modeling, which allows us to focus in on the most important controls, such as grain size and the amount of cement or ductile grains. Our ability to predict permeability in undrilled areas is now more often hampered by our inability to predict the variations in these controlling factors rather than by any lack of understanding of permeability itself.

[1]Present address: Center for Subsurface Modeling, Texas Institute for Computational and Applied Mathematics, University of Texas at Austin, U.S.A.

INTRODUCTION

Porosity and permeability are important parameters that help to define the commercial viability of an oil or gas accumulation. In particular, reservoir permeability is an important control on the flow rates that may be achieved from a well. As a result, the ability to predict permeability has important commercial significance.

Permeability measures the ability of a rock to allow fluids to move through its pore system. It is a key factor with respect to producing fluids from a reservoir. The controls on porosity are well understood, and methods of porosity estimation are becoming well established. In comparison, our understanding of the factors controlling permeability is less advanced. This chapter reviews the various methods that have been used to help minimize the uncertainty inherent in permeability prediction.

The data available for permeability prediction vary with the stage of a reservoir evaluation. At the wildcat stage, an assessment of permeability before drilling is essential to constrain the potential economic return. This usually will be based on regional porosity-permeability-depth trends together with sedimentological information; some burial history data may also be added. In appraisal and development, a detailed description of permeability is required. Direct measurements of reservoir characteristics from seismic reflection data, wireline logs, well tests, and core samples will be available. Prediction during these stages involves integration of permeability measurements with information on reservoir sedimentology, together with seismic reflection and wireline log data, to fill the gaps between wells and produce an overall reservoir description.

DETERMINATION OF PERMEABILITY

Permeability is the intrinsic characteristic of a material that determines how easily a fluid can pass through it. In the petroleum industry, the darcy is the standard unit of permeability, but millidarcys (1 md = 10^{-3} darcys) are commonly used. Permeabilities in clastic reservoir rocks may range from <0.1 md to >10 darcys. This intrinsic rock property is called absolute permeability when the rock is 100% saturated with one fluid phase.

The three main permeability measurement techniques are well testing, wireline tool analysis, and laboratory analysis of core samples.

Well Testing

Well testing can take various forms, but all involve the measurement of a flow rate for fluid moving into the well bore from the reservoir. The simplest test is a spinner survey, in which a turbine is moved up the well bore to record the location and velocity of any flow. Other forms of testing, such as the drill-stem test, involve taking measurements of pressure changes through time either before or after a restriction to flow. When these pressure data are combined with measurements of reservoir thickness, permeability can be calculated.

Well testing provides an average measurement of permeability across a certain reservoir interval. For oil or gas flows, well tests usually measure relative permeability, rather than absolute permeability, since more than one fluid phase is present.

Wireline Measurements

Many methods have been proposed for obtaining permeability measurements from wireline tool measurements. These include: (1) pressure/time measurement of formation fluids with the repeat formation test tool; (2) empirical correlation of permeability (from core analysis) with porosity and intergranular surface area (measured by wireline tools); (3) measurement of movable fluids with the nuclear magnetic resonance log; and (4) correlation of permeability with Stoneley wave velocity measured by acoustic logging tools.

The applications of these methods have been reviewed by Ahmed et al. (1991). Most of the methods are at best qualitative—capable of distinguishing high- and low-permeability zones. The exceptions are formation test measurements and the standard core-derived permeability vs. porosity regression method. The latter is valid only for formations similar to the calibrated formation.

Core Analysis

Core analysis allows direct measurement of porosity and permeability under controlled laboratory conditions. Measurements can be made at three scales: rotary sidewall core [samples <2.5 cm (1 in.) long], core plugs [samples 2.5–4 cm (1–1.5 in.) long], and whole core [samples ≤60 cm (2 ft) long].

Such measurements give an accurate representation of a particular core sample under specific laboratory conditions. Extrapolation to field conditions must be done with care.

Routine core analysis is normally carried out on core plugs taken every 30 cm (1 ft) through whole core. This provides data on porosity and air permeability (K_a). A correction is usually applied to the K_a values to give equivalent liquid permeability (K_L) (Klinkenberg, 1941).

Special core analysis (SCAL) may be performed on a selection of plugs from the reservoir interval to determine brine permeability (K_b). This may be measured over a range of confining pressures to determine K_b at overburden pressure. Other SCAL methods can determine a range of petrophysical parameters if required (e.g., capillary pressure, relative permeability, and formation factor).

CONTROLS ON PERMEABILITY

In clastic rocks, permeability is determined by the size of the pore throats present in the rocks and by the number of connected pores. Permeability prediction involves understanding how various geological factors affect these fundamental controls. In unconsolidated sands, the important factors are the grain size and sorting (Beard and Weyl, 1973). Rocks with coarser grain sizes will tend to have larger pore throats and, therefore, higher permeability. Rocks with poorer sorting will have smaller mean pore-throat diameters and, therefore, lower permeability than better sorted rocks with the same mean grain size. The presence of detrital clays will lead to smaller pore throats and less-connected pores, which reduces permeability. During burial, compaction reduces the size of pore throats and eventually blocks them off entirely, so again permeability is reduced. The rate of compaction and the rate of pore-throat blocking depends on the proportion of ductiles present; this also affects the permeability (Gluyas and Cade, this volume). The precipitation of cements similarly reduces the size and number of pore throats; different cement styles will reduce the permeability at different rates. Ethier and King (1991) illustrate a general understanding of these controls (Figure 1), but with little or no quantitative detail, the value of such trends is limited.

EMPIRICAL APPROACHES TO PERMEABILITY PREDICTION

Empirical techniques use a calibration data set (e.g., data from core samples) and multiple regression analysis to determine the relationship between rock property variables and reservoir quality. The calibrated regression relationships are then used to predict reservoir quality in different settings but within the range of the variables comprising the calibration data set. Dutton and Diggs (1992) and Bloch (1991) describe the most frequently used application of this approach, in which relationships between measured porosity and permeability (usually ambient helium porosity and single-phase gas permeability), and textural and mineralogical variables (usually measured on thin sections), are investigated. Commonly used variables are grain size, sorting, matrix clay content, volume of individual cements, total cement volume, and point-counted interparticle porosity.

A variation on the empirical approach is described by Ehrlich et al. (1991). Using the observation that, even in a single formation, permeability commonly varies by several orders of magnitude, they conclude that the configuration, rather than the absolute value, of porosity is the control on permeability. To characterize the pore system configuration, Ehrlich et al. (1991) make measurements of pores in two dimensions (on polished thin sections) and combine these with pore-throat size distribution data (from mercury porosimetry) to develop a simple pore system model. For selected data sets, a good relationship between the simple pore system model and measured permeability has been established. It is unlikely, however, that we would be able to predict confidently the pore type and pore-throat size distribution parameters in an undrilled sandstone.

In predicting permeability ahead of drilling, the criteria for success of any method must be that it establishes a quantitative link between measured permeability and another (or several other) rock parameter(s), and that those correlative parameters can themselves be predicted from a geological model. Many empirical approaches fail the second of these criteria.

APPLICATIONS OF THE EMPIRICAL APPROACH

In areas with sufficient well data (either core analysis, log, or well test data) to define significant regional or field porosity–permeability and porosity–depth regressions, the empirical approach described above can often be successfully used to predict porosity and permeability in areas away from well control.

This method is the one most commonly used in mature provinces, and gives good results provided there is not too much scatter in the data. However, the scatter is often such that the uncertainty in permeability prediction may cover several orders of magnitude.

Some of this scatter may be due to textural variation, controlled in turn by sedimentary facies and lithology. If sedimentological information is available (from core logs or reservoir models), lithofacies can be taken into account by plotting the poroperm values for each lithofacies separately. Often the regression relationships for a given lithofacies will be better than those for the whole data set, since variations in grain size, clay content, and so forth will be reduced. The combination of empirical relationships for each facies with a sedimentological reservoir model may then produce a reasonable description of reservoir permeability variations.

Further insight may be obtained through including mineralogical (e.g., from modal point counting), textural (e.g., grain size, sorting), and SCAL (e.g., critical pore-throat size, K_b) data in the regression analysis. In many cases, a few parameters will explain most of the variation in permeability (e.g., grain size, sorting, lithic content).

Example 1—Permeability Prediction While Drilling

Hogg et al. (1996) have recently presented a novel application of the empirical approach to permeability estimation. In the Triassic fluvial Sherwood Sandstone reservoir of the Wytch Farm field (onshore U.K.), permeability is controlled principally by porosity and grain size. Using previously drilled wells in the field, an empirical correlation was determined between porosity and permeability for a range of grain-size classes (Figure 2).

By measuring porosity (from logging-while-drilling density tool) and grain size (from sieve analysis of cuttings) during drilling, it is possible to estimate the permeability of the sandstone while drilling. This method was applied during drilling of extended-reach wells in order to ensure that a certain minimum productivity index (PI) is achieved before the well is stopped.

Test results from several extended-reach wells show that PI can be predicted with a high degree of accuracy using this technique (Figure 3). The main uncertainty is the density of the saturating fluid, which must be assumed when calculating porosity. An additional benefit is that the predicted permeability–depth plot can be used to optimize perforation intervals.

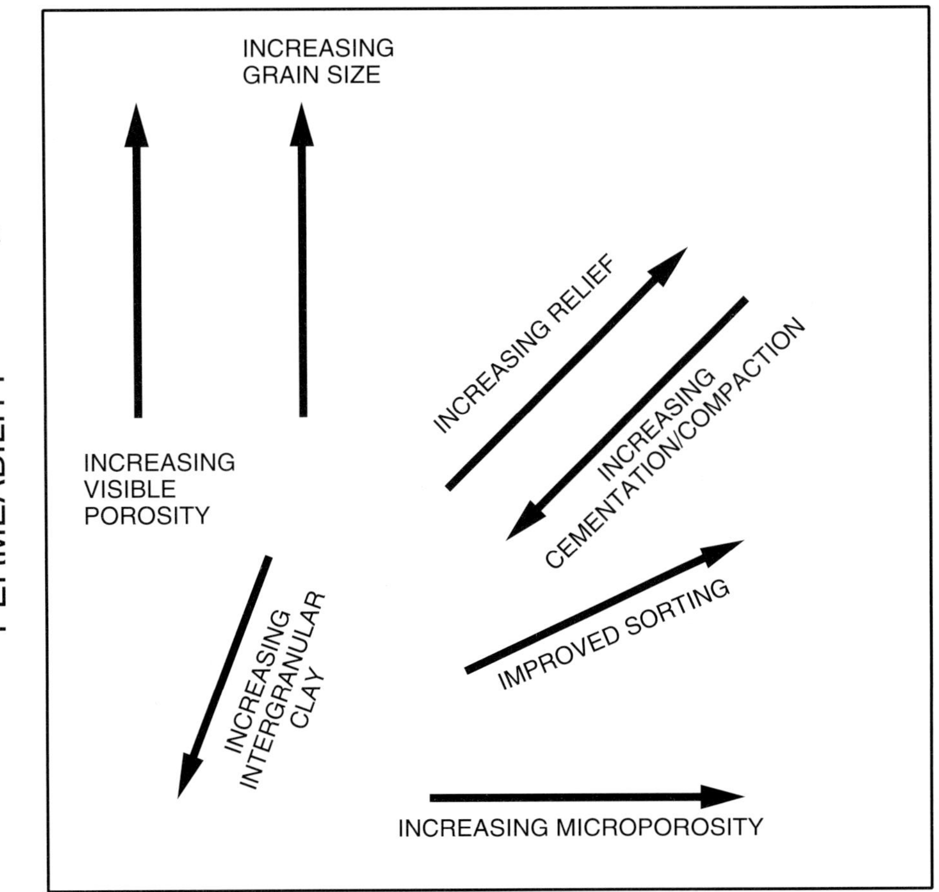

Figure 1. An interpretation of the effects of various controls on porosity and permeability (after Ethier and King, 1991).

Example 2—Mapping Permeability in Clyde Field, UKCS

The Clyde Field [United Kingdom Continental Shelf (UKCS)] contains oil in Upper Jurassic shallow marine sands. The field has been on production for some time, and infill drilling is currently in progress.

A study of the controls on porosity and permeability in the Clyde Field reservoir sandstone was performed in order to reduce the uncertainty associated with permeability prediction ahead of drilling infill wells and to improve the mapping of permeability between wells (J. Gluyas, 1995, personal communication).

Accurate prediction of permeability within the Clyde Field is difficult because in some reservoir zones permeability varies by up to 4 orders of magnitude for a given porosity. Porosity varies little within and between reservoir zones. Porosity and permeability display no correlation. There is, however, a good correlation between the maximum grain size and measured permeability from the core (Figure 4). For each reservoir layer that displays a grain size variation, there is a systematic and predictable trend across the field (e.g., Figure 5). Thus, the correlation between grain size and permeability can be used to predict permeability ahead of drilling or to map permeability in uncored areas for reservoir simulation purposes.

Despite the success of these empirical methods when dealing with a particular field, correlative techniques are limited by the need for pre-existing data. Also, since no insight is gained into the processes controlling permeability, there is no basis for extending predictions beyond the range of calibration data. Therefore, permeability prediction in areas with little or no well data requires another approach.

MODELING APPROACHES TO PERMEABILITY PREDICTION

The major factors controlling sandstone permeability are grain size, sorting, compaction, and cementation (Cade et al., 1994). Computer models may be used to help understand the effects of these parameters on permeability. The effects of compaction (vertical shortening of a rock volume) and cementation (various types of pore filling) have been modeled using a numerical representation of a real porous medium, a sphere-pack of like-sized, randomly packed grains (Bryant et al., 1993a, b; Cade et al., 1994). Using this approach, the porosity-permeability trends that result from the progressive application of various diagenetic processes, either on their own or in combination, can be understood. The effects of grain size on the porosity–permeability trends

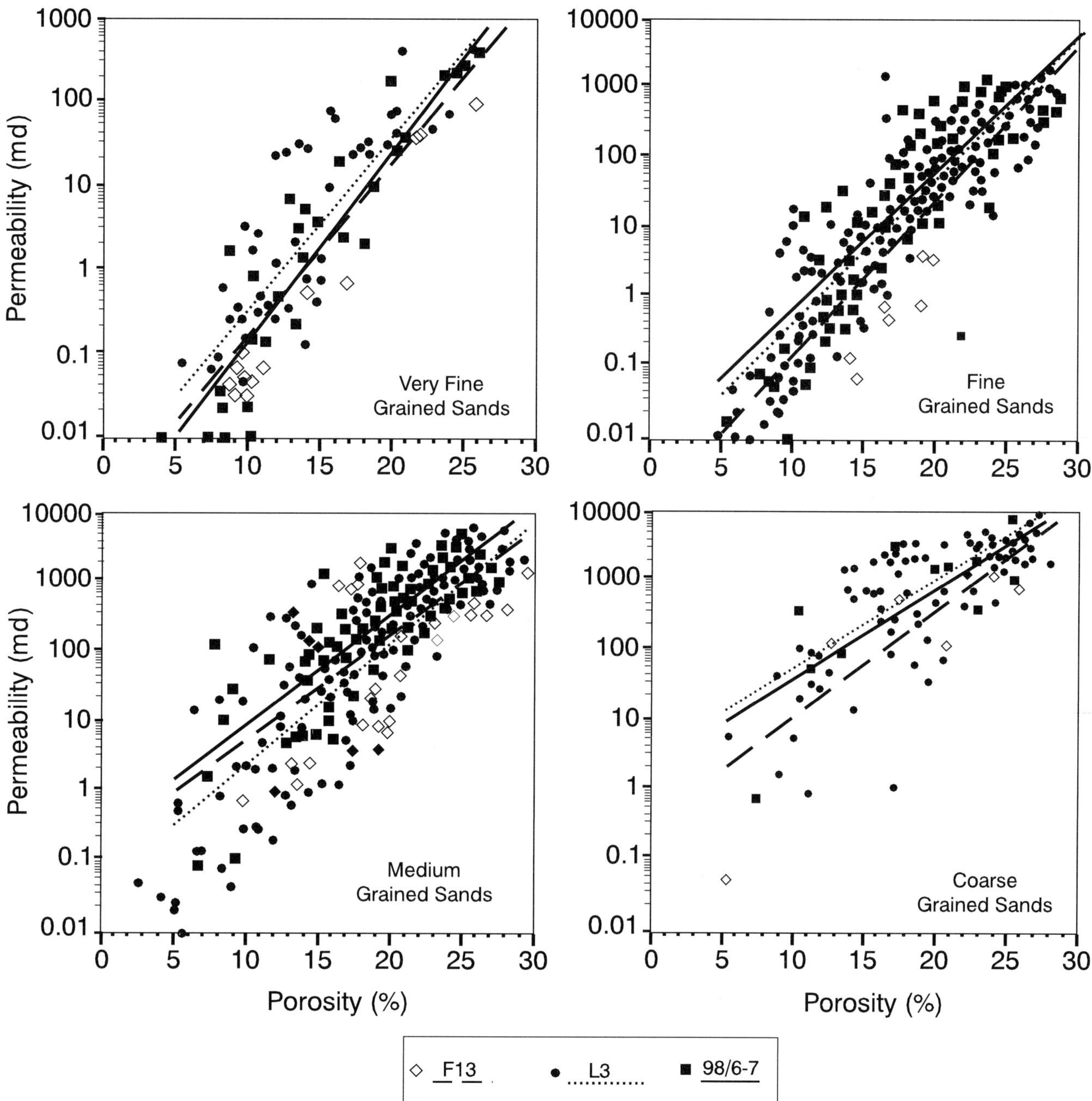

Figure 2. Variation in permeability with horizontal core plug porosity and grain size class in the Sherwood Sandstone Formation for three wells in the Wytch Farm field, onshore UK (after Hogg et al., 1996).

can also be modeled. The extension of this modeling to account for less than perfect sorting remains problematic (Panda and Lake, 1994, 1995). Clearly, sorting is of fundamental importance, but at present there is only an empirically based correction for sorting variation (Beard and Weyl, 1973). Despite this drawback, computer modeling will, in many cases, provide the basis for enhanced predictions of permeability in combination with predictions of texture, diagenetic modification, and porosity.

The "process-oriented" approach described by Bloch (1991) and Bloch and Helmold (1995) focuses on modeling diagenetic processes in an undrilled area, based on chemical and mathematical models, and the effects of those processes on reservoir quality. There are two important limitations to this approach: first, there is the uncertainty associated with the subsurface geological model, and how it impacts on the thermodynamics and kinetics of the models; second, there is the lack of a detailed quantitative understanding of how diagenetic processes control permeability. To date, the quantification of the impact of specific controls, particularly diagenetic controls, has been either formation specific (and therefore not widely applicable) or very general.

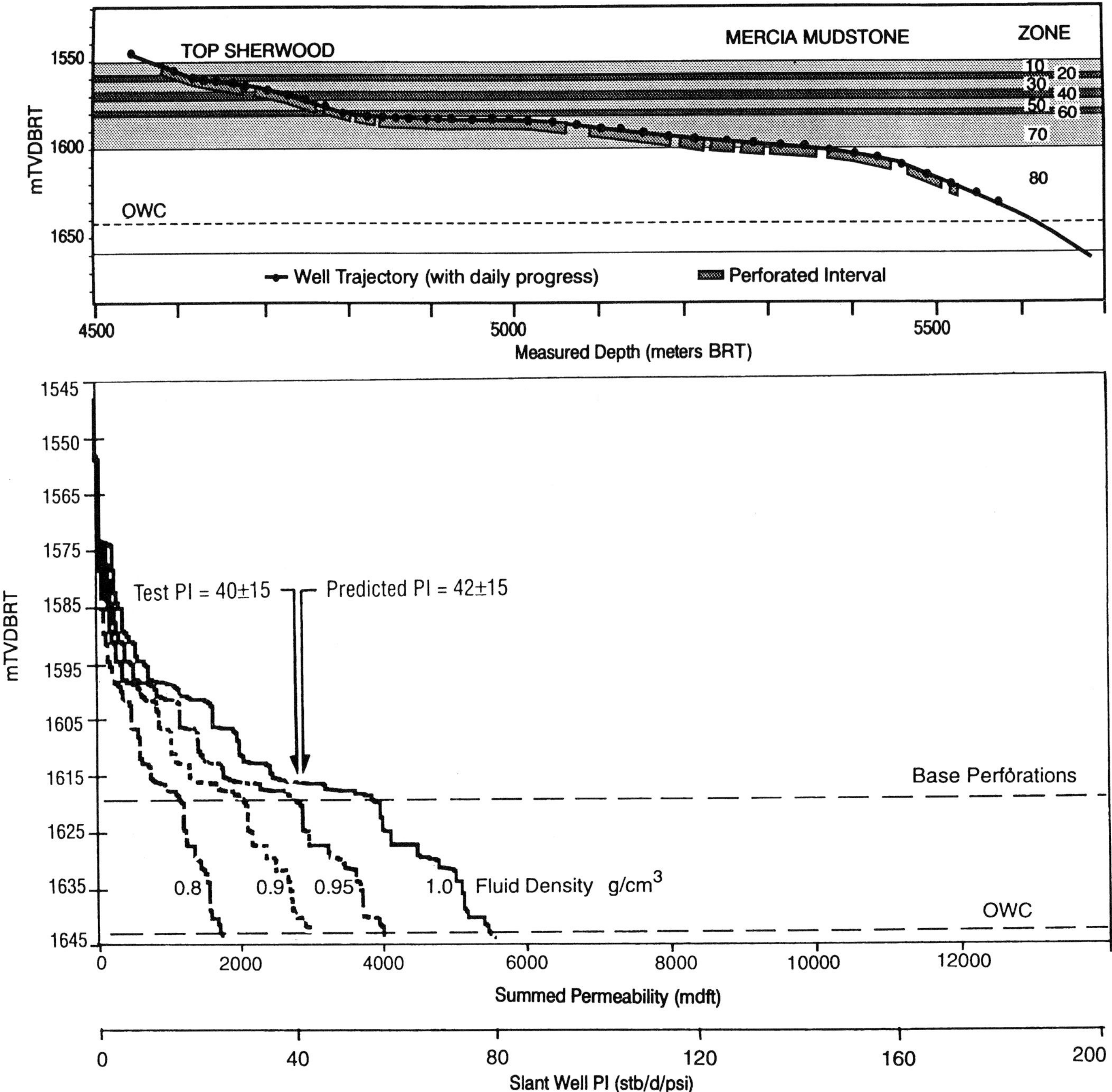

Figure 3. Predicted permeability–depth profiles and productivity index (PI) for well L98/6-F19 in the Wytch Farm field, onshore UK. Actual well trajectory and test results are shown for comparison (after Hogg et al., 1996). mTVDBRT = meters true vertical depth below rotary table; OWC = oil-water contact; brt = below rotary table; PI = productivity index; sbb/d/psi = stock tank barrels per day per psi.

APPLICATIONS OF THE MODELING APPROACH
Example 4—Frontier Exploration

Economic success in many frontier hydrocarbon basins is dependent on high oil production rates. Production rate is in turn controlled by the permeability of the reservoir rocks. In basins where potential reservoir targets are deep, reservoir quality is a key risk to discovering commercial hydrocarbon volumes. Conventionally, reservoir quality would be predicted based on comparison with analog basins or by calibration with nearby wells (see earlier discussion of empirical approaches). In frontier basins, however, few data may be available to help assess the depth to economic basement. In such cases, where it may be unclear which analogs are appropriate, a process-oriented approach to estimating the depth limit of effective permeability may be more appropriate.

Described below is a method to estimate the depth to economic basement that is based on geological models of porosity–depth relationships and porosity–permeability

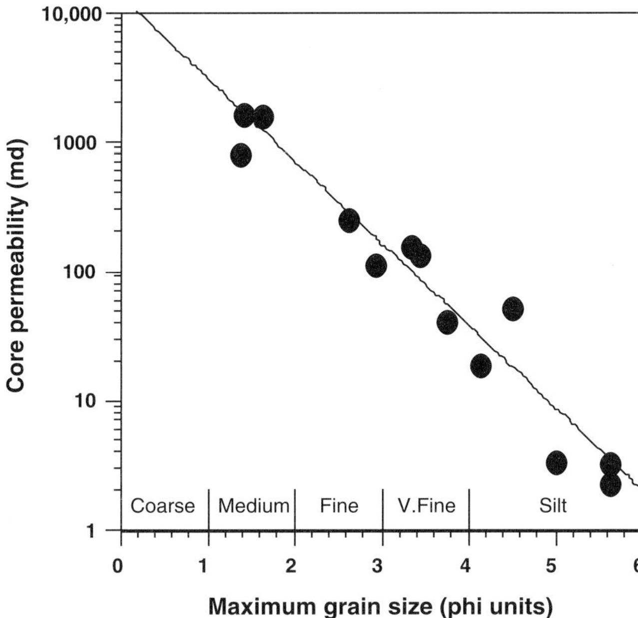

Figure 4. Correlation between maximum grain size in phi units ($-\log^2$ grain size in millimeters) and arithmetic mean core permeability for layer AL3 in Clyde Field, offshore UKCS.

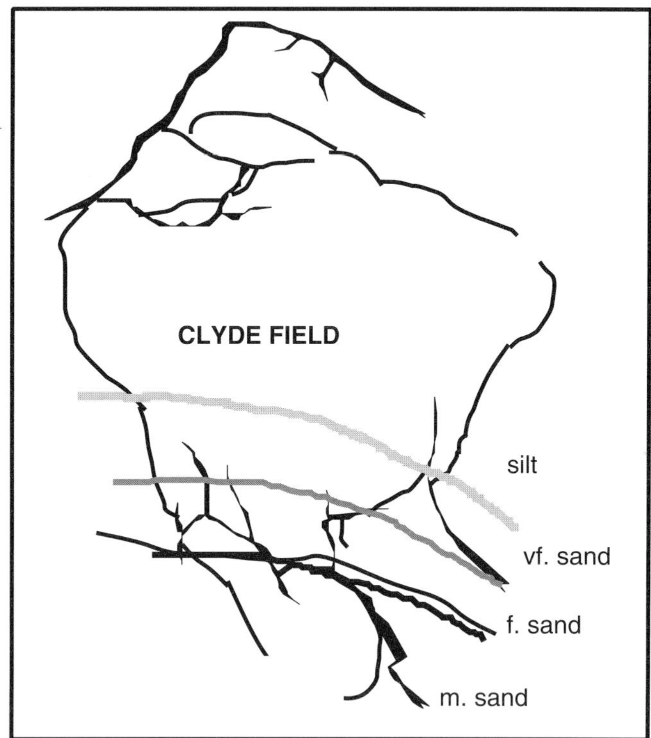

Figure 5. Variation of grain size within layer AL3 across Clyde Field, offshore UKCS. Grain size varies between medium sand in the south of the field to silt in the north. The grain size variations can be used to map permeability variations in this case. f = fine; m = medium; vf = very fine.

relationships (Bryant et al., 1993a; Gluyas and Cade, this volume). The method is illustrated by reference to a real example taken from a frontier basin in Southeast Asia.

To help in ranking prospects with different depths and containing sands with potentially different compositional maturity, "cutoff depths" below which reservoir quality will be insufficient to provide economic flow rates must be defined. Cutoffs have been calculated for a range of possible geological scenarios. The results provide a guide to what geological conditions will allow a given prospect to be economic. By assessing the chance of the necessary conditions occurring, the risk on reservoir quality can be estimated.

The likely range of geological variation was taken from the best and worst cases seen in nearby wells. Best case: Uncemented, coarse-grained pure quartz sand. Worst case: Very fine grained sand containing 25% ductile lithic grains and cemented by 20% quartz cement. Based on required flow rates and likely reservoir thickness, economic permeability cutoffs were given as 18 md for gas and 45 md for oil at surface conditions. At reservoir conditions, these are estimated to reduce by a factor of 3, because of overburden pressure and relative permeability effects.

The general method of calculating depth cutoffs is straightforward. Using the economic permeability cutoff as a starting point, we must calculate what the porosity of the rock was after it had been compacted but before any cements started to form (this porosity value is known as the compactional porosity and gives an upper limit to the porosity in normally pressured rocks). Using this porosity, the cutoff depth is calculated from compaction curves.

For uncemented rocks, the reservoir porosity (or "cutoff porosity") is equal to the compactional porosity. However, for rocks containing cement, the compactional

porosity is calculated by adding the volume of cement to the cutoff porosity.

The depth cutoffs were calculated using the following approach:

1. Construct the permeability-porosity curve for the relevant grain size, sediment composition (i.e., clean or ductile-bearing), and cement type.
2. Using the appropriate permeability cutoff (18 or 45 md), read the equivalent porosity (Figure 6).
3. Using the compactional porosity obtained above, calculate the cutoff depth using appropriate compaction curves (Figure 7).

The results of cutoff depth calculations for two most likely geological cases are presented in Tables 1 and 2. The values shown in the cutoff tables assume that the reservoirs are normally pressured. The cutoff depths will increase by ~550 m for every 1000 psi of overpressure (assuming overpressure was present before cements formed).

Prior to drilling prospects in this basin, reservoir effectiveness was perceived to carry a high risk, because ductile-rich sands were considered likely. Temperature gradients are such that quartz cements are likely to be present at depths greater than ~3500 m.

Since these original predictions were made, several wells have penetrated these potential reservoirs. The sandstones have proved to be ductile rich. The porosities are consequently low (15%–20% at 3000 m) and the

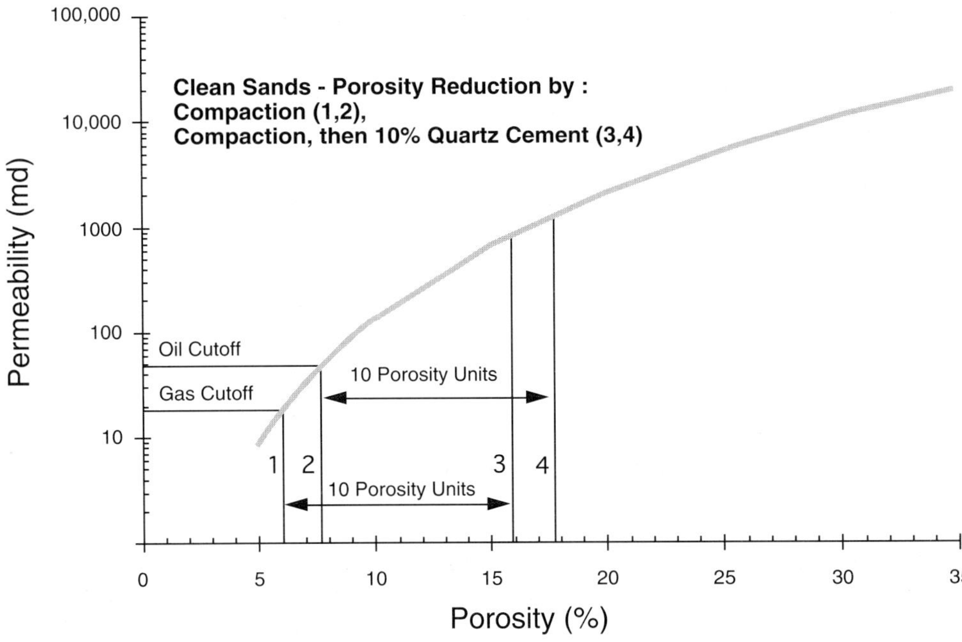

Figure 6. Example curve for calculating porosity cutoffs for a given permeability in clean sands in which porosity reduction is by compaction and/or quartz cementation. In this case, permeability cutoffs of 18 and 45 md (gas and oil) are equivalent to porosity cutoffs of 6% and 7.5%. In the case where the rocks contain quartz cement, the amount of cement (here 10 vol%) must be added to the cutoff porosity in order to determine the "compactional porosity," which is used to derive the depth cutoff for a given permeability (Figure 7).

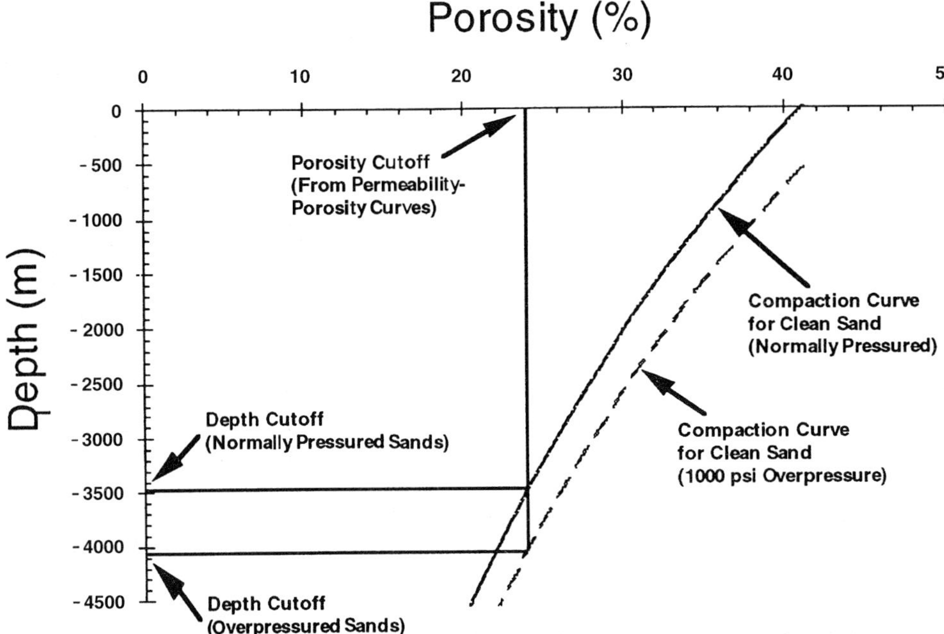

Figure 7. Example of how to determine depth cutoffs for a given permeability. Using the porosities determined in Figure 6, the equivalent cutoff depth may be found by reference to a clean sand compaction curve (Gluyas and Cade, this volume).

permeabilities are subeconomic below 3000–3500 m (Worden et al., 1996).

Example 5—Appraisal of Unconsolidated Reservoirs

Unconsolidated reservoirs are common in many parts of the world (e.g., Gulf of Mexico, North Sea, and West Africa). There are particular problems in determining the permeability of such reservoirs. Cores are difficult to obtain, and core analysis results are often unreliable due to "repacking" of grains. Also, dynamic testing is difficult, because the formations may collapse during flow stimulation. However, unconsolidated reservoirs are usually uncemented, so porosity and permeability are largely controlled by compaction.

Permeability may be accurately predicted if porosity and grain size are known or may be predicted.

The Harding Field (offshore UKCS) has an unconsolidated sandstone reservoir. During early appraisal, a knowledge of reservoir permeability was critical to determining the economic viability of the field. Two conflicting permeability estimates were available from different sources. Well-test data suggested a permeability of ~9–10 darcys, whereas core analysis data implied lower values of ~3–4 darcys. The choice of permeability value had implications for the time to water breakthrough and the optimum height in the reservoir at which to drill a horizontal well.

Using grain size measurements from sieve analysis of core samples (Table 3), together with porosity

Table 1. Case 1: Pure Quartz Sand, 10% Quartz Cement Porosity Reduction by Compaction, Then Quartz Cement.*

Grain Size, Sorting	Cutoff Permeability (K_L)	Cutoff Porosity (%)	Cutoff Depth (m)
vfs, mod. sorted	18 md	12	3000
	45 md	15	2200
fs, mod. sorted	18 md	7.5	>5000
	45 md	9.5	>5000
ms, mod. sorted	18 md	5	>5000
	45 md	6.5	>5000
cs, mod. sorted	18 md	3.5	>5000
	45 md	4.5	>5000

*cs = coarse sand; fs = fine sand; ms = medium sand; vfs = very fine sand.

derived from logs (estimated to be 34%–35%), the predicted porosity-permeability relationships of Bryant et al. (1993a) were used to estimate the likely permeability of the reservoir (Figure 8). This was then corrected to reservoir conditions (an overburden correction of ~1.5 was applied) to allow comparison with the test data. The permeability estimated in this way was 9–12 darcys (range due to grain size variations), which was much closer to the test results than to the core analysis data. The test data were therefore used for reservoir simulation, and the core analysis results were discarded.

Subsequently, it was discovered that the core analysis results had been in error. The analytical method had not been appropriate for such high-permeability samples, so that, in effect, the permeability of the test apparatus itself had been measured. More careful measurements were made, which confirmed the permeability modeling and well-test data.

CONCLUSIONS

Through recognizing the important controls on permeability in clastic rocks, namely, grain size, sorting, compaction, and cementation, it is possible to predict permeability in undrilled areas by application of geological models. In areas where there are existing core data, empirical methods that relate permeability to other predictable parameters (e.g., grain size variation) will give good results. However, in areas away from well control or in fields where the controls on permeability are complex, predictions based on geological models combined with permeability modeling results are likely to give better results. A combination of empirical and geological modeling approaches will often give the best results, even in areas where there is abundant data (see examples in Cade et al., 1994). The quantitative insight into the way in which different factors affect permeability, which has been provided by computer modeling (Bryant et al., 1993a; Panda and Lake, 1995), allows us to focus on the most important controlling factors. These are often grain size and the amount of cement or ductile grains. Our ability to predict permeability in undrilled areas is now more often hampered by our inability to predict these controls rather than by any lack of understanding of permeability itself. This shifts the emphasis back to the sedimentologists and geologists to better constrain their geological models so that the uncertainty in the possible range of permeabilities may be reduced.

Table 2. Case 2: Sand Containing 25% Ductiles, No Cement, Porosity Reduction Solely by Compaction.*

Grain Size, Sorting	Cutoff Permeability (K_L)	Cutoff Porosity (%)	Cutoff Depth (m)
vfs, mod. sorted	18 md	19.5	2650
	45 md	21	2350
fs, mod. sorted	18 md	17	3200
	45 md	18	2950
ms, mod. sorted	18 md	15.5	3600
	45 md	16.5	3350
cs, mod. sorted	18 md	14.5	3950
	45 md	15	3750

*cs = coarse sand; fs = fine sand; ms = medium sand; vfs = very fine sand.

Table 3. Grain Size Data Determined by Sieve Analysis of Unconsolidated Core Samples.

Sieve Data	9/23b-7 (Harding Central)	9/23b-8 (Harding South)
Mean grain size (µm)	243	240
Median grain size (µm)	243	238
Sorting	mod. well or well	mod., mod. well, or well

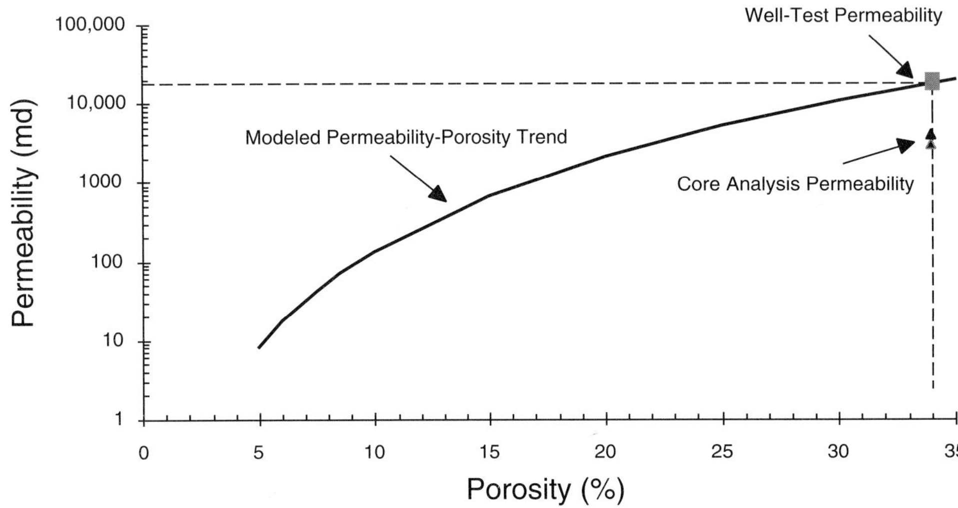

Figure 8. Modeled porosity–permeability curves for sands in the Harding Field (offshore UKCS) based on sieve data. Well-test data (corrected to surface conditions) and routine core analysis data are shown for comparison. The modeling results confirm that the test data are valid.

ACKNOWLEDGMENTS

Thanks to BP Exploration for permission to publish these data and ideas. Our understanding of permeability prediction has evolved over the past five years through interaction and discussions with many past and present colleagues. In particular, we would like to thank Craig Smalley, Tony Mitchell, Andrew Brayshaw, Sue Raikes, Dave Mellor, Harit Trivedi, Ed Warren, Andrew Hogg, Tim Primmer, Shona Grant, Jon Gluyas, Norman Oxtoby, Richard Worden, Kevin Schofield, and Mike Bowman.

REFERENCES CITED

Ahmed, U., S.F. Crary, and G.R. Coates, 1991, Permeability estimation: the various sources and their interrelationships: Journal of Petroleum Technology, v. 42, p. 578–587.

Beard, D.C., and P.K. Weyl, 1973, Influence of texture on porosity and permeability of unconsolidated sand: AAPG Bulletin, v. 57, p. 349–369.

Bloch, S., 1991, Empirical prediction of porosity and permeability in sandstones: AAPG Bulletin, v. 75, p. 1145–1160.

Bloch, S., and K.P. Helmold, 1995, Approaches to predicting reservoir quality in sandstones: AAPG Bulletin, v. 79, p. 97–115.

Bryant, S.L., C.A. Cade, and D.W. Mellor, 1993a, Permeability prediction from geological models: AAPG Bulletin, v. 77, p. 1338–1350.

Bryant, S.L., D.W. Mellor, and C.A. Cade, 1993b, Physically representative network models of transport in porous media: American Institute of Chemical Engineers Journal, v. 39, p. 387–396.

Cade, C.A., J. Evans, and S.L. Bryant, 1994, Analysis of permeability controls: a new approach: Clay Minerals, v. 29, p. 491–501.

Dutton, S.P., and T.N. Diggs, 1992, Evolution of porosity and permeability in the Lower Cretaceous Travis Peak Formation, East Texas: AAPG Bulletin, v. 76, p. 252–269.

Ehrlich, R., E.L. Etris, D. Brumfield, L.P. Yuan, and S.J. Crabtree, 1991, Petrography and reservoir physics III: physical models for permeability and formation factor: AAPG Bulletin, v. 75, p. 1579–1592.

Ethier, V.G., and H.R. King, 1991, Reservoir quality evaluation from visual attributes on rock surfaces: methods of estimation and classification from drill cuttings or cores: Bulletin of Canadian Petroleum Geology, v. 39, p. 233–251.

Gluyas, J.G., and C.A. Cade, this volume, Prediction of porosity in compacted sands, in J.A. Kupecz, J. Gluyas, and S. Bloch, eds., Reservoir quality prediction in sandstones and carbonates: AAPG Memoir 69, p. 19–28.

Hogg, A.J.C., A.W. Mitchell, and S. Young, 1996, Predicting well productivity from grain-size analysis and logging while drilling: Petroleum Geoscience, v. 2, p. 1–15.

Klinkenberg, L.J., 1941, The permeability of porous media to liquids and gases: Drilling and Production Practices, API, Dallas.

Panda, M.N., and L.W. Lake, 1994, Estimation of single-phase permeability from parameters of particle size distribution: AAPG Bulletin, v. 78, p. 1028–1039.

Panda, M.N., and L.W. Lake, 1995, A physical model of cementation and its effects on single-phase permeability: AAPG Bulletin, v. 79, p. 431–443.

Worden, R., M. Mayall, and J. Evans, 1996, The effect of lithic grains on porosity and permeability in Tertiary clastics, South China Sea: Journal of the Geological Society of London.

Ehrlich, R., M.C. Bowers, V.L. Riggert, and C.M. Prince, 1997, Detecting permeability gradients in sandstone complexes—quantifying the effect of diagenesis on fabric, *in* J.A. Kupecz, J. Gluyas, and S. Bloch, eds., Reservoir quality prediction in sandstones and carbonates: AAPG Memoir 69, p. 103–114.

Chapter 8

Detecting Permeability Gradients in Sandstone Complexes—Quantifying the Effect of Diagenesis on Fabric

Robert Ehrlich
Department of Geological Sciences, University of South Carolina
Columbia, South Carolina, U.S.A.

Mark C. Bowers
Conoco Incorporated
Houston, Texas, U.S.A.

Virginia L. Riggert
Amoco Production Co.
Denver, Colorado, U.S.A.

Christopher M. Prince
Department of Geological Sciences, University of South Carolina
Columbia, South Carolina, U.S.A.

ABSTRACT

Matrix permeability, the permeability associated with measurements on small samples, is controlled by depositional fabric and diagenesis. Prediction of matrix permeability requires: (1) specification of a fabric, (2) specification of the diagenetic state, and (3) a means to assess both factors in a sample set taken from a target basin. The data from the sample set can be used to extrapolate or interpolate within the basin or may be used to calibrate fabric response to basin history data (e.g., thermal history). The effects of fabric and diagenesis on the sample set can be determined using a combination of image analysis data and mercury porosimetry data.

Strong correlations exist between permeability and grain size of unconsolidated sands and gravels, with permeability increasing exponentially with increasing grain size. Permeability in clastic fabrics is controlled by networks of packing flaws, characterized by large pores connected by large pore throats. Such circuits comprise only a fraction of the porosity and represent the effective flow component of porosity. Diagenesis usually brings about permeability reduction, but preferentially affects the grains in close-packed arrangements that separate the networks of packing flaws. A methodology has been developed over the past decade that quantifies thin-section–based data precisely enough to estimate the effects of grain size and diagenesis on the rock fabric with respect to flow properties. Such rock physics data are necessary for permeability prediction as a function of basin position.

INTRODUCTION

Reservoir-scale permeability prediction requires knowledge of many properties at many scales and, therefore, requires a multidisciplinary team that includes petrologists, stratigraphers, basin modelers, and others. Because reservoir-scale permeability predictions are ultimately derived from matrix permeability measurements (usually associated with small volumes of rock), it is important to understand how matrix permeability varies within a basin. The objective of this chapter is to discuss how matrix permeability prediction can be used as a lead-in to reservoir-scale permeability prediction. We discuss methods for prediction of the highest permeability possible that might be encountered as a function of basin location, given a particular fabric, because only a small fraction of the porosity of a sandstone contributes to permeability (Ehrlich et al., 1991b; Prince et al., 1995). We show how the porosity components that most influence permeability can be identified, and how the rate of change of the size of associated pore throats with respect to depth (or basin location) can then be determined.

Prediction of matrix permeability (hereafter referred to simply as permeability) must take into account both the depositional fabric and the diagenetic modification of that fabric. Most of the permeability contrasts observed in a single core arise from grain size variation, because all samples have, to a first approximation, a common postdepositional history resulting in a common diagenetic state. Increasing diagenesis alters or obscures the relationship between depositional fabric and permeability, but it never completely erases it. Significant permeability contrasts observed in a single core are commonly associated with grain size variation because all samples have a common postdepositional history resulting in a common diagenetic state. Given rock of similar composition, diagenesis varies spatially as a response to gradients in pressure, temperature, and fluid chemistry. That is, individual components of diagenesis exist in the form of a diagenetic gradient; in theory, this component can be mapped over a basin. In our experience, such gradients are common in sandstones where reduction in permeability is largely due to factors such as quartz overgrowth development and/or compaction and pressure solution.

Intergranular pore shape is affected by the shapes of the bounding sand grains, as well as whether they are located in close-packed or loose-packed domains. Intergranular pore size is a function of grain size, modified according to whether the pore exists in close-packed (relatively smaller) or loose-packed (relatively larger) domains. Pore types can be defined as a population of pores with a characteristic size and shape. An objective quantitative porosity classification into pore types that is rapid and precise can be achieved by using image analysis procedures described in Ehrlich et al. (1991b). The automated classification is consistent with conventional classification, while easily capturing differences in size, shape, and type of porosity (intergranular, intragranular, and moldic). Image analysis breaks down porosity complexes into as many pore types as demanded by variations in depositional fabric (including grain size) or differential effects of diagenesis. Reservoirs commonly contain four to seven pore types, depending on grain size variability (Horkowitz, 1987; Ehrlich et al., 1991b; Bowers, 1992; Murray et al., 1994; Riggert, 1994), whereas individual samples are usually dominated by one or two pore types.

Permeability is strongly dependent on the efficiency of the intergranular porosity, of which moldic and intragranular porosity generally contribute little. Intergranular porosity covers a wide range of subtypes; the quantitative objective characterization of these subtypes is crucial to permeability prediction. Intergranular porosity falls into two types: that found in close-packed domains and that found in loose-packed domains (Graton and Fraser, 1935; Prince et al., 1995). These types can be expressed in a variety of ways, depending on grain size and sorting. Loose-packed porosity ("packing flaws") has large-scale spatial continuity and is associated with larger pore throat sizes. These two factors make loose-packed domains the major contributors to permeability (McCreesh et al., 1991; Anguy et al., 1994; Prince et al., 1995).

THE EFFECTIVE COMPONENT OF POROSITY

Permeability is independent of porosity in unconsolidated sands. Under progressive diagenesis, a rough relationship between porosity and permeability may arise. However, at the median porosity in a sample set, permeability commonly varies by more than 2 orders of magnitude. Among samples with the same permeability, porosity may vary by more than 10 porosity units (e.g., 10%–20%) (Figures 1A, 2A). Much of this scatter is due to the effect of variations in grain size, with porosity being preferentially reduced in finer grained sandstones.

Subsets of porosity are much more highly correlated with permeability than core (bulk) porosity, suggesting that some parts of the pore system do not support much

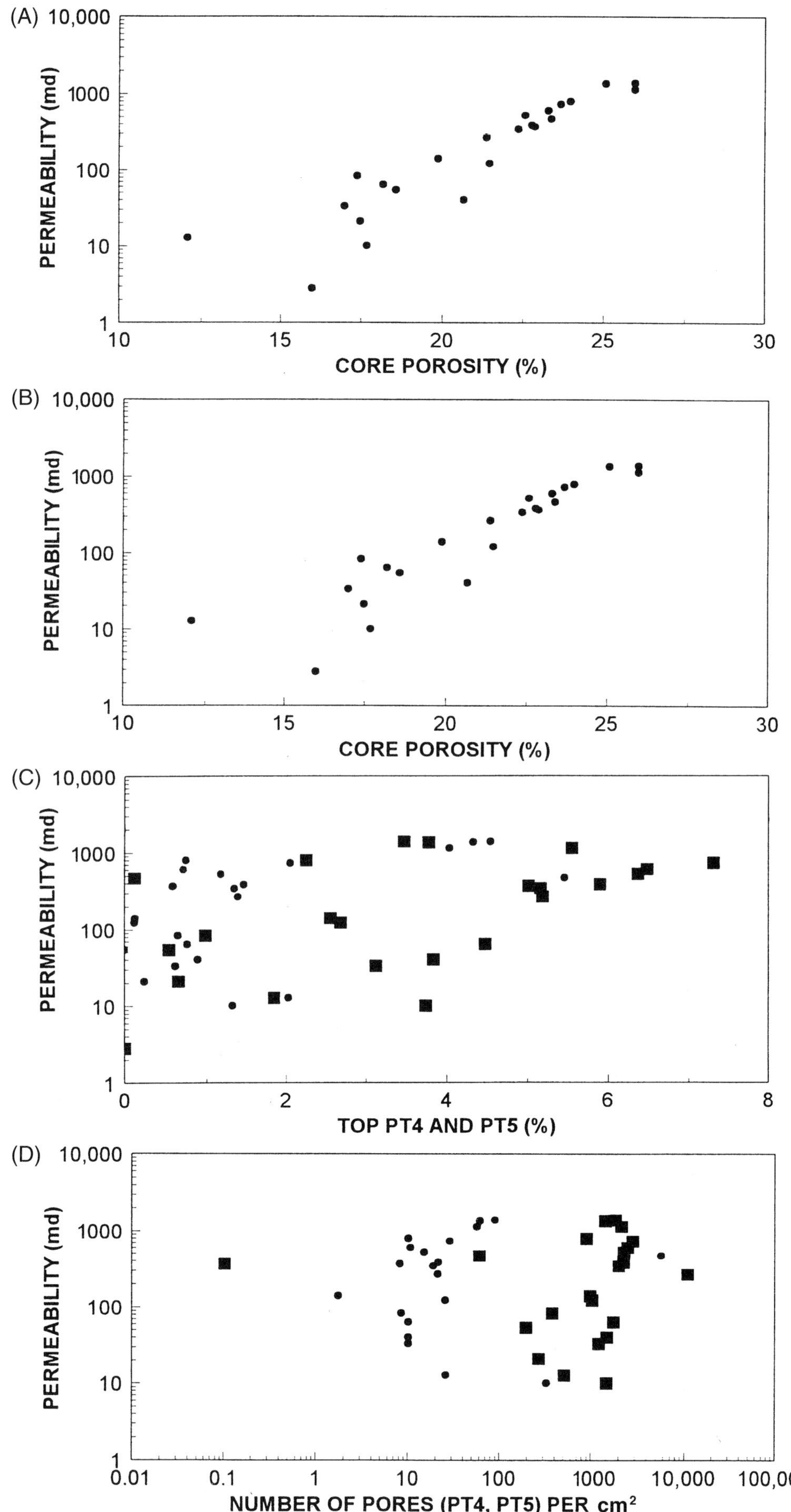

Figure 1. Relationships between permeability and subsets of porosity for Miocene sandstones, Gulf of Thailand: (A) core porosity, (B) total optical porosity (TOP), (C) TOP portion consisting of PT4 (pore type 4) and PT5, and (D) number of pores per unit area of PT4 and PT5. Squares represent PT4 and dots represent PT5.

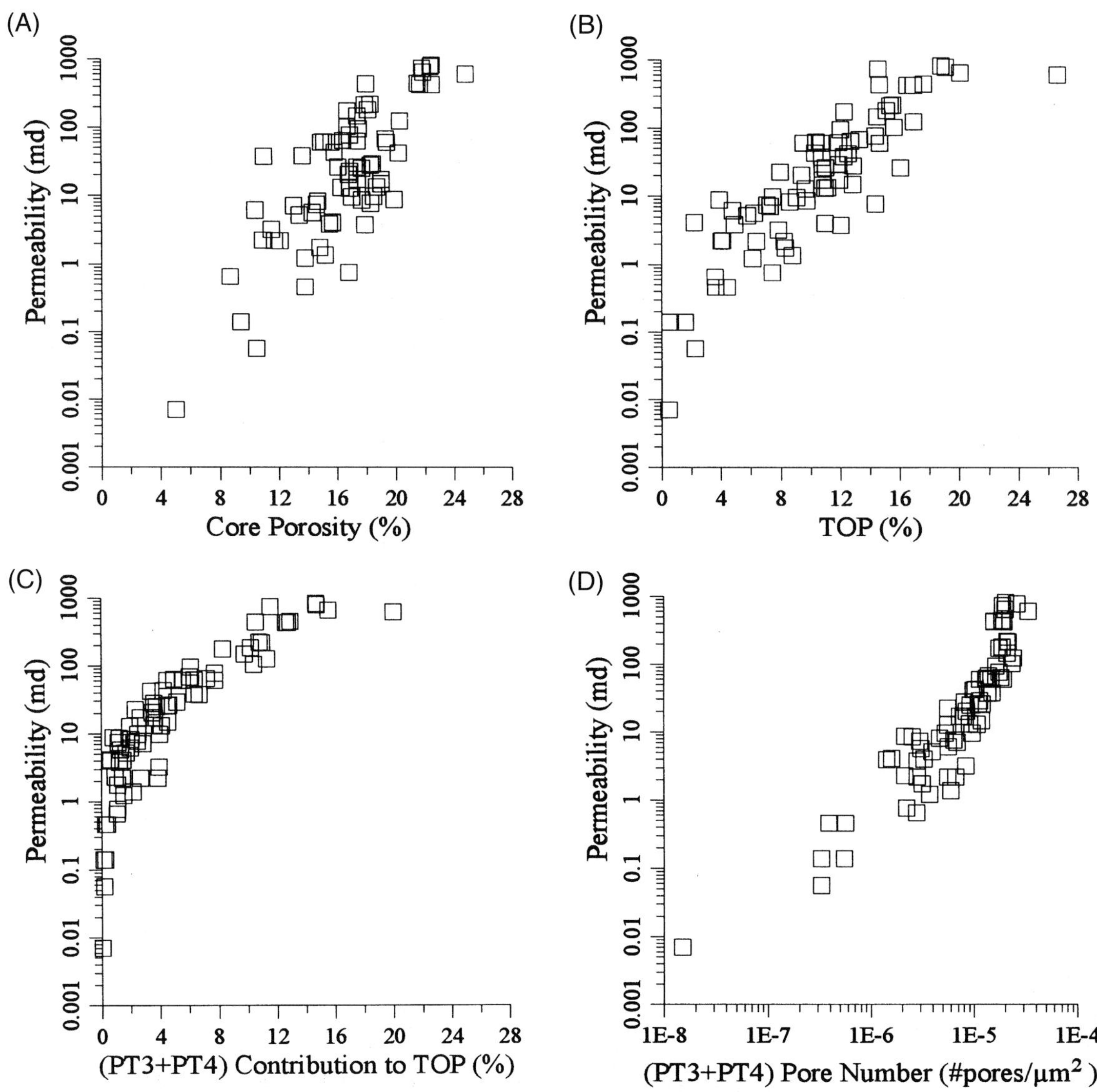

Figure 2. Relationships between permeability and subsets of porosity for Upper Carboniferous sandstones in Oklahoma: (A) core porosity, (B) total optical porosity (TOP), (C) TOP portion consisting of PT3 and PT4, and (D) number of pores per unit area of PT3 and PT4.

flow, even at the scale of matrix permeability. This can be illustrated by observing the increase in correlation between permeability and successive refinements of porosity (Figures 1, 2), using image analysis of petrographic thin sections. Digital image analysis at low magnification (<100×) will commonly yield a value of porosity [total optical porosity (TOP)] less than the measured value because small pores and small-scale roughness on pore walls cannot be resolved. Total optical porosity, however, always correlates more highly with permeability than does core porosity (Figures 1B, 2B); TOP can be subdivided into portions associated with each pore type. The amount of TOP associated with one or more pore types (the product of the relative proportion of a pore type and TOP) is more highly correlated with permeability (Figures 1C, 2C) than is either TOP or core porosity. The number of pores per unit area of these pore types is also highly correlated with

permeability (Figures 1D, 2D). The high correlation between certain pore types and permeability implies that such pores must be connected by relatively large throat sizes. The throat sizes associated with pores of each type can be quantified by relating the pore type data with mercury-injection porosimetry data.

PORE TYPES AND THROAT SIZE

The amount of porosity lying behind pore throats of various sizes can be determined from mercury-injection capillary pressure tests. McCreesh et al. (1991) found that, based on statistical analysis, different pore types tend to control different portions of the capillary pressure curves; that is, different pore types tend to fill in different pressure ranges. This can occur only if pores of like type are mutually adjacent, forming circuits characterized by a common throat size. Recently, Prince et al.

(1995) optically resolved such circuits using filtered Fourier transforms of large, high-resolution images; Riggert (1994) demonstrated the existence of these flow circuits by analysis of suites of capillary pressure curves.

The pore type–throat size relationship is based on a set of regression analyses in which the pore type proportions are used to predict the amount of saturation in successive pressure intervals from the mercury-injection curves (McCreesh et al., 1991). One equation is produced for each pressure interval that relates the pore type proportions to the mercury saturation. The set of equations is then used to calculate the distribution of throat sizes and mean throat size associated with each pore type.

PORE TYPES AND PERMEABILITY

Ehrlich et al. (1991a) demonstrated how the association between pore type and throat size can be used to model permeability. Using a modified Hagen-Poiseuille version of Darcy's Law, they showed that permeability is proportional to the product of the number of pores of each type per unit cross-sectional area and the fourth power of the associated pore throat size. The model is based on the assumption that the flow paths are relatively straight and parallel, and the number of effective throats is proportional to the number of pores. Pore throats at high angles to the pressure gradient are ineffective. With this model, the amount of permeability contributed by each pore type can be determined.

The modified Hagen-Poiseuille permeability model is effective over a range from <1 md to several darcys. In sandstones with permeabilities >20 md, commonly only one or two pore types contribute most of the permeability. The spatial rate of change of the throat size of such dominant pore types can be determined from a reference set of cores, and that function can be used to provide an estimate of the maximum permeability that may be encountered for such rocks.

GRAIN SIZE AND PERMEABILITY IN LITHIFIED SANDSTONE

Pore typing automatically takes into account both depositional and diagenetic effects, because any change in grain size and any diagenetic event affects the size and/or shape of a pore. Grain size and packing are the major depositional properties at perm plug scale. The relationship between grain size and permeability is, to our knowledge, only documented for unconsolidated sands (Shepard, 1989). Shepard stated that permeability is an exponential function of grain size, with the relationship proportional to grain size raised to a power ranging from about 1.3 to 2 in. well-sorted sands. Lithification commonly reduces the permeability at all grain sizes, but little is known concerning the relationship between permeability in consolidated sandstones.

Because investigations in lithified sandstones have not been published, some researchers have assumed that the relationship remains exponential after lithification. One problem in verifying this assumption is the determination of grain size in an indurated rock,

because three problems have had to be overcome: (1) many sandstones are not friable enough to disaggregate effectively; (2) acquisition of overgrowths bias direct measurements on quartz grains observed in thin section; and (3) measurements of grain size in thin section are generally biased by the fact that a grain may not be cut by the plane of section near its diameter. Therefore, the distance from grain center to grain boundary represents an apparent grain size, because the distance of magnitude is influenced by the location of the plane of section relative to the grain "equator," by overgrowth development, and by pressure solution. Prince et al. (1995) used a two-dimensional fast Fourier transform on thin-section–scale binary images to quantify the spatial fabric of sandstones. They pointed out that the center-to-center distances between pores are a good approximation of the distances between grain centers unaffected by the biases mentioned above.

A common assumption is that the grain size–permeability relationship observed in unconsolidated sandstones by Shepard (1989) also holds true for lithified sandstones. That is untrue, as shown by work done by C.M. Prince et al. (personal communication) on a small portion of their data for a Carboniferous sandstone (Perry Sandstone) from the Cherokee Basin, Oklahoma. Grain sizes range from ~100 to 250 µm, with permeabilities ranging from ~0.5 to 500 md. Results shown in Figure 3 indicate that extrapolations of the grain size–permeability relationships in unconsolidated sands should be used with caution when trying to characterize well-cemented sandstones in samples of low permeability (<10 md). This is especially true where diagenesis has produced patchy fabrics; no grain size–permeability relationship exists (Figure 3). The Perry Sandstone data also indicate the grain size–permeability relationship is exponential for samples with uniformly altered fabrics (e.g., rimming cements, quartz overgrowths), with an exponent much greater than that observed in unconsolidated sands (5 in the consolidated Perry Sandstones vs. 1.3–2.0 in unconsolidated sands) (Figure 3). The Perry Sandstone is similar to many Paleozoic sandstones we have studied; it is expected studies of other sandstones will verify this result.

PREDICTING PERMEABILITY

Permeability values observed among a set of samples may not include the maximum value likely to occur, because of incomplete sampling, incomplete core recovery, or the well bore missing the maximum development of porosity in a depositional subfacies. Given the Hagen-Poiseuille permeability model (Ehrlich et al., 1991b), however, permeability can be calculated for a series of "synthetic" rocks that can contain pore type proportions and porosities exceeding those in the sample set, but falling within plausible limits, as discussed in the example from the Pattani and Cherokee basins.

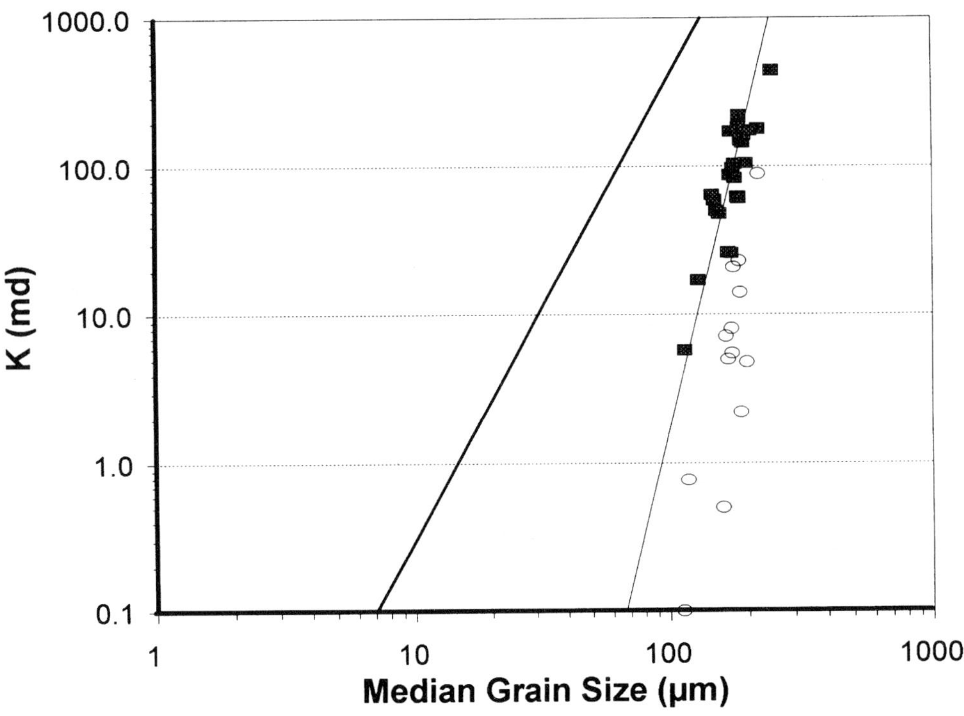

Figure 3. Relationship between grain size and permeability of the Perry Sandstone; open circles represent samples with patchy carbonate cement; squares represent samples with uniformly distributed quartz overgrowths. Slope of the regression line through the uniformly affected sample yields an exponent of ~5 compared with exponents in the range of 1.3 to 2 reported by Shepard (1989) (solid line to left of data).

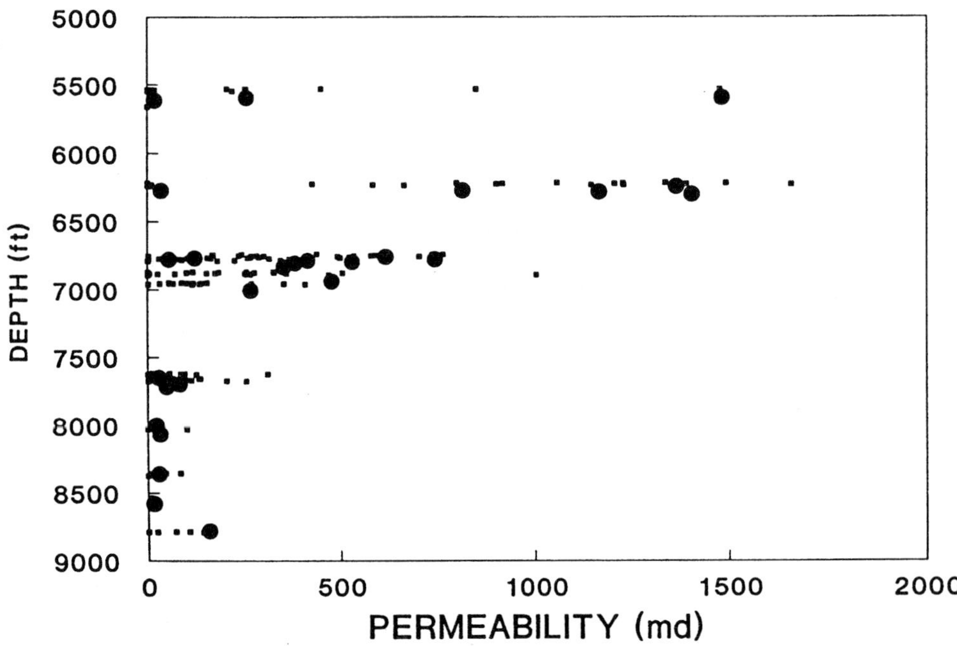

Figure 4. Relationship of permeability to depth in Miocene sandstones, Gulf of Thailand. Squares represent samples with permeability and porosity data only ($n =$ 197). Dots represent the subset of samples with mercury-injection and image analysis data.

Diagenetic changes are responsible for the changes in throat size associated with each pore type. Reduction in throat sizes associated with the loose-packed circuits in sandstones in the following examples can be of two types: (1) progressive development of quartz overgrowths can gradually reduce the throat size (the circuits remain intact) or (2) local patches of diagenetic carbonate or clay can plug the circuits. Both situations can be modeled, with the case involving intact circuits providing the most optimistic picture. As discussed in the following examples, the reduction in throat size associated with intact circuits can change smoothly as a function of depth, defining a diagenetic gradient, which in turn controls the maximum permeability that may be encountered.

Example 1: Satun Field, Pattani Basin, Gulf of Thailand

An extensive coring program in the Satun field, Pattani Basin in the Gulf of Thailand provided the opportunity to sample Miocene sands over a depth range of ~1000 m (Bowers et al., 1994). The sandstones share a

PORE TYPE 1

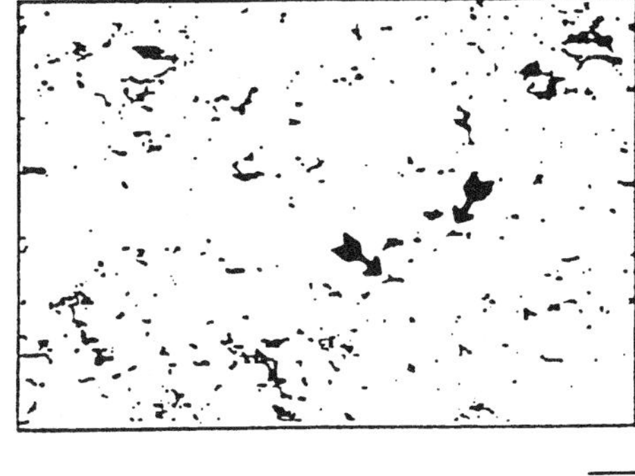

PORE TYPE 2

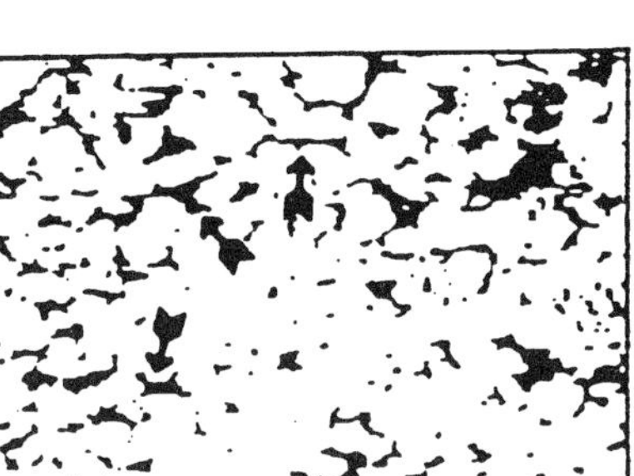

PORE TYPE 3

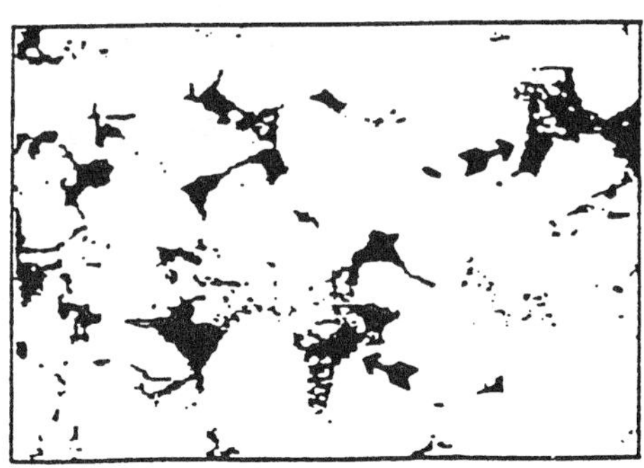

PORE TYPE 4

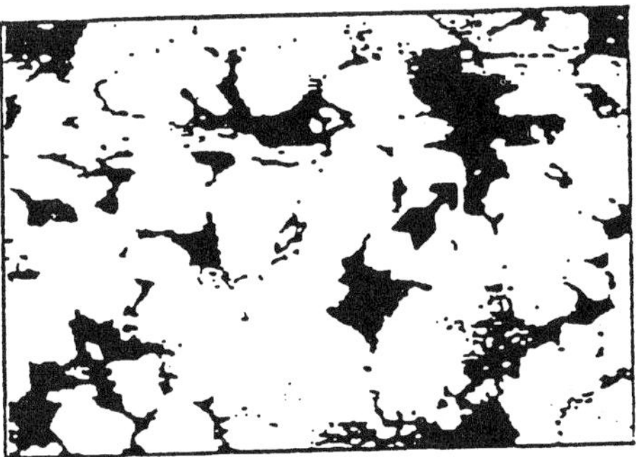

PORE TYPE 5

Figure 5. Pore types derived for the Miocene sandstones, Gulf of Thailand. Five pore types were derived, ranging in size from 19 to 160 μm in diameter. See text for a complete description of each pore type. Horizontal line equals 100 μm; arrows indicate an example of each pore type.

common depositional environment and have approximately the same grain size. Maximum permeability decreases with depth from >1 darcy (6000 ft) to <10 md (8500 ft) (Figure 4). The samples come from a "hot" basin where a high geothermal gradient (4.0–5.0°C/100 m) is associated with a diagenetic gradient as feldspars are progressively destroyed as a function of depth (Travena and Clark, 1986). Quartz overgrowth development and kaolinitization developed in step with feldspar dissolution. A hypothesis of the study was that detecting the diagenetic gradient would be reflected in the pore type–pore throat size relationship as a function of depth. As discussed below, this was the case, but in an unanticipated mode.

Image analysis data were linked with permeability and mercury-injection data by Bowers et al. (1994),

using the procedures described earlier in this chapter and detailed in Ehrlich et al. (1991a, b) and McCreesh et al. (1991). They derived five pore types (Figure 5). Pore type 1 (PT1; mean diameter of 19 μm) is the smallest pore type and occurs as cuspate to triangular shaped porosity elements in thin section. Pore type 2 (PT2; mean diameter of 37 μm) represents the surviving remnants of intergranular porosity bounded by quartz overgrowths. Pore type 3 (PT3; mean diameter of 53 μm) is associated with kaolinite, which can reduce the effective throat size. The two largest pore types are types 4 and 5 (PT4 and PT5) and have mean diameters of 76 μm and 160 μm, respectively. These two pore types occur as discrete patches of porosity surrounded by a more compact fabric and overgrown grains. Pore types 4 and 5 are associated with

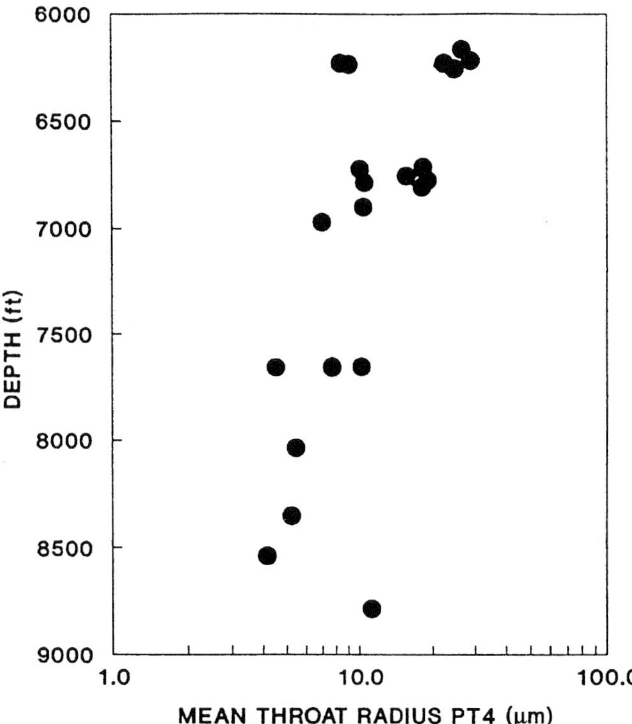

Figure 6. Relationship of mean throat radius of pore type 4 (PT4) with depth in Miocene sandstones, Gulf of Thailand. The throat radius associated with PT4 decreases log-linearly with depth.

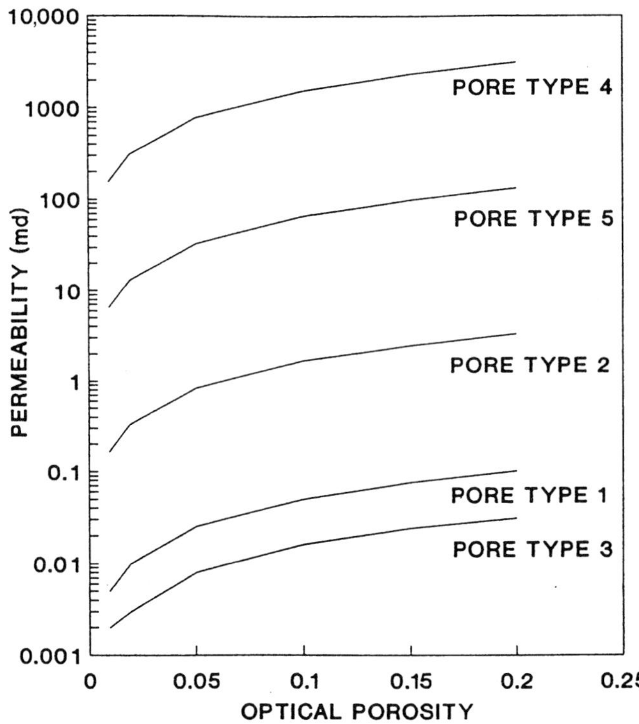

Figure 7. Relationship between optical porosity and permeability for rocks composed entirely of a single pore type in Miocene sandstones, Gulf of Thailand. The relationships were calculated from the permeability model assuming throat sizes associated with the most permeable zone (5560–6240 ft). Note that only PT4 and PT5 can account for values of permeability >1 md.

enhanced permeability (Figure 1C, D). In most samples, PT4 is the major carrier of permeability; PT5 was too low in abundance to significantly affect permeability in all but a few samples.

Permeability modeling using the Hagen-Poiseuille model shows that fabric dominated by PT1, PT2, and PT3 (conventional intergranular porosity) accounts for little permeability throughout the depth range. Low-permeability samples anywhere in the sampled depth range are dominated by these pore types. Therefore, the effectiveness of the conventional intergranular porosity has been impaired throughout the depth range. Pore types 1, 2, and 3 are associated with kaolinite throughout the depth range, indicating that the diagenetic gradient observed by Travena and Clark (1986) is not coincident with the gradient in the conventional intergranular porosity. The diagenetic gradient is reflected in the distribution of PT4 and PT5, however.

Pore types 4 and 5 are unusual in that they occur in thin section as patches of large pores completely surrounded by PT1, PT2, and PT3. Although isolated in section view, porosity of pore types 4 and 5 must be connected in the third dimension to account for the observed relationships between pore type and pore throat size. PT4 and PT5 contain little, if any, kaolinite, indicating that these pores were not in existence during the period of kaolinite formation. Samples with PT4 and PT5 also exhibit a less compacted fabric than the fabrics dominated by PT1, PT2, and PT3. These characteristics, coupled with the patchy distribution of the large intergranular pores associated with PT4 and PT5, led Bowers

et al. (1994) to interpret these pore types as representing the product of a late-stage dissolution of an early patchy carbonate cement. While no vestige of this cement occurs in the sampled rocks, carbonate cement is abundant in sandstones shallower in the sequence.

Throat sizes associated with PT4 decrease log-linearly over the sampled depth range (Figure 6), accounting for the observed reduction in maximum permeability with respect to depth. If a pore type is assumed to exist at any depth, the maximum permeability likely to occur at any given depth can be calculated. Using that assumption, the relative effectiveness of each pore type can be illustrated by creating "synthetic" rocks, each with the same porosity and containing a single pore type. Bowers et al. (1994) constructed such models containing throat sizes appropriate for shallow depths (Figure 7). They concluded that permeability values >1 darcy are possible only in the presence of PT4. Less than 10% of PT4 would ensure permeability values >100 md; PT4 permeability efficiency decreases with depth as its throat size decreases. Bowers et al. (1994) showed that similar values of PT4, yielding a permeability of 1 darcy at a depth of 6000 ft; would account for a permeability of ~100 md at 7000 ft, and only about 10 md at ~8000 ft (Figure 8). These values are in agreement with the maximum measured permeabilities over that depth range.

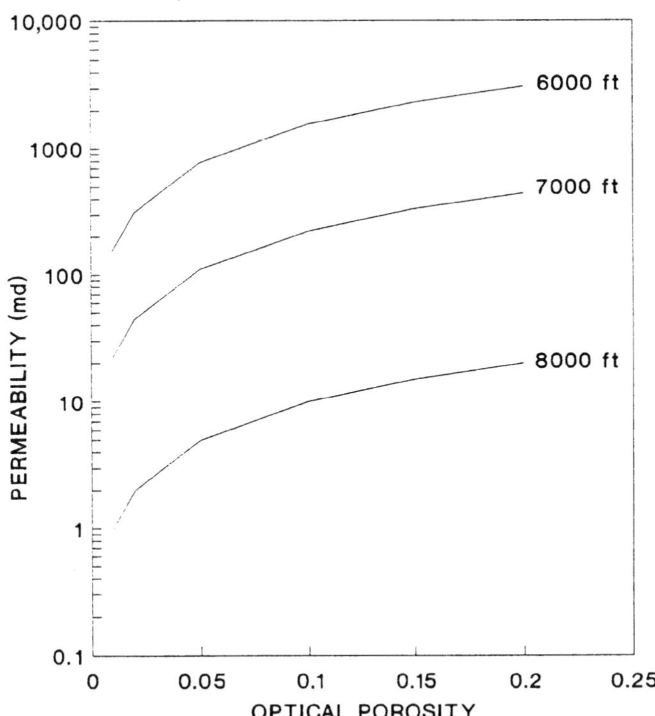

Figure 8. Relationship between optical porosity and permeability of modeled rocks composed of PT4 with throat radii appropriate for depths of 6000 ft, 7000 ft, and 8000 ft for Miocene sandstones, Gulf of Thailand.

Example 2: Upper Carboniferous Sandstones, Cherokee Basin, Oklahoma

An industry–university consortium was organized to evaluate measurement-while-drilling tools in vertical and deviated boreholes in the Cherokee Basin in North Central Oklahoma (Hutchinson, 1991). A by-product of this multidisciplinary investigation was porosity, permeability, density, and other analyses of more than 1000 plugs from a core taken in the vertical borehole. The core spanned 2700 ft of Permian and Upper Carboniferous sedimentary rocks. Of the plugs taken by the consortium, Riggert (1994) selected 73 samples, spanning >1000 ft in four Upper Carboniferous (Missourian and Virgilian) sandstones. The sandstones are medium to very fine grained, quartz-rich sandstones with subsidiary amounts of feldspar and lithic fragments. Patchy carbonate cement occurs in all samples and can be a major factor in permeability reduction. The samples come from a "cold" basin with a low geothermal gradient (<1.5°C/100 m).

Using image analysis, Riggert (1994) determined that four pore types were sufficient to account for essentially all of the petrographic variability in these sandstones. All pore types represent intergranular pores of various sizes and shapes and are illustrated in Figure 9. (Note: Pore type numbers refer to the relative sizing of optical porosity types within an individual reservoir. The largest pore type is identified by the largest pore type

number. A pore type with same number designation in one reservoir is not related to the same numerically labeled pore type in another reservoir.) Pore type 1 has a mean diameter of 14 µm and is characterized by compact intergranular pores. Pore type 2 has a mean diameter of 19 µm and is characterized by small elongated intergranular pores. Pore types 3 and 4 are the two largest pore types and have mean diameters of 39 µm and 79 µm, respectively; they are associated with loose-packed domains and are the primary agents for enhanced permeability. In some samples, the presence of these pore types does not ensure enhanced permeability because circuits associated with these pore types are blocked by carbonate cement. When this occurs, the pore throat radii associated with PT3 and PT4 decrease from 6–10 µm to <3 µm.

Using only samples with open circuits, permeability values are depth related, because the throat sizes of PT3 and PT4 decrease with depth. From shallowest to deepest, throat radii of these two pore types are ~10 µm in the Hoover Sandstone, ~8 µm in the Elgin Sandstone, ~6 µm in the Perry Sandstone, and ~5 µm in the Layton Sandstone. Assuming a maximum core porosity of 20% and a maximum relative proportion of PT3 of 50%, the maximum permeabilities for the sandstones in this sequence can be calculated (Figure 10A). From this relationship, the maximum permeability in the sandstone between the Elgin and Perry sandstones, the Tonkawa [sampled by the consortium, but not by Riggert (1994)] can be interpolated and compared with measurement (Figure 10B).

DISCUSSION AND CONCLUSIONS

Permeability is dependent on grain size, packing, sorting, and diagenetic state. In the absence of diagenesis, permeability prediction becomes an exercise in predicting depositional fabric with respect to basin location (i.e., facies distribution and burial history analysis). Holding depositional fabric constant, permeability varies in response to changes in diagenetic state.

Quantifying the diagenetic state relevant to permeability prediction is difficult to impossible at present because diagenesis is a combination of the effects of physical and chemical processes. Many geochemical changes involve the physical redistribution of phases. However, diagenetic processes invariably affect the porosity: pores change in size and shape, pore throat sizes change, or the relationship between pore sizes and pore throat sizes changes. Therefore, these characteristics (which are quantifiable) can be used for a direct characterization of diagenetic state (which is not quantifiable) for other rocks in the basin.

Most of these changes in the sizes of pores and pore throats can be detected by using the procedures of pore type determination described in Ehrlich et al. (1991b), and for relating pore types to pore throat size described by McCreesh et al. (1991). Building on this, permeability can be partitioned among pore types by using the methods described by Ehrlich et al. (1991a); the porosity elements responsible for enhanced permeability can be identified by this method. Examples

PORE TYPE 1

PORE TYPE 2

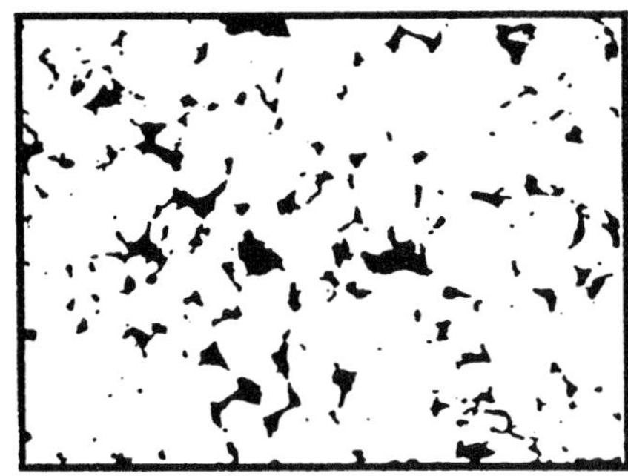

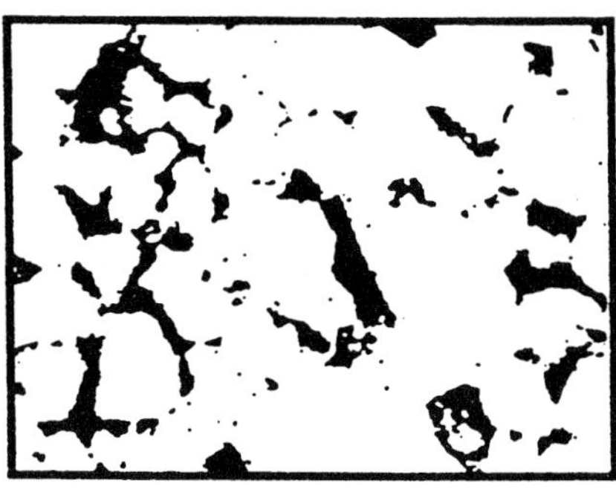

PORE TYPE 3

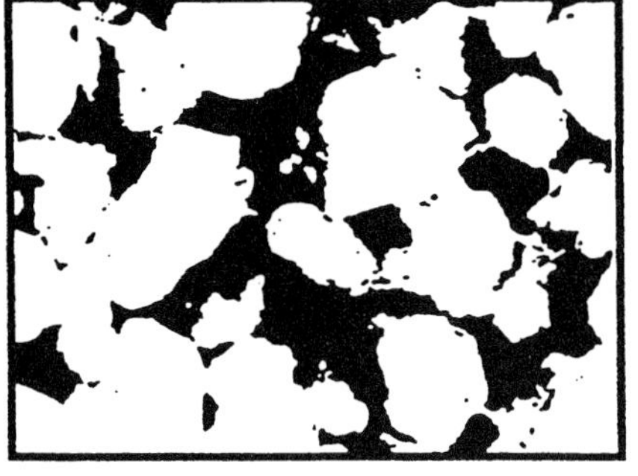

PORE TYPE 4

Figure 9. Pore types derived for the Upper Carboniferous sandstones of Oklahoma. Four pore types were derived, ranging in size from 14 to 79 μm in diameter. See text for a complete description of each pore type. Each view is 1075 × 832 μm.

discussed in this chapter showed that pore throat sizes associated with such pore types vary smoothly as a function of depth in two basins, one with a high geothermal gradient and one with a low geothermal gradient. The maximum permeability associated with these fabric types can be interpolated within the depth range. With care, extrapolation may be made to greater depths and laterally away from well control.

The results described here do not require that the maximum permeability be measured in available core, only that the pore types associated with permeability enhancement be present in a few samples. This means that the maximum measured permeability may be unrepresentative of what may be possible in a well bore; higher permeabilities may be encountered with additional drilling. On the other hand, the maximum likely permeability may be that which was measured; if that permeability is below the economic threshold permeability, reservoir quality may be too low for exploitation.

Our results until now describe changes with respect to depth. A logical next step is to attempt permeability prediction using data taken from reservoirs covering a wider areal extent. A potential shortcut is an attempt to relate basin history models (especially those incorporating heat flow over time) with the diagenetic gradient expressed by the changes in the pore throat size of the pore type that carries the majority of the permeability.

An unresolved aspect of permeability prediction is the degree of reduction of average permeability as a function of diagenesis. Decreased average permeability independent of grain size is generally the product of detrital mineral composition, burial depth, overpressure, temperature, and cement type. In the case of patchy carbonate cements, they progressively isolate portions of the loose-packed porosity that are responsible for much of the permeability in unaffected sandstones. The early stages of such mineralization may be benign; Prince et al. (1995) observed that there is a preference for mineralization of close-packed domains, accounting for the reduction of pore throat sizes compared to those in the loose-packed domains. However, given a great enough chemical potential, such mineralization can overcome the effect of fabric structure and reduce the permeability by blocking the loose-packed circuits.

In our experience, permeability reduction is commonly associated with the progressive nucleation and growth of patchy carbonate cement. We do not know at this time whether the degree of this kind of cementation has a large-scale spatial (basinal) component or whether it is essentially controlled by local factors. In the case of the Oklahoma sandstones, carbonate cementation becomes more pervasive with depth. Schmidt and McDonald (1979) report a tendency for cementation in some basins to decrease with depth, so

A

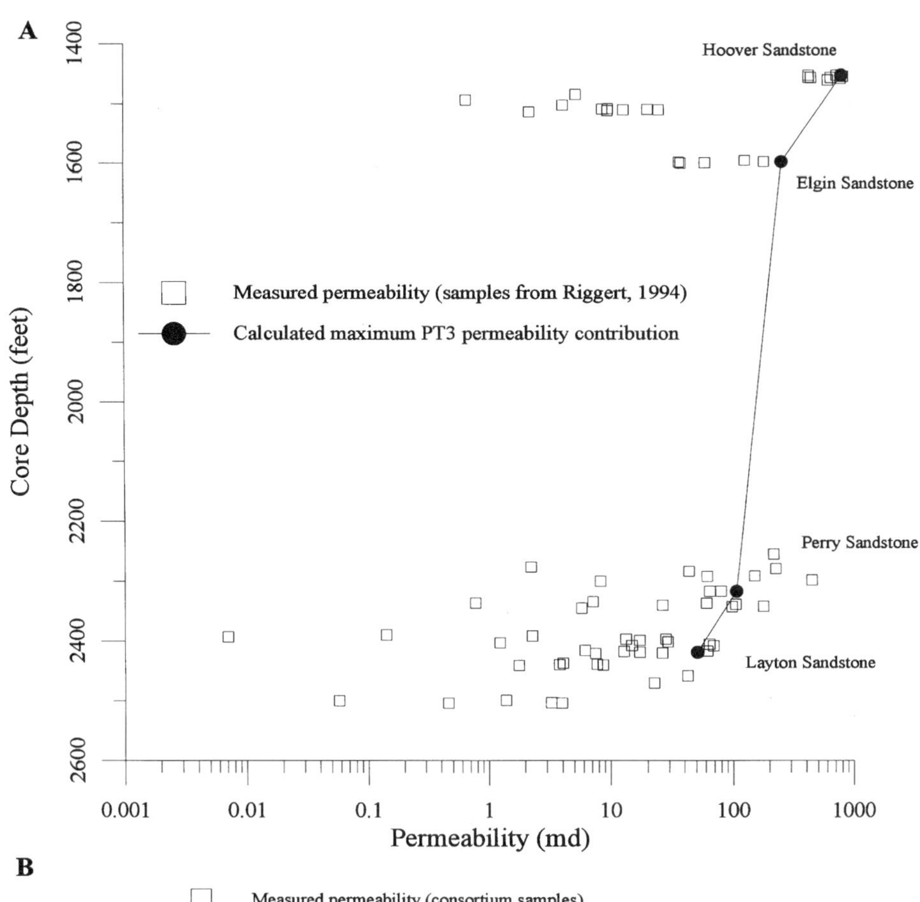

B

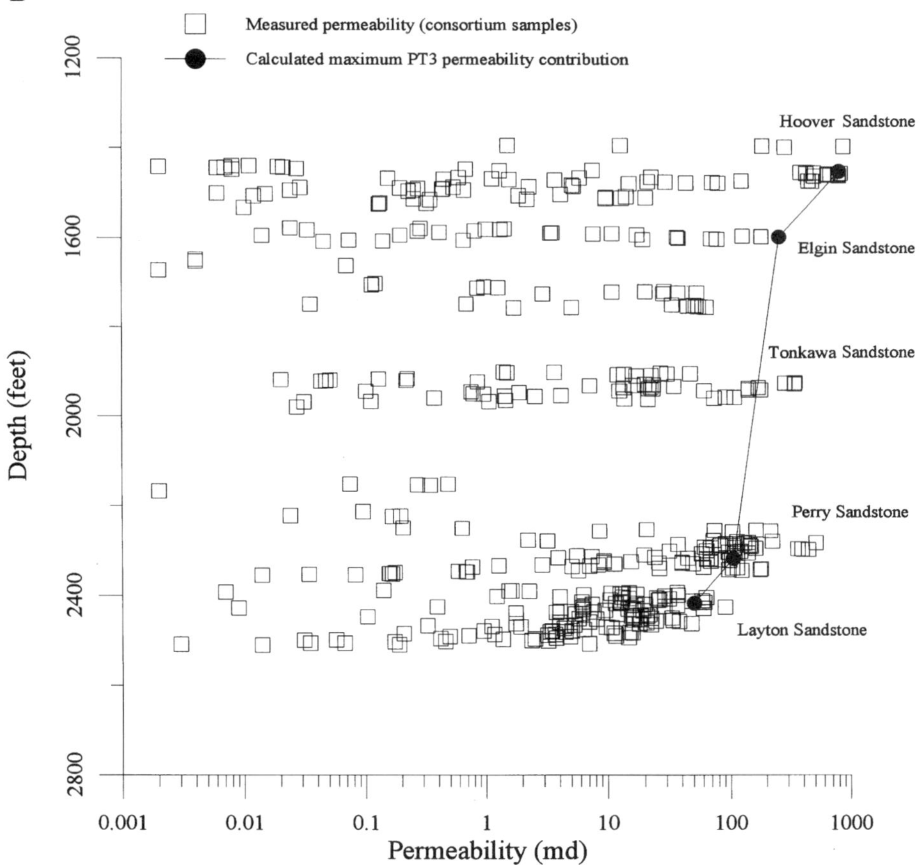

Figure 10. Relationship between permeability and depth, Upper Carboniferous sandstones, Oklahoma. Squares represent measured values; solid dots and the connecting line represent maximum permeability calculated according to the permeability model. (A) Samples taken by Riggert (1994); (B) samples taken by the industry–academic consortium in a similar depth interval ($n = 431$).

there is some hope that such cementation can be understood and quantified, permitting estimates of average permeabilities. In addition, as shown in the example from the Gulf of Thailand, relatively late stage dissolution of this cement can restore permeability after the conventional fabric is rendered ineffective by diagenesis. The possibility of such a restoration of porosity and permeability must be kept in mind in order not to exclude from consideration reservoirs whose conventional intergranular porosity would be predicted not to support high values of permeability.

The gradients described in this chapter imply a continuity of history from shallow to deep strata. Unconformities, faults, and other barriers may cause discontinuities in the rate of change of throat sizes and will make extrapolation more complicated, but still possible with the techniques described.

ACKNOWLEDGMENTS

This manuscript has been significantly improved by reviews from Jon Gluyas, Andy Bradshaw, Julie Kupecz, and Sal Bloch.

REFERENCES CITED

Anguy, Y., R. Ehrlich, C.M. Prince, V.L. Riggert, and D. Bernard, 1994, The sample support problem for permeability assessment in sandstone reservoirs, *in* J. M. Yarus and R.L. Chambers, eds., Stochastic modeling and geostatistics: AAPG Computer Applications in Geology 3, p. 37–54.

Beard, D.C., and P.K. Weyl, 1973, Influence of texture on porosity and permeability of unconsolidated sand: AAPG Bulletin, v. 57, no. 2, p. 348–369.

Bowers, M.C., 1992, The use of nuclear magnetic resonance, permeability and diffusion to characterize the porous microstructure of sandstones: Ph.D. thesis, University of South Carolina, Columbia, South Carolina, 152 p.

Bowers, M.C., R. Ehrlich, and R.A. Clark, 1994, Determination of petrographic factors controlling permeability using image analysis and core data, Satun Field, Pattani Basin, Gulf of Thailand: Marine and Petroleum Geology, v. 11, no. 2, p. 148–156.

Ehrlich, R., E.L. Etris, D. Brumfield, and L.P. Yuan, 1991a, Petrography and reservoir physics III: physical models for permeability and formation factor: AAPG Bulletin, v. 75, no. 10, p. 1579–1592.

Ehrlich, R., K.O. Horkowitz, J.P. Horkowitz, and S.J. Crabtree, 1991b, Petrography and reservoir physics I: objective classification of reservoir porosity: AAPG Bulletin, v. 75, no. 10, p. 1547–1562.

Evans, J.C., R. Ehrlich, D. Krantz, and W.E. Full, 1992, A comparison between polytopic vector analysis and empirical orthogonal function analysis for analyzing quasigeostrophic potential vorticity: Jour. Geophys. Res., v. 97, no. C2, p. 2365–2378.

Fraser, H.J., 1935, Experimental study of the porosity and permeability of clastic sediments: Jour. Geol., v. 43, no. 8, p. 910–975.

Full, W.E., R. Ehrlich, and J.E. Klovan, 1981, Extended QModel—objective definition of external end members in the analysis of mixtures: J. Math. Geol., v. 13, no. 4, p. 331–344.

Graton, L.C., and H.C. Fraser, 1935, Systematic packing of spheres with particular relation to porosity and permeability: Jour. Geol., v. 43, no. 8, p. 785–909.

Horkowitz, K.O., 1987, Direct and indirect control of depositional fabric on porosity, permeability, and pore size geometry: differential effect of sandstone subfacies on fluid flow, Cut Bank Sandstone, Montana: Ph.D. thesis, University of South Carolina, Columbia, South Carolina, 136 p.

Hutchinson, M.W., 1991, Comparisons of MWD, wireline and core data from a borehole test facility, 66th Annual Technical Conference and Exhibition: SPE Paper 22735, p. 741–754.

McCreesh, C.A., R. Ehrlich, and S.J. Crabtree, 1991, Petrography and reservoir physics II: relating thin section porosity to capillary pressure, the association between pore types and throat size: AAPG Bulletin, v. 75, no. 10, p. 1563–1578.

Murray, C.J., R. Ehrlich, E. Mason, and R. Clark, 1994, Evaluation of the diagenetic and structural influences on hydrocarbon entrapment in the Cardium Formation, Deep Basin, western Alberta: Bulletin of Canadian Petroleum Geology, v. 42, no. 4, p. 529–544.

Prince, C.M., R. Ehrlich, and Y. Anguy, 1995, Analysis of spatial order in sandstones II: grain clusters, packing flaws, and the small-scale structure of sandstones: Jour. Sed. Res., v. A65, no. 1, p. 13–28.

Riggert, V.L., 1994, Petrophysical relationships of pores and pore throats to spatial fabric elements in sandstones and their implications for fluid and electrical flow: Ph.D. thesis, University of South Carolina, Columbia, South Carolina, 192 p.

Schmidt, V., and D.A. McDonald, 1979, The role of secondary porosity in the course of sandstone diagenesis, *in* Aspects of Diagenesis: SEPM Special Publication 26, p. 175–207.

Shepard, R.G., 1989, Correlations of permeability and grain size: Groundwater, v. 27, no. 5, p. 633–638.

Travena, A.S., and R.A. Clark, 1986, Diagenesis of sandstone reservoirs of Pattani Basin, Gulf of Thailand: AAPG Bulletin, v. 70, p. 299–308.

Cabrera-Garzón, R., J.F. Arestad, K. Dagdelen, and T.L. Davis, 1997, Geostatistical simulation of reservoir porosity distribution from 3-D, 3-C seismic reflection and core data in the Lower Nisku Formation at Joffre Field, Alberta, *in* J.A. Kupecz, J. Gluyas, and S. Bloch, eds., Reservoir quality prediction in sandstones and carbonates: AAPG Memoir 69, p. 115–125.

Chapter 9

Geostatistical Simulation of Reservoir Porosity Distribution from 3-D, 3-C Seismic Reflection and Core Data in the Lower Nisku Formation at Joffre Field, Alberta

Raúl Cabrera-Garzón

John F. Arestad

Kadri Dagdelen

Thomas L. Davis
Department of Geophysics, Colorado School of Mines
Golden, Colorado, U.S.A.

ABSTRACT

Rock properties such as lithology and porosity can be obtained from comparative P- and S-wave traveltimes or velocities measured from multicomponent (3-D, 3-C) seismic reflection data. A 3-D, 3-C seismic reflection data survey was acquired by the Colorado School of Mines Reservoir Characterization Project at Joffre field, Alberta, to map the complex porosity distribution in a shelf carbonate reservoir. Velocity ratio analysis, of compressional velocity to shear velocity (Vp/Vs), indicates a linear correlation with porosity in the Devonian Nisku reservoir. Vertical porosity distribution at wells and horizontal porosity distribution derived from seismic reflection data are used to map 3-D porosity distribution using geostatistical methods. The results show enhanced mapping of porosity distribution and better definition of the lateral limits of the reservoir. These results will assist in reservoir simulation of this field.

INTRODUCTION

An accurate determination of the spatial distribution of porosity is key to understanding and predicting petroleum reservoir performance. The information that can be used to characterize porosity distribution is diverse. For instance, well information provides good vertical resolution; however, it gives poor horizontal resolution due to the large separation between wells. On the other hand, seismic reflection data provide high horizontal resolution but lower vertical resolution than does well information. Geostatistical tools are useful in relating different types of rock property measurements, such as wireline logs, core measurements, and seismic reflection data, to provide models that describe the spatial distribution of the properties being estimated. Significant porosity differences occur in the lower Nisku interval at Joffre field, where a 3-D, 3-C seismic reflection data survey was acquired to provide P- and split shear wave data [(fast) S1 and (slow) S2] . Shear wave splitting is considered to occur due to differential horizontal stress, fracturing, and pore shape elongation. Conventional P-wave seismic reflection data have been ineffective for porosity characterization of this reservoir. Geostatistics is used to analyze the diverse geophysical data, to develop useful relationships among the different types of information, and to characterize porosity distribution.

POROSITY PREDICTION FROM SEISMIC REFLECTION DATA

The motivation for using seismic reflection data to characterize the spatial distribution of porosity (or other physical properties) comes from the ability to provide useful relationships between the seismic reflection data and physical properties. As will be shown later, porosity–seismic data relationships have been developed both theoretically and experimentally, mainly for sand and sand-clay models, but little work has been done on carbonate reservoirs.

Traditionally, velocity–porosity relationships have been estimated from regression methods that fit linear trends for certain intervals. More recently, the use of statistical techniques has provided better results in relating porosity to seismic reflection attributes, particularly for describing interwell porosity from surface seismic reflection data. Doyen (1988) applies geostatistical techniques to relate transit times from surface seismic reflection to porosity measurements from wells, and compares the results to those derived from linear regression. Scerbo and Mazzotti (1991) apply cokriging methods to relate seismic velocities to porosity. This approach provides better results than those provided by kriging methods. However, the results still show the smoothing and hole effects imposed by the original kriging method.

More recent approaches use simulation techniques that provide porosity models that describe the spatial distribution of this property. Such models are strongly supported by both statistical models that correlate porosity and seismic reflection attributes and by the information itself (Deutsch and Journel, 1992).

The first approaches to estimate porosity from seismic reflection data have considered changes in compressional and shear velocity due to this property. Experimental relationships among velocity, porosity, and clay content have been described by Wyllie et al. (1956, 1962), Eberhart-Phillips et al. (1989), Klimentos (1991), Marion et al. (1992), and Mavko and Nolen-Hoeksema (1992).

Davis et al. (1992) provide results that were derived from detailed three-dimensional, multicomponent (3-D, 3-C) seismic reflection data. Such results show the potential to relate anisotropy and porosity to Vp/Vs ratios and shear velocity differences.

Vernik and Nur (1992) presented work relating petrophysics to porosity and velocity. They developed a petrophysical classification of siliciclastics to predict lithology and porosity from seismic velocities. They presented results for Vp/Vs vs. porosity. In their work they fit linear and polynomial trends to the laboratory data; their results show an increase of Vp/Vs with porosity. The fits for arenite and clean arenite are polynomial, whereas the fits for the case of shale and wackestones are linear (Figure 1).

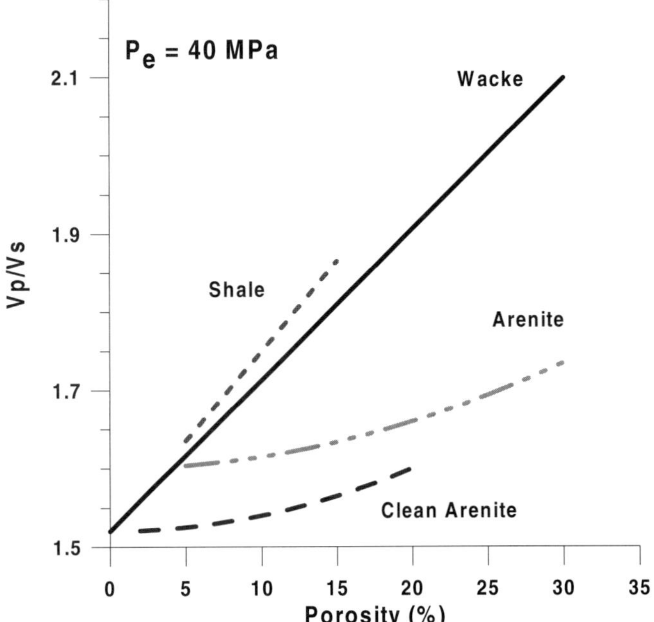

Figure 1. Vp/Vs ratios for saturated rocks vs. porosity (after Vernik and Nur, 1992). P$_e$ = effective pressure, MPa = megapascals.

Berge et al. (1995) established an excellent agreement between compressional and shear velocities from laboratory and theoretical predictions of velocity from bounded methods. Figure 2 shows the Vp/Vs–porosity relationship estimated from their numerical results, which is good for the case of wet samples. On the other hand, a Vp/Vs–porosity relation cannot be estimated for the dry samples. As in Vernik and Nur's (1992) Vp/Vs–porosity relationship, these Vp/Vs results (for the wet case) also increase with increasing porosity.

Sarmiento (1994) discussed the usefulness of the Vp/Vs ratios as a tool for identifying lithology, and also established that Vp/Vs vs. Vp plots of the Nisku reservoir are not constant relationships but vary with porosity. Therefore, Vp/Vs ratios can be used not only for lithology discrimination, but also for porosity mapping. From wireline log data, Sarmiento proposed the values 2.0 for anhydrite and 1.87 for dolomite for the Nisku Formation. One of the biggest concerns when analyzing the Vp/Vs1 data was that high Vp/Vs ratios were expected in the NE part of the seismic area and lower Vp/Vs values in the SW, according to the knowledge of anhydrite distribution in the field. However, the trend from the Vp/Vs1 ratio map shows the opposite relationship.

If we consider that porosity increases with Vp/Vs ratio in the dolomite zone, and that porosity is almost zero due to anhydrite plugging, then the increasing trend that we observe in our data is valid if a superposition of effects is considered. Figure 3 illustrates the idea that validates the use of Vp/Vs not only for lithology discrimination but also for porosity estimation.

Also to be considered is that the values given by Sarmiento (1994) are for pure anhydrite and pure dolomite. In the Nisku reservoir, anhydrite has

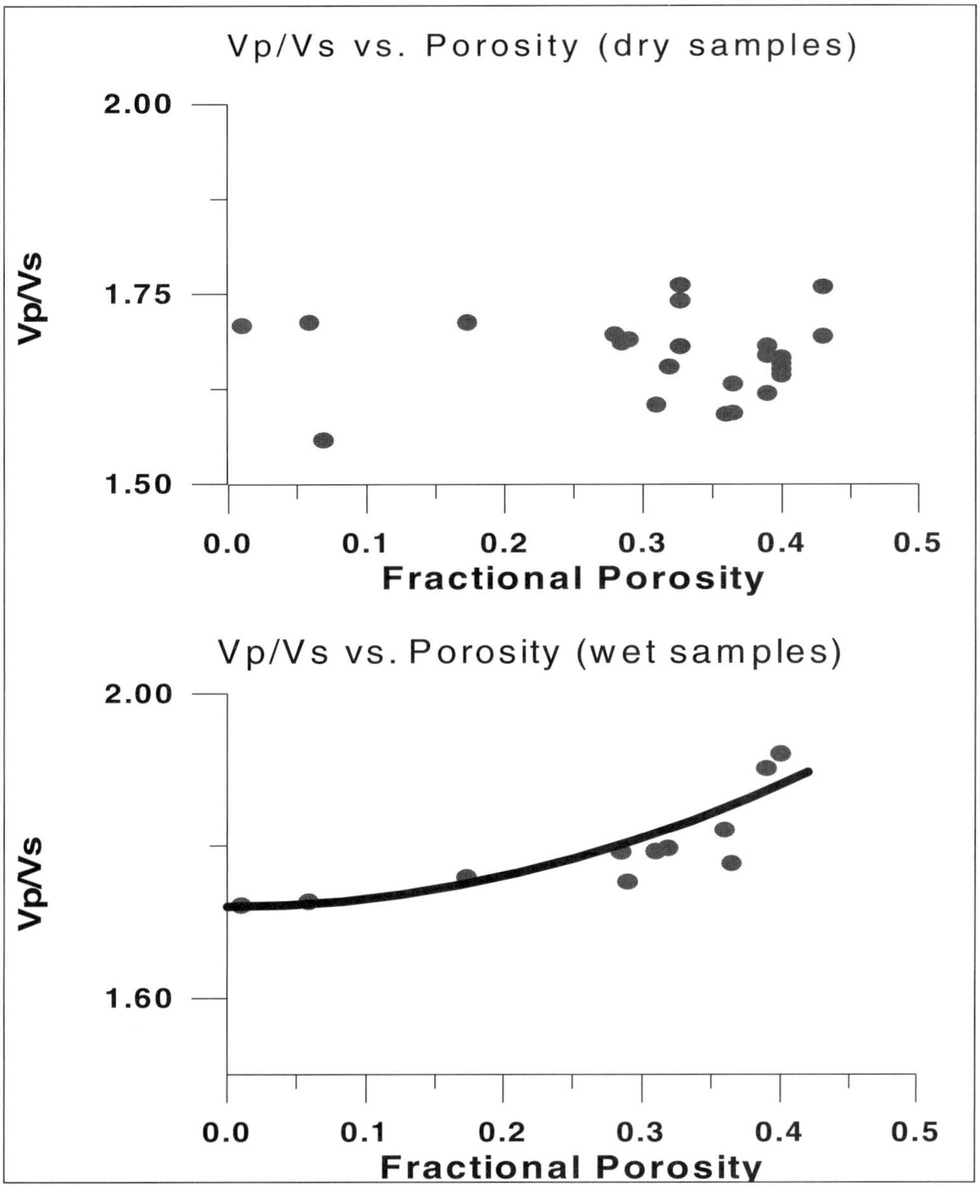

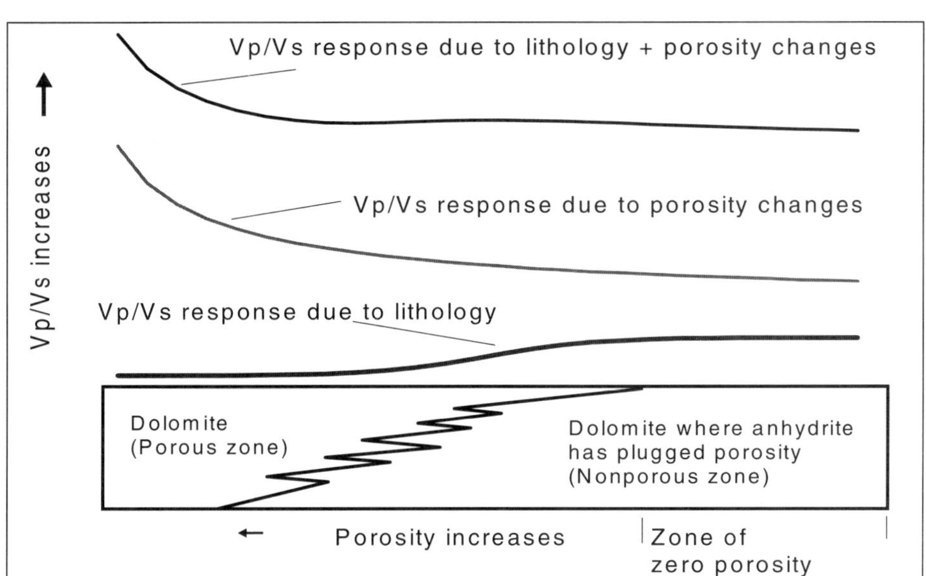

Figure 2. Vp/Vs vs. porosity for sandstone analogs. Dry and wet cases (data from Berge et al., 1995).

Figure 3. Qualitative interpretation of Vp/Vs ratio changes seen as a combination of responses due to lithology and porosity changes.

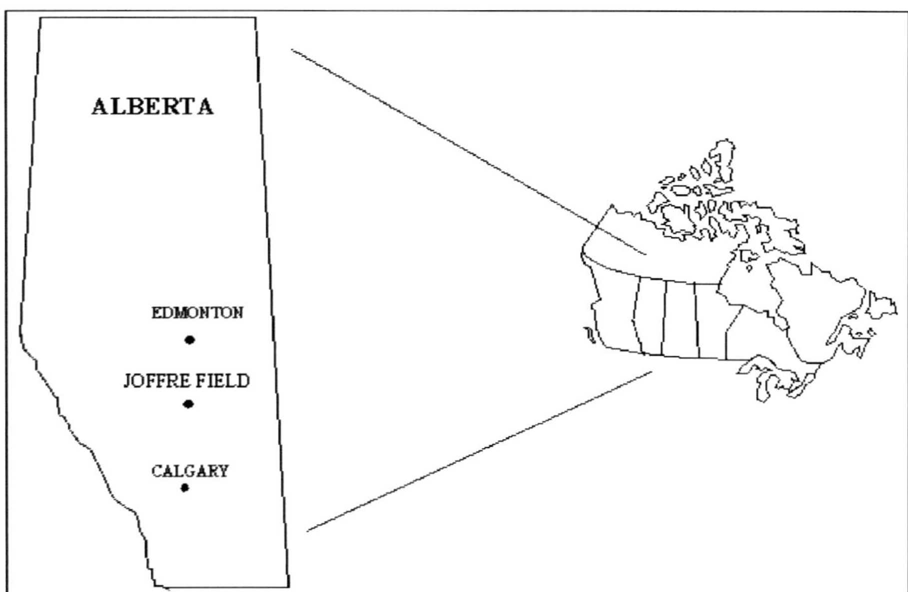

Figure 4. Location of the Joffre field study area (after Sarmiento, 1994).

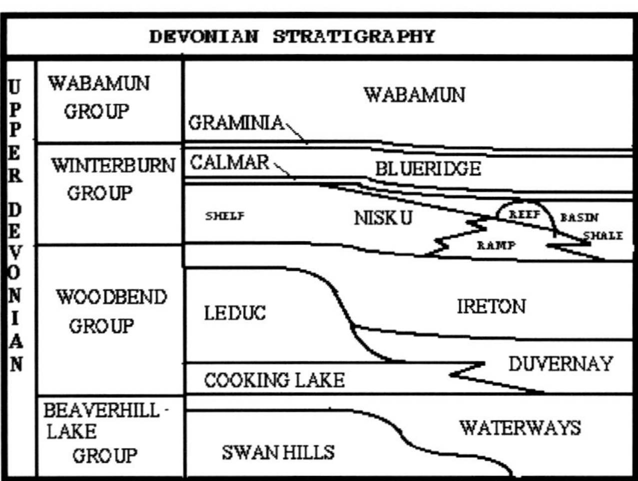

Figure 5. Generalized stratigraphy of the Upper Devonian in South-Central Alberta (after Watts, 1987).

plugged porosity in dolomite rock; thus, the non-porous zone will consist of a mixture of dolomite and anhydrite. If this is true for the reservoir, the difference of Vp/Vs ratios between dolomite and dolomite+anhydrite will be very small, and Vp/Vs changes due to porosity will predominate. Another important point to justify the use of Vp/Vs1 ratios for porosity estimation is that reservoir production and pressure data support the conclusion that Vp/Vs ratio variations over the Nisku reservoir are due to porosity, and not changing reservoir fluids (Arestad, 1995).

Reservoir pressures within the limits of the 3-D, 3-C survey were above the bubble point pressure at the time of data acquisition, showing that no free reservoir gas is located in the survey area (Al-Bastaki et al., 1995).

JOFFRE FIELD

Joffre field is located on the inner Nisku shelf region of South-Central Alberta, between Calgary and Edmonton, at the western edge of the Bashaw Complex (Figure 4). The field covers townships 39 and 38 and ranges 27 and 26, for an areal extent of 45–50 mi^2 (116.5–130.5 km^2). The geological model of the Nisku interval has been continuously changed as knowledge of the field has increased. This area is located in the central part of the Phanerozoic Western Canada Sedimentary Basin. The Devonian sedimentary section consists of marine carbonates, shales, and evaporites (Al-Bastaki et al., 1995). The Upper Devonian in the Western Canada Sedimentary Basin has been divided into four groups, from oldest to youngest: the Beaverhill Lake, Woodbend, Winterburn, and Wabamun groups (Figure 5). The Nisku Formation of the Winterburn Group consists of two units in the Joffre area: an upper unit of dolomite interbedded with anhydrite, and a lower, open marine dolomite with vuggy porosity and minor anhydrite. Arestad (1995) established that the stratigraphic zonation found in the open marine unit is present, with local variations, in most of the cored wells. The reservoir portion of the lower Nisku beneath the seismic reflection data area has an almost constant thickness of 22 m.

GEOPHYSICAL DATA

The well information used for this study describing the Nisku interval is restricted to an area of 6.5 × 5.5 km, slightly larger than the 3-D seismic data set. The area contains a total of 44 wells, including core samples, porosity–permeability measurements, and wireline logs (Figure 6).

A total of twenty-three wells were selected for petrophysical measurements. All of these wells are located inside the well control area for this study. The core

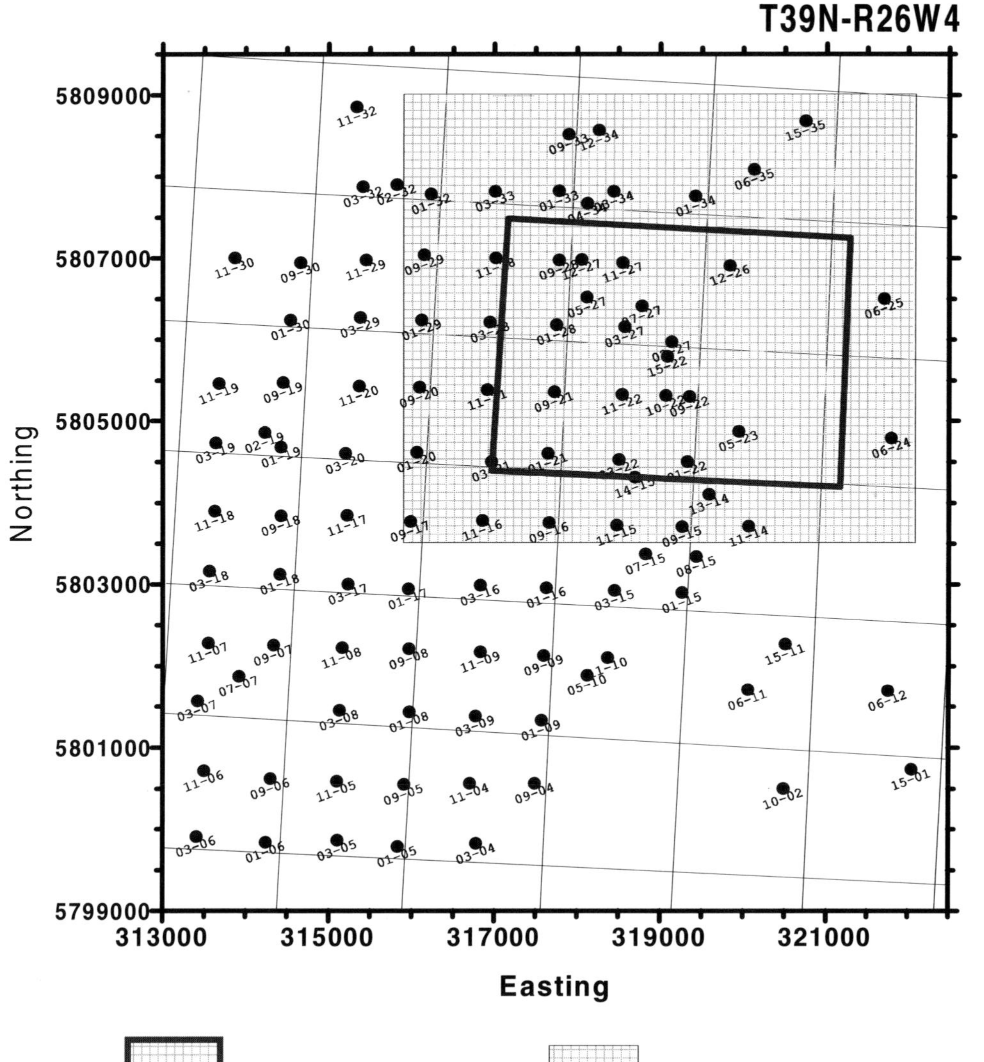

Figure 6. Location map showing the limits of the 3-D survey and the well control area.

samples were 3.5–4 in. (7.62–10.16 cm) in diameter and 5–11 in. (12.5–28 cm) long. To measure porosity and permeability, complete cores were used rather than small plugs, due to the moldic and vuggy nature of the porous Nisku reservoir rock.

Figure 7 shows porosities and permeabilities vs. depth for well 09-21-39-26 to show how these properties are distributed through the reservoir interval.

The seismic reflection data used in this study are part of the result of a 3-D, 3-C survey acquired over the northeastern edge of the Nisku reservoir by the Colorado School of Mines Reservoir Characterization Project. Conventional seismic reflection data have failed to characterize the complex diagenetic dolomite reservoir. In general, the acoustic impedance contrast between the Nisku reservoir and surrounding rock is very small. Additionally, strong interbed multiples and converted waves (P-SV) can interfere with the Nisku event on stacked data (Davis, 1992); thus, variations in the Nisku reflection event are usually not reliable indicators of porosity development or of reservoir quality. Therefore, 3-D, 3-C seismic reflection data technology

has been applied to the field to improve reseroir characterization. The use of this technology allows the recording of compressional as well as shear wave data. Anisotropic media (like carbonate reservoirs, which present azimuthal fracturing, and elongated pore shape porosity) create splitting and polarization of the shear waves (Martin and Davis, 1987). The results are a fast (S1) and a slow (S2) shear wave data set.

The compressional S1 and S2 shear data sets were used by Arestad (1995) to generate velocity ratio maps, amplitude maps, and time structure maps at several intervals or times. The seismic data studied in this work consist of a map of velocity ratios (Vp/Vs1) for the D1 to mid-Ireton interval (Figure 8). The map covers an area of ~4 km × 3 km, with a bin size of 30 m × 30 m. The velocity ratio map was computed utilizing interval traveltimes from both the compressional and shear wave data sets, using VSP data to tie the P- and S-wave reflections originated from equal depths. Arestad (1995) gives a detailed description of the timing analysis and calculations for Vp/Vs mapping.

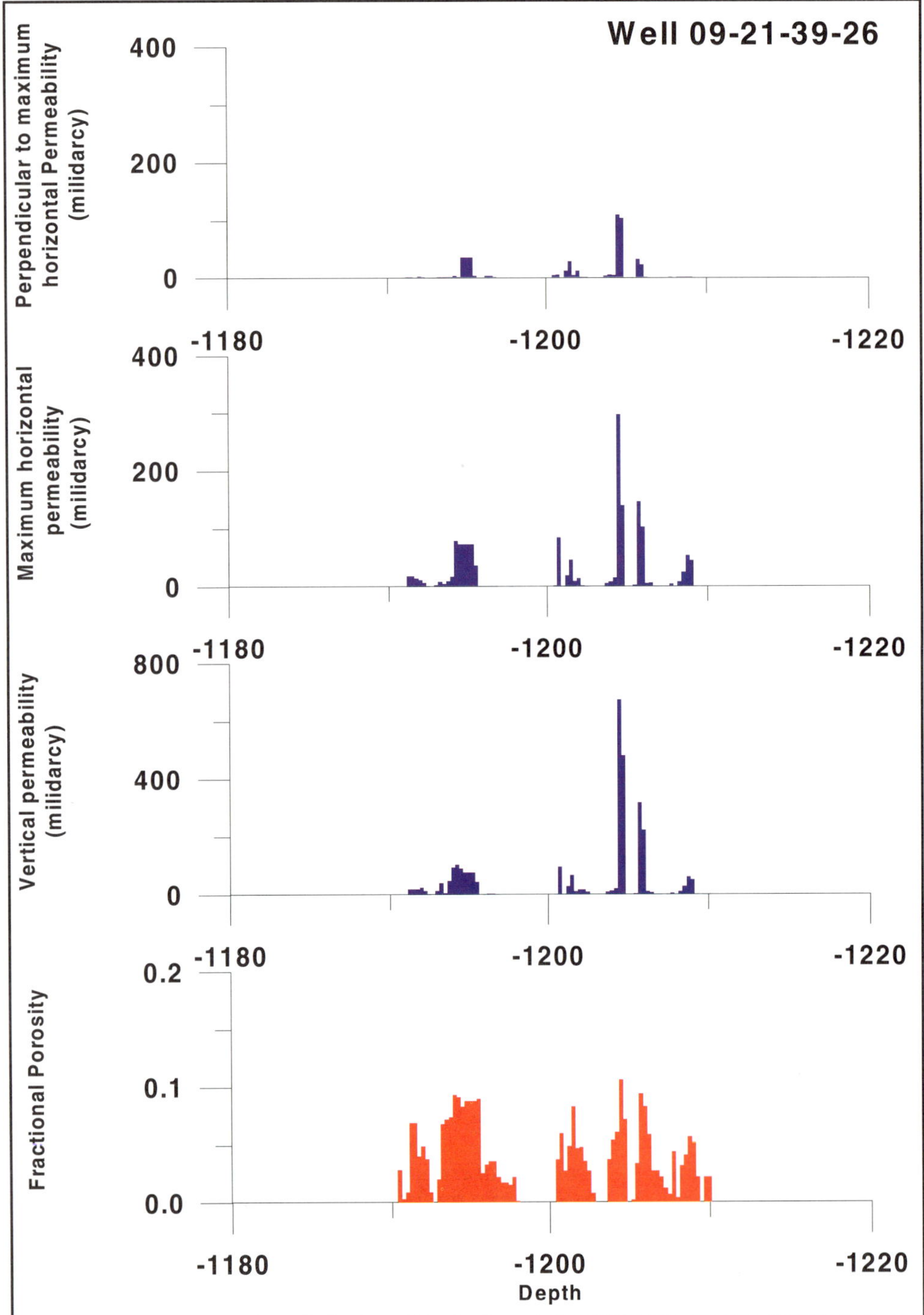

Figure 7. Porosity and permeability distribution for well 09-21-39-26.

GEOSTATISTICAL ANALYSIS

The 3-D estimation of porosity using a sequential Gaussian simulation requires the knowledge of both the probability distribution and the 3-D covariance or variogram model for the physical property (Deutsch and Journel, 1992). Other techniques such as cokriging require variogram models for primary and secondary data (porosity and seismic) and the cross-variogram between variables. For this work, neither the variogram model nor the cross-variogram could be determined from core porosity and seismic data. Therefore, the cokriging technique could not be used to estimate porosity distribution.

The histogram of the core porosity is shown in Figure 9. The histogram is not normally distributed. The median is the parameter that better indicates the high of the population. Porosity values range from ~0% to 20%, and most of the values are concentrated in the interval 0%–6%. The next step in calculating statistical parameters is to estimate the 3-D variogram model from core porosity. A variogram model is represented by a function that varies with increasing distance. The maximum of the curve is called the sill, and its magnitude represents the variance of the data. The range is the distance at which the sill is reached; it represents the correlation length of the parameter being

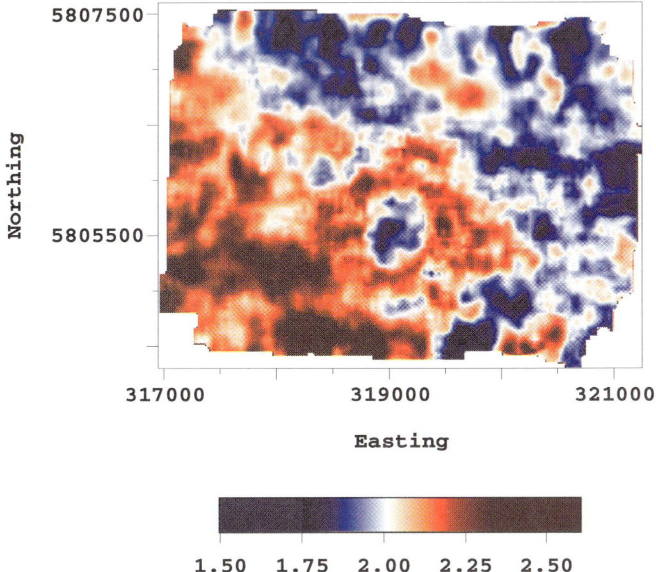

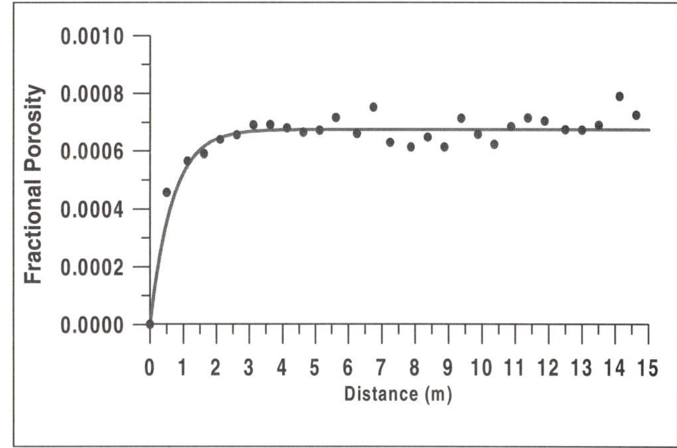

Figure 8. Velocity ratio (Vp/Vs1) map of the Wabamun (D1) to mid-Ireton interval.

Figure 10. Vertical core porosity variogram; the variogram model is exponential.

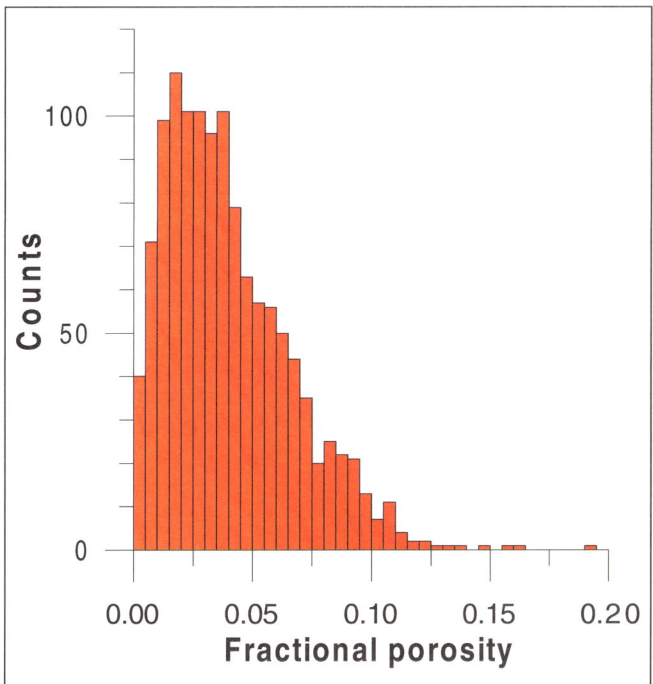

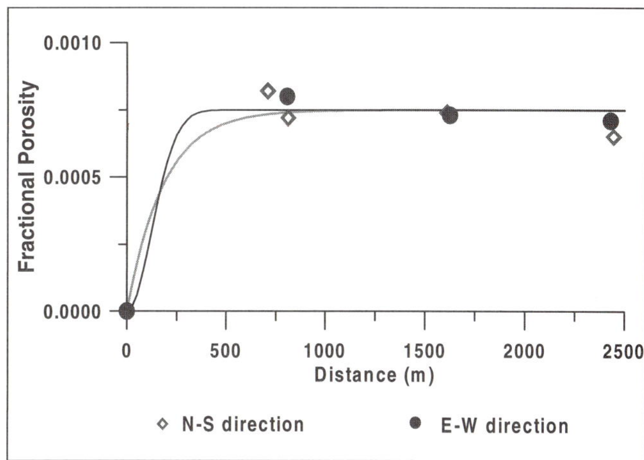

Figure 11. Horizontal core porosity variograms. Note the lack of definition due to well separation.

Figure 9. Histogram of core porosity measured at 23 well locations within the lower Nisku.

wells. The computed variograms cannot be modeled because the sill has been reached before the first lag distance. Two variogram models are plotted along the computed variogram to show how the range (or distance of maximum continuity) can be represented for any model (Figure 11).

It is clear that we need to estimate the horizontal variogram model from other data. Thus, by establishing the relationship between core porosity and seismic reflection information (in the form of Vp/Vs or amplitude of shear wave maps), we can transform seismic reflection data into horizontally distributed porosity data. Considering that we have a high-resolution porosity description in the vertical direction that is to be compared with high-resolution horizontal seismic reflection data, we need to obtain an average (or mean) porosity at each well location. Once the mean porosities were calculated, the seismic reflection values surrounding each well were extracted from the data set to calculate a mean seismic reflection data attribute value. Data within a 90-, 150-, and 250-m radius around each well were considered. Then, the mean

measured. Along the vertical direction, the sample interval is small enough to obtain a high-resolution variogram, which allows modeling of the Z-component of the variogram model. The result of this computation is shown in Figure 10. On the other hand, computation of horizontal variograms for the N-S and E-W directions shows a lack of information due to the large spacing (~0.5 mi; 0.805 km) distance between

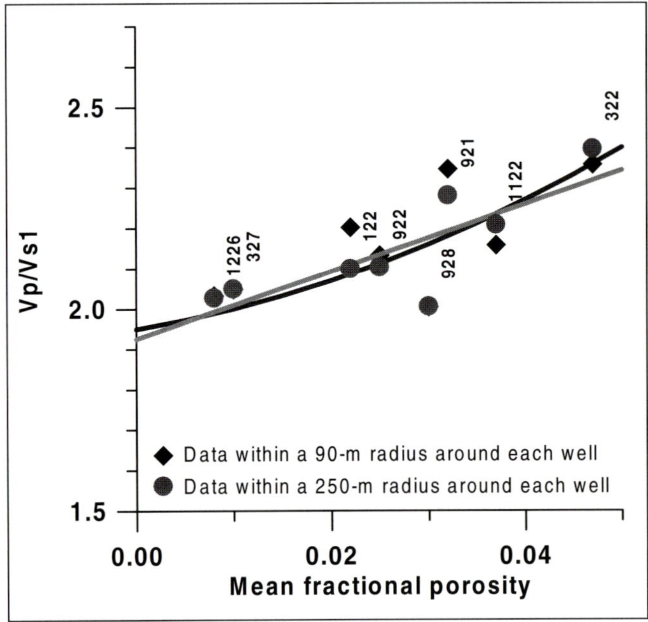

Figure 12. Vp/Vs1 vs. mean porosity. Linear and exponential fits are plotted.

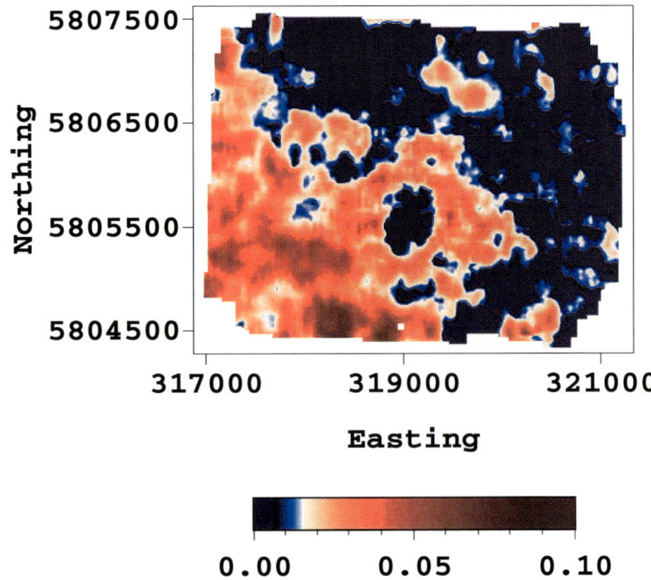

Figure 13. Mean porosity map derived from Vp/Vs1 ratios (linear fit). The scale shows fractional mean porosity.

Vp/Vs1 ratios were compared to the mean porosities in the form of crossplots. Plots for radii of 90 and 250 m are shown in Figure 12.

The Vp/Vs1 vs mean porosity crossplot indicates an increase of porosity with increasing Vp/Vs1. Linear and exponential models were fitted to the data to derive a mathematical relationship between Vp/Vs1 and mean porosity. There are no significant differences between these two models over the examined interval; therefore, the linear model is chosen to transform Vp/Vs1 ratios into mean porosity values. Figure 13 shows a map of the estimated porosities from Vp/Vs1. If Vp/Vs1 also indicates lithology, limits of the reservoir due to anhydrite plugging can also be interpreted from this map.

Variograms were calculated from the porosity map derived from Vp/Vs1 for eight directions. The results show that horizontal variation of porosity is geometrically isotropic within a distance of 600 m. Two directions (N68W and N22E) are shown in Figure 14.

Assuming that the spatial variability structures of the porosity from seismic reflection and core data are the same, the variogram analysis of the seismic reflection data-derived mean porosity map defines the near-lag portion of the curve that is not defined in the horizontal core porosity variogram. The sills of both horizontal variograms (core porosity and porosity from seismic reflection data) can be related by the following scaling relationship:

$$\frac{\sigma_s^2}{m_s^2} = \frac{\sigma_c^2}{m_c^2} \tag{1}$$

where σ_c = core porosity standard deviation, m_c = core porosity mean, σ_s = porosity from seismic reflection data

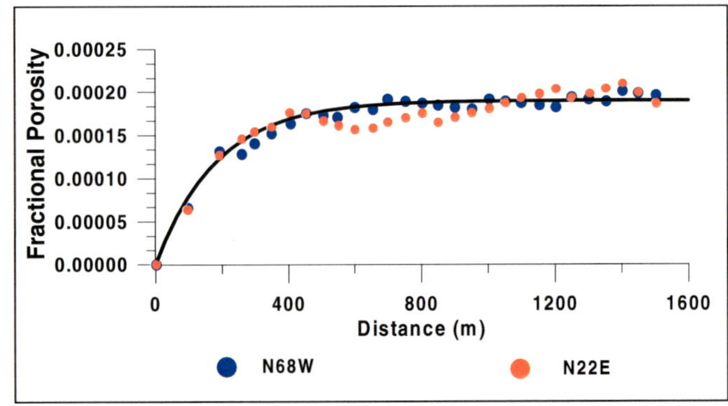

Figure 14. Horizontal variogram model and computed variograms from porosity from seismic reflection data (porosity from Vp/Vs1 ratios).

standard deviation, m_s = porosity from seismic reflection data mean; with the computed scaling factors being

$$\frac{\sigma_s^2}{m_s^2} = \frac{0.00068}{0.0485^2} = 0.289 \tag{2}$$

and

$$\frac{\sigma_c^2}{m_c^2} = \frac{0.00019}{0.026^2} = 0.281 \tag{3}$$

The combination of previous results from core data with the results obtained from seismic reflection data gives the 3-D variogram model, which is defined by the following parameters: lateral range N-S/E-W = 600 m, vertical range (depth) Z = 3–4 m, dip direction = 0.5° W, horizontal geometrical anisotropy ratio = 1.0, and vertical geometrical anisotropy ratio = 0.005–0.0066.

Once the 3-D variogram model was defined, the porosity distribution was computed for the 2-D and 3-D

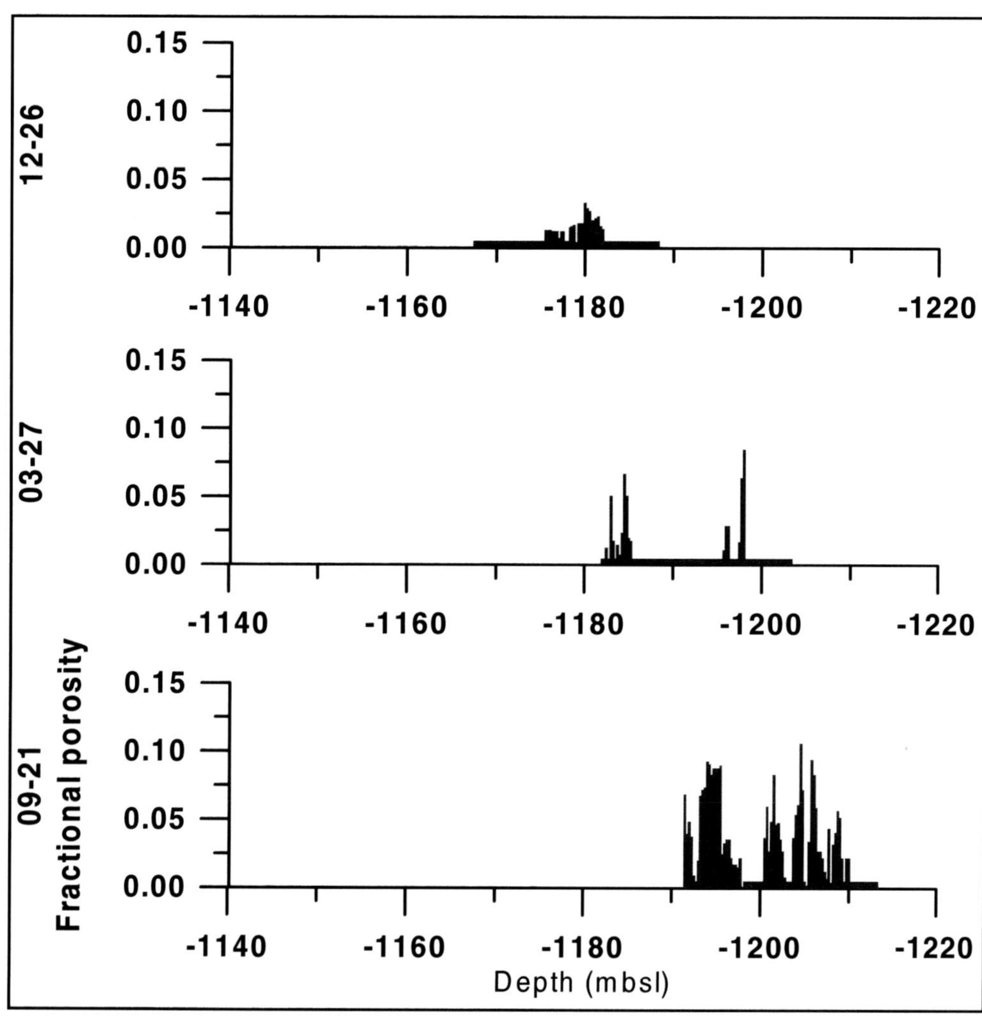

Figure 15. Distribution of core porosity for the wells used in 2-D porosity simulation.

cases. Figure 15 shows the porosity distribution at the three well locations used for the 2-D case. Figure 16 shows the result of the conditional simulation.

Finally, the sequential Gaussian technique is used to estimate a 3-D porosity distribution. The horizontal grid defined was 100 m, whereas the vertical sampling was 0.5 m. According to the lateral extension defined by the well control, the number of cells calculated for each simulation is approximately 300,000. A change in the grid size involves a significant change in the number of cells of the model and, consequently, a significant change in the number of computations needed to perform the simulation.

Figure 17 shows the result of a 3-D realization. As expected, the simulation of the volume of porosity distribution shows connectivity zones, trends of porosity horizons, and distribution of low- and high-porosity zones. Some of the characteristics of the 3-D porosity simulation are the enhancement of the connectivity of high- and low-porosity zones, the preservation of the dip angle of the porous lower Nisku interval, and a clear definition of the lateral limits of the reservoir zone.

An oblique slice that runs parallel to the lower Nisku interval is shown in Figure 18. This slice shows how high-porosity values are concentrated in the southwest portion of the model, as indicated by the well and seismic reflection data.

CONCLUSIONS

Geostatistical techniques have been useful to derive porosity distribution from a limited amount of core data when integrated with seismic reflection data (in the form of Vp/Vs1 ratios). The improvement on the porosity model can be evaluated by mapping the model into a petroleum reservoirsimulation grid. The results of history matching and reservoir performance prediction will indicate quantitatively the amount of information gained by using estimation techniques.

Oil reservoirs with geological settings similar to those of Joffre field exist around the world; therefore, the use of geostatistical techniques and 3-D, 3-C seismic reflection data are of importance to reservoir characterization in general, and to the implicit goal of improved hydrocarbon recovery.

Geostatistics can be applied to the study of spatial and temporal relationships among porosity, permeability, and fluid saturation related to observed changes in seismic reflection data attributes. In this case, reproduction of the spatial dependence of several variables is critical. Therefore, conditional simulation algorithms can be generalized to join simulation of several variables.

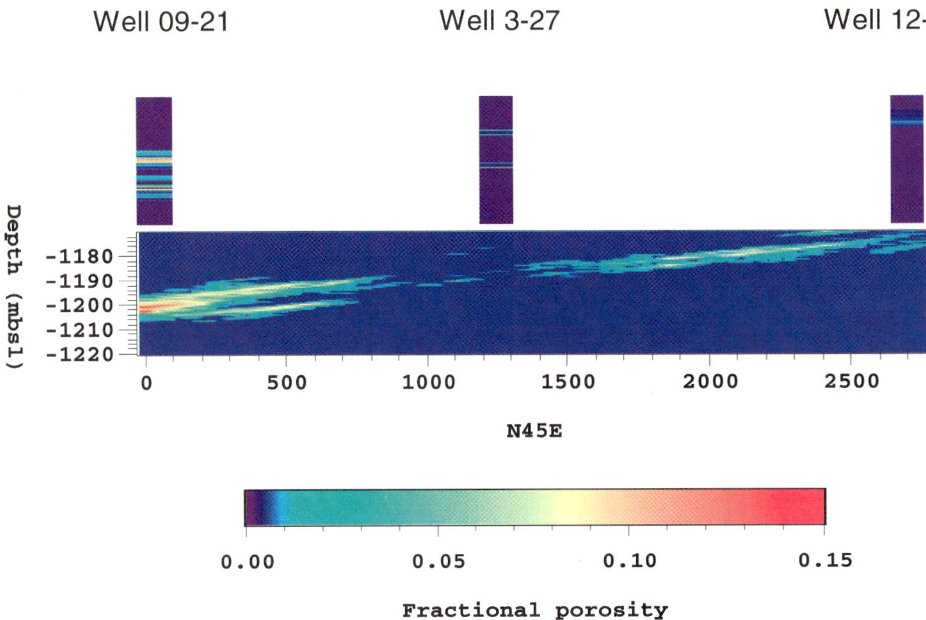

Figure 16. Sequential Gaussian conditional simulation of porosity for profile SW 45 NE.

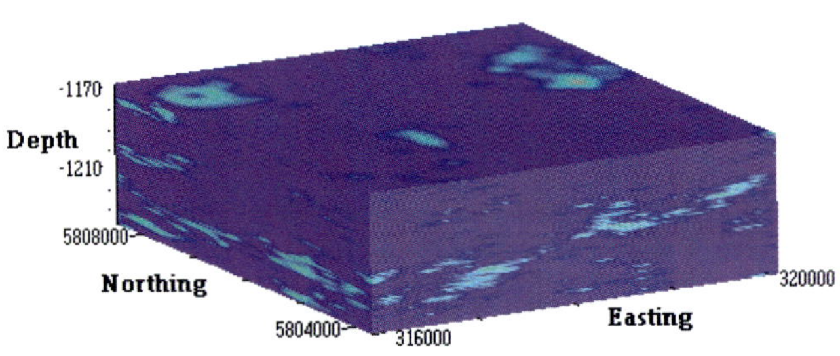

Figure 17. Three-dimensional conditional simulation of porosity (from the lower Nisku interval) using a sequential Gaussian technique. Lighter colors represent higher porosity values.

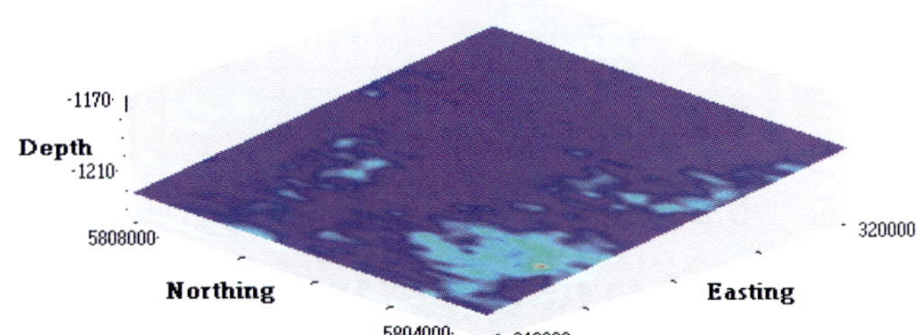

Figure 18. Oblique slice from the 3-D porosity simulation (from the lower Nisku interval). Lighter colors represent higher porosity values.

ACKNOWLEDGMENTS

We want to express appreciation for direction and financial support for this research to the RCP (Reservoir Characterization Project)—Phase V Industry sponsors and to the Mexican Petroleum Institute. We also thank an anonymous reviewer, Robert Kendall, Julie Kupecz, and the RCP research team members for their comments and suggestions on this work.

REFERENCES CITED

Al-Bastaki, A.R., J.F. Arestad, K. Bard, R. Cabrera-Garzón, B. Mattocks, and M.R. Rolla, 1995, Multidisciplinary multicomponent reservoir characterization, Joffre field, South-Central Alberta, Canada: Colorado School of Mines Reservoir Characterization Project—Phase V Sponsor Meeting Notes, April 6, 1995, 256 p.

Arestad, J.F., 1995, An integrated multicomponent three-dimensional seismic characterization of Joffre field, Alberta, Canada: Ph.D. thesis, Colorado School of Mines, Golden, Colorado, 293 p.

Berge, P.A., B.P. Bonner, and J.G. Berryman, 1995, Ultrasonic velocity–porosity relationships for sandstone analogs made from fused glass beads: Geophysics, v. 60, no. 1, p. 108–119.

Davis, T.L., 1992, Reservoir Characterization Project—Phase V: CSM Proposal 3788, 24 p.

Deutsch, C.V., and A.G. Journel, 1992, Geostatistical software library and user's guide: New York, Oxford University Press, 340 p.

Doyen, P.M., 1988, Porosity from seismic data, a geostatistical approach: Geophysics, v. 53, no. 10, p. 1263–1275.

Eberhart-Phillips, D., D.-H. Han, and M.D. Zoback, 1989, Empirical relationships among seismic velocity, effective pressure, and clay content in sandstones: Geophysics, v. 54, no. 1, p. 82–89.

Klimentos, T., 1991, The effects of porosity-permeability-clay content on the velocity of compressional waves: Geophysics, v. 56, no. 12, p. 1930–1939.

Marion, D., A. Nur, H. Yin, and D. Han, 1992, Compressional velocity and porosity in sand-clay mixtures: Geophysics, v. 57, no. 4, p. 554–563.

Martin, M.A., and T.L. Davis, 1987, Shear wave birefrigence: a new tool for evaluating fractured reservoirs: The Leading Edge, v. 6, no. 10, p. 22–28.

Mavko, G., and R. Nolen-Hoeksema, 1994, Estimation of seismic velocities at ultrasonic frequencies in partially saturated rocks: Geophysics, v. 59, no. 2, p. 252–258.

Sarmiento, V., 1994, Petrophysical relationships from wireline logs for seismic calibration of the Devonian Nisku and Wabamun formations, Joffre field, Alberta: Master's thesis, Colorado School of Mines, Golden, Colorado, 97 p.

Scerbo, F., and A. Mazzotti, 1991, Geostatistical estimates of porosity from seismic data: Bollettino di Geofisica Teorica ed Applicata, v. 33, no. 130–131, p. 85–110.

Vernik, L., and A. Nur, 1992, Petrophysical classification of siliciclastics for lithology and porosity prediction from seismic velocities: AAPG Bulletin, v. 76, no. 9, p. 1295–1309.

Watts, N.R., 1987, Carbonate sedimentology and depositional history of the Nisku Formation (within the Western Canadian Sedimentary Basin) in South Central Alberta: GSPG Second International Symposium on the Devonian System, p. 87–152.

Wyllie, M.R.J., A.R. Gregory, and L.W. Gardner, 1956, Elastic wave velocities in heterogeneous and porous media: Geophysics, v. 21, no. 1, p. 41–70.

Wyllie, M.R.J., L.W. Gardner, and A.R. Gregory, 1962, Studies of elastic wave attenuation in porous media: Geophysics, v. 27, no. 5, p. 569–580.

Chapter 10

Zempolich, W.G., and L.A. Hardie, 1997, Geometry of dolomite bodies within deep-water resedimented oolite of the Middle Jurassic Vajont Limestone, Venetian Alps, Italy: analogs for hydrocarbon reservoirs created through fault-related burial dolomitization, *in* J.A. Kupecz, J. Gluyas, and S. Bloch, eds., Reservoir quality prediction in sandstones and carbonates: AAPG Memoir 69, p. 127–162.

Geometry of Dolomite Bodies Within Deep-Water Resedimented Oolite of the Middle Jurassic Vajont Limestone, Venetian Alps, Italy: Analogs for Hydrocarbon Reservoirs Created Through Fault-Related Burial Dolomitization

William G. Zempolich[1]
Lawrence A. Hardie
Department of Earth and Planetary Sciences, The Johns Hopkins University
Baltimore, Maryland, U.S.A.

ABSTRACT

The Middle Jurassic Vajont Limestone of the Venetian Alps, Italy, is predominantly composed of resedimented ooids that were deposited in slope and basin settings. The Vajont Limestone has been partly replaced by massive dolomite that can be mapped at both regional and local scales. Dolomite bodies that are present within or are associated with the Vajont Limestone include: (1) a large-scale wedge, ~25 km long, 10–15 km wide, and ≥400–500 m thick (50–94 km^3), located on the hanging wall of the Alpine-aged, thrust-based Mt. Grappa–Visentin anticline. This dolomite body is located within the axis of the anticline and crosscuts the stratigraphic section where subvertical to vertical faults penetrate the crest of the anticline; (2) Isolated, rootless plume-shaped bodies, 100–200 m wide and >300 m high (≥2 × 10^{-2} km^3), which penetrate a footwall syncline within an Alpine-aged thrust sheet. These dolomite "plumes" possess extensively brecciated cores and exhibit sharp to gradational transitions with surrounding Lower to Middle Jurassic basinal limestone; (3) Isolated dolomite "towers" that have partly replaced Cretaceous-age synsedimentary fault breccia. These bodies are found in overlying basinal strata (i.e., the Fonzaso Formation, the Ammonitico Rosso, and the Biancone Formation), but emanate from the underlying dolomitized Vajont; and (4) Small-scale wedge-shaped dolomite bodies on the scale of meters found along small faults and fractures. The connection between these dolomite bodies and Alpine-aged faults and fractures clearly indicates that dolomitization was a late burial process.

---

[1]Present address: Mobil Oil Company, Dallas, Texas, U.S.A.

It is proposed that during the Alpine deformation event, convection-driven fluids derived from Late Tertiary seawater were circulated through subaqueous Alpine-aged faults and fractures and paleosynsedimentary breccias, thus creating the multitude of dolomite bodies now found in the Vajont and other Mesozoic basinal sediments. Paleogeographic, tectonic, and hydrologic systems, similar to the one proposed for dolomitization of the Vajont, appear to be active in modern subaqueous thrust zones of the Caribbean and Northwest Pacific Coast.

Potential reservoir attributes of Vajont dolomite bodies include their large size and medium to coarsely crystalline replacement fabric that is characterized by significant amounts of partial moldic, intercrystalline, and vug pore space. Visual estimates of porosity within dolomitized grainstone and packstone range up to 10% to 15%, with inferred permeabilities of 1–100 md. Permeability of Vajont dolomite replacement fabrics is enhanced through recrystallization and the formation of touching-vug networks (inferred permeabilities ≥100 md).

Results of this study indicate that (1) massive replacement dolomitization in thermotectonic (i.e., burial) settings may be much more important than previously thought, and (2) significant reservoirs may be hosted in otherwise tight basinal limestones as the result of late-stage burial dolomitization. Consequently, the geometries of the Vajont dolomite bodies may provide analogs for reservoir characterization and new exploration plays in the subsurface. Exploration methods for analogous dolomite reservoirs in the subsurface may include the mapping of dolomitization fronts using core and log data and seismic reflection identification of crosscutting dolomite bodies. The focus of such efforts should be placed on anticlinal and synclinal structures within buried fold and thrust belts, and along zones of deep-seated tectonic fractures and faults within intracratonic basins.

INTRODUCTION

Dolomites constitute some of the best-quality reservoirs for oil and gas, due to several unique properties, much as intercrystalline pore space resulting in high permeability and resistance to burial compaction. Therefore, the prediction of dolomite body geometries is of paramount importance in reservoir exploitation. A key to understanding dolomite distribution lies in understanding its origin and timing.

The origin of massive replacement dolomite has remained one of the major unresolved problems of sedimentology and sedimentary geochemistry for more than a century (van Tuyl, 1916; Morrow, 1982a, b; Land, 1985; Machel and Mountjoy, 1986; Hardie, 1987). In the last two to three decades the favored interpretations for the origin of massive dolomites have centered on "early" low-temperature replacement of limestones and lime sediments. Early low-temperature models involve surface or near-surface marine waters such as refluxing sabkha brines or coastal mixing zone brackish waters

and thus circumvent the magnesium "supply" problem (Morrow, 1982a, b). Hardie (1987) has pointed out some of the serious weaknesses and uncertainties in these low-temperature models for dolomitization, and has argued that attention be turned to the many alternative ways by which massive dolomites can be made. Along these lines, it is valuable to compile well-documented, clear-cut case histories of as many different modes of dolomitization as can be identified.

In this regard, the occurrence of massive replacement dolomite in the Vajont limestone is particularly notable because of the deep-water slope and basin setting in which the Vajont sediments were deposited (Bosellini et al., 1981). Features that make the Vajont area especially valuable for field and laboratory study of reservoir development and prediction include:

1. The Vajont area is massively, but not completely, dolomitized, so that dolomitization fronts (Wilson et al., 1990) can be mapped and provide direct clues to the subsurface pathways that dolomitizing

fluids may have taken. Outcrops are abundant and well exposed, allowing mapping at scales that range from centimeters (reservoir scale) to kilometers (exploration scale).

2. The part of the Vajont Limestone that hosts the dolomite bodies covers an area of >100 km^2, so that a regional-scale fluid flow system capable of producing 50–100 km^3 of dolomite must have been involved.

3. The Vajont consists of bedded grainstones and packstones composed of shallow-water ooids (and some skeletal remains of shallow-water organisms) that were resedimented in deep water by slope processes (Bosellini et al., 1981; Zempolich, 1995). This unusual oolite is an integral part of a thick, conformable Jurassic succession of deep-water sediments that were never exposed to shallow-water syndepositional diagenetic processes. Thus, shallow-water and land-surface-related dolomitization processes can be ruled out.

4. The ooids, skeletal grains, and their intergranular cements are beautifully preserved in the undolomitized parts of the Vajont area so that it is possible to document in detail their predolomitization petrography and isotope geochemistry; thus, the changes produced by dolomitization can be identified and measured.

5. Exposures of partly and massively dolomitized Vajont sediments in road cuts reveal clearly that dolomitization fluid pathways were controlled by fractures and faults of Late Tertiary age (Alpine orogeny), indicating that the dolomitization was of "late" burial (synfaulting to postfaulting) origin. Through an understanding of the timing of Vajont dolomitization, the potential exists for the "prediction" of reservoirs in other similarly deformed carbonate strata.

Although a number of workers have put forward evidence and arguments for "late" elevated-temperature dolomitization during burial [Jodry, 1969; Zenger, 1976, 1983; Mattes and Mountjoy, 1980; Broomhall and Allen, 1985; Gregg, 1985; Barrett, 1987; Lee and Friedman, 1987 (and the discussion of this paper by Kupecz et al., 1988, and the reply by Lee and Friedman, 1988); Aulstead et al., 1988; Zenger and Dunham, 1988; Machel and Anderson, 1989; Cervato, 1990; Wilson et al., 1990; Kupecz and Land, 1991; Mountjoy and Halim-Dihardja, 1991; Zempolich and Hardie, 1991a, b; Amthor et al., 1993; Dix, 1993; Coniglio et al., 1994; Miller and Folk, 1994; Montañez, 1994; Mountjoy and Amthor, 1994; Yao and Demicco, 1995; Zempolich, 1995], burial dolomitization remains a controversial process believed by many sedimentologists to be of little importance in the origin of ancient massive dolomites (Blatt, 1982; Morrow, 1982b; Wilkinson and Algeo, 1989) and limited to the enhancement of preexisting or poor reservoirs (Sun, 1995). However, most, if not all, of the kinetic problems of dolomite formation that plague low-temperature systems essentially disappear at the elevated temperatures of burial (Hardie, 1987), making massive burial dolomitization a

likely, if not common, process. Through a comparative study of the Vajont limestone and dolomite, a much clearer understanding of burial dolomitization processes has been developed, including the role of deeply circulated subsurface fluids, in the origin of regional-scale and isolated dolomite bodies and the evolution of porosity through dolomitization. These results lead to prediction of dolomite reservoir geometries that can be created through tectonic and burial diagenetic processes.

THE VAJONT LIMESTONE AND ITS GEOLOGIC SETTING

Deposition of the Vajont Limestone is closely associated with the breakup of Pangea, during which time prolific oolite was deposited along the margins of the Tethys Ocean in the circum-Mediterranean region (Bosellini, 1989; Zempolich, 1995). In the Early Jurassic, Europe and northern Africa began to separate, and by the Late Jurassic an extensive transform zone was present (Weissert and Bernoulli, 1985). The breakup of Pangea established a horst-and-graben tectonic setting along the southern Tethyan margin, and led to the structural definition of local platforms and basins (Figure 1). The Trento Platform, the most landward horst block of the Southern Alps, was bounded to the west by the Lombardy Basin, and to the east by the Belluno Basin, which separated the Trento Platform from the Friuli Platform (i.e., the stable foreland). Thick sequences of these Mesozoic platform and basin carbonates, now partly to completely dolomitized, are extensively exposed in the Venetian Alps (Figures 2, 3).

Age dating of the Vajont Limestone is problematic. Stratigraphic-age constraints from formations located below and above the Vajont area suggest a general age range of Bajocian to Callovian (Casati and Tomai, 1969; Bosellini et al., 1981). The biostratigraphic study of Casati and Tomai (1969) suggests an age assignment (in part) of Upper Bajocian–Lower Bathonian for the Vajont limestone, based on overlapping ranges of the foraminiferal zones *Protopeneropolis striata* and *Trocholina*. New age constraints provided by nannofossil and ammonite data collected during the present study suggest that the Vajont Limestone was deposited during the latest Aalenian to the earliest Bajocian (Zempolich, 1993, 1995).

The Vajont Limestone is a particularly interesting carbonate deposit because it is a thick sequence (≤600 m along the platform margins) predominantly composed of shallow-water oolitic sand and biogenic skeletal debris that was redeposited by gravity flow processes in slope and basin environments (Bosellini et al., 1981; Zempolich, 1995). Depositional units include meter-scale debris flows and turbidites and bedded hemipelagic mudstone. Paleogeographic reconstructions suggest that the Vajont Limestone is an eastward-thickening wedge with a depositional area in excess of 100 km along strike and 50 km across strike (Figure 4). Vajont ooids were derived from the western edge of the

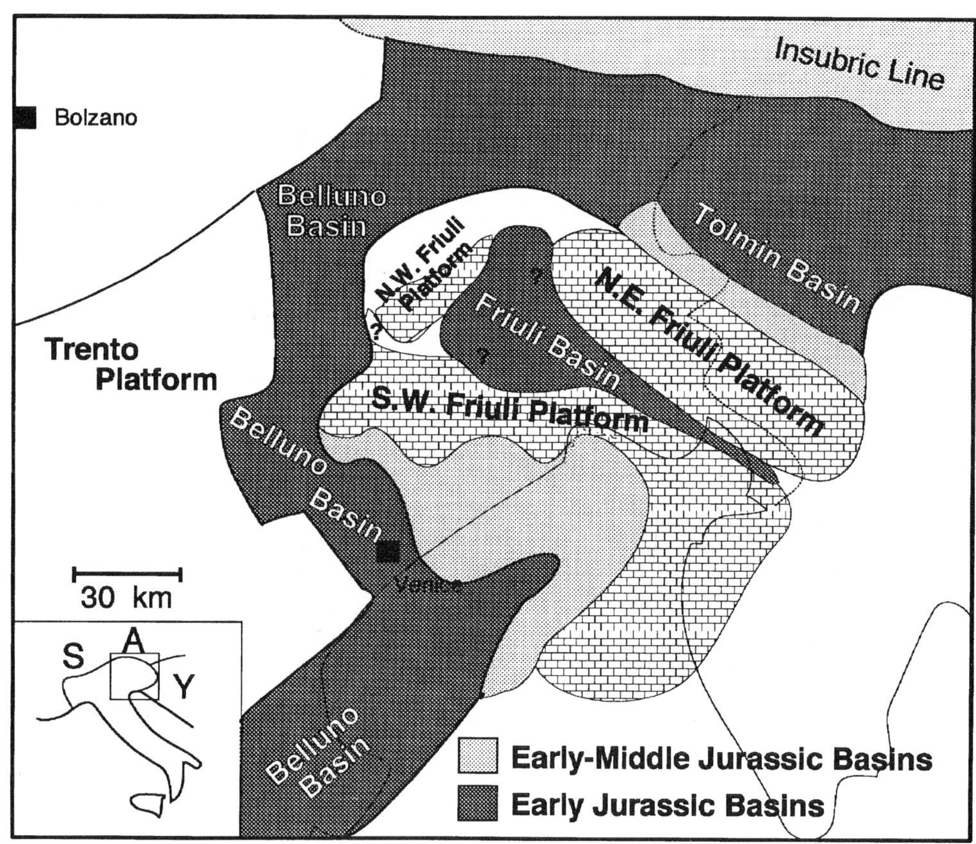

Figure 1. Early and Middle Jurassic paleogeography of the Venetian Alps (modified from Bosellini et al., 1981; Cati et al., 1987). The study area is divided into several carbonate platform and basin domains, including the Trento Platform, Belluno Basin, Friuli Platform, and Tolmin Basin. The Friuli Platform is subdivided into several platforms and basins based on the interpretation of seismic reflection data from the Friuli Plain and Po Basin (Cati et al., 1987).

Friuli Platform and were deposited as a carbonate slope apron in the Belluno Trough (Zempolich, 1995). The Vajont Limestone thins basinward and onlaps parts of the Trento Platform to the west. Well penetrations in the Po Plain and northern Adriatic Sea (nonproductive) suggest that the Vajont sediment is present to the south in the subsurface (Bosellini et al., 1981; Cati et al., 1987).

Within the Belluno Basin, the Vajont Limestone overlies dense, chert-rich micritic limestone and shale belonging to the Igne Formation (Toarcian–Aalenian; Figures 2, 3, 5, and 6). The Fonzaso Formation (Callovian–Lower Kimmeridgian?) overlies the Vajont Limestone and contains cherty, skeletal-rich turbidites and debris flows. In the central Belluno Basin, the Fonzaso Formation grades upward into nodular, micritic red limestone belonging to the Upper Ammonitico Rosso (Kimmeridgian–Tithonian), which in turn grades into thick, hemipelagic white limestone of the Biancone Formation (Tithonian–Cretaceous). Toward the east, the Fonzaso Formation and Ammonitico Rosso grade into the Soccher Formation (Lower Kimmeridgian–Cretaceous), which contains resedimented shallow-water carbonate and hemipelagic limestone. Along the western margin of the Friuli Platform, the Soccher Formation directly overlies the Vajont Limestone (e.g., Mt. Sestier section; Zempolich, 1995) and passes upward from thin-bedded peloidal/skeletal grainstone to thick skeletal-rich beds to massive coral and *Ellipsactinia* (hydrozoan) reefs and back-reef *Nerinacea* gastropod grainstone (Cellina Limestone; Upper Oxfordian–Lower Kimmeridgian). The progradation of Upper Jurassic slope and reef

margin sediments over the Vajont area along the eastern Belluno Basin indicates that Vajont sediments found here and to the west were deposited in periplatform, slope, and basinal settings (Zempolich, 1995). Along the western edge of the Belluno Basin, slope and basinal sediments of the Vajont limestone and Fonzaso Formation onlap downfaulted blocks and margins of the east Trento Platform (Bosellini et al., 1981).

DOLOMITE FIELD RELATIONSHIPS

Regional and Stratigraphic Distribution of Dolomite

Regional field mapping has established that dolomitization is mostly confined to slope and basinal facies of Jurassic and Cretaceous sediments in the central and western Belluno Basin (Zempolich, 1991a, b; 1995).

Toward the east, the source of the resedimented ooids, only Vajont limestone is found in periplatform areas adjacent to the western margin of the Friuli Platform (e.g., Mt. Sestier); dolomite bodies are conspicuously absent (Figures 2, 3, 5, and 6). Toward the west, dolomite first occurs in the central Belluno Basin at the Vajont Dam and Col Visentin localities (Figures 2, 3, 5, and 6). At the Vajont Dam, dolomitization of slope and basin facies has resulted in formation of an isolated, rootless dolomite plume ≥300 m high and ~100–200 m wide. At Col Visentin, dolomitization occurs along small faults and fractures within a predominantly limestone section. Other occurrences of Vajont dolomite within the

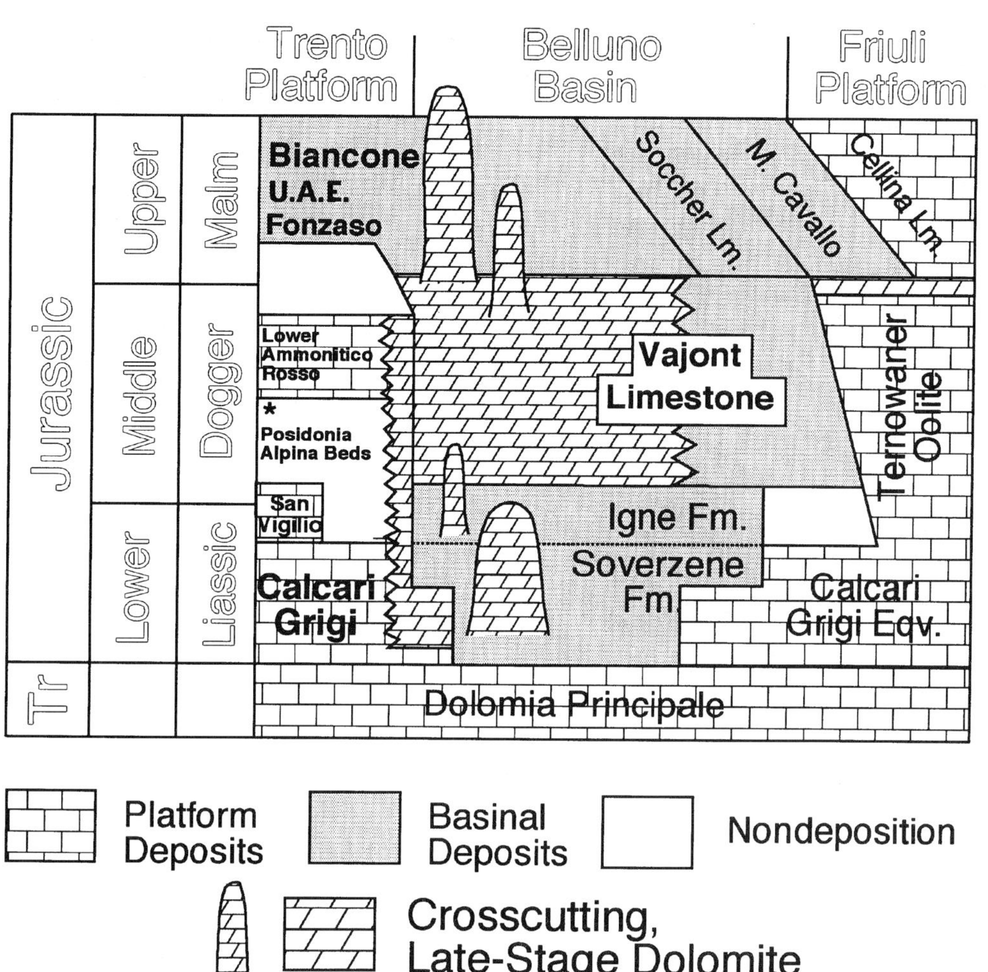

Figure 2. Stratigraphic age relationships and distribution of dolomite within the Mesozoic area of the Venetian Alps (stratigraphy modified from Bosellini et al., 1981). Massive replacement dolomitization is predominantly confined to the Middle Jurassic Vajont Limestone in the central and western portions of the Belluno Basin. Dolomite also occurs as plume-shaped bodies within the underlying Lower Jurassic Soverzene and Igne formations, and as dolomite "towers" in overlying Upper Jurassic–Lower Cretaceous strata following paleosynsedimentary dikes. Dolomitization has also affected the eastern margin of the South Trento Platform. These crosscutting relationships indicate that dolomitization occurred during or following the Early Cretaceous.

central Belluno Basin are found: (1) along the intersection of the Piave Graben and the mouth of the Vajont Canyon; (2) in the subsurface of the Piave Graben (i.e., the Belluno 1 well); and (3) at Villanova. The total extent and geometry of these last three bodies is poorly known due to limited exposure and data.

Within the western Belluno Basin, dolomite bodies span the entire Jurassic basinal succession and climb upward into the lower Cretaceous section (Figures 2, 3, 5, and 6). At Val Zoldo, an isolated, rootless dolomite plume ≥300 m high and ~100 m wide penetrates the Soverzene, Igne, and Vajont Limestone formations. At the San Boldo, Col dei Moi, and Val Sassuma sections, massive dolomitization (thickness >400 m over 20–25 km) has affected the Igne, Vajont Limestone, and Fonzaso formations that are now exposed in the crest of the Mt. Grappa–Visentin anticline. At Val Sassuma and Mt. Tomatico, dolomitization continues higher in the stratigraphic section along vertically oriented paleo-synsedimentary breccia, and locally replaces the Fonzaso, Ammonitico Rosso, and Biancone formations.

Massive dolomite is also found at the boundary between the western Belluno Basin and the eastern edge of the South Trento Platform, where an abrupt transition takes place from limestone platform facies to

a thin belt of dolomitized platform facies to completely dolomitized basinal facies (Figures 2, 3, and 5). At the platform margin, meter-scale occurrences of replacement dolomite are found along fractures and faults that penetrate platform strata (e.g., Upper Pliensbachian reef sediments, Mt. Grappa; Zempolich 1993, 1995). Platform strata associated with other large structural features such as the Seren Graben and other north-south–trending paleolineaments (Figures 2, 3) that comprise the eastern platform/basin boundary fault are also massively dolomitized, making stratigraphic correlations difficult (e.g., Dolomie Selcifere, Calcari Grigi, Toarcian–Aalenian) (Masetti, 1971; Trevisiani, 1991).

Structural and Crosscutting Relationships of the Dolomite Bodies

Regional and local detailed field mapping of dolomite bodies within Jurassic basinal sediments of the Belluno Basin indicates that dolomite bodies are linked to fracture and fault systems associated with Alpine deformation that was imposed on the Southern Alps during the Late Paleogene to Neogene (Figures 3, 7, and 8). The dolomite bodies include: (1) an extensive wedge-shaped body (~20–25 km long, 10–15 km wide, and 400–500 m thick) that has replaced the Vajont and other

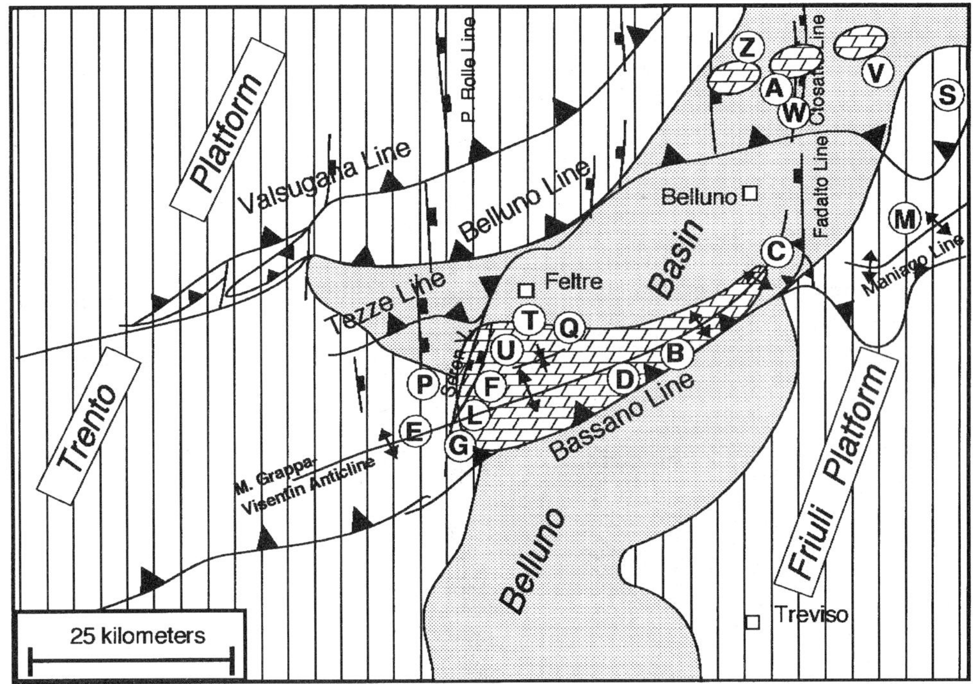

Figure 3. Tectonic map of the southern Alps (modified from Doglioni, 1990). Mesozoic platform and basin strata were thrusted toward the south during the Alpine orogeny (Late Oligocene to Recent) (Massari et al., 1986; Doglioni, 1990). Study localities are marked by circled letters. Southern study localities include: B = Passo San Boldo; C = Col Visentin; D = Col dei Moi; E = Grigno; F = Fontana Secca; G = Mt. Grappa; L = Valpore di Cima; M = Mezzamonte; P = Ponte Serra; T = Mt. Tomatico; U = Val Sassuma. Northern study localities include: A = Villanova; S = Mt. Sestier; V = Vajont Dam/Canyon; W = Belluno 1 exploration well; Z = Val Zoldo/Igne. Geographic distribution of dolomite bodies is noted by dolomite shade pattern. Most dolomite bodies occur within Jurassic and Cretaceous basinal sediments of the Belluno Basin. A major dolomite body, ~25 km × 15 km in area and ≥400 m in thickness, is present in the southern study area and is located within the crest of the Mt. Grappa–Visentin anticline. Isolated plume-shaped dolomite bodies, which are ~200–300 m in width and ≥300 m in height, occur in the northern study area and are hosted within footwall synclines of the Belluno thrust sheet.

basinal sediments present in the core of the Mt. Grappa–Visentin anticline (Figures 9, 10); (2) large, "rootless" dolomite plumes (hundreds of meters thick and high) that have penetrated upward into the Vajont Limestone from underlying Lower Jurassic strata (Figures 11–13); (3) large cylindrical and elliptical shaped plumes that penetrate upward from dolomitized Vajont Limestone through Upper Jurassic and Lower Cretaceous strata along Cretaceous-age synsedimentary breccias (Figure 14); and (4) smaller meter-scale dolomite bodies found along faults and fractures (Figure 15).

DOLOMITE BODIES AND REACTION FRONTS

Within the Belluno Basin, both large-scale (kilometer-scale) and small-scale (meters to hundreds of meters) dolomite bodies are present. Dolomitization fronts (Wilson et al., 1990) are noted in outcrop by an easily recognized and distinct transition from brown dolomite to blue limestone. In the southern study area (i.e., the massive dolomite wedge located within the hanging wall of the Mt. Grappa–Visentin anticline), transitions from partially dolomitized to completely

dolomitized limestone occur over distances of several tens of centimeters to hundreds of meters. In the northern study area, where small-scale dolomite plumes and fault-related dolomite bodies penetrate upward through the stratigraphic section, the dolomitization fronts are relatively sharp and occur over distances of centimeters to tens of centimeters.

Large-Scale Dolomite Bodies

Areal Distribution

In the southern study area, a massive areally extensive wedge of replacement dolomite is found within the Vajont Limestone and other basinal sediments on the hanging wall of the Bassano Line (Figures 3, 5, 7–10). The hanging wall is the southward-dipping limb of the M. Grappa–Visentin anticline, which trends N60–80°E. This dolomite body is >400 m thick at Passo di San Boldo and Col dei Moi and extends laterally 20–25 km, paralleling the overthrust from northeast to southwest. To the northeast, where the anticline wraps around toward the north-northeast, the dolomite exposures are reduced to several small occurrences (meter-scale) at Col Visentin. To the southwest, the Vajont limestone is completely

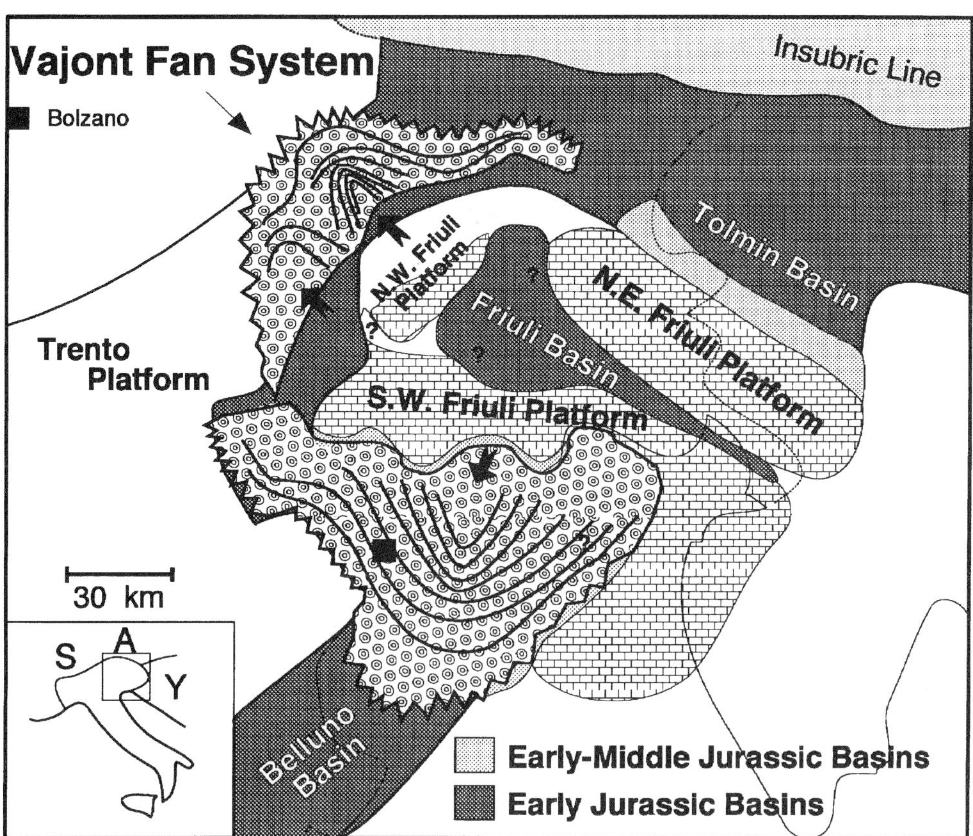

Figure 4. Distribution of the Vajont Limestone based on both outcrop and subsurface data (modified from Bosellini et al., 1981; Cati et al., 1987). The Vajont limestone thickens toward the Friuli Platform and attains a maximum thickness of ≤600 m near the platform margin. Toward the west, the Vajont limestone thins and onlaps portions of the Trento Platform.

dolomitized as far as the border fault associated with the eastern margin of the South Trento Platform (Figure 3). Mapping to the north of this front (Quero-Vas locality) suggests that the dolomite front wedges out over 10–15 km to the north-northwest, with the wedge thinning downward through the stratigraphic section.

In the central part of the Mt. Grappa–Visentin anticline, the dolomite front penetrates the crest of the anticline and passes upward through the Vajont Limestone and into the overlying Fonzaso Formation at Passo di San Boldo (Figures 9, 10). At Passo di San Boldo, a flower structure is recognized in the crest of the anticline by the presence of large-scale vertical to subvertical faults that were formed during regional transpression (Doglioni, 1990). These faults and related fractures are reflected in the present-day topography as small canyons oriented parallel to the crest of the anticline (Figures 9, 10). It is in these fault-controlled canyons that the dolomite fronts can be observed penetrating upward into the Fonzaso Formation. Elsewhere, the upper dolomitization front is typically found near the top of the Vajont Limestone or higher, where the stratigraphy is transected by vertically oriented faults or breccia (e.g., paleosynsedimentary breccia in Upper Jurassic and Lower Cretaceous strata; Val Sassuma and Mt. Tomatico). Offset of the stratigraphic section along these faults is minor, and dolomitization fronts can be observed at the top of the section where remnants of unaltered Vajont limestone are preserved in the east wall of the San Boldo Canyon. The Vajont Limestone is entirely dolomitized from Passo di San Boldo to Col dei

Moi and out into the leading edge of the anticline, where it disappears into the subsurface. Faulting apparently has controlled dolomitization, because lateral contacts between dolomite and limestone are commonly abrupt. Passing northward away from the crest of the anticline and into the Belluno thrust sheet, the limestone–dolomite contact stratigraphically drops within the upper 100 m of the Vajont limestone. This upper contact can be viewed in both the east and west hills on either side of the pass at Col dei Moi. At the base of Col dei Moi, limited exposure of the Vajont–Igne contact indicates that dolomitization has also affected the uppermost portion of the underlying Igne Formation. From here, the dolomite body disappears downward into the subsurface. Total thickness of the dolomite body along the Mt. Grappa–Visentin anticline may exceed 400–500 m where dolomitization of the underlying Igne and Soverzene formations has occurred.

Continuing toward the southwest along the Mt. Grappa–Visentin anticline, the Vajont Limestone is massively dolomitized in the vicinity of the eastern margin of the South Trento Platform (Figures 2, 3, and 5). In this region, both platform and basinal strata are complexly faulted due to Tertiary uplift. Also found in association with massive Vajont dolomite are isolated dolomite and limestone breccia bodies that penetrate upward through the Fonzaso, Ammonitico Rosso, and Biancone formations at the Val Sassuma and Mt. Tomatico localities (Figures 2, 5, and 14). These vertically oriented dolomite breccias are ellipsoidal in shape, penetrate 100–200 m into the Upper Jurassic–Lower Cretaceous

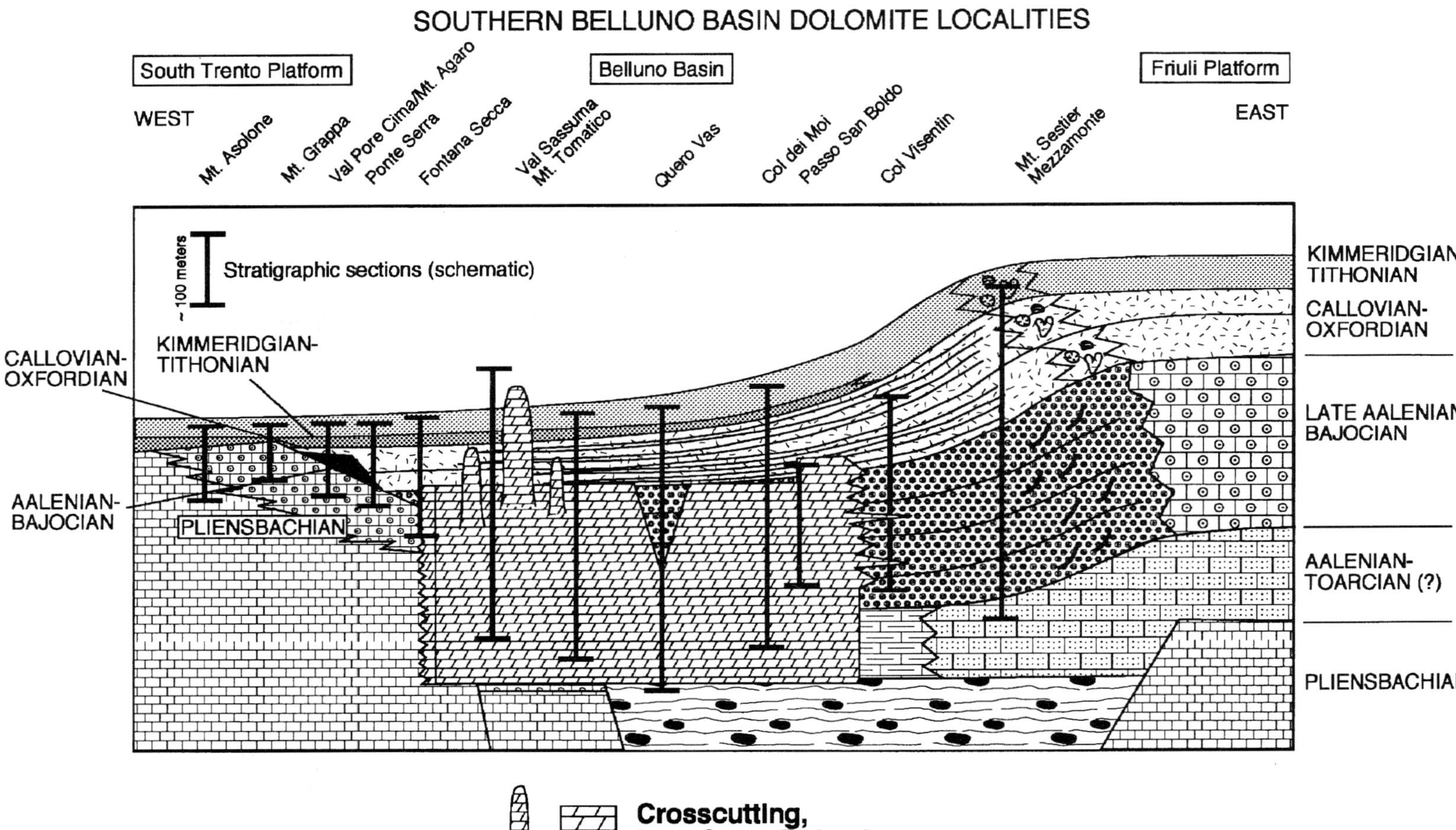

Figure 5. Paleogeographic depositional profile and stratigraphic cross section showing distribution of dolomite within the dominantly limestone Venetian Alps (southern study area). Outcrop localities, including sampling traverses and measured sections, are noted by black bars (Figure 3). Massive replacement dolomitization is predominantly confined to slope and basin facies of the Vajont Limestone, beautifully exposed in the core of the Mt. Grappa–Visentin anticline.

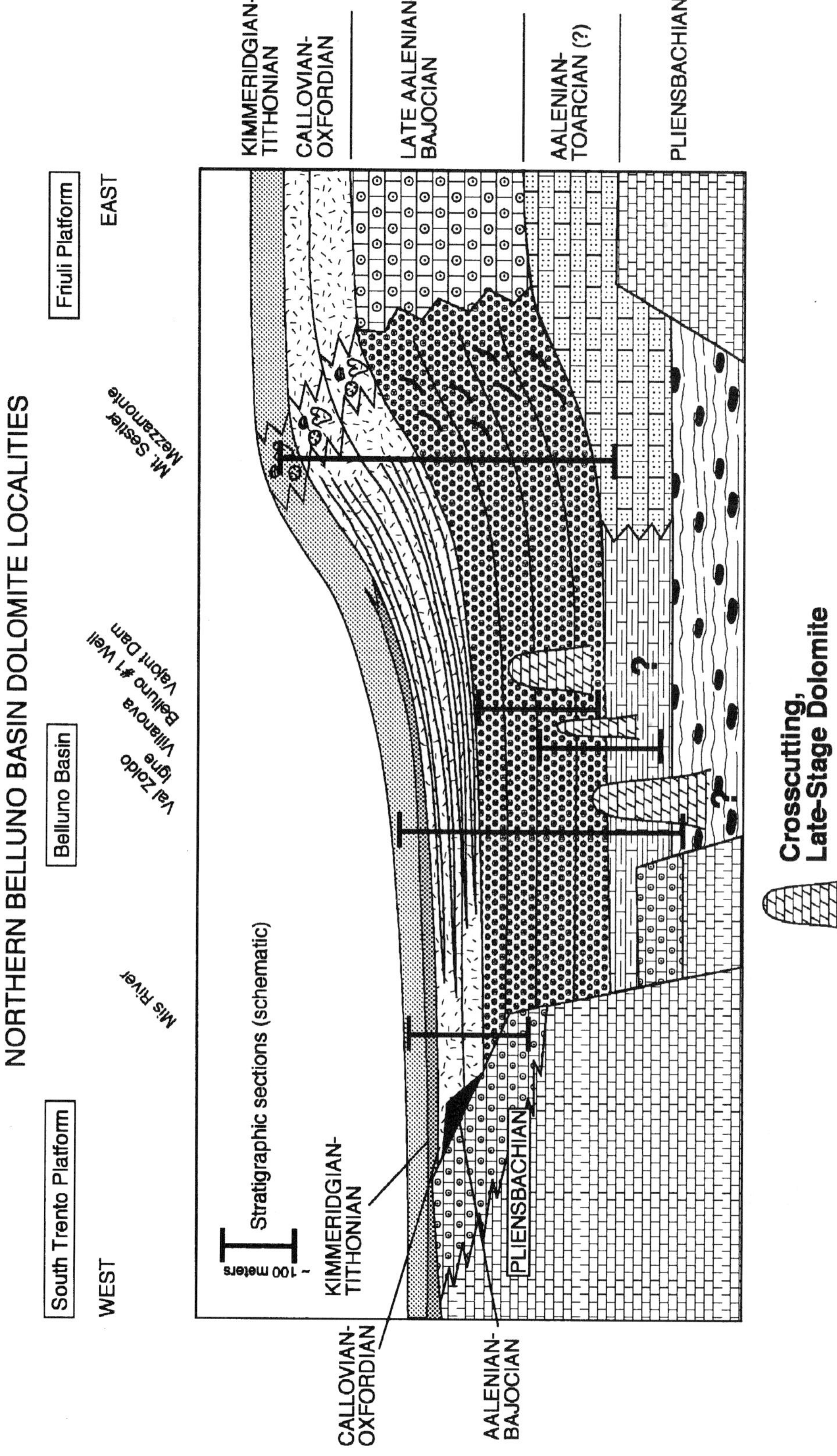

Figure 6. Paleogeographic depositional profile and stratigraphic cross section showing distribution of dolomite within the dominantly limestone Venetian Alps (northern study area). Outcrop localities, including sampling traverses and measured sections, are noted by black bars (Figure 3). Replacement dolomitization is confined to dolomite plumes that are hosted within slope and basin facies of the Soverzene, Igne, and Vajont Limestone formations.

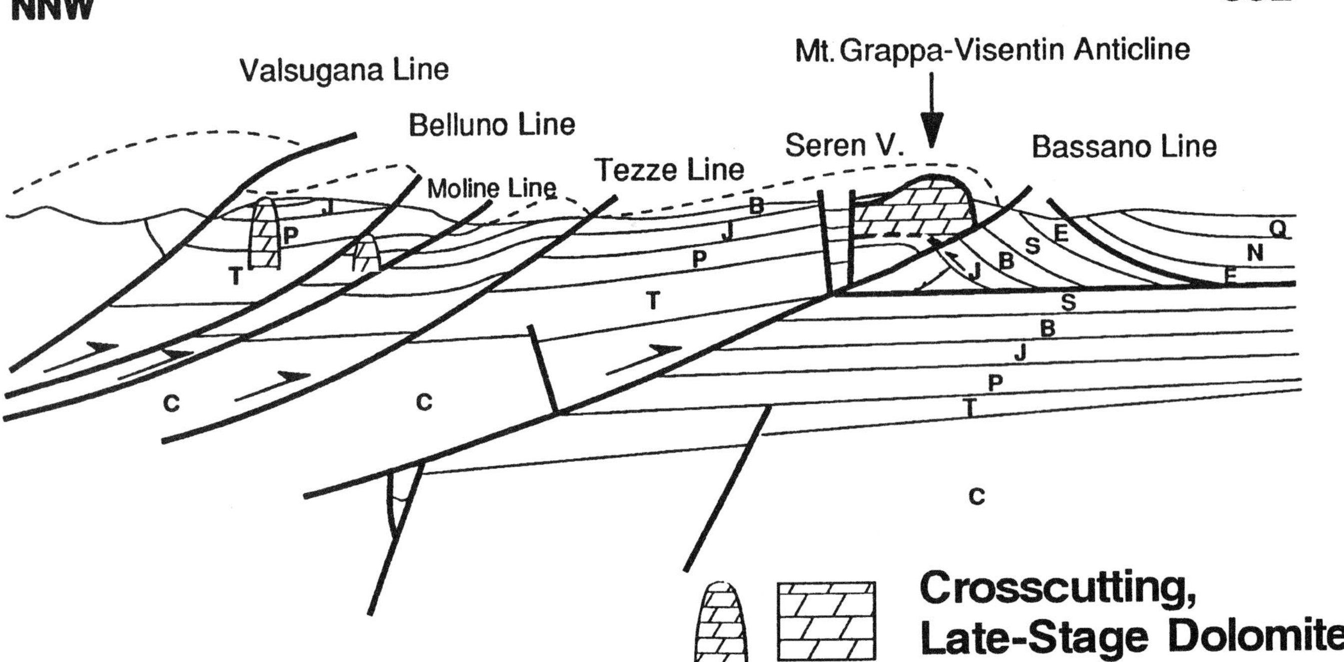

Figure 7. Distribution of dolomite within the present-day thrust and fold structural configuration of the Venetian Alps (structural interpretation modified from Doglioni, 1990). Dolomite bodies include a major wedge of dolomite located within the crest and hanging wall of the Mt. Grappa–Visentin anticline. Isolated dolomite plumes are found in footwall synclines located along the trailing edge of the Belluno thrust sheet. Movement of the thrusts can be accurately dated by analysis of sedimentation events and patterns in the Venetian foredeep (Massari et al., 1986; Doglioni, 1990). These data indicate that thrusting was initiated by the late Oligocene. Doglioni (1990) suggests that some thrusting may have begun even earlier, as indicated by onlap relationships of early Eocene sediment. Extensive uplift and subaerial exposure of the growing fold belt began during the early middle Miocene and continued into the Pliocene. T = Late Permian–Middle Triassic, B = Early Cretaceous (Biancone Formation), C = crystalline basement, E = Paleogene, J = Jurassic, N = Neogene, P = Late Triassic (Dolomia Principale), Q = Quaternary, S = Late Cretaceous (Scaglia Rosso Fm.).

section, and pass upward into undolomitized limestone breccia. The original brecciation of Jurassic–Lower Cretaceous limestone is interpreted by Doglioni (1990) and Masetti (1990, personal communication) to have occurred during formation of Cretaceous-age synsedimentary dikes. The presence of relic limestone breccia above dolomitized breccia indicates that the dolomitizing fluids originated from below the Upper Jurassic to Cretaceous section and ascended along the breccia, which was more permeable than surrounding bedded, chert-rich micritic limestone.

To the northeast, the massive dolomite body found in the core of the Mt. Grappa–Visentin anticline rapidly thins to several-meter-thick occurrences of dolomite at Col Visentin. Here, small-scale dolomite reaction fronts are found in association with minor faults (Figures 3, 5). Calculated volumes of this massive dolomite wedge range from 50 to 94 km³.

Isolated Dolomite Plumes

Areal Distribution

In the northern study area, several isolated dolomite bodies hundreds of meters in height and width are found at Vajont Canyon and Val Zoldo. These dolomite bodies are located on the trailing edge of the Belluno thrust sheet in or near the axis of an east-west–oriented footwall syncline (Figures 3, 7, and 8). These isolated plume-shaped bodies are oriented upward through the stratigraphic section and are cored by hydrothermal breccia. The occurrence of a succession of such isolated plume-shaped bodies along the same structural trend suggests that the dolomite bodies were formed by flow of Mg-bearing fluid along the axis of the east-west footwall syncline.

Vajont Canyon

Remnants of a massive plume-shaped dolomite body, >300 m high and several hundred meters wide, are found on both the north and south walls of the Vajont Canyon just to the west of the Vajont Dam (Figures 11, 12). The dolomite body is distinguished by a distinct color change from brown (dolomite) to blue (limestone) in both the north and south canyon walls. Along the north canyon wall, the vertically oriented dolomite body is discordant, with bedded limestone lying at low-angle dip, and has the shape of a simple upward-oriented plume (Figure 11). Discordance is

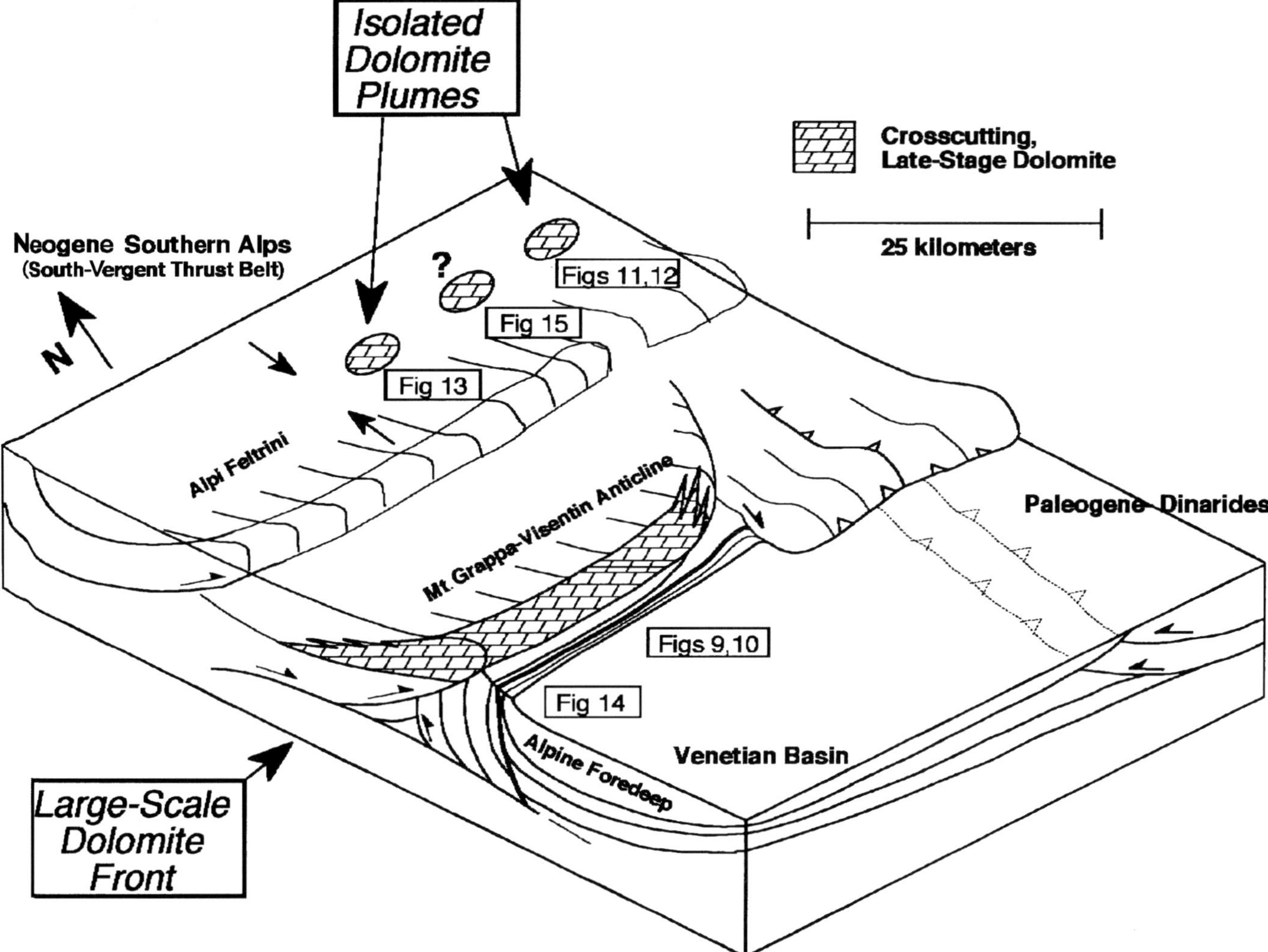

Figure 8. Geographic distribution of dolomite within anticlines and synclines of the Venetian Alps (structural interpretation modified from Massari et al., 1986; Doglioni, 1990). Major dolomitization is related to fault and fracture systems that are associated with anticlines and synclines formed during Tertiary compression (figure numbers refer to exemplary photographic plates or sketches of dolomite bodies). The major dolomite front located within the crest and hanging wall of the Mt. Grappa–Visentin anticline is ~25 km long, 10–15 km wide, and ≥400 m thick, and represents 50–94 km^3 of dolomite. Toward the northeast, the dolomite front thins and occurs within several meter-scale beds of Vajont limestone. In the central portions of the Mt. Grappa–Visentin anticline, the dolomite front climbs upward through the stratigraphic section. Toward the southwest, the dolomite front broadens and affects the eastern margin of the South Trento Platform. To the north, isolated dolomite plumes are located within footwall synclines of the Belluno thrust sheet. Each of these bodies represents ~2.4 × 10^{-2} km^3 of dolomite.

characterized by the abrupt disappearance of limestone bedding planes at the contacts between limestone and dolomite on each side of the plume. The presence of bedded limestone over the top of the plume marks the upper limits of the dolomite–limestone contact along the north canyon wall. At the base of the south canyon wall, the vertical margin of the dolomite plume is sharply discordant, with bedded limestone now lying at low-angle dip (Figure 12). Moving upward from the base, the dolomite body turns toward the west, becoming concordant with bedded limestone and eventually pinching out. Thus, the top of the dolomite body along both the north and south walls of the Vajont Canyon

appears to be confined to the upper Vajont limestone. Along the Vajont River, at the base of both north and south canyon walls, the dolomite plume disappears into the subsurface.

At the edges of the main dolomite plume, wedge-shaped apophyses of dolomite and associated limestone–dolomite transitions emanate from the main dolomite body and follow bedding planes and fractures into surrounding unaltered limestone (Zempolich, 1995). Dolomite–limestone transitions are narrow bands that range in thickness from several centimeters to several meters. At the center of the dolomite plume, the replacement dolomite is extensively brecciated and cemented

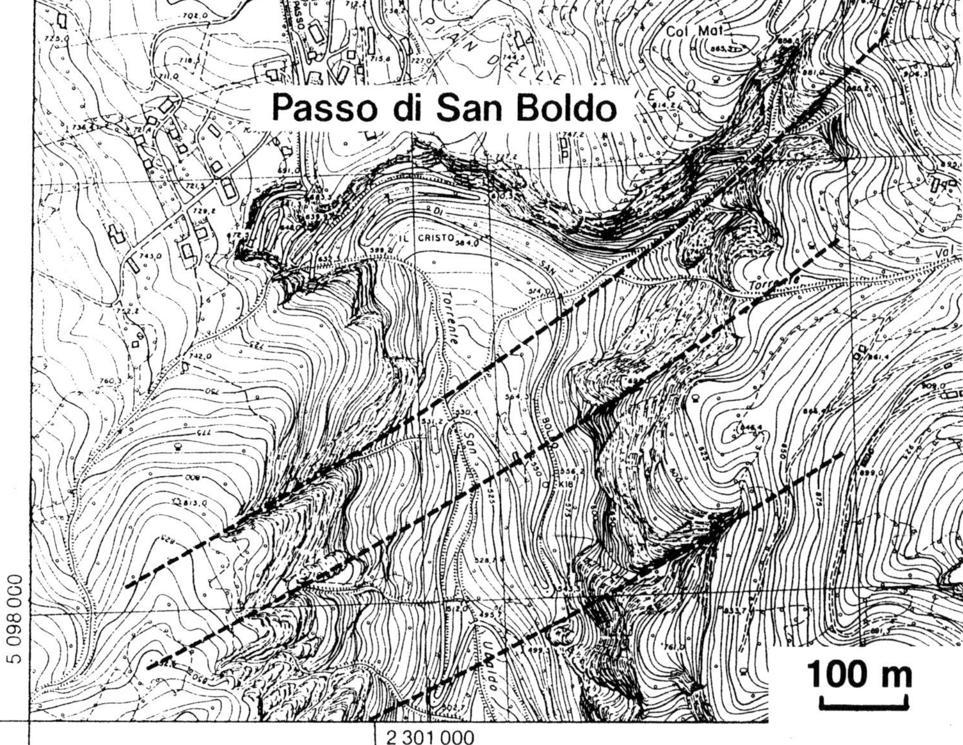

Figure 9. Dolomite in outcrop at Passo di San Boldo, Mt. Grappa–Visentin anticline. Top photograph: View looking northward into the crest of the Mt. Grappa–Visentin anticline. Tunnels and road climb through cliffs (≥400 m) composed of Vajont dolomite. This outcrop is part of an extensive wedge of dolomite that is hosted within the crest and hanging wall of the anticline (see Figures 3 and 7). Lower inset: Topographic map of the Passo di San Boldo area (5-m contour interval). Large-scale faults and fractures (dashed lines; Figure 10) are oriented parallel to the axis of the anticline, which trends N60°E. The dolomite front climbs upward along faults and fractures through the stratigraphic section at this locality; it has also affected the Upper Jurassic Fonzaso Formation. Present-day drainages accentuate the faults and fractures, which minimally offset the stratigraphic section ~10 m).

by thick linings of baroque dolospar cement, which completely fills interclast pore space. The angular nature of the dolomite clasts indicates that replacement dolomitization preceded brecciation and baroque dolomite cementation. Moreover, breccia and baroque dolomite cement are only found within the interior of the dolomite plume. These petrographic relationships demonstrate that brecciation and baroque dolomite cementation were the final diagenetic events associated with replacement dolomitization (Zempolich, 1995).

Dolomite is also found at the mouth of the Vajont Canyon (east wall of the Piave Graben), located ~0.5 km to the west of the main dolomite body just described. While similar replacement dolomite and breccia are present at this locality, the full extent of this dolomite body is unknown due to normal faulting during the late Tertiary, and downdropping of the stratigraphic section into the Piave Graben. However, this dolomite, together with small occurrences of replacement dolomite outcropping along the west wall of the Piave

Figure 10. Vertical to subvertical faults within the crest of the Mt. Grappa–Visentin anticline at Passo di San Boldo (view is toward the west wall of the pass; Figure 9). Arrows point to large faults and fractures that dissect cliffs composed of Vajont dolomite (≥400 m). Faults and fractures are oriented parallel to the anticline, which trends N60°E. The dolomitization front continues along the south limb of the anticline, where it disappears into the subsurface and forms the hanging wall of the Bassano thrust (left part of photograph; Figures 3 and 7).

Graben (e.g., at Villanova and Longarone), suggests the presence of a second dolomite body (plume?), much of which might be buried beneath the Piave Graben. The occurrence of dolomitized Vajont limestone in the Piave Graben is confirmed by an exploration well (AGIP Belluno 1 well), located just to the south of these outcrops, that penetrated and cored the Vajont Limestone directly beneath the Dolomia Principale Formation, which forms the hanging wall of a buried overthrust (AGIP, personal communication, unpublished well results). Exposure of Vajont dolomite along graben walls and within buried thrusts of the Piave Graben indicates that dolomitization of the Vajont occurred before down-faulting of segments of the Belluno and Moline thrust sheets (Figures 3, 7) into the Piave Graben during the Late Tertiary.

Val del Zoldo/Igne

The upper section of a large dolomite plume is found within the chert-rich micritic Soverzene and Igne formations near the town of Soffranco (Figure 13). This plume is located within the same footwall synclinorium as the plume described at Vajont Canyon, but is on the opposite side of the Piave Graben. Large clasts of dolomitized Soverzene and Igne carbonate, fractured chert clasts, dolomitized geopetal silt, and baroque dolospar characterize the hydrothermal breccia found in the center of this body (Zempolich, 1995). The brecciated core of this body is ~100 m wide and ≥200 m high. Along the edges of the body, thin "fingers" of dolomite breccia (tens of centimeters to several meters thick) follow bedding

planes for ≤10 m before grading into cherty argillaceous micrite. Angular dolomite clasts and the presence of dolomitized geopetal silt indicate that replacement dolomitization preceded and overlapped brecciation, and preceded the precipitation of baroque dolomite cement. These petrographic relationships are similar to those found in the breccia at the Vajont Dam locality and carry the same implication; that is, the brecciation and baroque dolomite cementation were the final diagenetic events associated with replacement dolomitization.

Directly above the main body of the dolomite plume, replacement dolomitization can be followed upward from the brecciated dolomite core along small fractures and faults (Figure 13). Replacement dolomitization continues along these pathways stratigraphically upward through the Soverzene and Igne formations and into the overlying Vajont Limestone. A replacement dolomite halo is present within nonbrecciated Vajont lithologies exposed in the cliff above the plume and in nearby outcrops lacking breccia located behind the cliff along trails leading west from the town of Igne. Examination of limited exposures of the Fonzaso and Ammonitico Rosso formations suggests that they were not affected by dolomitization at this locality. Near the town of Igne, undolomitized Igne and Vajont bedded limestone is exposed. These field relationships indicate that replacement dolomitization was restricted to the plume-shaped dolomite breccia body found within the Soverzene, Igne, and Vajont strata at Soffranco, and

(A)
LS
LS
LS
DOL
LS
DOL
DOL
LS
DOL
DOL
LS
DOL
(B)
8 m

suggest that dolomitizing fluids ascended from depth. Reaction fronts and textural transitions between initial replacement dolomite at Val del Zoldo/Igne are similar to those observed at the Vajont Dam.

Small-Scale Dolomite Occurrences

Meter- to decameter-scale dolomite bodies occur as isolated reaction fronts, or occur in association with large-scale dolomite bodies. These small-scale dolomite bodies provide important petrographic and geochemical evidence of fluid movement, the mechanism by which precursor limestone was replaced, and the formation of pore space through dolomitization (Zempolich, 1995). Small-scale dolomite bodies occur as: (1) small wedges (10–30 m) found parallel and sub-parallel to fault, fracture, and bedding planes; and (2) strata-bound beds. Small-scale reaction fronts between dolomite and limestone are seen in outcrop as sharp fronts, on the scale of centimeters or less, and transitional fronts over a distance of several meters, grading from zones of completely dolomitized rock to partially dolomitized limestone to unaltered limestone. Detailed analysis of closely spaced samples across these fronts indicates that Vajont dolomite textures and compositions show progressive textural and compositional maturity with increasing proximity to fluid conduits (Kupecz and Land, 1994; Zempolich, 1995). A particularly well defined example of a transitional front is found at Villanova. Here, a small dolomite body ~20–30 m wide is exposed along a road cut that dissects a small rollover anticline in the Vajont limestone. Dolomite-to-limestone transitions at this locality occur along a small fault that cuts obliquely across the bedding and within bedded limestone. The dolomite front on the south side of the body is wedge-shaped, narrowing upward along the fault plane (Figure 15), pinching out obliquely beneath undolomitized bedded limestone where the fault soles out into a bedding plane. The front on the north side of the body is a simple gradation from partially dolomitized limestone to unaltered limestone within an individual bed, and has been described in detail by Zempolich (1995).

Figure 11. (A) Large dolomite plume exposed along the north wall of the Vajont Canyon (tunnel located on the right side of dolomite body is ~8 m high; downward and upward dimensions of this photo montage are distorted by the camera angle). Dolomite (DOL) is dark brown in outcrop and crosscuts bedded limestone (arrows), which is light blue (LS). The plume is 200–300 m wide and >300 m high. It is bounded above by bedded limestone and disappears below into the subsurface. (B) Schematic of (A) depicting the large dolomite plume exposed along the north wall of the Vajont Canyon. Dolomite in stipple pattern. Tunnel located on the right side of dolomite body is ~8 m high.

STRATIGRAPHIC CONSTRAINTS ON THE TIMING OF DOLOMITIZATION

The occurrence of crosscutting relationships between dolomite bodies and the host Jurassic and Cretaceous strata places constraint on the timing of dolomitization. These data indicate that dolomitization occurred during or following the Early Cretaceous and was focused along structural features related to paleolineaments and Tertiary-age deformation.

Lower Jurassic Dolomite Bodies

Dolomite bodies hosted in the Lower Jurassic Soverzene and Igne formations occur as "rootless" plumes, and pass into dolomite found in the overlying Vajont Limestone. At the Val del Zoldo locality (Figures 2, 6, and 13), the large, partly brecciated dolomite body penetrates upward through cherty, dark micritic limestones of the Soverzene and Igne formations and into the overlying Vajont Limestone. Some brecciation of Soverzene limestone is attributed to the occurrence of synsedimentary (Lower Jurassic) growth faults and slumps, examples of which are widespread in the western Belluno Basin and Alpi Feltrine (Masetti and Bianchin, 1987). Dolomitization of the Igne Formation is also observed in the hanging wall of the Mt. Grappa–Visentin anticline and underlies a thick occurrence of dolomitized Vajont limestone (Figures 2, 3, 5, and 7). At these localities, crosscutting relationships of the Lower and Middle Jurassic dolomite bodies indicate that dolomitization must have occurred during or following the Middle Jurassic.

Upper Jurassic Dolomite Bodies

Bodies of dolomitized Vajont limestone located in fault zones cut stratigraphically upward into the overlying Upper Jurassic section [e.g., the Mt. Grappa–Visentin anticline (Figures 2, 3, 5, and 7)]. Such dolomite crosscutting relationships indicate that dolomitization of the Vajont Limestone must have occurred during or following the Late Jurassic. Furthermore, the large dolomite wedge of dolomitized Vajont limestone associated with crestal faults within the Mt. Grappa–Visentin anticline suggests that dolomitization may be related to Tertiary deformation and the formation of the Venetian Alps thrust belt during the late Oligocene–Recent.

Lower Cretaceous Synsedimentary Breccia

At Val Sassuma and Mt. Tomatico, massive Vajont dolomites can be traced upward into the Fonzaso, Upper Ammonitico Rosso, and Biancone formations (Figures 2, 6, and 14). These vertically oriented breccias are roughly columnar in shape, penetrate 100–200 m into the Upper Jurassic–Lower Cretaceous section, and pass upward into limestone breccia. The original brecciation of Jurassic–Lower Cretaceous limestone is interpreted by Doglioni (1990) to have occurred during formation of Cretaceous-age synsedimentary dikes in association with extensional tectonics. At these

(A)

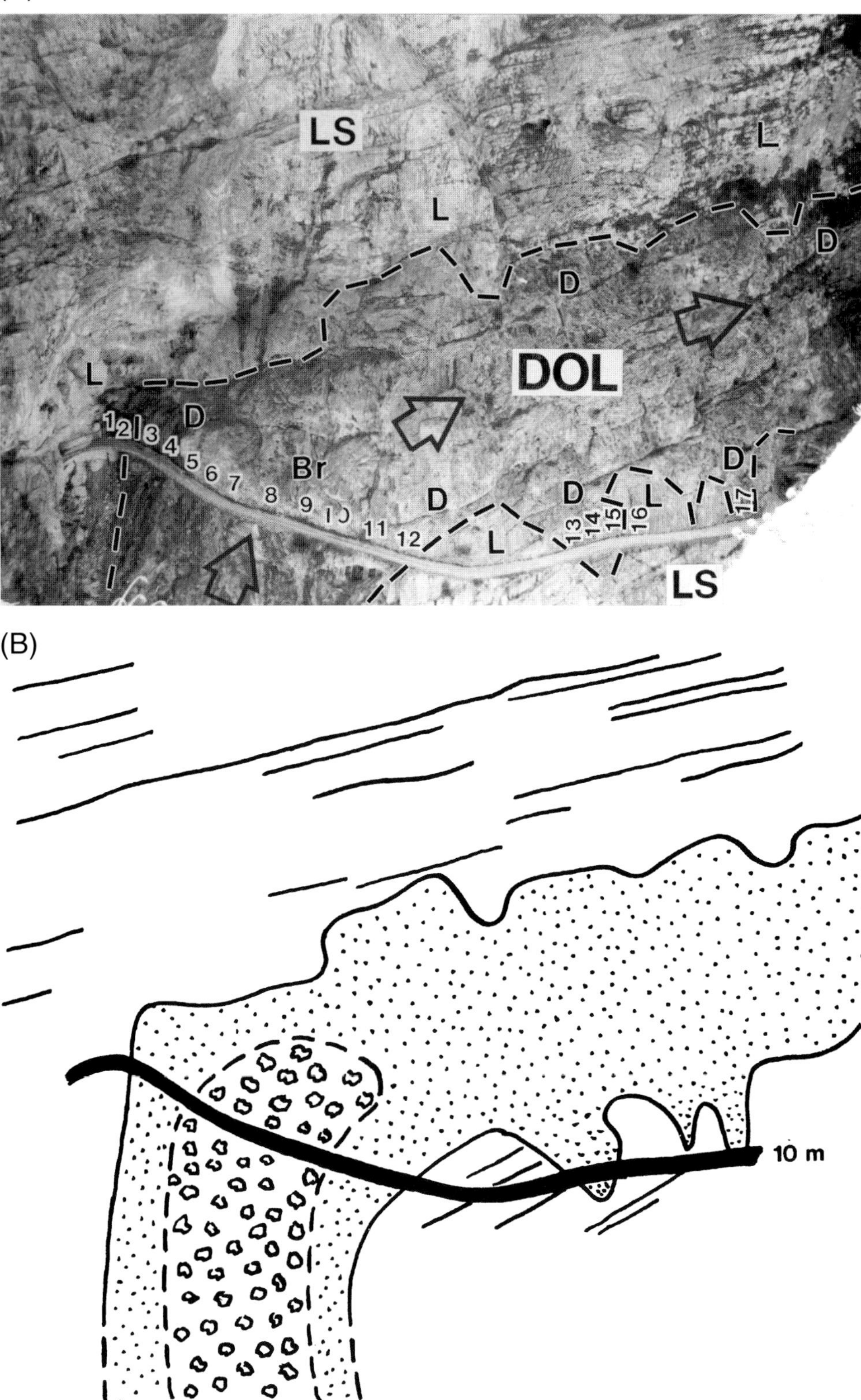

Figure 12. (A) Large dolomite plume exposed along the south wall of the Vajont Canyon (the road that is visible crossing the dolomite body is ~10 m wide; downward and upward dimensions of this photo montage are distorted by the camera angle). Dolomite (DOL) is dark brown in outcrop and crosscuts bedded limestone (dashed lines), which is light blue (LS). The dolomite plume is cored by hydrothermal breccia (Br), which is composed of replacement dolomite clasts and baroque dolomite cement. The plume is 200–300 m wide and >300 m high. Toward the top, the dolomite body becomes concordant with bedded limestone and eventually pinches out. At the base, the body disappears below into the subsurface. Numbers refer to detailed sampling that was conducted along the dam access road, which crosses the plume (Zempolich, 1995). Arrows point in the direction of probable fluid flow during dolomitization. (B) Schematic of the large dolomite plume exposed along the south wall of the Vajont Canyon. Dolomite in stipple pattern; breccia core noted by clast pattern. The road that is visible crossing the dolomite body is ~10 m wide.

localities, dolomite crosscutting stratigraphic relationships indicate that dolomitization of these synsedimentary breccias must have occurred during or following the Early Cretaceous.

Collectively, crosscutting stratigraphic relationships of dolomite bodies observed throughout the study area indicate that massive dolomitization of Vajont and other Jurassic and Lower Cretaceous basinal strata occurred during or following the Early Cretaceous. To provide a further constraint on the timing of dolomitization, the relationship between Vajont dolomite bodies and Tertiary structural elements is examined below.

RELATIONSHIP BETWEEN DOLOMITE BODIES AND ALPINE DEFORMATION

Dolomite bodies distributed within Mesozoic basinal sediments are exposed within hanging-wall anticlines and footwall synclines that were formed during late Oligocene–Recent thrusting (Figures 3, 7, and 8). The specific distribution of dolomite along these structural features suggests that extensional and transpressive faulting in the axes of these anticlines and synclines controlled the circulation of dolomitizing fluid. If this is true, dolomitization of Mesozoic basinal sediments must have occurred sometime between the initiation of thrusting and regional epierogenic uplift (i.e., subaerial exposure) of the sequence.

The timing of initial thrusting and the uplift history in the southern Alps has been accurately dated by studies of pre- and synorogenic sediments that were deposited and shed into the surrounding foredeeps. Deformation of the southern Alps began during the Late Cretaceous, and progressed from to east to west due to the oblique convergence of Europe and Adria during the Tertiary (Massari et al., 1986). To the west of the study area, initial deformation is recorded by the deposition of Late Cretaceous and Early Tertiary flysch in both the Lombard and western Venetian basins (Gaetani and Jadoul, 1979; Massari et al., 1986). To the east, thrusting related to late Eocene compression (Dinaric fold belt) deposited flysch in the eastern Venetian basin (Doglioni, 1990). Major deformation of the study area (i.e., the central Venetian Alps) occurred during the later phases of regional transpressive deformation (Late Oligocene–Recent) and is characterized by a series of south-vergent thrusts (trending N60–80°E) that involve crystalline basement to the north (Doglioni, 1990). Major thrusts, from south to north, include the Bassano-Maniago, Tezze, Belluno, and Valsugana (Figure 3). However, onlap relationships and angular unconformities between early Eocene flysch and late Oligocene molasse in the Belluno syncline (i.e., the trailing edge of the Mt. Grappa–Visentin anticline) suggest that initial detachment and thrusting may have begun earlier (Doglioni, 1990).

Prior to major thrusting and development of the Venetian foredeep, marine siliciclastic and carbonate shelf sediments of late Oligocene to early middle Miocene age were deposited regionally across the study area under the influence of the Dinaric fold belt to the east (Chattian to Langhian cycle; Massari et al., 1986). Major thrust movement and loading in the study area began at least by early middle Miocene time (Serravalian) and led to foreland subsidence and thick accumulation of hemipelagic marls and mudstones, an intermediate basin-fill sequence, and fan-delta and alluvial deposits (Serravalian to Recent cycle; Massari et al., 1986). This synorogenic sedimentary sequence marks a change in the polarity of sedimentation of the Venetian Basin and records the first major movement of thrust sheets toward the south. Uplift and denudation of these thrust sheets in the Late Tertiary can be accurately dated by the inclination and alteration of molasse sedimentation patterns, which were shed off of the growing anticlines. Progressive inclination of sediment packages and formation of angular unconformities occur along both dip and strike sections (Massari et al., 1986; Doglioni, 1990). These data indicate that major thrusting and uplift took place rapidly from the early middle Miocene to the late Pliocene (~10 Ma).

The spatial association of Vajont dolomite with these Tertiary thrust features (Figures 3, 7, and 8) suggests that dolomitization of Mesozoic basinal sediments may have occurred within these thrust sheets sometime between the early Eocene/late Oligocene and the late Oligocene/early middle Miocene (in agreement with stratigraphic crosscutting relationships exhibited by dolomite bodies, which indicates that dolomitization postdated the Lower Cretaceous). During this time range, the study area, and specifically the Mesozoic basinal succession, was still buried beneath several kilometers of section and was located beneath coastal and marine environments. The entire region experienced major uplift during the early middle Miocene to late Pliocene. If uplift and surficial expression of the Mt. Grappa–Visentin anticline and other compressional structures occurred by the middle Miocene (Massari et al., 1986), structural constraints place the timing of dolomitization somewhere between the initiation of compressional tectonics during the early Eocene, and the uplift and exposure of the advancing thrust sheets during the early middle Miocene. Dolomitization must have been completed prior to extensive uplift because (1) significant topographic expression of the anticline would have initiated meteoric recharge, and theoretically would have shut down the subsurface circulation of Mg-bearing fluid; and (2) metastable dolomite replacement textures and fronts are beautifully preserved in these outcrops, which suggests that dolomitization preceded uplift.

In summary, field and stratigraphic relationships indicate that dolomitization of Mesozoic-age sediments in the Venetian Alps is mostly confined to slope and basin facies contained in the Belluno Basin. Within these basinal strata, the majority of dolomite occurs as massive and isolated bodies within the Vajont Limestone and as isolated bodies beneath and above Vajont dolomite. Structural and stratigraphic crosscutting relationships collectively suggest that late-stage dolomite bodies within the Vajont and other basinal sequences were formed following the Lower Cretaceous. The spatial distribution of these dolomite bodies within otherwise tight basinal strata and their relationship to Tertiary-aged compressional structures, other paleolineaments, and (paleo)synsedimentary faults suggest that dolomitizing fluids were focused along zones of structurally enhanced porosity and permeability during Tertiary deformation. The timing of these Alpine structural events indicates that dolomitization occurred sometime between the early Eocene and the early middle Miocene during initial compression and prior to rapid uplift of the region during the early middle Miocene to the Pliocene.

These conclusions about the dolomitization of the Vajont and other basinal strata of the Belluno Basin are quite different from those reached by Cervato (1990) for the dolomite bodies in the nearby Lessini Mountains

(A)
Vj
Ig
Sz
(B)
Truck

Figure 14. Dolomitized Vajont limestone and Upper Jurassic–Lower Cretaceous dolomite breccia, Val Sassuma, Mt. Grappa–Visentin anticline. (A) Panorama of the south wall of Val Sassuma depicting the Mesozoic basinal stratigraphic succession composed of the Vajont Limestone (Vj), the Fonzaso Formation (Fz), and the Biancone Formation (B). The Ammonitico Rosso (2–4 m thick) is found just above the Fonzaso Formation. The Vajont Limestone, which is 400–500 m thick at this location, is massively dolomitized. The dolomite front is located near the contact between the Vajont Limestone and the Fonzaso Formation, except where the front climbs upward through the stratigraphic section (B–D) (Figures 2, 6). In these instances, partial dolomitization of paleo-synsedimentary breccia has formed erosion-resistant "towers" (arrows in C, closeup in B and D) [Doglioni (1990) credits these synsedimentary breccias as having originally formed during the Cretaceous]. Higher in the stratigraphic section, these paleosynsedimentary breccias are composed of limestone.

Figure 13. (A) Large dolomite plume exposed along Val Zoldo, across from the village of Soffranco (truck is shown for scale; upward dimensions of this photo montage are distorted by the camera angle). Dolomite plume is cored by hydrothermal breccia that is composed of clasts of dolomitized carbonate, chert, and baroque dolomite cement. Dolomite "fingers" protrude from the main dolomite body and penetrate into surrounding limestone along select beds and bedding planes. The breccia core is ~100 m wide and ≥200 m high. The plume penetrates upward through the Soverzene (Sz) and Igne (Ig) formations; a replacement dolomite halo is present within the Vajont Limestone (Vj) in the cliff above. (B) Schematic of (A) depicting the large dolomite plume exposed at Val Zoldo, across from the village of Soffranco. Dolomite in stipple pattern; breccia core noted by clast pattern. Dolomite fingers protrude from the main dolomite body and penetrate into surrounding limestone along select beds and bedding planes. Truck is shown for scale.

Figure 15. Small-scale dolomite–limestone reaction front, Villanova locality. This mineralogic transition forms the left side of the dolomite wedge observed. Dolomite fronts (brown; DOL) emanate from a fault (F) and propagate (open arrows) toward the left into unaltered limestone (light blue; LS) and toward the right into the core of the dolomite wedge. Arrows along the fault point toward the probable direction of fluid flow.

(southern Trento Platform). In the latter area, where dolomite is found predominantly in platform strata, Cervato attributes the dolomitization to the hydrothermal circulation of seawater related to the emplacement of magmatics within the southern Trento Platform during the Tertiary. The absence of magmatics in the present study area and the presence of a thick sequence of platform limestone (central Trento Platform) between Tertiary volcanics to the southwest and the Belluno Basin suggest that dolomitization of the Vajont and other basinal limestone was unrelated to the hydrothermal dolomitization of platform sequences further to the southwest (Cervato, 1990).

PETROGRAPHY AND GEOCHEMISTRY

Limestone Components

The undolomitized Vajont ooids typically display cortices that are composed of radially oriented, small

subequant to bladed calcite crystals (Zempolich, 1995). Radial calcite fabrics, nonluminescence, enriched ^{13}C and ^{18}O isotopic compositions (average δ^{13}C = +2.13‰ and δ^{18}O = –3.12‰), low covariant Sr-Mg contents, low Mn-Fe contents, and an absence of neomorphic texture collectively suggest that Vajont ooids were originally composed of radial low-Mg calcite and underwent little diagenesis prior to dolomitization (Zempolich, 1995). These data suggest that Vajont ooids were redeposited in the Belluno Basin as relatively pristine, mineralogically stable low-Mg calcite.

Aragonitic and high-Mg calcite skeletal grains in the Vajont limestone that were deposited along with radial ooids in gravity flows are now replaced by low-Mg calcite (Zempolich, 1995). These grains exhibit a spectrum of fabric-retentive and fabric-destructive neomorphic fabrics, and possess enriched ^{13}C and ^{18}O isotopic compositions similar to radial ooids. This suggests that original metastable components were altered to low-Mg calcite early in the diagenetic history of the limestone.

Intergranular pores in resedimented ooid grainstone were first cemented by pore-lining, nonluminescent, equant low-Mg calcite cement. Nonluminescent equant calcite cement occurs as thin isopachous linings in resedimented grainstone, as intraskeletal pore fill, and as isopachous linings in skeletal molds formed through the dissolution of original aragonite (Zempolich, 1995). Isotopic compositions of nonluminescent equant calcite are enriched with respect to other calcite cements and fall within the field defined by radial calcitic ooids. Analogous, equant low-Mg cement has been described in modern slope and basin settings by Schlager and James (1978). Based on these data, isopachous nonluminescent equant cement is interpreted as an early marine precipitate in slope settings. Furthermore, its occurrence in the ooid grainstones as a pore-lining phase in primary intergranular voids, intraskeletal pore space, and within skeletal molds indicates that precipitation began soon after deposition of carbonate in slope settings and continued during shallow burial diagenesis.

Late diagenetic calcite fabrics include banded luminescent equant calcite that overlies nonluminescent equant calcite and fills remaining intergranular pore space, coarse luminescent calcite that fills molds of skeletal grains, and fracture-filling luminescent calcite that crosscuts all previously described fabrics. Progressive depletion in oxygen values from banded luminescent calcite to mold-filling luminescent calcite to fracture-filling luminescent calcite suggests progressive cementation in a burial environment (Zempolich, 1995). Importantly, late calcite cement occluded the majority of intergranular, intragranular, and moldic porosity that remained after early cementation and dissolution in slope and shallow burial environments.

In summary, early and late diagenesis of Vajont sediments in basinal settings resulted in the formation of a relatively impermeable and mineralogically stable (low-Mg calcite) volume of rock (Zempolich, 1995).

Limestone–Dolomite Reaction Fronts

Dolomite bodies within the Vajont Limestone in both the southern and northern study areas exhibit a spectrum of replacement and recrystallized dolomite fabrics and variable Ca-Mg compositions that illustrate the initial nonmimetic replacement of limestone and progressive stabilization of intermediate dolomite phases (e.g., Kupecz et al., 1993; Kupecz and Land, 1994). These textures and compositions are distributed over centimeter- to meter-scale transitions from partially dolomitized limestone to completely dolomitized limestone and exhibit a concomitant increase in the degree of neomorphism and recrystallization with increasing proximity to fluid conduits (i.e., fractures, faults, and bedding planes). The inherent metastability and evolution of these initial and intermediate dolomite fabrics, as defined by petrographic and geochemical study, has been explored by Zempolich (1995).

Transitions from limestone to dolomite occur over several tens of centimeters to tens and hundreds of meters in relationship to faults and fractures (Figure 15). Macroscale replacement fabrics include the gross retainment of lithoclastic grains and sedimentary structures through variations in the size of replacement dolomite rhombohedra (Zempolich, 1995). Microscale dolomite textures, which record the initial step-by-step replacement of limestone by dolomite and the neomorphism and recrystallization of initial replacement dolomite fabrics, are distributed across dolomite–limestone transitions. These petrographic data define the mechanism by which limestone was progressively replaced by dolomite, and by which initial replacement fabrics were progressively recrystallized.

Replacement Dolomite

Initial replacement fabrics are found toward the periphery of dolomite reaction fronts within partially dolomitized limestone (Figure 16). Initial replacement dolomite is composed of calcian dolomite that contains inclusions of relic calcite (Zempolich, 1995). Two types of initial replacement styles are exhibited by the Vajont dolomite: *intragranular replacement*—initial dolomitization begins with the selective dolomitization of ooids and other grains within oolite where intergranular calcite cement has completely occluded pore space; and *intergranular replacement*—initial dolomitization begins with the dolomitization of ooids and intergranular matrix (i.e., carbonate mud) preferentially along grain peripheries (Zempolich, 1995). Field and petrographic data suggest that these different replacement styles are dependent on the degree of cementation within the precursor limestone fabric and original carbonate mud content. Recrystallized replacement fabrics are found in completely dolomitized limestone nearest to faults and fractures.

Cathodoluminescent petrography and microprobe analysis of initial replacement fabrics indicate that replacement and recrystallized dolomite found in both the northern and southern study areas luminesces a homogeneous dull red color, and that individual crystals are not compositionally zoned. Cathodoluminescence also reveals that replacement dolomite crosscuts ooid grains, pore-lining nonluminescent equant calcite cement, and banded-luminescent equant calcite spar cement (Zempolich, 1995). The widespread uniformity in luminescence and lack of compositional zoning suggests that initial replacement by calcite-inclusion–rich, nonstoichiometric dolomite, and neomorphism of these phases to more stoichiometric compositions, was the product of one progressive dolomitization event (Zempolich, 1995). Postdolomitization processes include the recrystallization, dedolomitization, and dissolution of initial calcite-inclusion–rich replacement fabrics.

These petrographic observations are important for several reasons. First, a replacement origin for dolomite in the Vajont limestone is inferred by the pervasive retainment of ooid ghosts in both dolomitized matrix and lithoclasts (Figure 16). These observations indicate that precursor limestone was not wholly dissolved, and later reprecipitated as dolomite in voids. Second, a late replacement origin for the dolomite is indicated by cathodoluminescent study that indicates replacement dolomitization occurred sometime after calcite cementation in burial settings.

Baroque Dolomite Cement

Baroque dolomite cement is found in association with replacement dolomite along both large-scale and small-scale fractures within the Mt. Grappa–Visentin anticline, and as massive pore fill within brecciated cores of dolomite plumes located in the northern study area (Figures 11, 12). Baroque dolomite cement was not observed within limestone or along fractures within undolomitized limestone. This suggests that faults and fractures were the conduits by which dolomitizing fluid circulated, and that baroque dolomite cement was a final pore-filling phase that precipitated after replacement dolomitization.

Regional Stable Isotopic Geochemistry

Compositions of replacement dolomite exhibit a wide range of $\delta^{18}O$ and a relatively narrow range of $\delta^{13}C$ values that overlap the Middle Jurassic marine carbonate compositions (Zempolich, 1995). Regional $\delta^{13}C$ and $\delta^{18}O$ compositions of replacement dolomite and baroque dolomite cement are summarized in Figure 17. The ^{18}O of replacement dolomite in northern dolomite localities is depleted relative to replacement dolomite located along the Mt. Grappa– Visentin anticline. Compositions of baroque dolomite cement exhibit depleted ^{18}O compositions relative to replacement dolomite and Middle Jurassic marine carbonate, and possess variable carbon compositions. The ^{18}O of baroque dolomite cement appears to be uniformly depleted throughout the region. These data, in addition to fluid inclusion data (T_h = 80–132°C, mean = 125°C, n = 12), suggest that replacement dolomitization and baroque dolomite cementation occurred at elevated temperature, and that dolomite replacement in the northern study area took place at higher temperatures than that of the southern study area (Zempolich and Hardie, 1991a, b; Zempolich, 1995).

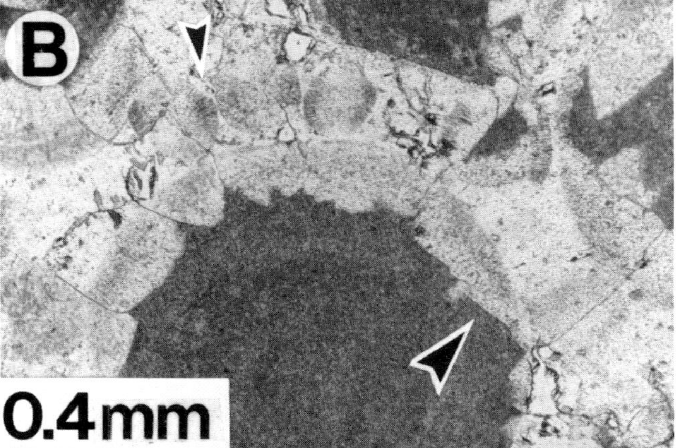

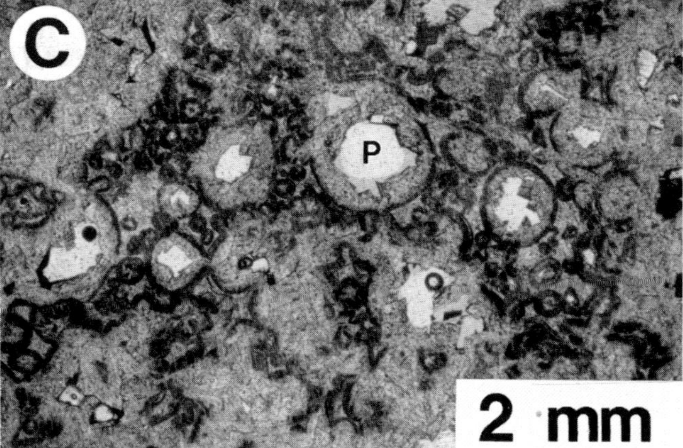

Figure 16. Dolomite replacement textures (plane light and cross-polarized light photomicrographs. (A, B) Partly replaced oolitic limestone. Replacement rhombohedra have preferentially nucleated within intergranular matrix and along the periphery of ooids. Precursor ooid structures are defined by a greater density of calcite inclusions within replacement rhombohedra (arrows). The dolomite–limestone contact between radial ooid cortices and replacement rhombohedra is sharp. (C) Replacement dolomite (partially recrystallized) with moldic pores "P". Complete replacement dolomitization of oolite results in formation of ≤10%–15% porosity.

Trace Elements

A characteristic of Vajont replacement dolomite in both the southern and northern study areas is a low concentration of Sr, Fe, and Mn (e.g., Col Visentin: Sr = 62.6 ppm, Fe = 93.2 ppm, and Mn = 27.5 ppm; Vajont Dam: Sr = 32.1 ppm, Fe = 92.1 ppm, and Mn = 42.6 ppm) (Zempolich, 1995). These values are similar to or are much lower than estimates of "marine" dolomite (Sr = 50–850 ppm, Fe = 10–2000 ppm, Mn = 5–275 ppm) (Al-Aasm and Veizer, 1982; Major, 1984; Saller, 1984; Aissoui, 1988; Dawans and Swart, 1988; Vahrenkamp and Swart, 1990); "deep marine" dolomite (Fe = 2100 ppm, Mn = 590 ppm) (Lumsden, 1988); late-stage recrystallized and burial dolomites (Sr = 35–147 ppm, Fe = 287–5115 ppm, Mn = 0.1–1069 ppm) (Montañez and Read, 1992; Montañez, 1994); dolomites of various depositional settings (average Fe = 2790 ppm, Mn = 245 ppm) (Weber, 1964); and dolomites of hydrothermal brine origin (Gregg, 1985; Gregg and Shelton, 1989). This comparison suggests that Vajont trace element compositions are not compatible with dolomite replacement, neomorphism, or recrystallization involving fluids enriched in Sr, Fe, and Mn (i.e., burial fluids or hydrothermal brine). While recrystallization of replacement dolomite and loss of Sr, Fe, and Mn through time is a possibility (Kupecz et al., 1993), most Vajont replacement dolomite exhibits petrographic evidence of initial replacement crystal fronts and engulfment of dissolution-resistant precursor calcite (Zempolich, 1995). The retention of these microfabrics suggests that geochemical compositions of replacement dolomite were emplaced during initial dolomitization and not through recrystallization (Zempolich, 1995).

However, modeling of isotopic, trace element, and fluid inclusion data collected from both limestone and dolomite components indicate that Vajont stable isotopic compositions and trace element concentrations are compatible with initial dolomite replacement and neomorphism or recrystallization by seawater derived fluid at elevated temperature (Zempolich, 1995). If correct, these data and models may suggest that circulation of seawater at temperatures ≤100°C may have caused dolomitization along faults and fractures within the Mt. Grappa–Visentin anticline (southern study area), and that circulation of seawater and/or modified seawater at temperatures ≤200°C may have caused dolomitization along faults and fractures within synclines in the northern study area. Moreover, $^{87}Sr/^{86}Sr$ values of replacement dolomite from southern ($^{87}Sr/^{86}Sr$ = 0.707104–0.707570; N = 9) and northern study localities ($^{87}Sr/^{86}Sr$ = 0.707040–0.708180; N = 2) overlap model ranges that utilize early Eocene and late Oligocene to early Miocene seawater values (Zempolich, 1995). Collectively, these results suggest that (1) dolomitization of the Mt. Grappa–Visentin anticline occurred by the circulation of Early Tertiary seawater at temperatures of 35–100°C concomitant with initial early Eocene compression and (2) dolomitization of the northern dolomite localities occurred by the circulation of Early to Middle Tertiary seawater or modified seawater at temperatures ≤200°C concomitant with initial early Eocene or late Oligocene to early middle Miocene compression.

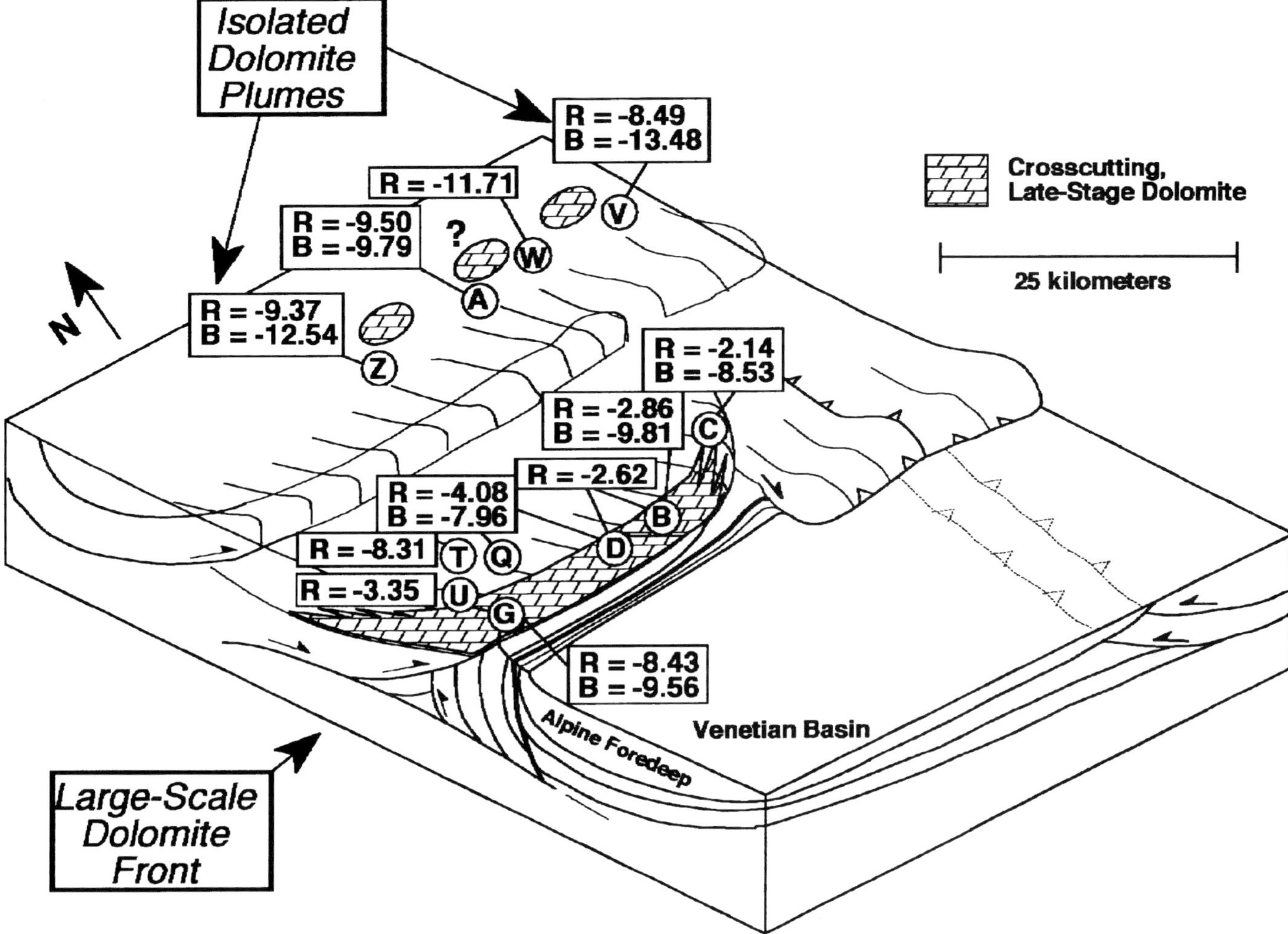

Figure 17. Regional distribution of Vajont dolomite and oxygen isotopic compositions. Compositions of replacement dolomite in the northern study area are more depleted in [18]O relative to those of replacement dolomite in the southern study area. Compositions of baroque dolomite cement in the northern study area are slightly depleted in [18]O relative to those of baroque dolomite cement in the southern study area (Zempolich, 1995). These and other geochemical and petrographic data suggest that dolomitizing fluids in the north circulated at higher temperatures (75–175°C) than did those in the south (35–100°C). B = baroque dolomite cement; R = replacement dolomite.

EVOLUTION OF POROSITY AND RESERVOIR QUALITY THROUGH DOLOMITIZATION

The complete dolomitization of Vajont oolite results in the replacement of calcite ooids and the formation of partly oomoldic and intercrystalline pores (e.g., Figures 16, 18). Partial oomoldic fabric forms through the complete replacement of ooid grains by medium to coarsely crystalline dolomite. Intercrystalline porosity develops as a result of the replacement of fine oolite and mud matrix. Visual estimates of porosity within dolomitized oolite range to 10%–15% in thin section. This is in agreement with a theoretical 13% increase in porosity through the volume-for-volume replacement of calcite by dolomite (Weyl, 1960). The distribution of moldic pores within grain interiors and the development of some intercrystalline porosity suggest that the pores in initial replacement dolomite fabrics were relatively isolated, and that the permeability developed at this stage of dolomitization was relatively low (inferred permeabilities of ~1–100 md; Lucia, 1995).

The macroscale rearrangement of moldic pores to form separate-vug and touching-vug pore space (Lucia, 1995) is first observed as a progression from moldic pores to separate vugs (nonfabric-selective pores) in completely replaced grainstone and packstone (Figure 18). The retainment of ooid ghosts around the margins of separate vugs indicates that enlargement of moldic pores occurred through local pore migration and crystal rearrangement (Zempolich, 1995). Next, a transition from separate-vug fabric to touching-vug fabric indicates that continued dissolution and recrystallization caused separate vugs and crystalline material to migrate and align, thereby forming alternations of touching-vug and dense recrystallized fabrics (e.g., Figure 18E, F). The development of interconnected pores in touching-vug fabric suggests that permeability has been increased without a change in the total porosity of the matrix (inferred permeabilities $\geq$100 md; Lucia, 1995). The retainment of ooid ghosts in these fabrics is less common and indicates a progressive-phase separation between pores and crystalline material. This is a characteristic of the sintering process (Barrett, 1987; Zempolich, 1995).

The significance of progressive textural modification in Vajont dolomite is that while reservoir-grade porosity may be formed through the initial replacement of limestone by dolomite ($\leq$10%–15%), reservoir-grade permeability is created through the recrystallization of intial replacement dolomite and pores. Given the rock volumes of the large-scale dolomite wedge and dolomite plumes (i.e., 50–94 km^3 and $\geq$2 × 10^{-2} km^3, respectively) found in the present study area, such pore space and enhanced permeability could potentially form significant economic hydrocarbon accumulations. For example, assuming 10% porosity, the large-scale dolomite wedge (Mt. Grappa–Visentin anticline) may contain $\leq$3 to 6 billion bbls of pore space, while the typical dolomite plume (e.g., Vajont Canyon) may contain $\geq$ 12.5 million bbls of pore space. Importantly, the large size of Vajont dolomite bodies and the formation and redistribution of porosity through late-stage replacement dolomitzation and recrystallization illustrate that significant dolomite reservoirs may be created through massive, late-stage, fault-related burial dolomitization.

THE ORIGIN OF THE VAJONT DOLOMITE

Field, petrographic, and geochemical data point to dolomitization of the Vajont Limestone by regional-scale circulation of Tertiary seawater within anticlines and synclines that were formed during Tertiary Alpine deformation. Theoretical and laboratory circulation models and fluid flow patterns observed in modern thrust zones are consistent with this interpretation, as discussed below.

Physiochemical Factors

As reviewed by Hardie (1987), a number of physiochemical factors influence the formation of dolomite in sedimentary and burial environments. These factors include thermodynamics, kinetics, mass transfer, and the nature of the precursor host rock. Current models of dolomitization, such as mixing-zone, tidal flat, evaporative-brine, and schizohaline, have inherent weaknesses with regard to one or more of these factors, the more serious of which are related to thermodynamics, kinetics, and the mass transfer of Mg. These problems are easily overcome at elevated temperature and within flow regimes capable of circulating large amounts of Mg-bearing fluid (Hardie, 1987; Wilson, 1989; Wilson et al., 1990).

With regard to the Vajont dolomitization, we can draw the following conclusions about the physiochemical factors involved:

1. The widespread occurrence of reaction fronts between dolomite and precursor limestone at all localities indicates that dolomitizing fluids were oversaturated with respect to dolomite and undersaturated with respect to calcite. Initial replacive dolomitization of limestone, therefore, most probably occurred through the general reaction:

$$2\ CaCO_3(cal) + Mg^{2+}(aq) <\!\!-\!\!> \\ CaMg(CO_3)2(dol) + Ca^{2+}(aq) \tag{1}$$

2. Field and petrographic data show that dolomite fronts moved out and away from fluid conduits (i.e., faults and fractures).
3. The ^{18}O compositions of replacement dolomite in southern localities are relatively enriched compared with those of northern dolomite localities, and suggest that dolomitization occurred at moderate temperatures ($\leq$100°C).
4. The ^{18}O compositions of replacement dolomite in northern localities are depleted relative to Middle Jurassic marine carbonate and, together with fluid inclusion data, suggest that dolomitization occurred at more elevated temperatures ($\geq$125°C).

These factors indicate that the dolomitizing fluids were introduced to the Vajont basinal limestones along fractures and faults at elevated temperatures, and that diffusion of Mg through relatively nonporous limestone resulted in the formation of massive replacement dolomite (Zempolich and Hardie, 1991a, b; Zempolich, 1995).

Flow Volumes and Delivery of Magnesium

To get some measure of the mass transfer requirements, the volumes of fluid necessary for the dolomitization of the large wedge within the Mt. Grappa–Visentin anticline (50–94 km^3 dolomite) was calculated. With seawater as the dolomitizing fluid, the calculation yields 2.74 × 10^4 km^3 at 35°C and 2.53 × 10^5 km^3 at 100°C (Zempolich, 1995). Such large volumes demand that dynamic transport of Mg from an external source must have occurred. For scale, the fluid volume calculated for the seawater case at 100°C and a dolomite rock volume of 50 km^3 is equal to the volume of a small sea, 1 km deep and 165 × 165 km in area. Although smaller, the calculated volumes of the fluid necessary for the formation of a representative dolomite plume (2.36 × 10^{-2} km^3 dolomite) found in the northern study area (e.g., seawater test case at 200–300°C = 6.6–12.9 km^3 seawater) (Zempolich, 1995) are impressive, and demand the dynamic flow of Mg-bearing fluid to promote dolomitization in the northern study area.

What remains to be explained is what kind of Mg-bearing fluid was involved, where massive quantities of this fluid were generated, and how the fluid was transported to the network of large- and small-scale fluid conduits.

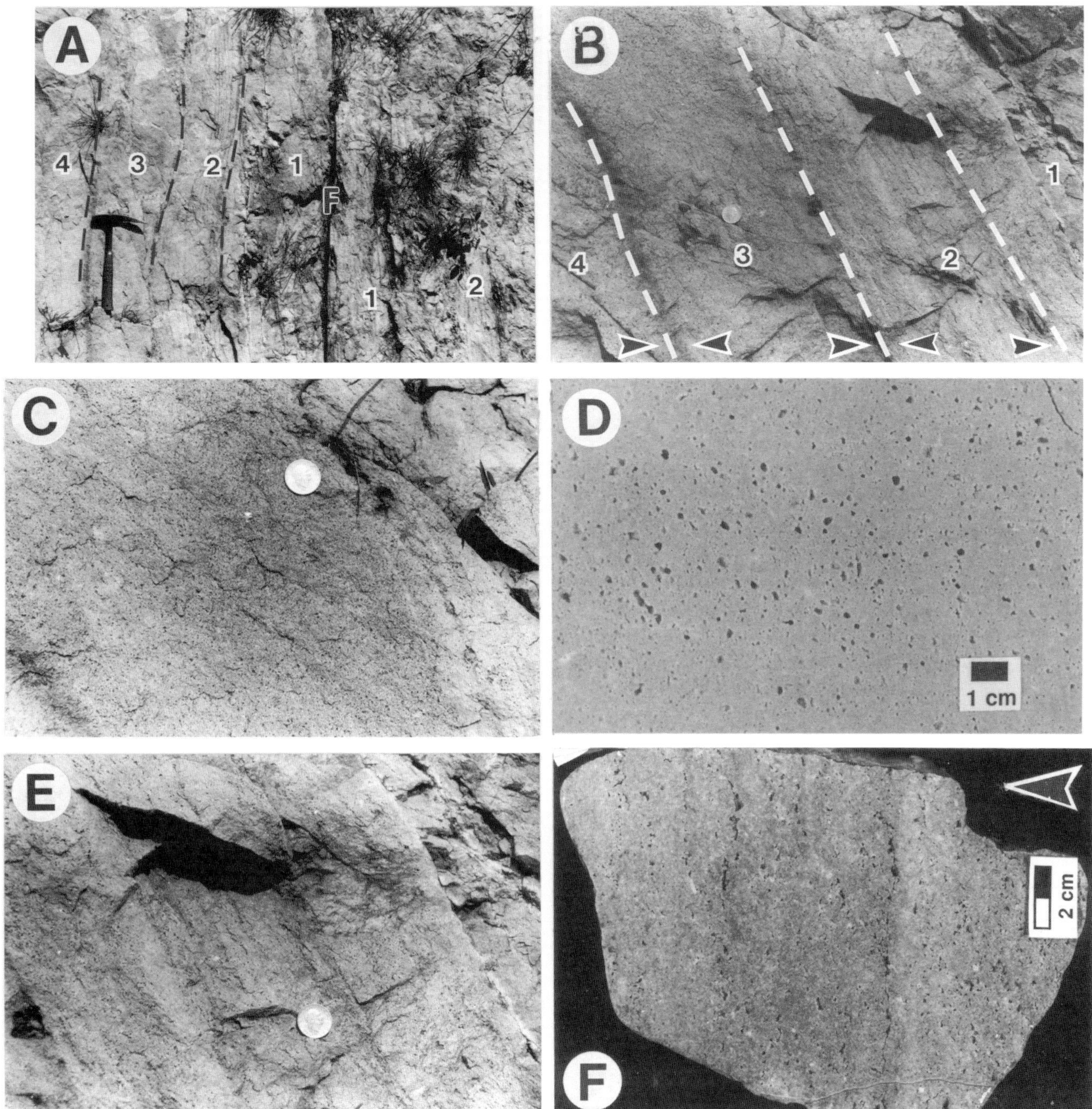

Figure 18. Dolomite textures and associated pores formed through the progressive replacement of limestone and the progressive recrystallization of intermediate replacement fabrics (Villanova section). (A, B) Close-up photographs of the left side of the reaction front that is depicted in Figure 15 (note hammer). The reaction front is separated into discrete zones of dolomite textures and pores oriented subparallel to the fault. Starting from the fault marked "F" these include: (1) densely crystalline dolomite with minor amounts of pore space; (2) recrystallized dolomite with touching-vug pores; (3) recrystallized and neomorphic dolomite with separate-vug and moldic pores (Figure 16C); (4) initial replacement dolomite in partially dolomitized limestone. (C) Close-up field photograph of moldic and separate-vug fabric (3). (D) Polished-slab photograph of moldic and separate-vug fabric (3). Visual estimates of porosity in both slab and thin section range to 10%–15%, inferred permeabilities range from 1–100 md (e.g., Lucia, 1995). (E) Close-up field photograph of touching-vug fabric (2). Bands of touching-vug pores are located ~1–2 cm apart and are oriented parallel to the reaction front. (F) Polished-slab photograph of touching-vug fabric (2) (sample oriented so that reaction front is to the left). Vugs are ≤1 cm in length and are laterally interconnected. Visual estimates of porosity in both slab and thin section range to 10%–15% and are similar to those estimated for moldic and separate-vug fabric. Inferred permeabilities are ≥100 md (Lucia, 1995).

Models for Dolomitization of the Vajont Limestone

Dolomite distribution in the Belluno Basin is confined to Mesozoic slope and basinal sediments. The dolomitization of such a thick sequence of relatively impermeable lithologies is unusual in that primary depositional porosity and permeability were negligible. This presents a major problem for the transportation of large quantities of Mg-bearing fluid to the sites of reaction because primary large-scale fluid conduits, such as permeable siliciclastic sands or carbonates, are absent in these deep-water sediments.

A number of hydrologic models have been developed or invoked to explain the occurrence of dolomite in shallow platform settings. These models have been based largely on the occurrence of dolomite in Holocene environments and have been applied to ancient dolomites by analogy with facies and paleogeographic settings. Shallow dolomitization models include mixing zone (Hanshaw et al., 1971; Badiozamani, 1973; Hanshaw and Back, 1980; Sandford, 1987), reflux (Adams and Rhodes, 1960; Sears and Lucia, 1980; Simms, 1984; Whitaker et al., 1994), tidal pumping (Carballo et al., 1987), and evaporative pumping (McKenzie, 1981; Ruppel and Cander, 1988). These hydrologic and depositional models predict the occurrence of dolomite in shallow shelf and platform margin settings (Kaufman, 1994), and so cannot explain the dolomitization of the Vajont Limestone, which was deposited as a thick sequence of carbonate gravity flows in slope and basin settings of the Belluno Basin.

Other hydrologic models and settings for dolomitization include contemporaneous dolomitization of deep marine sediments by cool ocean water (Baker and Burns, 1985; Lumsden, 1985, 1988; Mullins et al., 1985), topographic-driven flow (Garven and Freeze, 1984; Garven, 1985; Gregg, 1985; Barrett, 1987; Ge and Garven, 1989; Yao and Demicco, 1995), compaction-driven flow (Jodry, 1969; Mattes and Mountjoy, 1980; Bethke, 1985), and thermally driven flow (Elder, 1965; Kohout et al., 1977; Simms, 1984; Aharon et al., 1987; Wilson et al., 1990; Kaufman, 1994).

Deep-marine sedimentary dolomitization is an unlikely explanation for massive dolomitization of the Vajont limestone because (1) dolomite bodies crosscut basinal stratigraphy, (2) Vajont replacement dolomite displays relatively coarse textures and depleted oxygen compositions that are quite different from the fine-grained disseminated dolomite that characterizes these occurrences (Lumsden, 1988), and (3) deep-water dolomite is a volumetrically minor component (average 0.5%) of modern deep-water sedimentary cover (Lumsden, 1988).

Topographic-driven flow may result in the long-term or transient flow of fluid in basins through the development of sufficient recharge and hydrostatic head in neighboring uplift areas (Garven and Freeze, 1984; Garven, 1985; Ge and Garven, 1989). However, such a model is unlikely for dolomitization of the Vajont limestone because: (1) uplift of the Venetian Alps and the formation of a possible recharge area during the middle Miocene (Massari et al., 1986;

Doglioni, 1990) included the basinal sedimentary section that is now dolomitized (i.e., Mt. Grappa–Visentin anticline, Belluno thrust); (2) initial thrust movement occurred synchronous with deposition of early Eocene to middle Miocene marine siliciclastics, marls, and carbonates in the Venetian Basin (as far north as Cortina); (3) dolomite bodies are aligned parallel to and are hosted within structural axes formed during initial compression of the Venetian Alps and prior to significant uplift; and (4) basin-scale topographic flow emanating from the Appenine Mountains to the south and migration through the Po Basin into the study area would have produced isotopic trends opposite to those observed (i.e., depleted ^{18}O values in the south, enriched ^{18}O values in the north).

The compaction and dewatering of shales is an unlikely source of dolomitizing fluids because shale is a volumetrically minor component of Late Paleozoic and Mesozoic sediments of the area, and because of the enormous volume of Mg-bearing fluid, which must be accounted for by mass balance calculations. Kohout convection (Simms, 1984) would predict the occurrence of dolomite in platform margin to periplatform settings, which is a dolomitization pattern that is unsupported by field, sedimentologic, and stratigraphic evidence.

Considering the discussion above and the collective field, petrographic, and geochemical evidence that suggests that dolomitization of the Vajont Limestone occurred by the large-scale circulation of fluid at elevated temperature along faults and fractures, dolomitization most likely occurred through the circulation of hydrothermal fluids at depth. Considering the low Sr, Fe, and Mn concentrations in replacement dolomite and a lack of associated Mississippi Valley-type mineralization, dolomitizing fluids must also have been low in Sr, Fe, Mn, and other metals, yet were capable of transporting large quantities of Mg. Geochemical data and modeling (see below) suggest that hydrothermal dolomitization most likely occurred through the circulation of seawater or modified seawater at depth (Zempolich, 1995).

Thermal Convection

As argued above, the fluid volumes that are required to produce both the extensive dolomite body in the southern study area and the narrow isolated dolomite plumes in the northern study area require that large-scale fluid transport must have occurred. Given that the fluid inclusion and geochemical data indicate that dolomitization of the Vajont limestone occurred at elevated temperature, it is likely that massive volumes of Mg-bearing fluid were delivered to the dolomitization sites by thermal convection, and that the geometry of these convective cells was dependent on temperature and availability of fluid conduits.

Through study of dolomitization patterns in the Latemar buildup, an isolated Late Triassic carbonate platform penetrated by rift-related Late Triassic volcanics, Wilson et al. (1990) have proposed several thermal-convective flow models to explain the occurrence of a massive mushroom-shaped dolomite body 1–2 km^2 in diameter. Critical to the generation of convective flow in these models is the presence of a heat source and a supply of Mg. Using other field and geochemical data, Wilson et al. (1990) conclude that dolomitization was most likely caused by the thermal convection of Late Triassic seawater driven by local elevated temperatures due to a volcanic intrusion. Convective models similar to the one proposed for the Latemar buildup (e.g., Kaufman, 1994) are unsuitable explanations for dolomitization of the Vajont limestone because (1) volcanic intrusions are not known in either the southern or northern study areas, (2) dolomite bodies are distributed within basinal rather than platform strata, (3) dolomite bodies are associated with compressional structures, and (4) multiple dolomite plumes are found along the same structural trend. Thus, if the Vajont dolomite geometries were produced by the thermal convection of Mg-bearing fluid, other convection models must be called upon to explain the unusual dolomite geometries that are now found in basinal strata contained within compressional structures.

Vajont Dolomite Geometries and Theoretical Convective Flow Patterns

Convection models that may be applicable to dolomite geometries observed in the Vajont and other Mesozoic basinal sediments have been explored by Elder (1965, 1967, 1977). Elder's models were developed through two-dimensional numerical simulations and scaled laboratory experiments to simulate the circulation and mass discharge of fluid in geothermal and volcanic areas, rifts, and oceanic rises (Figure 19). These scaled models rely on linear, basal heat sources, and approximate the physical dimensions of zones of fracture-enhanced permeability that are found in the faulted crests and troughs of anticlines and synclines that are host to Vajont dolomite bodies (Zempolich, 1995). The theoretical convective flow patterns of these models predict the occurrence of multiple isolated plumes and large-scale flow geometries that approximate the dolomite geometries that are observed in the northern and southern study areas (Figure 20).

Isotopic, trace element, and fluid inclusion evidence suggests that dolomite bodies in the northern and southern areas evolved under different thermal regimes (Zempolich, 1995). As dolomite bodies in the south and north are hosted in a similar succession and thickness of basinal strata, it is unlikely that this temperature difference arose from differences in burial history. Temperature differences due to an extraneous heat source, such as the intrusion of magma, can be ruled out because such intrusions are not present in the study areas. Thus, the temperature differences must have resulted from some other thermotectonic perturbation.

The difference in temperature regimes may be related to the depth that structural deformation reached and to patterns of fluid circulation. For example, in the southern study area, balanced reconstructions of Doglioni (1990) suggest that the Mt. Grappa–Visentin anticline detached along a shallow

decollement (≤5 km deep) within the Mesozoic section, whereas thrusts in the northern study area detached along fairly deep decollements (5–10 km) within the Mesozoic section and Paleozoic basement. Initial detachment and thrusting took place prior to the early middle Miocene during development of the Venetian Alps foredeep (Massari et al., 1986; Doglioni, 1990). Within this subaqueous thrust system, anticlines and synclines were dissected by numerous vertical-subvertical faults (Figures 9, 10). Assuming a normal thermal gradient of ~30°C/km, ambient temperatures at the depth of the decollement horizons (5 and 10 km) would be ~150° and 300°C. These observations suggest that fluids, circulating downward along faults and fractures within anticlines and synclines, may have been heated by conduction to temperatures approaching 150°C and 300°C, respectively. These postulated temperature differences are consistent with geochemical data and a change in the geometry of the dolomite bodies from narrow isolated plumes in the north to a broad dolomite wedge in the south (Zempolich, 1995). Therefore, it is postulated that vertical to subvertical faults within anticlines and synclines produced hydrologic conduits that connected overlying Tertiary seawater with deeply buried Mesozoic basinal sediments, thereby creating large-scale convective hydrologic systems, which enabled the massive dolomitization of otherwise tight basinal sequences (Figure 20).

In summary, massive dolomitization of the Vajont Limestone by the convective circulation of Early to Middle Tertiary seawater is suggested by a consistency among dolomite and structural field relationships, petrographic data, geochemical data, and theoretical hydrologic models. It is proposed that the delivery of Mg-bearing fluid and massive replacement dolomitization was promoted by a combination of: (1) large-scale fluid flow along Tertiary compressional structures that provided the main plumbing by which an Mg-rich reservoir (seawater) was put in communication with a deep heat source. The ensuing generation of thermal-convection cells within anticlines and synclines ultimately controlled the overall shape and distribution of large-scale dolomite bodies; and (2) small-scale fluid flow emanating from large-scale flow systems. Small-scale fluid flow along faults, fractures, and bedding planes controlled the propagation and orientation of reaction fronts. In this manner, vast quantities of Mg were delivered to relatively impermeable basinal sediments of the Belluno Basin at elevated temperature and at a broad range of scales. Kinetic barriers involved in the formation of dolomite were overcome by the elevated temperature and the high Mg/Ca ratio of seawater (Hardie, 1987).

MODERN HYDROLOGIC AND STRUCTURAL ANALOGS

The proposed model for massive dolomitization of the Vajont Limestone depends on the thrusting of thick sequences of limestone in a marine environment, the

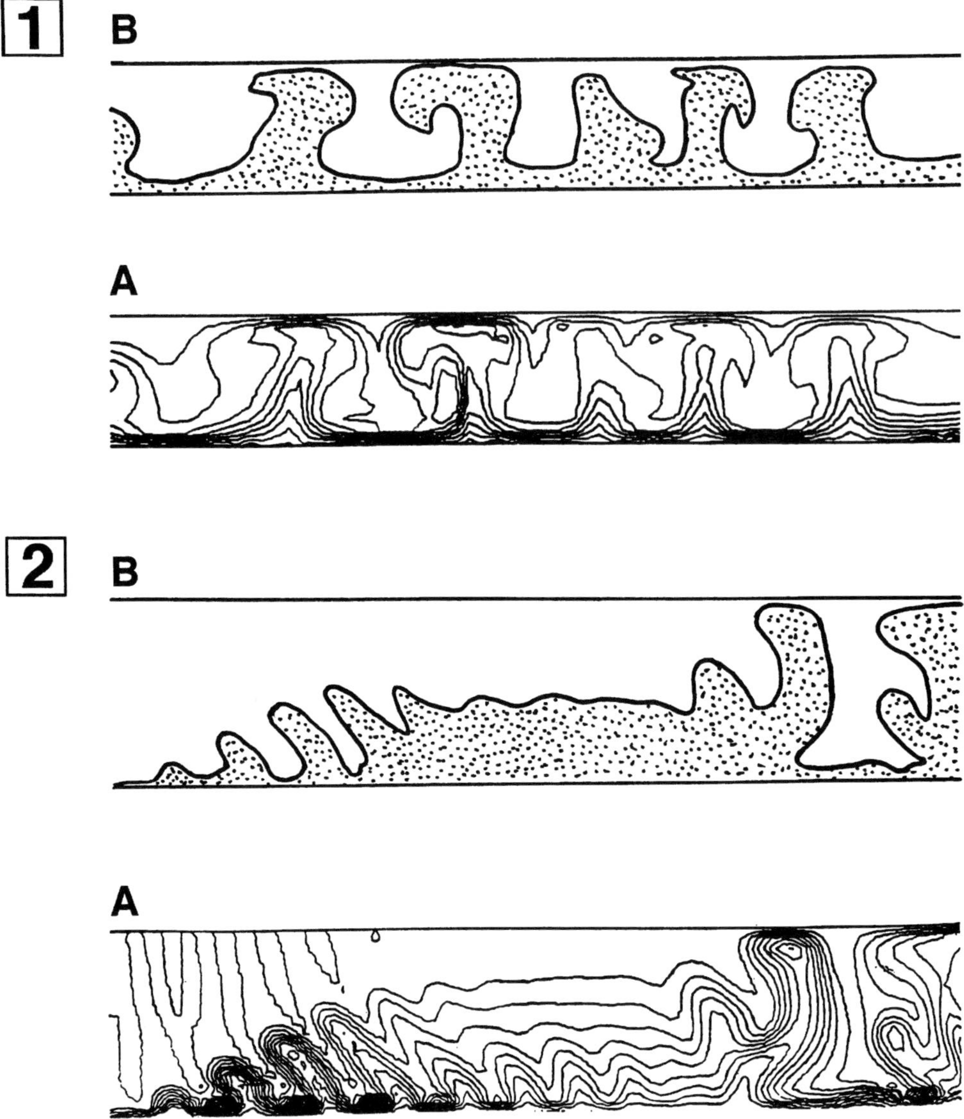

Figure 19. Theoretical convective fluid flow models for the circulation of heated fluid in two-dimensional porous mediums (modified from Elder, 1977). 1: Fluid flow model with restricted upward outflow. The thermal interface is uniformly heated from below, and surface discharge is localized within the middle of the upper surface. (A) Theoretical isotherms. (B) Hypothetical dolomite geometries resulting from convective fluid flow (A), assuming that the kinetic inhibitions of dolomitization are overcome at elevated temperature. It is postulated that similar convective flow of Mg-bearing fluid, first downward and then upward along fractures within the axes of synclines, may have given rise to the multiple occurrence of dolomite plumes that are now present in the northern study area (compare with Figures 3, 8, 11–13, and 20). 2: Fluid flow model with enhanced lateral outflow (to the right). The thermal interface is uniformly heated from below. Regional fluid flow is from left to right. (A) Theoretical isotherms. (B) Hypothetical dolomite geometry resulting from convective fluid flow (B), assuming that the kinetic inhibitions of dolomitization are overcome at elevated temperature. It is postulated that similar large-scale convective flow of Mg-bearing fluid may have given rise to the large, thickening-westward dolomite wedge that is now present along the Mt. Grappa–Visentin anticline in the southern study area (compare with Figures 3, 8–10, and 20).

development of a deeply rooted fault and fracture network through the crests and troughs of thrust-related anticlines and synclines, and the generation of large-scale convective fluid flow and diffusion of Mg through limestone matrix. It is proposed that large-scale fluid flow systems developed along fracture sets located within anticlinal and synclinal structures, and that these fracture networks controlled the overall geometry and distribution of dolomite bodies. Many of these structural and hydrologic features are present in modern subduction or transpressive compressional zones where carbonate and siliciclastic sediments are

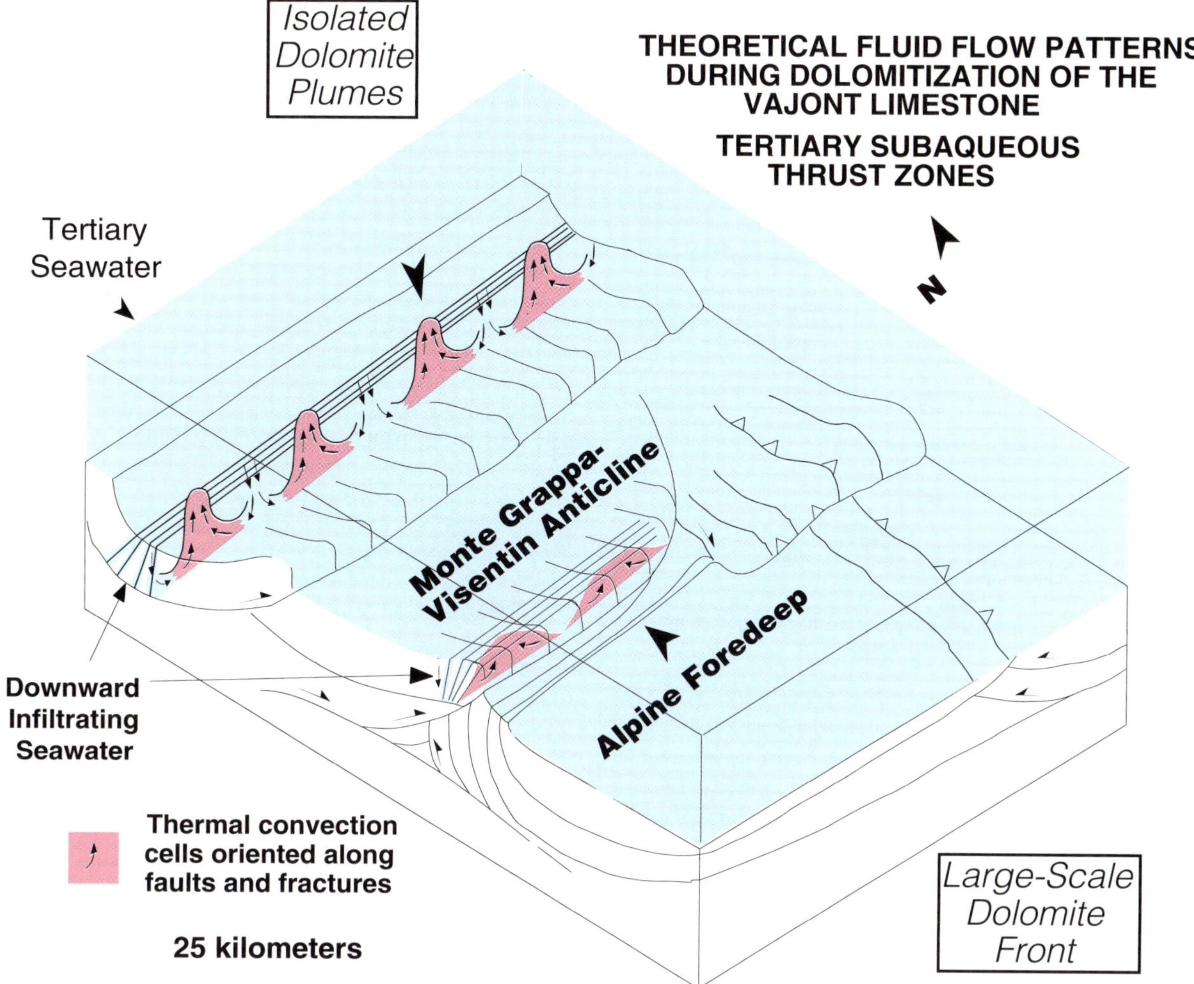

Figure 20. Proposed flow pathways of dolomitizing fluids and their relation to structural features of the Venetian Alps thrust belt (fluid flow patterns after Elder, 1977; present-day structural configuration after Doglioni, 1990). It is proposed that massive dolomitization of the Vajont Limestone and other Mesozoic basinal sequences was a consequence of the thermal convection of Early to Middle Tertiary seawater along faults and fractures that were formed during initial tectonic deformation and that breached the Tertiary sea floor. Structural and crosscutting stratigraphic relationships of late-stage dolomite suggest that dolomitization occurred sometime during the early Eocene/late Oligocene or late Oligocene/early middle Miocene, and prior to significant uplift of the basinal succession that occurred during the early middle Miocene to Pliocene. Convective fluid circulation is postulated to have developed as a result of extensive fracturing and faulting within the axes of synclines and anticlines and the downward infiltration of seawater. At depth, the seawater was presumably heated by conductive heat flow and then driven upward along the fracture and fault network by buoyant forces (compare with Figures 3, 17, 21, and 22).

deformed into a series of thrusted anticlines and synclines. For example, shallow- and deep-water carbonate of Mesozoic and Tertiary age is found in an extensive thrust zone located along the southeast Bahamas–Hispaniola collision zone (Ditty et al., 1977; Austin, 1983; Mullins et al., 1992). This submerged thrust zone is composed of a series of anticlines and synclines at subsea depths of ~1000–3000 m. Interestingly, gas-hydrate zones and "groundwater seeps" identified by high seismic data reflectivities (Austin, 1983; Mullins et al., 1992) are present in some, but not all, of the crests and hanging walls of anticlines (Figure 21). These features are distributed in the same structural position as the large-scale dolomite front within the Mt. Grappa–Visentin anticline. Moreover, crestal positions of these anticlines are extensively dissected by large-scale vertical to subvertical "keystone" faults (Figure 22) (Austin, 1983). The buried geometry of these anticlines and their associated faults are remarkably similar to the structural setting and fault pattern recognized in the Mt. Grappa–Visentin anticline. Moreover, the subsea sedimentologic setting of the southeast Bahamas–Hispaniola collision zone is similar to the Tertiary foredeep formed during initial thrusting of the Venetian Alps. These data, together with previously presented field, petrographic, and geochemical data, support the idea

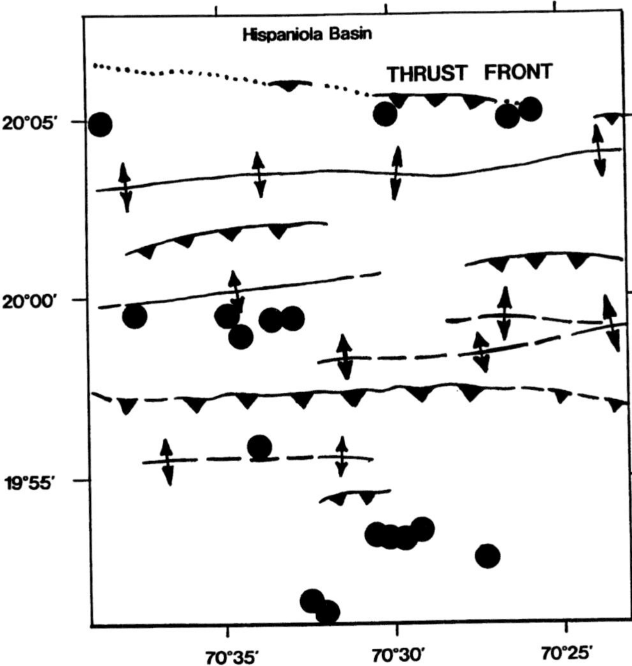

Figure 21. Subaqueous thrust belt (~1000–3000 m subsea) north of Hispaniola (modified from Mullins et al., 1992). Shallow- and deep-water carbonates of Mesozoic and Tertiary age are thrusted into a series of anticlines and synclines due to regional transpression. "Groundwater seeps" (circles) and gas-hydrate zones, which are identified by high seismic data reflectivities, are present along and parallel to some of the crests and hanging walls of anticlinal thrusts. These structural and hydrostratigraphic settings are thought to be analogous to those that promoted the dolomitization of the Vajont Limestone and other Mesozoic basinal sediments during deformation of the Venetian Alps during the Tertiary.

that thrusting of the Venetian Alps during the Tertiary and the concomitant thermal circulation of seawater through faults and fractures within the rising fold belt was responsible for the occurrence of bodies of massive dolomite within the Vajont Limestone.

Recent study has identified that fluid expulsion and venting along vertical strike-slip faults and crestal faults in thrust-related anticlines of the Cascadia accretionary prism is significant (Kulm et al., 1986; Ritger et al., 1987; Carson et al., 1990; Sample et al., 1993; Tobin et al., 1993). Flow rates of 100 m/yr have been described for these systems, suggesting that large-scale flow has developed in response to the dewatering of prism sediments and expulsion of fluid through vertical channelized flow. Calcite cements sampled from siliciclastic sediments outcropping along these fault zones possess depleted oxygen compositions, enriched to depleted carbon compositions, and radiogenic Sr compositions that are thought to represent precipitation at temperatures as great as 100°C (calcite cement $\delta^{18}O$ = –4 to –13‰, $\delta^{13}C$ = –1 to –25‰, $^{87}Sr/^{86}Sr$ = 0.70975–0.71279) (Sample et al., 1993).

The origin and composition of these cements has been attributed to the complex interaction of deeply derived interstitial pore fluid with clays, thermogenic methane, and marine water (Sample et al., 1993). Despite the marine influence observed in these cements and the distribution of cements in the crests of anticlines, the origin of these fluids has been solely attributed to the dewatering of the accretionary complex. An alternative explanation for the occurrence of channelized fluid flow within these structures would be the thermal-driven circulation of seawater through these extensive fault and fracture systems. In such a scenario, the modification of some seawater would presumably occur through reaction with siliciclastic sediment and organics. If correct, channelized fluid flow along faults and fractures, and the chemical modification of seawater in the Cascadia accretionary prism, would be analogous to that proposed for the dolomitization of the Vajont limestone, including the relationship of dolomite bodies to structure and large-scale fluid flow and marine to nonmarine compositions of replacement dolomite.

The similarity between the Mt. Grappa–Visentin anticline and Belluno thrust with these modern structures supports the proposed hydrostratigraphic model that involves the convection of seawater along linear zones of high permeability. The southeast Bahamas–Hispaniola collision zone and Cascadia accretionary prism may be modern structural and hydrostratigraphic analogs for the tectonic deformation and the dolomitization of the Vajont and other Mesozoic basinal sediments during the Tertiary.

POTENTIAL DOLOMITE RESERVOIR ANALOGIES

Field mapping, petrography, and geochemistry of the Vajont dolomite reveal a strong relationship between hydrothermal dolomitization and tectonism. The enhanced porosity and permeability within these dolomite bodies suggests that these bodies may well represent exhumed dolomite reservoirs created in tectonically deformed carbonate provinces. Dolomite reservoir geometries illustrated in this study include: meter-scale dolomite bodies located parallel to fracture networks; multiple isolated dolomite plumes, several hundred meters in width and height, located along structural trends; and large-scale dolomite bodies, kilometers to tens of kilometers in scale, encompassing the crests of anticlines. Accordingly, the recognition of these dolomite geometries and this style of dolomitization in subsurface settings may define new exploration targets in the search for oil and gas, and/or provide analog geometries for reservoir characterization (Figure 23).

Subsurface examples of ellipsoidal, areally restricted dolomite bodies apparently associated with tectonic lineaments are postulated to exist in several basins. For example, in Paleozoic strata of the Michigan Basin, small-scale ellipsoidal dolomite reservoirs are aligned NW-SE in association with the dominant NW-SE fracture network imposed on the basin during Appalachian orogenesis (Prouty, 1983). Dolomite isopleths suggest that these bodies thin away rapidly from

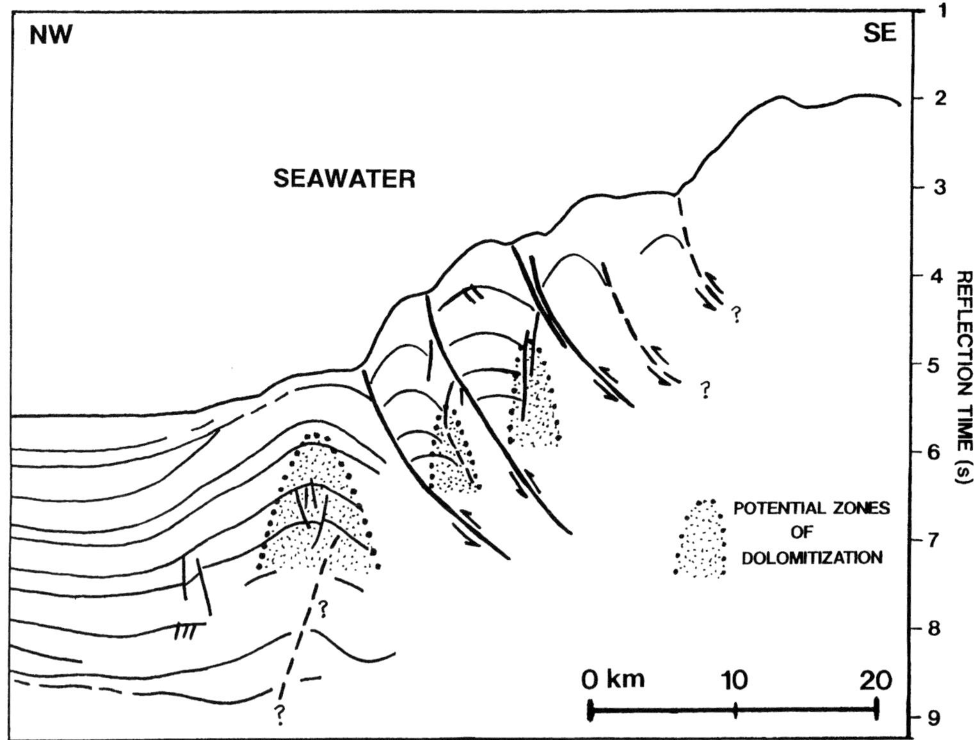

Figure 22. Schematic cross section of the southeast Bahama–Hispaniola collision zone (modified from Austin, 1983). Subaqueous thrusts are composed of shallow- and deep-water carbonates of Mesozoic–Tertiary age and are found at present-day subsea depths of 1000–3000 m. Numerous vertical to subvertical faults dissect anticlines and synclines that were formed during regional transpression. It is postulated here that extensive fracturing and faulting within the axes of these submerged structures may allow for the downward infiltration of seawater. Potential zones of dolomitization may exist in the cores of these anticlines and synclines (stippled pattern) due to the thermal-convective circulation of seawater upward and along these extensive fracture and fault networks (compare with Figures 3, 21).

tectonic lineaments. Dolomite types include fine-crystalline dolomite and late coarse-crystalline and baroque textures (Zempolich, 1984, personal observation). Plume-shaped isolated dolomite reservoirs are also found in Paleozoic rocks in the Williston Basin (R.D. Perkins, 1991, personal communication). By analogy with dolomite bodies observed in the Vajont Limestone, it is possible that these dolomite bodies were produced by the convection of heated fluid(s) along deep-seated tectonic fractures and faults. If correct, these basin and outcrop examples would predict that significant dolomite reservoirs may be hosted in carbonate strata that have been deformed in peri- and intracratonic tectonic settings and subjected to hydrothermal dolomitization processes. Such hydrothermal dolomite bodies may be difficult to recognize in sequences that have also been affected by early shallow dolomitization processes.

The delineation of hydrothermal dolomite bodies in the subsurface may include (1) the mapping of dolomitization fronts using dolomite abundances calculated from well logs and core, and (2) the identification of thermal diagenetic fluids and textural trends using petrographic and geochemical techniques. Once the hydrothermal process has been delineated, the search for new exploration targets may be concentrated on anticlinal and synclinal structures within buried fold and

thrust belts, and along zones of deep-seated tectonic fractures and faults within intracratonic basins. Dolomite plumes may be identified by seismic reflection data methods due to the disruption of bedded limestone by crosscutting dolomite fronts and by formation of breccia cores. The seismic reflection data expression of such dolomite bodies would, theoretically, consist of a "rootless," oriented chaotic zone (several hundred meters high) interspersed within layered reflectors (i.e., undolomitized bedded limestone).

SUMMARY AND CONCLUSIONS

Through field and laboratory study of Vajont limestone and dolomite, a number of inferences can be made as to the formation of massive replacement dolomite and formation of dolomite reservoirs through late-stage fault-related, burial dolomitization. Field distribution of dolomite bodies and petrographic and geochemical data collectively suggest that massive replacement dolomitization occurred as a result of the circulation of hot Mg-bearing fluids piped into the Vajont and other Mesozoic basinal sediments along a master network of faults and fractures. The faulting and fracturing of Mesozoic basinal sediment is related to Alpine thermotectonics, which

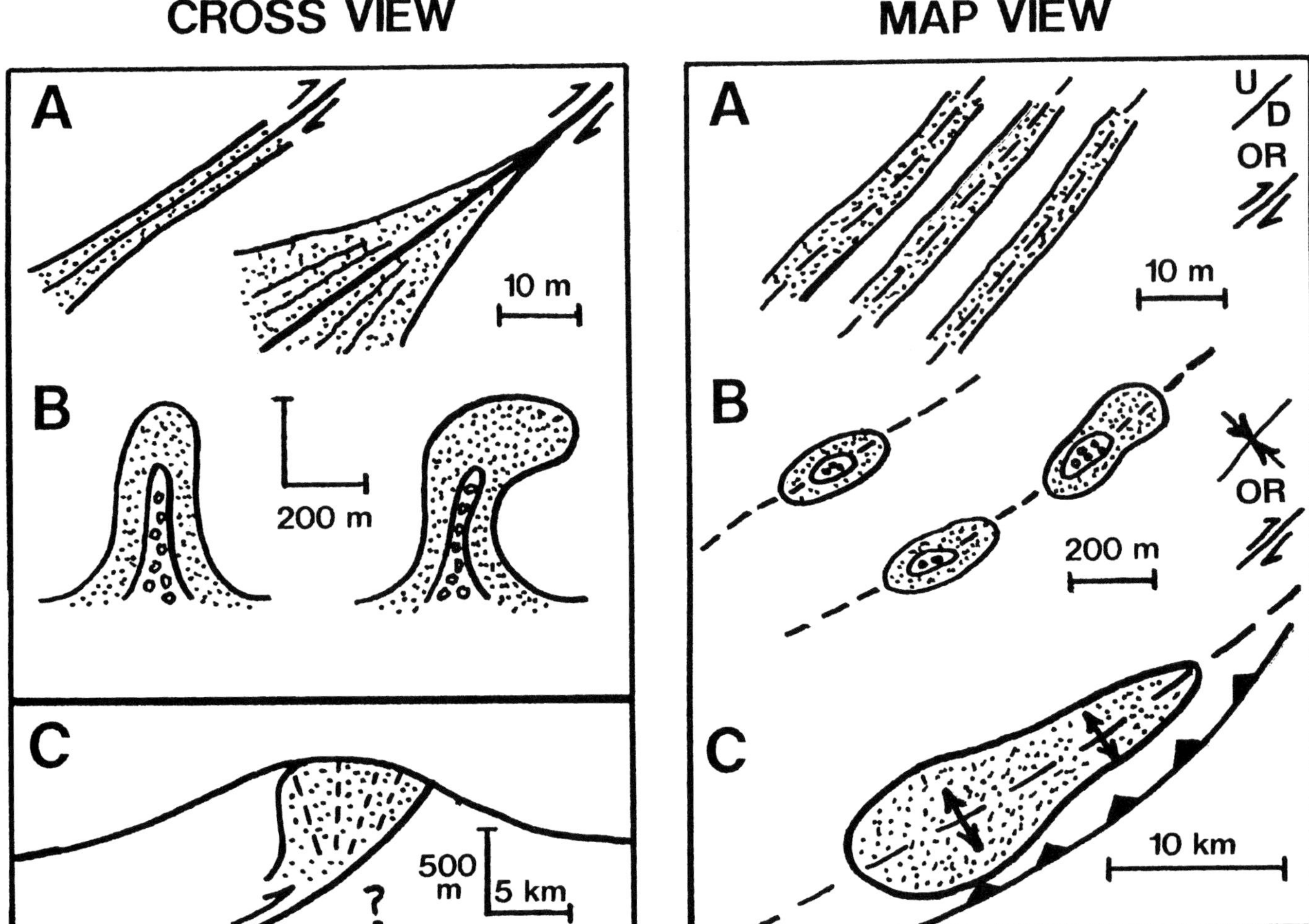

Figure 23. Summary of Vajont dolomite bodies that are found in outcrop of the Venetian Alps. The diagram on the left is a schematic cross view of these bodies; the diagram on the right is a schematic map view of these bodies if projected into the subsurface. Vajont dolomite bodies are potential analogs for dolomite reservoirs created in subsurface settings due to the formation of porous and permeable bodies in otherwise-tight deep-water limestone. Potential reservoir geometries that were created in the Vajont Limestone by massive replacement dolomitization and recrystallization in association with tectonism and the hydrothermal circulation of Mg-bearing fluid include: **(A)** small-scale dolomite wedges (meters to tens of meters in width) oriented parallel to subparallel with faults and fractures; **(B)** multiple isolated dolomite plumes (200–300 m wide, 300–400 m high) cored by dolomite breccia and located along structural trends; and **(C)** large-scale dolomite bodies (10–20 km long, 5–10 km wide, and ≥400 m thick) located in the crests of major anticlines.

formed a series of thrust-related anticlines and synclines in the study area during early Eocene? and late Oligocene to Recent times. These faults and fractures opened up porous and permeable pathways within an otherwise tight sequence of basinal limestone. Dolomitization is postulated to have occurred concomitant with initial thrusting during early Eocene and/or late Oligocene to early middle Miocene time while the study area was still inundated by seawater. Dolomitization was completed prior to rapid uplift and subaerial exposure of the region during the middle Miocene to Pliocene. Rapid uplift following initial deformation and dolomitization preserved metastable dolomite textures and compositions across limestone–dolomite transitions.

It is proposed that circulation of seawater was driven by both large- and small-scale transport processes that controlled the shape and distribution of dolomite bodies, reaction fronts, and replacement styles. Large-scale fluid movement involved the thermal convection of Tertiary seawater through anticlines and synclines. In addition to Tertiary structures, dolomitizing fluids also utilized inherited structural elements such as paleolineaments and paleosynsedimentary breccia. Convection cells were developed parallel to the axes of these structures through extensive subvertical to vertical faults and fractures. In the southern study area, fluid convection resulted in the formation of a large-scale dolomite body that is ~25 km long, 10 × 15 km wide, and ≥400–500 m thick. In the northern study area, fluid convection resulted in the formation of multiple rootless dolomite plumes that are >300 m high and ~100–200 m wide.

Replacement of limestone by dolomite occurred by the microscale dissolution of precursor limestone and precipitation of dolomite. Complete dolomitization formed porous moldic and intercrystalline fabrics with porosities up to 10% to 15%, and inferred permeabilities of 1–100 md. Recrystallization progressively amalgamated moldic and intercrystalline pores and dolomite to form separate-vug, touching-vug, and dense crystalline fabrics with inferred permeabilities ≥100 md. Consistency in the development of dolomite textures, dolomite composition, and porosity with respect to limestone–dolomite transitions suggests that the massive replacement of limestone by dolomite, and the formation of reservoir-grade porosity and permeability, occurs through a predictable pattern of replacement and recrystallization.

Dolomite geometries in the northern and southern study areas are consistent with theoretical circulation models that predict the formation of large-scale flow systems and the multiple occurrence of isolated plumes due to the thermal convection of fluid. The proposed thermotectonic model for the formation of massive replacement dolomite in the Vajont Limestone may have modern analogs in active thrust zones of the southeast Bahamas and the Pacific Northwest.

The geometry, size, and distribution of dolomite bodies within the Vajont Limestone and other Mesozoic basinal sediments indicate that late-stage thermotectonic dolomitization is an important process by which massive replacement dolomite may form. Moreover, these examples illustrate that both large- and small-scale dolomite reservoirs may be created through late-stage dolomitization. Similar bodies in the subsurface may prove to be attractive exploration targets.

ACKNOWLEDGMENTS

Daniele Masetti, Carlo Doglioni, and Alfonso Bosellini of the University of Ferrara provided logistical support that made this study possible. Dmitri Sverjensky, Grant Garven, Owen Phillips, and Saki Olsen (Johns Hopkins University) provided help and instruction on many of the geochemical and hydrologic concepts that were evaluated during the course of this study. K.C. Lohmann and Jim Burdett (University of Michigan) provided carbon and oxygen isotopic analyses, and Lynn Walters and Ted Huston (University of Michigan) provided trace element (ICP) analyses. Tim Denison and Mobil Oil Corporation provided Sr isotopic analyses. Special thanks are extended to AGIP for providing access to core from the Belluno 1 well, and to ENEL for access to the Vajont Dam area. This study benefited from the support and help of many family members, fellow students, and friends, including Michele Claps, Paul A. Dunn, Linda A. Hinnov, Joseph B. Paul, and Lyndon A. Yose. We would like to thank J.A. Kupecz and J.R. Markello for providing critical review of this manuscript. This study was made possible by grants from the American Association of Petroleum Geologists, the Geological Society of America, Sigma Xi, Mobil Oil Corporation, The Johns Hopkins University Balk Fund, and the National Science Foundation (Grant #EAR910510).

REFERENCES CITED

Adams, J.E., and M.L. Rhodes, 1960, Dolomitization by seepage refluxion: AAPG Bulletin, v. 44, p. 1912–1921.

Aharon, P., R.A. Socki, and L. Chan, 1987, Dolomitization of atolls by sea water convection flow: test of a hypothesis at Niue, South Pacific: Journal of Geology, v. 95, p. 187–203.

Aissaoui, D.M., 1988, Magnesian calcite cements and their diagenesis: dissolution and dolomitization, Mururoa Atoll: Sedimentology, v. 35, p. 821–841.

Al-Aasm, I.S., and J. Veizer, 1982, Chemical stabilization of low-Mg calcites: an example of brachiopods: Journal of Sedimentary Petrology, v. 52, p. 1101–1109.

Amthor, J.E., E.W. Mountjoy, and H.G. Machel, 1993, Subsurface dolomites in Upper Devonian Leduc Formation buildups, central part of Rimbey-Meadowbrook reef trend, Alberta, Canada: Bulletin of the Canadian Society of Petroleum Geology, v. 41, p. 164–185.

Aulstead, K.L., R.J. Spencer, and H.R. Krouse, 1988, Fluid inclusion and isotopic evidence on dolomitization, Devonian of western Canada: Geochimica et Cosmochimica Acta, v. 52, p. 1027–1035.

Austin, J.A., 1983, OBC 5-A: overthrusting in a deep-water carbonate terrane, *in* A.W. Bally, ed., Seismic expression of structural styles: AAPG Studies in Geology Series 15, v. 3, p. 167–172.

Badiozamani, K., 1973, The dorag dolomitization model—application to the Middle Ordovician of Wisconsin: Journal of Sedimentary Petrology, v. 43, p. 965–984.

Baker, P.A., and S. Burns, 1985, Occurrence and formation of dolomite in organic-rich continental margin sediments: AAPG Bulletin, v. 69, p. 1917–1930.

Barrett, M.L., 1987, The dolomitization and diagenesis of the Jurassic Smackover Formation, southern Alabama: Ph.D. thesis, The Johns Hopkins University, Baltimore, Maryland, 362 p.

Bethke, C.M., 1985, A numerical model of compaction-driven groundwater flow and heat transfer and its application to the paleohydrology of intracratonic sedimentary basins: Journal of Geophysical Research, v. 90B, p. 6817–6828.

Blatt, H., 1982, Sedimentary petrology: San Francisco, Freeman & Co., 564 p.

Bosellini, A., 1989, Dynamics of Tethyan carbonate platforms, *in* P.D. Crevello, J.L. Wilson, J.F. Sarg, and J.F. Read, eds., Controls on carbonate platform and basin development: SEPM Special Publication, p. 3–13.

Bosellini, A., D. Masetti, and M. Sarti, 1981, A Jurassic "Tongue of the Ocean" infilled with oolitic sands: the Belluno Trough, Venetian Alps, Italy: Marine Geology, v. 44, p. 59–95.

Broomhall, R.W., and J.R. Allen, 1985, Regional caprock-destroying dolomite on the Middle Jurassic to Early Cretaceous Arabian shelf: Society of Petroleum Engineers, SPE 13697, p. 157–163.

Carballo, J.D., L.S. Land, and D.L. Miser, 1987, Holocene dolomitization of supratidal sediments by active tidal pumping, Sugarloaf Key, Florida: Journal of Sedimentary Petrology, v. 57, p. 153–165.

Carson, B., E. Suess, and J.C. Strasser, 1990, Fluid flow and mass flux determinations at vent sites on the Cascadia margin accretionary prism: Journal of Geophysical Research, v. 95, p. 8891–8898.

Casati, P., and M. Tomai, 1969, Il Giurassico ed il Cretacio del versante settentrionale del Vallone Bellunese e del Gruppo del M. Brandol: Riv. Italiana Paleontologia e Stratigrafia, v. 75, p. 205–341.

Cati, A., D. Sartorio, and S. Venturini, 1987, Carbonate platforms in the subsurface of the northern Adriatic area: Mem. Soc. Geol. It., v. 40, p. 295–308.

Cervato, C., 1990, Hydrothermal dolomitization of Jurassic–Cretaceous limestones in the southern Alps (Italy): relation to tectonics and volcanism: Geology, v. 18, p. 458–461.

Coniglio, M., R. Sherlock, A.E. Williams-Jones, K. Middleton, and S.K. Frape, 1994, Burial and hydrothermal diagenesis of Ordovician carbonates from the Michigan Basin, Ontario, Canada, in B. Purser, M. Tucker, and D. Zenger, eds., Dolomites—a volume in honour of Dolomieu: International Association of Sedimentologists Special Publication 21, p. 231–254.

Dawans, J.M., and P.K. Swart, 1988, Textural and geochemical alternations in Late Cenozoic Bahamian dolomites: Sedimentology, v. 35, p. 385–403.

Ditty, P.S., C.J. Harmon, O.H. Pilkey, M.M. Ball, and E.S. Richardson, 1977, Mixed terrigenous carbonate sedimentation in the Hispaniola Caicos turbidite basin: Marine Geology, v. 24, p. 1–20.

Dix, G.R., 1993, Patterns of burial- and tectonically controlled dolomitization in an Upper Devonian fringing-reef complex: Leduc Formation, Peace River Arch area, Alberta, Canada: Journal of Sedimentary Petrology, v. 63, p. 628–640.

Doglioni, C., 1990, The Venetian Alps thrust belt, in K.R. McClay, ed., Thrust tectonics: London, Chapman and Hall, p. 319–324.

Elder, J.W., 1965, Physical processes in geothermal areas, in W.H.K. Lee, ed., Terrestrial heat flow: American Geophysical Union Monograph Series No. 8, p. 211–239.

Elder, J.W., 1967, Steady free convection in a porous medium heated from below: Journal of Fluid Mechanics, v. 27, p. 29–48.

Elder, J.W., 1977, Thermal convection: Journal of the Geological Society of London, v. 133, p. 292–309.

Gaetani, M., and F. Jadoul, 1979, The structure of the Bergamasc Alps: Rend. Acc. Naz. Lincei, v. 66, no. 5, p. 411–416.

Garven, G., 1985, The role of regional fluid flow in the genesis of the Pine Point deposit, Western Canada Sedimentary Basin: Economic Geology, v. 80, p. 307–324.

Garven, G., and R.A. Freeze, 1984, Theoretical analysis of the role of groundwater flow in the genesis of stratabound ore deposits 2: quantitatve results: American Journal of Science, v. 284, p. 1125–1174.

Ge, S., and G. Garven, 1989, Tectonically induced transient groundwater flow in foreland basin: Int. A.G.U. Monograph Series, No. 48, I.U.G.G., v. 3, p. 145–157.

Gregg, J.M., 1985, Regional epigenetic dolomitization in the Bonneterre dolomite (Cambrian), southern Missouri: Geology, v. 13, p. 503–506.

Gregg, J.M., and K.L. Shelton, 1989, Minor- and trace-element distributions in the Bonneterre Dolomite (Cambrian), southeast Missouri: evidence for possible multi-basin fluid sources and pathways during lead-zinc mineralization: Geological Society of America Bulletin, v. 101, p. 221–230.

Hanshaw, B.B., and W. Back, 1980, Chemical mass-wasting of the northern Yucatan Peninsula by groundwater dissolution: Geology, v. 8, p. 222–224.

Hanshaw, B.B., W. Back, and R.G. Deike, 1971, A geochemical hypothesis for dolomitization by groundwater: Economic Geology, v. 66, p. 710–724.

Hardie, L.A., 1987, Dolomitization: a critical review of some current views: Journal of Sedimentary Petrology, v. 57, p. 166–183.

Jodry, R.L., 1969, Growth and dolomitization of Silurian reefs, St. Clair County, Michigan: AAPG Bulletin, v. 53, p. 957–981.

Kaufman, J., 1994, Numerical models of fluid flow in carbonate platforms: implications for dolomitization: Journal of Sedimentary Research, v. A64, p. 128–139.

Kohout, F.A., H.R. Henry, and J.E. Banks, 1977, Hydrogeology related to geothermal conditions of the Floridan Plateau, in The geothermal nature of the Floridan Plateau: Florida Bureau of Geology Special Publication 21, p. 1–41.

Kulm, L.D., et al., 1986, Oregon subduction zone: Venting, fauna, and carbonates: Science, v. 231, p. 561–566.

Kupecz, J.A., and L.A. Land, 1991, Late-stage dolomitization of the Lower Ordovician Ellenberger Group, West Texas: Journal of Sedimentary Petrology, v. 61, p. 551–574.

Kupecz, J.A., C. Kerans, and L.S. Land, 1988, Discussion: Deep-burial dolomitization in the Ordovician Ellenberger Group Carbonates, West Texas and Southeastern New Mexico: Journal of Sedimentary Petrology, p. 908–910.

Kupecz, J.A., and L.A. Land, 1994, Progressive recrystalization and stabilization of early-stage dolomite: Lower Ordovician Ellenberger Group, west Texas, in B. Purser, M. Tucker, and D. Zenger, eds., Dolomites—a volume in honour of Dolomieu: International Association of Sedimentologists Special Publication 21, p. 255–279.

Kupecz, J.A., I.P. Montañez, and G. Gao, 1993, Recrystallization of dolomite with time, in R. Rezak and D. Lavoie, eds., Carbonate microfabrics, frontiers in sedimentology: New York, Springer-Verlag, p. 187–194.

Land, L.S., 1985, The origin of massive dolomite: Journal of Geological Education, v. 33, p. 112–125.

Lee, Y.I., and G.M. Friedman, 1987, Deep-burial dolomitization in the Ordovician Ellenberger Group carbonates, West Texas and southeastern New Mexico: Journal of Sedimentary Petrology, v. 57, no. 3, p. 544–557.

Lee, Y.I., and G.M. Friedman, 1988, Reply: deep-burial dolomitization in the Ordovician Ellenberger Group carbonates, West Texas and southeastern New Mexico: Journal of Sedimentary Petrology, v. 58, p. 910–913.

Lucia, F.J., 1995, Rock fabric/petrophysical classification of carbonate pore space for reservoir characterization: AAPG Bulletin, v. 79, p. 1275–1300.

Lumsden, D.N., 1985, Secular variations in dolomite abundance in deep marine sediments: Geology, v. 13, p. 766–769.

Lumsden, D.N., 1988, Characteristics of deep-marine dolomite: Journal of Sedimentary Petrology, v. 58, p. 1023–1031.

Machel, H.G., and J.H. Anderson, 1989, Pervasive subsurface dolomitization of the Nisku Formation of Central Alberta: Journal of Sedimentary Petrology, v. 59, p. 891–911.

Machel, H.G., and E.W. Mountjoy, 1986, Chemistry and environments of dolomitization—a reappraisal: Earth Science Reviews, v. 23, p. 175–222.

Major, R.P., 1984, The Midway Atoll coral cap: meteoric diagenesis, amplitude of sea level fluctuation, and dolomitization: Ph.D. thesis, Brown University, Providence, Rhode Island, 133 p.

Masetti, D., 1971, Sedimentologia e paleogeografia del Giurassico tra Brenta e Piave: Ph.D. thesis, University of Ferrara, Italy, 104 p.

Masetti, D., and G. Bianchin, 1987, Geologia del Gruppo della Schiara (Dolomiti Bellunesi). Suo inquadramento nella evoluzione giurassica del margine orientale della Piattaforma di Trento: Mem. Ist. Geol. Min. Univ. Padova v. 39, p. 187–212.

Massari, F., P. Grandesso, C. Stefani, and P.G. Jobstraibizer, 1986, A small polyhistory foreland basin evolving in a context of oblique convergence: the Venetian basin (Chattian to Recent, Southern Alps, Italy): International Association of Sedimentologists Special Publication 8, p. 141–168.

Mattes, B.W., and E.W. Mountjoy, 1980, Burial dolomitization of the Upper Devonian Miette buildup, Jasper National Park, Alberta: SEPM Special Publication 28, p. 259–297.

McKenzie, J., 1981, Holocene dolomitization of calcium carbonate sediments from the coastal sabkhas of Abu Dhabi, U.A.E.: a stable isotope study: Journal of Geology, v. 89, p. 185–198.

Miller, J.K., and R.L. Folk, 1994, Petrographic, geochemical and structural constraints on the timing and distribution of postlithification dolomite in the Rhaetian Portoro ("Calcare Nero") of the Portovenere area, La Spezia, Italy, in B. Purser, M. Tucker, and D. Zenger, eds., Dolomites—a volume in honour of Dolomieu: International Association of Sedimentologists Special Publication 21, p. 187–202.

Montañez, I.P., 1994, Late diagenetic dolomitization of Lower Ordovician, Upper Knox carbonates: a record of the hydrodynamic evolution of the Southern Appalachian Basin: AAPG Bulletin, v. 78, p. 1210–1239.

Montañez, I.P., and J.F. Read, 1992, Fluid-rock interaction history during stabilization of early dolomites, Upper Knox Group (Lower Ordovician), U.S. Appalachians: Journal of Sedimentary Petrology, v. 62, p. 753–778.

Morrow, D.W., 1982a, Diagenesis 1. Dolomite—Part 1: the chemistry of dolomitization and dolomite precipitation: Geoscience Canada, v. 9, p. 5–13.

Morrow, D.W., 1982b, Diagenesis 2. Dolomite—Part 2: dolomitization models and ancient dolostones: Geoscience Canada, v. 9, p. 95–107.

Mountjoy, E.W., and J.E. Amthor, 1994, Has burial dolomitization come of age? Some answers from the Western Canada Sedimentary Basin, in B. Purser, M. Tucker, and D. Zenger, eds., Dolomites—a volume in honour of Dolomieu: International Association of Sedimentologists Special Publication 21, p. 203–229.

Mountjoy, E.W., and M.K. Halim-Dihardja, 1991, Multiple phase fracture and fault-controlled burial dolomitization, Upper Devonian Wabamun Group, Alberta: Journal of Sedimentary Petrology, v. 61, p. 590–612.

Mullins, H.T., N. Breen, J. Dolan, R.W. Wellner, J.L. Petruccione, M. Gaylord, B. Andersen, A.J. Melillo, A.D. Jurgens, and D. Orange, 1992, Carbonate platforms along the southeast Bahamas–Hispaniola collision zone: Marine Geology, v. 105, p. 169–209.

Mullins, H.T., S.W. Wise, L.S. Land, D.I. Siegel, P.M. Masters, E.G. Hinchey, and K.R. Price, 1985, Authigenic dolomite in Bahamian periplatform slope sediment: Geology, v. 13, p. 292–295.

Prouty, C.E., 1983, The tectonic development of the Michigan Basin intrastructures, in R.E. Kimmel, ed., Tectonics, structure, and karst in northern Lower Michigan: Michigan Basin Geological Society 1983 Field Conference, p. 36–81.

Ritger, S., B. Carson, and E. Suess, 1987, Methane-derived authigenic carbonates formed by subduction-induced pore water expulsion along the Oregon/Washington margin: Geological Society of America Bulletin, v. 98, p. 147–156.

Ruppel, S.C., and H.S. Cander, 1988, Dolomitization of shallow-water carbonates by seawater and seawater-derived brines: San Andres Formation (Guadalupian), West Texas, in V. Shukla and P.A. Baker, eds., Sedimentology and geochemistry of dolostones: SEPM Special Publication 43, p. 245–262.

Saller, A.H., 1984, Petrologic and geologic constraints on the origin of subsurface dolomite, Enewetak Atoll: an example of dolomitization by normal seawater: Geology, v. 12, p. 217–220.

Sample, J.C., M.R. Reid, H.J. Tobin, and J.C. Moore, 1993, Carbonate cements indicate channeled fluid flow along a zone of vertical faults at the deformation front of the Cascadia accretionary wedge: Geology, v. 21, p. 507–510.

Sandford, W.E., 1987, Assessing the potential for calcite dissolution in coastal saltwater mixing zones: Ph.D. thesis, Pennsylvania State University, State College, Pennsylvania, 103 p.

Schlager, W., and N.P. James, 1978, Low-magnesian calcite limestones forming at the deep-sea floor, Tongue of the Ocean, Bahamas: Sedimentology, v. 25, p. 675–702.

Sears, S.O., and F.J. Lucia, 1980, Dolomitization of northern Michigan Niagara reefs by brine refluxion and freshwater/seawater mixing, *in* D.H. Zenger, J.B. Dunham, and R.L. Ethington, eds., Concepts and models of dolomitization: SEPM Special Publication 28, p. 215–235.

Simms, M., 1984, Dolomitization by groundwater-flow systems in carbonate platforms: Transactions of the Gulf Coast Association of Geological Societies, v. 34, p. 411–420.

Sun, S.Q., 1995, Dolomite reservoirs: porosity evolution and reservoir characteristics: AAPG Bulletin, v. 79, p. 186–204.

Tobin, H.J., J.C. Moore, M.E. MacKay, D.L. Orange, and L.D. Kulm, 1993, Fluid flow along a strike-slip fault at the toe of the Oregon accretionary prism: implications for the geometry of frontal accretion: Geological Society of America Bulletin, v. 105, p. 569–582.

Trevisani, E., 1991, Il Toarciano-Aaleniano nei settori centro-orientali della Piattaforma di Trento (Prealpi Venete): Riv. Italiana Paleontologia e Stratigrafia, v. 97, p. 99–124.

Vahrenkamp, V.C., and P.K. Swart, 1990, New distribution coefficient for the incorporation of strontium into dolomite and its implications for the formation of ancient dolomites: Geology, v. 18, p. 387–391.

van Tuyl, F.M., 1916, The origin of dolomite: Iowa Geological Survey Annual Report, v. 25, p. 251–422.

Weber, J.N., 1964, Trace element composition of dolostones and dolomites and its bearing on the dolomite problem: Geochimica et Cosmochimica Acta, v. 28, p. 1817–1868.

Weissert, H.J., and D. Bernoulli, 1985, A transform margin in the Mesozoic Tethys: evidence from the Swiss Alps: Geologische Rundschau, v. 73, p. 665–679.

Weyl, P.K., 1960, Porosity through dolomitization: conservation-of-mass requirements: Journal of Sedimentary Petrology, v. 30, p. 85–90.

Whitaker, F.F., P.L. Smart, V.C. Vahrenkamp, H. Nicholson, and R.A. Wogelius, 1994, Dolomitization by near-normal seawater? Field evidence from the Bahamas, *in* B. Purser, M. Tucker, and D. Zenger, eds., Dolomites—a volume in honour of Dolomieu: International Association of Sedimentologists Special Publication 21, p. 111–132.

Wilkinson, B.H., and T.J. Algeo, 1989, Sedimentary carbonate record of calcium-magnesium cycling: American Journal of Science, v. 289, p. 1158–1194.

Wilson, E.N., 1989, Dolomitization of the Triassic Latemar buildup, northern Italy: Ph.D. thesis, The Johns Hopkins University, Baltimore, Maryland, 272 p.

Wilson, E.N., L.A. Hardie, and O.M. Phillips, 1990, Dolomitization front geometry, fluid flow patterns, and the origin of massive dolomite: the Triassic Latemar buildup, Northern Italy: American Journal of Science, v. 290, p. 741–796.

Yao, Q., and R.V. Demicco, 1995, Paleoflow patterns of dolomitizing fluids and paleohydrology of the southern Canadian Rocky Mountains: evidence from dolomite geometry and numerical modeling: Geology, v. 23, p. 791–794.

Zempolich, W.G., 1993, The drowning succession in Jurassic carbonates of the Venetian Alps, Italy: a record of supercontinent breakup, gradual eustatic rise, and eutrophication of shallow-water environments, *in* R.G. Loucks and J.F. Sarg, eds., Carbonate sequence stratigraphy: recent developments and applications: AAPG Memoir 57, p. 63–105.

Zempolich, W.G., 1995, Deposition, early diagenesis, and late dolomitization of deepwater resedimented oolite: the Middle Jurassic Vajont limestone of the Venetian Alps, Italy: Ph.D. thesis, The Johns Hopkins University, Baltimore, Maryland, 659 p.

Zempolich, W.G., and L.A. Hardie, 1991a, Massive burial dolomitization: the Jurassic Vajont oolite of northeast Italy, *in* A. Bosellini, R. Brandner, E. Flügel, B. Purser, W. Schlager, M. Tucker, and D. Zenger, eds., Dolomieu Conference on Carbonate Platforms and Dolomitization Abstracts: Ortisei, Italy, p. 298.

Zempolich, W.G., and L.A. Hardie, 1991b, Massive burial dolomitization: the Jurassic Vajont oolite of northeast Italy (abs.): Geological Society of America, Abstracts with Programs, p. 411.

Zenger, D.H., 1976, Dolomitization and dolomite "dikes" in the Wyman Formation (Precambrian), northeastern Inyo Mountains, California: Journal of Paleontology, v. 46, p. 457–462.

Zenger, D.H., 1983, Burial dolomitization in the Lost Burro Formation (Devonian) East-Central California, and the significance of late diagenetic dolomitization: Geology, v. 11, p. 519–522.

Zenger, D.H., and J.B. Dunham, 1988, Dolomitization of Siluro–Devonian limestones in a deep core (5350 m), southeastern New Mexico: SEPM Special Publication 43, p. 161–173.

Gluyas, J.G., and T. Witton, 1997, Poroperm prediction for wildcat exploration prospects: Miocene Epoch, Southern Red Sea, *in* J.A. Kupecz, J. Gluyas, and S. Bloch, eds., Reservoir quality prediction in sandstones and carbonates: AAPG Memoir 69, p. 163–176.

Chapter 11

Poroperm Prediction for Wildcat Exploration Prospects: Miocene Epoch, Southern Red Sea

Jon G. Gluyas[1]
*BP Exploración de Venezuela S.A., Edificio Centro Seguros de Sud America
Caracas, Venezuela*

Trevor Witton
*BP Exploration
London, England, United Kingdom*

ABSTRACT

Prior to BP Exploration's drilling the well Antufash-1 in the Yemeni waters of the Southern Red Sea, reservoir quality was estimated to be poor; it was dry, plugged, and abandoned. The Miocene sandstones encountered were tight, with a mean porosity of 4% in the cored section and a permeability of only 0.07 md.

The prediction of low quality for the reservoir section of Antufash-1 was based on very few core analysis data. The diagenetic history of potential reservoir sands in the Antufash acreage was calculated from data on depth to prospect, burial and thermal history of the area, reservoir sand provenance, and depositional environment.

An initial assessment, using limited local well data, led to the conclusion that only at depths <0.5 km was it reasonable to expect high reservoir quality (>100 md). However, at depths >1.5 km, permeability was likely to be as low as 10 md. Throughout this depth range, the chances of halite cementation were also reasoned to be high. The rapid deterioration of reservoir quality with depth was attributed to the instability of the original volcaniclastic detritus. Such detritus was predicted to have converted to a mixture of zeolites and smectitic clay soon after deposition. The reactivity of the assemblage was also predicted to have been exaggerated by the high thermal gradients in the area.

The recommendation was to avoid large parts of the license area known to have received input of volcaniclastic sediment, and to develop prospects in the few areas thought to have had arkosic sand input. These sands, it was reasoned, would suffer less degradation of reservoir quality. The Antufash-1 well successfully proved the existence of such arkosic sands in the basin, and their diagenetic history was as predicted. Unfortunately, the sandstones were tight. Halite cement filled, as predicted, all remaining porosity.

---

[1]Present address: Monument Oil and Gas plc, London, United Kingdom.

INTRODUCTION

In 1990, BP Exploration was awarded a 100% working interest in the Antufash Concession, located off the west coast of Yemen in the Southern Red Sea (Figure 1). In 1992 the commitment well Antufash-1 was drilled in the northern part of the license. The well results demonstrated that the sandstones encountered were fully cemented, and only minor gas shows were recorded.

Earlier exploration drilling elsewhere in the Southern Red Sea had resulted in a few very minor oil and condensate discoveries. The general perception was that trap risk was relatively low, but significant risks were associated with oil source and reservoir. Indeed, even if a working source were present, oil charge was predicted to be limited due to the small prospect drainage areas. However, the quality of the likely reservoir interval remained the key risk.

The few sandstones that had been encountered were of very low permeability. The aim of our study was to determine, ahead of the drill bit, the controls on reservoir quality, and to develop a methodology for either mapping reservoir quality prior to drilling or estimating reservoir quality at the prospect location prior to drilling.

DATABASE

Most of the reservoir quality data available at the time of prospect evaluation were limited to qualitative descriptions of cuttings, petrographic point-count data for cuttings, and a few petrographic descriptions of core chips from wells scattered across the Red Sea. A few scanning electron microscope (SEM) photomicrographs were available for some of the core chips.

Core analysis data (porosity and permeability) were available for two old wells in the Antufash license (Al Meethag 1 and Al Meethag 2; Table 1).

GEOLOGICAL BACKGROUND

Basin Development

The Red Sea occupies a long (2000 km) linear rift of late Oligocene age 180–50 km wide (Hughes et al., 1991). The conjugate margins are bounded by a series of large fault terraces with ≤2500 m of relief. Within the sea itself, there are three physiographic elements: (1) shallow shelfal areas, narrow north of 21°N but wider to the south; (2) a main trough 600–1000 m deep occupying most of the sea area to the north of 21°N; and (3) a narrow axial trough, ~2000 m deep and 5–30 km wide (Coleman, 1993).

The crust beneath the axial trough has been determined as oceanic on the basis of magnetic anomalies (Girdler and Styles, 1978), with the oldest crust thought to have been formed ~5 Ma. There remains much doubt as to the nature of the crust beneath the shelfal areas. It may be oceanic crust formed during the Oligo–Miocene (Hall, 1989), thinned continental crust (Egloff et al., 1991), or a bit of both (Cochran, 1983). Sea-floor spreading is still active within the Red Sea area.

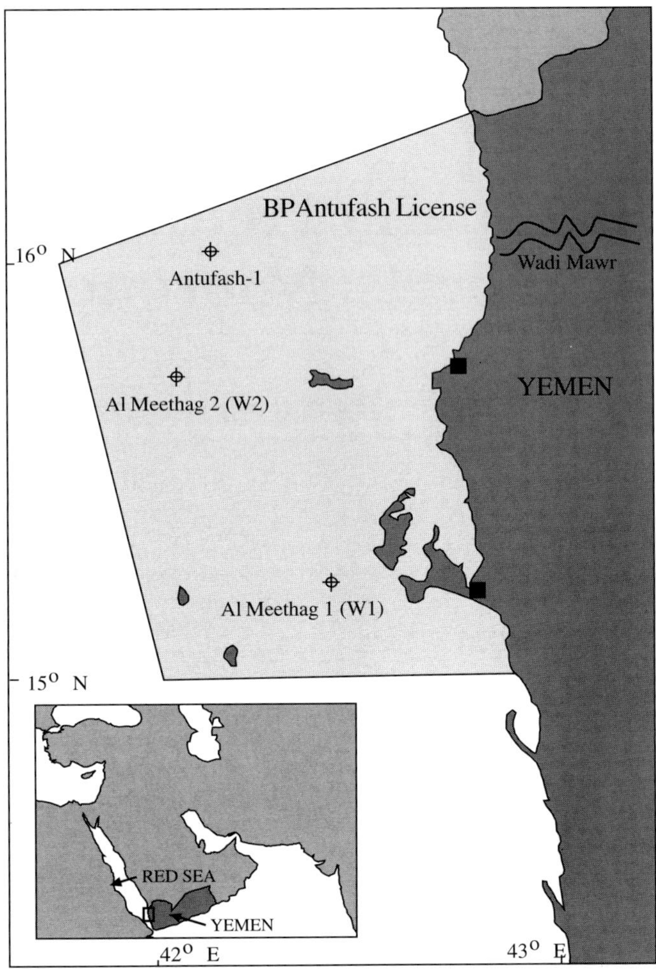

Figure 1. Location map for the Antufash license in the Yemeni Red Sea.

The thin continental crust and active sea-floor spreading has resulted in heat flows, typically >200 mWm^2 (megawatts per square meter) in the axial trough (Coleman, 1993), and almost everywhere greater than the world average of 55 (mWm^2). As a consequence, thermal gradients are also high, from ~73°C km^{-1} at the basin center to ~45°C km^{-1} at the basin margin. Volcanism associated with the rifting continued throughout the Miocene (Davison and Rex, 1980).

Stratigraphy

Four megasequences have been identified (Figure 2; Hughes and Beydoun, 1992; Mitchell et al., 1992). These, with their approximate ages, are described below.

Prerift Pre-Late Oligocene, 26 Ma or Older

The oldest megasequence comprises minor marine deposits generated by occasional flooding by the Indian Ocean into the incipient rift in the Mesozoic. The Oligocene is dominated by regional flood volcanism.

Table 1. Core Porosity and Permeability Data for Regional Wells in the Antufash License.

Well	Average Depth (m subsea)	Number of Plugs	Mean Porosity (%)	1σ (%)	Mean Permeability (md)	1σ (md)
Al Meethag 1 (W1)	1417	51	28.5	5.3	34.7	77.0
Al Meethag 2 (W2)	1540	29	21.6	3.7	11.5	19.0

Synrift Early to Middle Miocene, 26–16 Ma ("Infra-Evaporite")

In quick succession, the rift phase was typified by uplift, high rate of extension, and subsidence as Arabia and Africa separated. Transgression occurred from the north, and widespread marine conditions were established. By the middle Miocene, the environment remained shallow marine, but water circulation was restricted.

Postrift Middle to Late Miocene, 16–5 Ma ("Evaporite")

Following development of a silled basin, thick evaporite deposits were developed during lowstand drawdown. Uplift of the rift shoulders resulted in deposition of thick clastic wedges along the basin margin. Intermittent connection with the Indian Ocean and periods of anoxia led to the development of potential petroleum source rocks (El-Anbaawy et al., 1992; Cole et al., 1995). The massive salt at the base of this megasequence also began to move at this time due to the gravity loading in the coastal areas (Heaton et al., 1993; Davidson et al., 1994, 1995). Cyclic deposition of source reservoir and seal parasequences occurred in the salt withdrawal basins.

Axial Rift Pliocene–Recent, 5 Ma–Present ("Supra-Evaporite")

Eustatic sea level fall accentuated erosion on the basin margins. Spreading continued with rapid subsidence of the axial trough. New oceanic crust was formed in the south, and open marine conditions were established with the Indian Ocean; major carbonate deposition occurred. The continued clastic deposition at the basin margins and the resultant salt movement accentuated earlier developed structural traps.

Reservoir Development

A combination of seismic reflection data mapping and information obtained from existing wells revealed that reservoir potential was likely to be best developed within upper Miocene highstand progradational systems and associated lowstand systems tracts (Crossley et al., 1992). In both tracts, basin-fringing alluvial/fluvial systems were predicted to be the most likely reservoirs. Some marine influence was likely to have occurred during maximum flood.

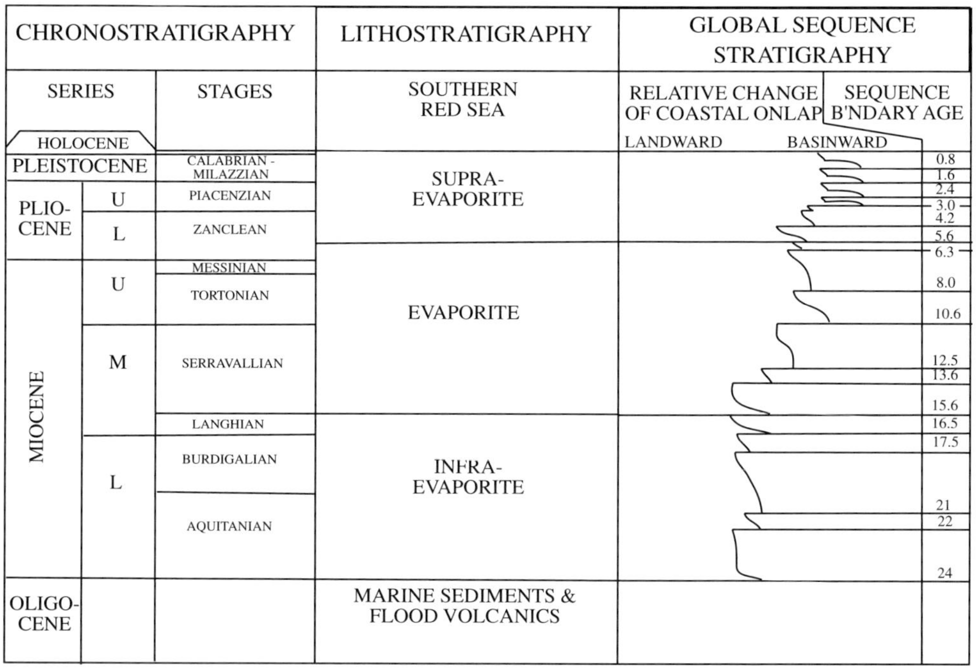

CHRONOSTRATIGRAPHY		LITHOSTRATIGRAPHY	GLOBAL SEQUENCE STRATIGRAPHY	
SERIES	STAGES	SOUTHERN RED SEA	RELATIVE CHANGE OF COASTAL ONLAP	SEQUENCE B'NDARY AGE
			LANDWARD BASINWARD	
HOLOCENE				
PLEISTOCENE	CALABRIAN - MILAZZIAN	SUPRA-EVAPORITE		0.8 / 1.6
PLIO-CENE U	PIACENZIAN	SUPRA-EVAPORITE		2.4 / 3.0 / 4.2
PLIO-CENE L	ZANCLEAN			5.6 / 6.3
MIOCENE U	MESSINIAN	EVAPORITE		8.0
MIOCENE U	TORTONIAN	EVAPORITE		10.6
MIOCENE M	SERRAVALLIAN			12.5 / 13.6 / 15.6
MIOCENE M	LANGHIAN			16.5 / 17.5
MIOCENE L	BURDIGALIAN	INFRA-EVAPORITE		21 / 22
MIOCENE L	AQUITANIAN			24
OLIGO-CENE		MARINE SEDIMENTS & FLOOD VOLCANICS		

Figure 2. Stratigraphy of the Red Sea area (R. Jones, 1994, personal communication).

Sand Provenance/Composition

The last major uplift in the Red Sea area began during the middle Miocene and continues today (Davidson et al., 1994). As a consequence, the present-day geological maps of the circum Red Sea area are taken to represent the potential provenance area for Upper Miocene sediments (Sudan, 1963; U.S. Geological Survey, 1963; Kazmin, 1973; Merla, 1979). In broad terms there were two very different provenance terrains during the mid-Miocene:

- Pre-Cambrian acid and acid-intermediate metamorphic and igneous granites and gneisses with minor Jurassic and Cretaceous sandstones (Tawila Formation).
- Pre-Cambrian basic metamorphic and igneous rocks, Oligo–Miocene volcanics, and minor Jurassic limestones.

The following depositional sand compositions were estimated using descriptions of SEM preparations from core and cuttings obtained from the two Al Meethag wells (Figure 1).

These wells showed basic/volcanic-derived and acid-derived deposits as follows: quartz 30 ± 20% and 50 ± 20%, respectively; feldspar 30 ± 20% and 30 ± 20%, respectively; various rock fragments 40 ± 20% and 20%, respectively.

The rock fragments include volcaniclastic grains, mafic mineral grains, and a little glauconite.

In the Antufash acreage, as evidenced by the Al Meethag wells, much of the provenance appears to have been from the basic metamorphics and volcanics. Only in the area west of Wadi Mawr (Figure 1) is this basic/volcanic input likely to have been diluted. This wadi drains along a Jurassic transfer fault and taps into an area that may have shed large quantities of arkosic Tawila Sandstone during the middle Miocene.

EVALUATION OF RESERVOIR QUALITY

Method

Two approaches were used to calculate reservoir quality for the Miocene sandstones of the Southern Red Sea. First, the limited poroperm data that did exist were analyzed in terms of the controls on porosity and permeability, using the methodology of Cade et al. (1994). The results from this analysis were compared with the qualitative petrographic descriptions.

Second, reservoir quality data were evaluated using the broad geological data available for the area. The methods for porosity and permeability synthesis are given below.

Correct prediction of porosity requires that the volume losses due to compaction and cementation are quantified. Permeability prediction further requires knowledge of the grain size and sorting characteristics, cement types, and their distribution.

Porosity loss due to compaction was calculated using the methodology of Gluyas and Cade (this volume). Cement types and volumes were calculated on the basis of a BP Exploration in-house regional diagenesis study (Primmer, 1993; Primmer et al., this volume), in which the links between sand mineralogy at deposition, depositional environment, burial, and thermal history were quantified.

Grain size and sorting data were adopted from the existing well information. Permeabilities were calculated from the cement data and estimates of grain size and sorting, using the sphere pack modeling approach of Cade et al. (1994).

Data Analysis

The reservoir intervals of the two Al Meethag wells contain quartz-poor, feldspar- and volcaniclastic-rich, fine- to medium-grained sandstones. Their diagenetic history is complex, with calcite, dolomite, chlorite, smectite, zeolite, quartz, illite, and halite cements. Given that the sediments are at most 15 m.y. old and even now buried only to 1.5–1.7 km, all of these processes must have occurred in a short geological time and at shallow depth.

An attempt to construct an empirical porosity depth plot proved futile. The problems encountered are illustrated in Figures 3 and 4. In short, there are too few data from which to draw any valid conclusions as to how, or if, porosity varies with depth in this basin.

Porosity and permeability data from conventional core analysis for the two Al Meethag wells are plotted in Figure 5. Plotted on the same graph are modeled curves for the porosity-to-permeability relationship in similar grain size (fine- to medium-grained) clean, compacted, and/or quartz cemented sandstones (Evans et al., this volume). Most of the data from the two wells describe two distinctly different prolate clusters of reasonably similar permeability range but significantly different porosity range. The outliers to these two trends are medium/coarse grained sandstones, carbonate cemented sandstones, and, in one instance, a fractured core analysis plug.

The poroperm data for both wells lie well below the modeled clean sand lines. The steep porosity-to-permeability gradient is indicative of a sand with a large proportion of poorly interconnected porosity: either intragranular secondary porosity or microporosity trapped between clay fibers and plates. The similarity of poroperm gradients in the two wells was taken to indicate that the process controlling permeability evolution in both was similar. This relationship did not hold for porosity. The inferred importance of clay in controlling the permeability of these sandstones is fully supported by the petrographic descriptions.

In order to explain the porosity difference between the wells, a process is needed to reduce porosity with only a minor (relative to the clays) effect on permeability. At the high porosities seen in these cores, two processes could have been responsible: compaction and/or syntaxial quartz precipitation (Cade et al., 1994).

There is insufficient difference in burial depth of the two sandstones to account for the porosity difference in terms of compaction alone, even when the 13 MPa overpressure in well W1 is taken into account (Robinson and Gluyas, 1992). It is possible that quartz

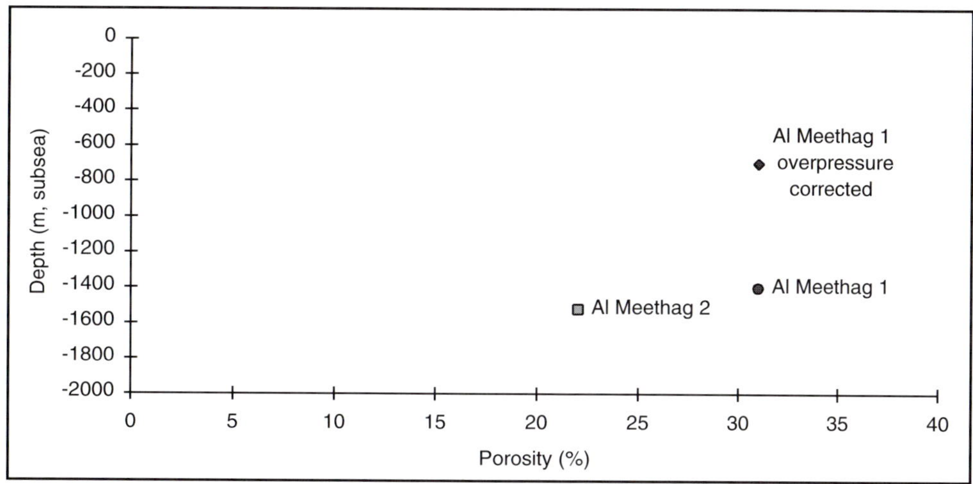

Figure 3. Porosity depth plot for cored intervals from Antufash License. Al Meethag 1 is plotted twice, at its current burial depth and at its hydrostatic equivalent burial depth. The Miocene sands in Al Meethag 1 are overpressured by 9 MPa (1300 psi); 1 MPa is ~80 m of burial in a hydrostatic system at these burial depths (Gluyas and Cade, this volume). Data are averages for wells; individual plug data are plotted in Figure 5. Circle = Al Meethag 1; square = Al Meethag 2; diamond = Al Meethag 2 overpressure corrected.

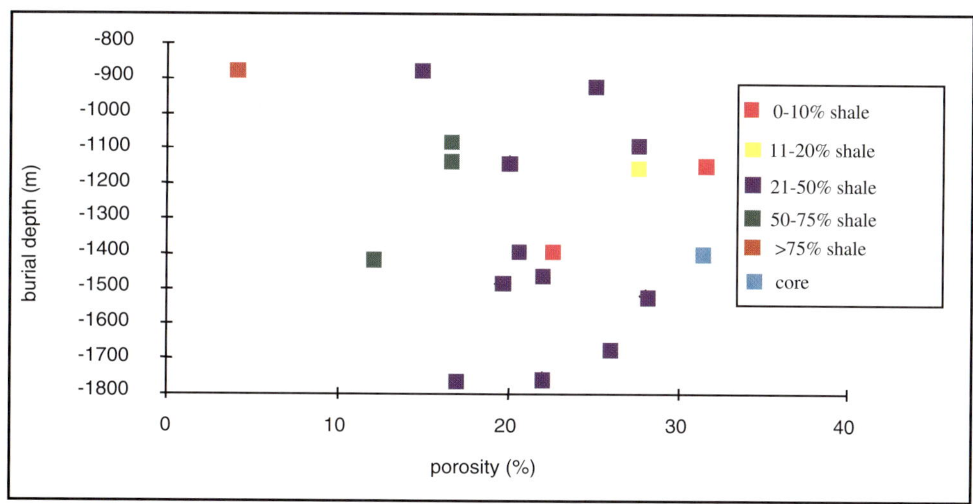

Figure 4. Porosity–depth plot for log data from intervals in Al Meethag 1. Porosity and shale percentages were calculated from a combination of neutron density and resistivity logs.

cementation may account for much of the difference. This suggestion is supported by the qualitative descriptions of the petrography of the sandstones from the two wells. Quartz cement was described from W2 but not from W1. Modeling data (see the following section) also lend some support to this suggestion.

No equivalent quantitative reservoir quality data were available for the sandstones derived from the Pre-Cambrian acid igneous and gneiss terrains.

Data Synthesis—Modeled Poroperm Evolution

The following criteria were used to construct a semiquantitative diagenetic history for the Miocene sandstones (Figures 6, 7).

Volcaniclastic Sandstones

- Volcaniclastic sandstones are likely to react in situ at temperatures below 25°C to produce aluminium and iron smectites, zeolites (clinoptilolite), and chlorite. By 75°C, the same assemblage can further react to produce higher temperature zeolites at the expense of aluminum smectite. At 100°C, laumonite is likely to be the stable zeolite alongside albite and quartz and the persistent chlorite (Bloch and Helmold, 1995; Primmer et al., this volume).
- At temperatures >70°C, burial rates exceeding ~100 m.y.$^{-1}$ (meters per million years), and heating rates exceeding 2°C m.y.$^{-1}$, quartz is likely to be an important cement phase during open system diagenesis (Gluyas et al., 1993).

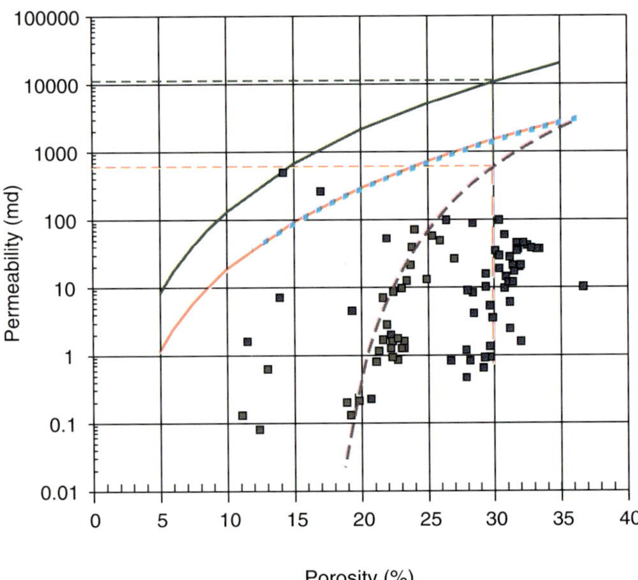

Figure 5. Porosity and permeability data from the cored intervals of Antufash License wells. Al Meethag 1 (blue) high porosity; Al Meethag 2 (brown) lower porosity. The 900-md outlier is from a fractured plug; the remaining outliers are from thin, medium-grained sandstones.

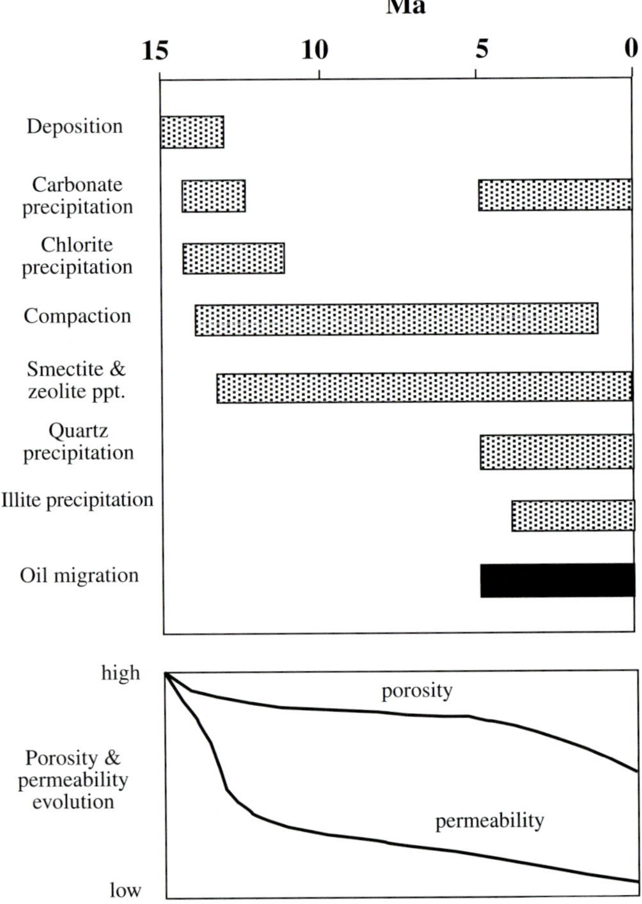

Figure 6. Synthesized diagenetic history for volcaniclastic sandstones in the Antufash License.

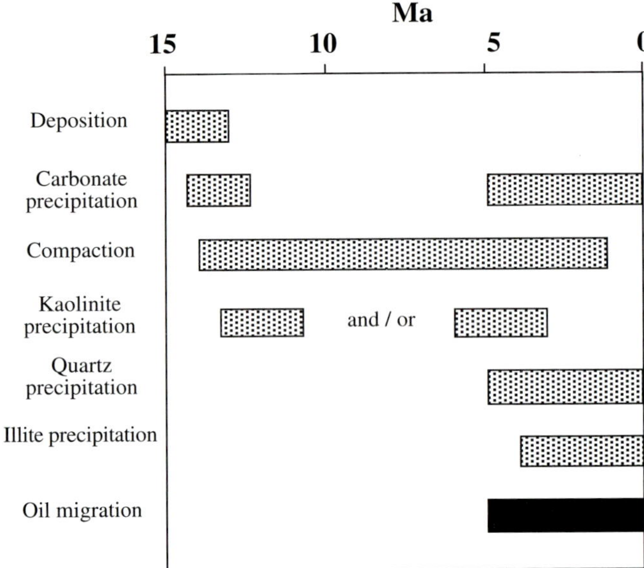

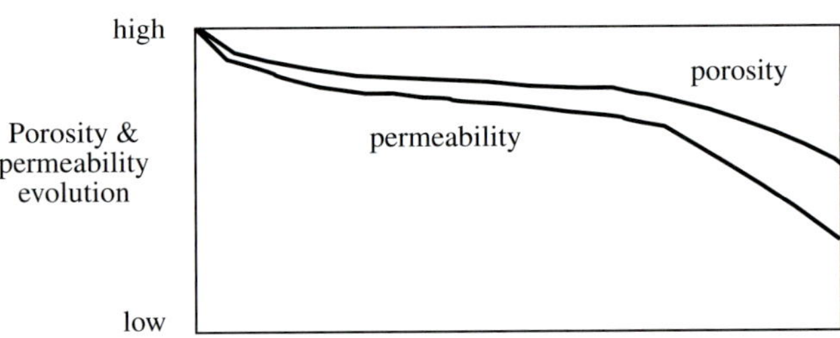

Figure 7. Synthesized diagenetic history for arkosic sandstones in the Antufash License.

- There are sufficient components for illite to form, although significant quantities are unlikely to exist at temperatures below about 100°C (Small et al., 1992).
- In a depositional system containing some marine influence, a little early diagenetic carbonate is to be expected (Bjørkum and Walderhaug, 1990); some of this cement is likely to have been dissolved and reprecipitated during the later stages of diagenesis. Some decarboxylation carbonate may have also been added (Gluyas and Coleman, 1992).
- In sequences interbedded with evaporites, there is a possibility that any residual porosity will have been filled by halite and other evaporite minerals. This point is speculative. We do not yet have information that would allow us to describe the process or timing of such cementation.
- Finally, we made the assumption that of the components required for silicate mineral cementation, only silica is likely to have been imported to the sands in quantities large enough to appreciably affect porosity (Gluyas and Coleman, 1992). This point could be considered controversial given the current debate in the literature with respect to the sources of silica for quartz cementation. However, we imply no scale of transport here; import could mean derivation from local silica sources, such as nearby pressure dissolution seams, or more distant sourcing from unspecified sources. Other elements such as potassium and

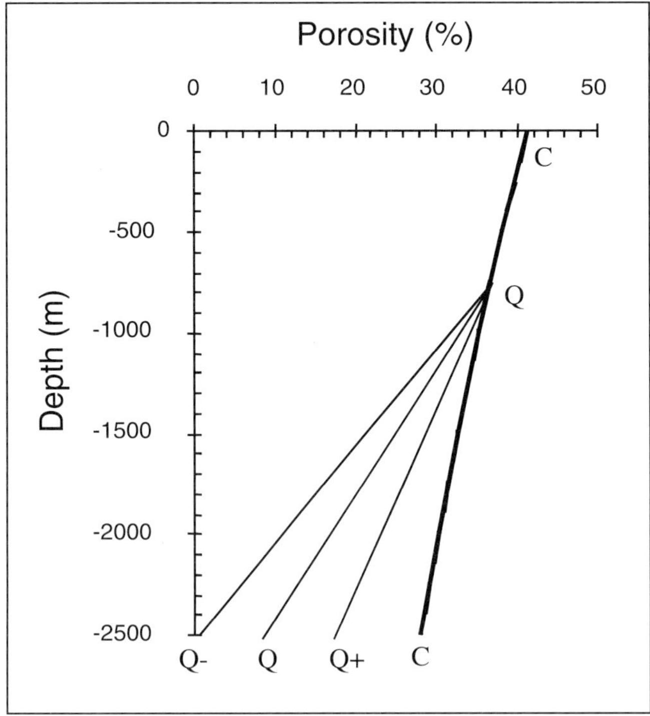

Figure 8. Synthesized and simplified porosity–depth relationship for all types of Miocene sandstones of the Antufash License. Curve C-C is a pure compaction curve for a rigid grain hydrostatically pressured sandstone; curve Q-Q is the expected porosity decline of 16% km⁻¹ hung from a depth equivalence of ~70°C (800 m) at the time of silicate cementation (quartz, clays, and/or zeolites). Q+ and Q– are the potential ±5% km⁻¹ variance on the expected value.

aluminum are likely to have been supplied internally (Gluyas and Leonard, 1995).

Arkosic Sandstones

The diagenesis of arkosic sandstones is likely to have been very different (Figure 7). The most common low-temperature product is likely to have been kaolinite, precipitation of which could have accompanied ingress of undersaturated water of near-surface, meteoric, or connate origin (Gluyas, 1985; Bjørkum et al., 1993). In an open system, quartz is likely to have precipitated once the sandstones exceeded 70°C. By 100°C, illite will have been the most likely clay phase to precipitate (Small et al., 1992).

The presence of carbonate and evaporite cement is likely to be common to both the volcanic and arkosic sourced sandstones, since both cements would have been supplied from largely outwith the sandstone.

Estimating Porosity and Permeability—Antufash-1

The location for Antufash-1 is shown in Figure 1. The area was a poorly explored anticline/diapir fairway comprising upper-middle Miocene reservoirs. The prospect lay above a well-defined NNW-SSE–trending salt-cored anticline with four-way dip closure throughout the Pliocene and Miocene sections. Multiple reservoirs were expected to be present in transgressive sands

of the upper-middle Miocene sections. Four such reservoirs were included in the volumetric calculations. The seal was expected to be salt. Depth to crest of the uppermost prospective horizon was estimated at 850 m, and gross reservoir thickness was calculated at 450 m. The surface temperature is 25°C, and the present thermal gradient is 60°C km⁻¹. The reservoir was expected to be normally pressured. The sands were estimated to be fine grained and well sorted. The sediment source area was thought to be along Wadi Mawr (Figure 1), which drains dominantly granitic terrain and is likely to have yielded arkosic sands.

Modeled Porosity

Using the above criteria, the effects of compaction and quartz cementation were modeled. The likely effect of carbonate cement on bulk porosity was assumed to be small, by analogy with the Al Meethag wells, while the potential for evaporite plugging of porosity was estimated to be large.

The modeled porosity–depth curve for either arkosic or volcaniclastic sands in the Antufash acreage is shown in Figure 8. In order to generate such a porosity–depth curve, several simplifying assumptions were made. Those that might have introduced a systematic error in the porosity estimate are:

- Formation of overpressuring during burial, leading to a low porosity estimate
- Conversion of labile volcaniclastic grains to ductile "clay clasts," which are more susceptible to compaction than rigid grains, leading to an overestimation of porosity

The sensitivity of the porosity estimate at 1 km burial, errors in pressure, or ductile grain content are examined in Table 2.

The porosity gradient associated with cementation, 16 ± 5% km⁻¹, is based on the empirical observation that subregional porosity gradients resulting from quartz cementation covary with thermal gradients (Rønnevik et al., 1983) (Table 3; Figure 9).

Modeled Permeability

In addition to the data required for porosity calculation, data on grain size, sorting, and cement mineralogy were required for the permeability estimate. Grain size data were taken from the Al Meethag wells and, for want of hard data, sorting was assumed to be moderate. Two cases were run for the cement mineralogy. For the volcaniclastic sands, a case based on pervasive, pore-lining smectite and/or zeolites was calculated, while the arkosic sand calculation was based on a pore structure comprising clean, "grain-lined" pores with randomly scattered kaolinite-filled pores.

Modeled curves of the porosity to permeability relationship for both the arkosic and volcaniclastic cases are shown in Figures 10 and 11.

Using the porosity/permeability relationships and porosity-to-depth relationships, it is possible to determine depth equivalencies for the permeability cutoffs (100 md, 10 md) (Table 4).

Table 2. Sensitivity of Porosity to Compaction as a Function of Burial Depth, Overpressure, and Ductile Grain Content.

Burial Depth (m subsea)	Ductile Grain Content (%)	Overpressure (MPa)	Porosity (%)
1000	0	0	35
1000	10	0	32
1000	20	0	30
1000	30	0	29
1000	50	0	26
1000	0	1	36
1000	10	1	32
1000	20	1	31
1000	30	1	29
1000	50	1	27
1000	0	2	36
1000	10	2	33
1000	20	2	31
1000	30	2	30
1000	50	2	27
1000	0	5	38
1000	10	5	35
1000	20	5	33
1000	30	5	32
1000	50	5	30
1000	50	10	35

Prospect-Specific Estimates of Porosity and Permeability

The estimated depth to top reservoir was 850 m. At this shallow depth, the risk on reservoir effectiveness was low for arkosic and volcaniclastic sandstones. Both types of sandstones were likely to have porosities ~36% and permeability >2 darcys.

ANTUFASH-1—WELL RESULTS

Antufash-1 was drilled to 2062 m in December 1992. The terminal depth was in middle Miocene halite and anhydrite, having penetrated Pleistocene, Pliocene, and upper Miocene sequences. Hydrocarbon gas peaks were encountered while drilling through mudstone intervals with source potential. However, other than these minor shows, the well was dry.

Thin sands were encountered in the upper part of the middle Miocene section (1200 m), and thicker sandstones were found at 1700 m within the middle Miocene section. Both sequences were ~200 m deeper than expected. The sandstones at both 1200 m and 1700 m were largely tight. Rig-site core examination led to the conclusion that much of the cement was halite. This was later confirmed from detailed petrography on samples cut from the core without using water as a lubricant. The cored intervals were at hydrostatic pressure.

Cores 1 and 2, cut from the shallower sandstones at 1200 m (Table 5), comprise a mixture of mudstones with a few fine- to medium-grained sandstone and disruptive, authigenic evaporite mineral layers and dikes. These intervals were interpreted to have been deposited in a marginal marine environment on the basis of sedimentary structures and the presence of fish debris.

Core 3, cut from deeper sandstones at 1700 m, comprises sandstone (90%) and mudstones rich in organic matter (10%). The sandstones occur as 1–1.75 m upward-fining, coarse to fine, cross-laminated sandstones. Mud-clast lags are common in the basal parts of the beds. Planar bedded sandstones with current ripple tops are also present. The sediments of core 3 are interpreted to have been deposited in high-energy fluvial channels and overbank areas.

Both the marine and fluvial sequences are comparable to those encountered in the Al Meethag wells.

Due to the extensive halite cementation, plugs for core analysis and thin-section preparation were cut

Table 3. Porosity Gradients and Thermal Gradients—Sandstones Around the World.

Sandstone	Location	Porosity Gradient (% km^{-1})	Thermal Gradient (°C km^{-1})	Reference
Brae	North Sea	9	33	Gluyas, 1985 (sandstone A)
Brent	North Sea	7–8	30	Gluyas, 1985 (sandstone B)
Stø	Barents Sea	16	60	Rønnevik et al., 1983
Sihapas	Sumatra	20	60+	Gluyas and Oxtoby, 1995
Garn	Haltenbanken	8	30	Ehrenberg, 1990
—	E. Pacific	12.8	35	Bjørlykke et al., 1989
—	E. Pacific	8.5	25	Bjørlykke et al., 1989
—	Gulf Thailand	11	49	Bjørlykke et al., 1989
San Joaquin	California	6.4	35	Bjørlykke et al., 1989
Frio	Texas Gulf Coast	6.7	32.2	Loucks et al., 1984
Frio	Texas Gulf Coast	4.8	38.3	Loucks et al., 1984
Jackson	Texas Gulf Coast	7.5	20	Loucks et al., 1984
Queen City	Texas Gulf Coast	6.1	20	Loucks et al., 1984
Mungaroo	NW Shelf, Australia	5.0	18	Gluyas et al., 1993

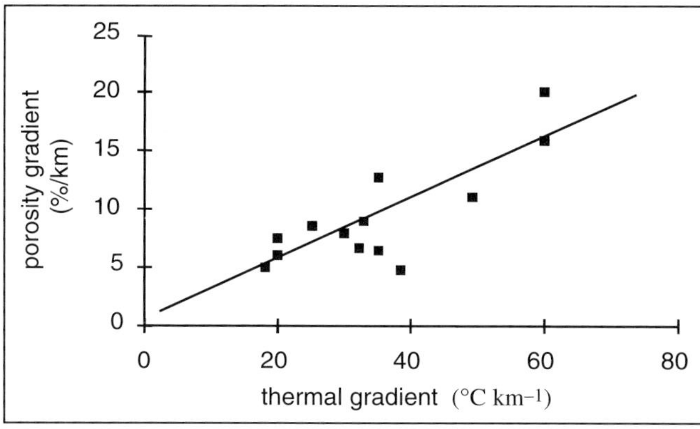

Figure 9. Empirical relationship between porosity gradient (due to quartz cementation) and thermal gradient for sandstones worldwide (Table 3).

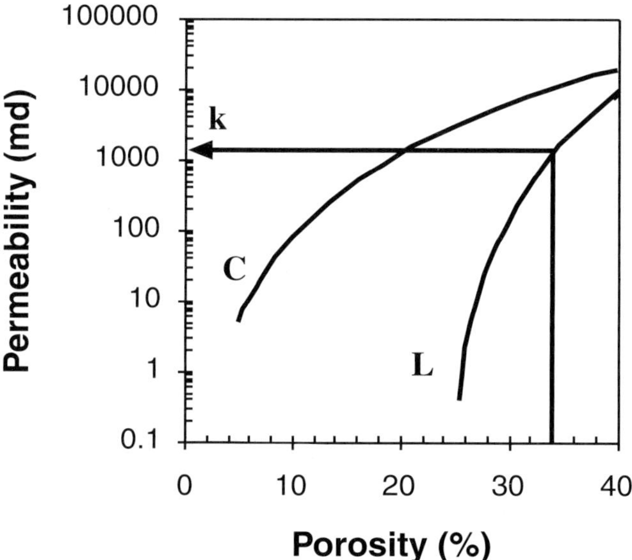

Figure 10. Modeled porosity-to-permeability relationship for volcaniclastic sandstones of the Antufash License. C = compaction/quartz cementation curve for fine (200 µm), moderately sorted sandstone. L = poroperm relationship for pore-lining clay cemented sandstone. Porosity of 36% is derived from Figure 8 (850 m burial). k = predicted permeability. The system contains about 10% pore-lining clay.

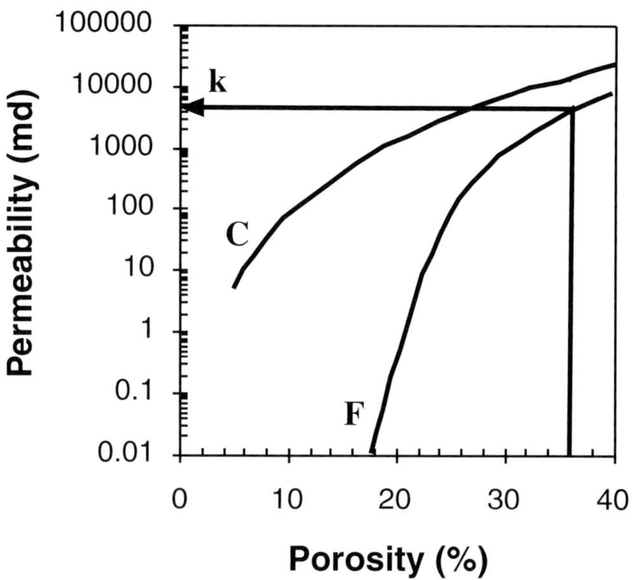

Figure 11. Modeled porosity-to-permeability relationship for arkosic sandstones of the Antufash License. C = compaction/quartz cementation curve for fine (200 µm), moderately sorted sandstone. F = poroperm relationship for pore-filling clay cemented sandstone. Porosity of 36% derived from Figure 8 (850 m burial). k = predicted permeability. The system contains about 10% pore-filling clay.

using diesel as a lubricant. A few of the plugs were cleaned of halite; their porosity and permeability were measured during and after the cleaning process. Five porosity and two permeability measurements were made on cores 1 and 2; 36 pairs of measurements were made on core 3. Unless otherwise stated, the remainder of the discussion centers on core 3.

Before cleaning, the average porosity of the sandstones was 1.5% and permeability was <1 md. Data are presented in Table 6 for the five samples subject to cleaning.

The sandstones of Antufash-1 are highly feldspathic. The major portions are: quartz plus polycrystalline quartz (42%), total feldspar (35%), and lithic fragments (22%). The feldspar is largely orthoclase and extensively altered. Microcline is also abundant. The rock fragments are largely of igneous origin, composed of quartz, feldspar, and mica aggregates. Most samples contain a few percentages of metamorphic rock fragments and volcaniclastic rock fragments.

The sandstones contain an average of 38% diagenetic components. More than half of this cement is halite, and much of the remainder is dolomite. Some feldspar appears to have undergone a late diagenetic alteration to mica. A few percentages each of siderite, calcite, quartz, anhydrite, kaolinite, and feldspar are present with trace amounts of pyrite and anatase (Figure 12; Table 7).

Detrital Mineralogy

The sandstone mineralogy of Antufash-1 is within the range of mineralogy expected for a granitic source. This supports the hypothesis of a Wadi Mawr source for the sediment.

The predicted composition/granitic source and Antufash-1 compositions of quartz are 50 ± 20% and 42%, respectively; feldspar, 30 ± 20% and 35%, respectively; and lithic fragments, 20 ± 20% and 23%, respectively.

Diagenetic Mineralogy

The predicted and actual mineral parageneses are similar (Table 8). Early diagenetic carbonate was followed by feldspar dissolution with quartz and kaolinite precipitation, with late diagenetic "mica" and more carbonate. Halite was the last mineral to precipitate.

Table 4. Depth and Porosity Criteria for Effective Reservoir (for Oil).

Sand	Permeability Cutoff (md)	Porosity (%) Equivalent to Permeability	Depth (mss)* Equivalent to Permeability
Arkosic	10	14	2200
Volcaniclastic	10	30	1100
Arkosic	100	20	1800
Volcaniclastic	100	32	1000

* mss = meters subsea.

Table 5. Cores Cut in Antufash-1.

m (BRT)*	Top	Base
Core 1	1196	1206.73
Core 2	1211	1222.70
Core 3	1761	1770.70

*BRT = below rotary table.

Reservoir Quality

A correct prediction of diagenetic sequence is still a long way from a quantitative prediction of porosity and permeability. Table 9 shows that the abundance of quartz was overestimated, while that of early diagenetic carbonate (dolomite) was underestimated. The likely presence of pervasive halite was indicated, and the reservoir effectiveness risked accordingly. However, no attempt was made to quantify the halite volume prior to drilling the well.

The predicted and actual poroperms are shown in Figure 13 and Table 9. In the very broadest of terms, the prediction of poroperm was perfect insofar as the probability of halite plugging pores was estimated to be high. On a halite-minus porosity basis, the predictions failed. For a prospect depth of ~1800 m, the estimated porosity for a hydrostatically pressured, arkosic sand reservoir was ~20 ± 2.5%. This compares with 27 ± 3% in Antufash-1. Permeability for this same sandstone

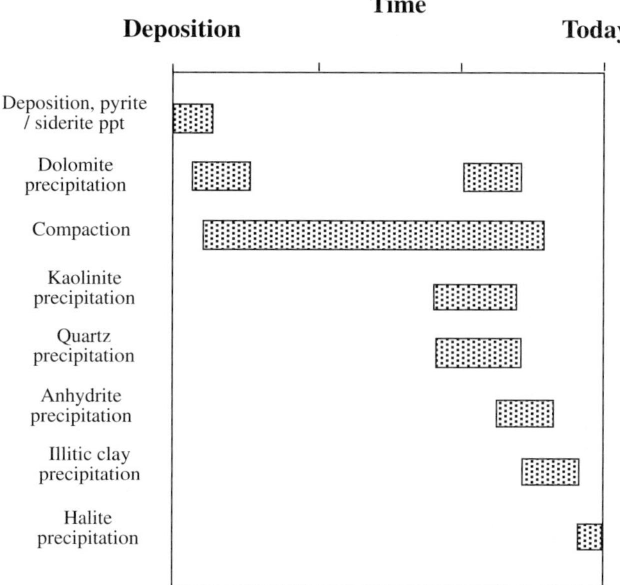

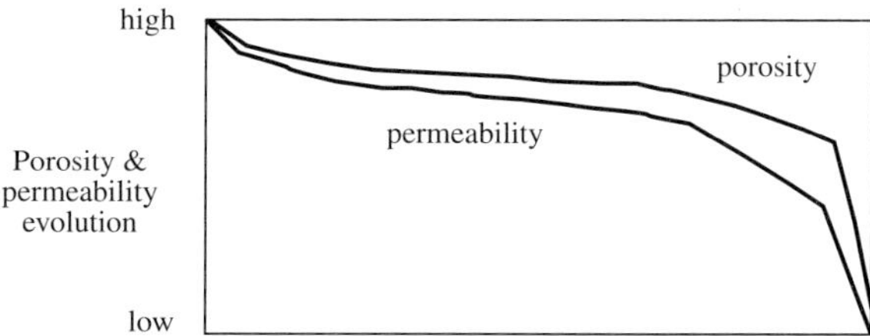

Figure 12. Paragenetic sequence for the Miocene sandstones of Antufash-1.

was calculated at 100 md; this is less than the 2700 ± 1900 md encountered in Antufash-1. If the comparison of halite-minus porosity (predicted) vs. halite-minus porosity (from six cleaned plugs) is a valid (and it may not be), our poroperm predictions were low.

There are a several reasons why predicted and actual halite-minus porosities differ: laboratory dissolution of halite may not replicate the porosity prior to halite cementation; the six samples selected for halite removal may not be representative of the sandstone;

Table 6. Reservoir Quality of Sands After Partial and Complete Dissolution of Halite.

Plug Number	Porosity (%) Unclean	Permeability (md) Unclean	Porosity (%) Partly Clean	Permeability (md) Partly Clean	Porosity (%) Clean	Permeability (md) Clean
6	1.1	0.11	11.5	564	24.4	709
13	0.6	0.17	10.4	646	25.5	1530
15	1.7	0.67	14.7	2040	28.6	5080
24	0.8	0.04	14.1	1930	29.7	3260
35	0.3	0.02	6.5	7.8	20.8	14
Average	0.90	0.20	11.44	1038	25.8	2119
st dev (1σ)	0.53	0.27	3.29	900	3.5	2052
av (without plug 35)	1.05	0.25	12.68	1295	27.1	2645
st dev (without plug 35)	0.48	0.29	2.06	799	2.5	1941

Table 7. Point-Count Mineralogy of Sandstones From Core 3 of Antufash-1.

	Depth (mKB)*												
	1761.3	1762.2	1764.3	1764.9	1765.8	1766.4	1767.3	1767.3	1767.6	1768.8	1769.1	1769.7	1770.6
Grain size	csL	msL	msU	msU	msU	msU	csU	csU	msU	msU	msU	msL	msL
Sorting	mws	vws	ws	ws	mws	ws	ws	mws	ws	ws	ws	ws	ws
Monoquartz	15.0	22.5	19.0	13.5	16.0	11.5	7.0	8.5	15.5	19.5	4.5	19.5	13.5
Polyquartz	12.5	9.5	7.0	15.5	10.0	10.5	10.5	17.0	12.5	10.5	6.0	5.5	8.0
Feldspar	19.0	16.0	22.5	14.5	19.5	16.5	9.5	17.0	15.5	17.0	3.0	21.0	14.5
Igneous/meta fragments	9.0	8.5	8.5	10.0	14.5	14.5	5.0	13.5	13.0	14.0	3.0	11.5	14.5
Volcanic fragments	1.5	1.0	0.5	0.5	1.0	0.5	0.5	—	1.0	1.0	—	0.5	1.0
Sedimentary rock fragments	t	1.5	1.5	1.0	3.5	1.0	34.5	6.0	1.5	3.5	50.0	0.5	1.5
Heavy minerals	—	t	t	—	0.5	t	—	t	t	t	—	t	0.5
Mica	0.5	3.5	1.5	1.5	3.0	2.0	1.5	1.5	1.5	t	—	3.5	6.0
Matrix clay	0.5	0.5	t	—	0.5	—	—	—	—	—	0.5	—	—
Organic matter	—	0	—	—	—	—	—	—	—	—	0.5	—	—
Quartz cement	3.5	2.5	1.5	1.0	0.5	1.0	t	1.0	3.0	2.0	—	5.0	3.0
F'eldspar cement	1.0	1.5	1.0	0.5	t	1.0	t	1.0	1.0	0.5	—	1.0	0.5
Halite	27.5	20.5	22.0	23.0	21.5	29.0	6.5	21.5	22.5	21.5	1.5	13.5	18.0
Anhydrite	0.5	1.5	4.5	t	t	1.0	2.0	0.5	t	0.5	5.5	1.0	3.0
Dolomite	5.0	8.0	6.0	14.0	7.5	4.5	20.5	9.0	6.5	5.0	23.0	9.0	9.5
Kaolinite	2.5	2.0	2.0	3.0	1.0	3.5	1.5	2.0	1.0	1.5	1.0	2.0	1.5
Illitic clay	1.5	0.5	1.5	1.0	1.0	3.0	0.5	1.5	5.5	3.5	—	6.0	4.5
Pyrite	t	0.5	0.5	—	—	—	0.5	t	—	t	0.5	—	—
Anatase	0.5	t	0.5	1.0	t	0.5	—	t	—	—	—	0.5	0.5

*mKB = meters Kelly Bushing (a measure of depth from the rig floor).

Table 8. Predicted and Actual Diagenetic History for Antufash-1.

Diagenetic Sequence	Predicted Composition —Granitic source	Antufash-1 Composition
Earliest	calcite precipitation	dolomite precipitation
	kaolinite ± quartz precipitation quartz, feldspar dissolution	quartz, kaolinite, feldspar precipitation, and dissolution
	illite ± ferroan dolomite precipitation	dolomite, and then anhydrite + mica precipitation
Latest	halite precipitation	halite precipitation

Table 9. Quantitative Effect of Cementation on Porosity.

Diagenetic Minerals	Predicted Abundance —Granitic Source	Antufash-1 Abundance
Early diagenetic carbonate	10% reduction on net:gross, no effect elsewhere	average 9%, range 2–23%
Quartz	minor	2%, range 0.5–3.5%
Kaolinite	minor	2%, range 0–6%
Illite/mica	minor	2%, range 0–6%
Late diagenetic carbonate	minor	minor
Halite	not quantified	2%, range 1.5–25%

halite cementation in nature may not have occurred when predicted with respect to porosity evolution.

DISCUSSION

Estimates of reservoir porosity and permeability are a fundamental part of a prospect evaluation. The most commonly used method of predicting a parameter such as porosity (or permeability) is to take an existing data analog, or perhaps a global data set, and plot porosity against something that correlates with it and is predictable. The first thing to attempt is a correlation with depth. If a porosity-to-depth correlation is acceptable, the process may go no farther. However, if no other parameters are added in, such as "shalyness" or reservoir age or thermal maturity or ... (Schmoker and Gautier, 1988).

The empirical approach can work well, but does, however, suffer from some drawbacks. The very nature of the approach allows prediction of the average. Anomalies, sandstones that are uncommonly porous at depth or well-cemented at shallow depths, go unrecognized. More fundamentally, in the case of the Antufash area, there were too few data on which to construct any sort of empirical porosity–depth plot. By considering diagenetic process, we were able to recognize that the two Al Meethag wells with their porous but impermeable volcaniclastic sandstones could hardly be called

candidate reservoirs. The need to avoid volcaniclastic sandstones led to work directed at identifying adjacent areas onshore, which might have yielded a sand with greater chemical stability. The perception of reservoir effectiveness as a function of sediment composition was incorporated within the prospect-specific risks for the Antufash License prior to drilling.

The diagenetic sequence encountered in Antufash-1 was close to that predicted, but mineral volume estimates were incorrect. Because almost all of the porosity in the Antufash sands was plugged by halite, which could have limited precipitation of other phases (no porosity left to fill), a quantitative comparison of predicted vs. actual poroperm for Antufash-1 is difficult.

CONCLUSIONS

The approach used here for prediction of reservoir quality ahead of drilling Antufash-1 was an attempt at quantification of the sandstone reservoir quality as a function of depositional characteristics and burial history, conditioned to existing well data. Many assumptions were made in order to generate the prediction of reservoir quality. We have justified our assumptions on the basis of empirical observations.

The absolute numerical success of this approach for Antufash-1 is difficult to assess. However, the discipline imposed on the poroperm prediction methodology by

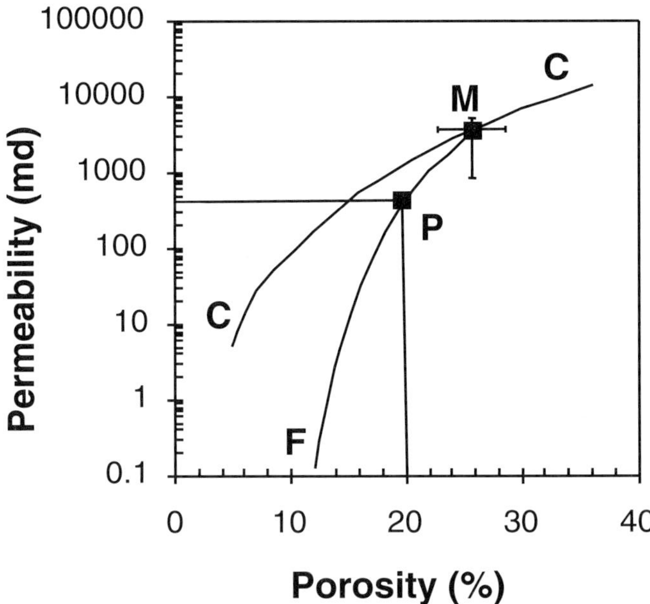

Figure 13. Poroperm data for halite cemented and cleaned sandstones from Antufash-1. C = compaction/quartz cementation curve for fine (200 mm), moderately sorted sandstone. F = poroperm relationship for pore-filling clay cemented sandstone with about 10% pore-filling clay. Porosity of 20% derived from Figure 8 (1800 m burial). M = measured poroperm of the Antufash-1 sandstones after halite dissolution; P = predicted permeability.

the need to understand how diagenetic processes were contriving to reduce reservoir quality had an important outcome that would have been unavailable to a "guessed prediction" of reservoir quality. The well was drilled where the volcaniclastic component of the sand was expected to be low. Thus, we created an intervention to mitigate against the high risk of encountering porous but impermeable reservoir.

ACKNOWLEDGMENTS

The authors thank BP Exploration for allowing publication of this paper. The paper was improved by the comments and suggestions of Keith Myers and Bob Jones. We also thank David Lawrence and John Aggett, who logged the cores from Antufash-1, and subsequently performed the petrographic analysis. To get this far with poroperm prediction would not have been possible without the efforts of an outstanding "Reservoir Quality Prediction" team at BP Research between 1991 and 1994; we thank: Chris Cade, Shona Grant, Andrew Hogg, Mark Hopkins, Norman Oxtoby, Tim Primmer, Craig Smalley, Ed Warren, and Richard Worden. David G. Roberts and Richard Heaton are thanked for their thorough reviews of this paper.

REFERENCES CITED

Bjørkum, P.A., R. Knarud, and M. Bergan, 1993, How important is the late Cimmerian unconformity in controlling the formation of kaolinite in sandstones of the North Sea (examples from the Snorre field)?, *in* A.D. Harbury and A.G. Robinson, eds., Diagenesis and basin development: AAPG Studies in Geology 36, p. 261–269.

Bjørkum, P.A., and O. Walderhaug, 1990, Geometrical arrangement of calcite cemented layers in shallow marine sandstones: Earth-Science Reviews, v. 29, p. 145–161.

Bjørlykke, K., M. Ramm, and G. Saigal, 1989, Sandstone diagenesis and porosity modification during basin evolution: Geologische Rundschau 78, p. 243–268.

Bloch, S., and K.P. Helmold, 1995, Approaches to predicting reservoir quality in sandstones: AAPG Bulletin, v. 78, p. 97–115.

Cade, C.A., I.J. Evans, and S.L. Bryant, 1994, Analysis of permeability controls—a new approach: Clay Minerals, v. 29, p. 491–501.

Cochran, J.R., 1983, A model for the development of the Red Sea: AAPG Bulletin, v. 67, p. 41–69.

Cole, G.A., M.A. Abu-Ali, H.I. Halpern, W.J. Carrigan, R. Savage, R.J. Scolaco, and S.H. Al-Sharidi, 1995, The source rock geochemistry of the Midyan and Jaizan basins of the Red Sea, Saudi Arabia: Monam, Bahrain, Gulf Petrolink Selected Papers from the Middle East Geoscience Conference, Bahrain, April 25–27, 1994, p. 307–319.

Coleman, R.G., 1993, Geological evolution of the Red Sea: Oxford Monographs on Geology and Geophysics, No. 24: New York, Oxford University Press, 186 p.

Crossley, R., C. Watkins, M. Raven, D. Cripps, A. Carnell, and D. Williams, 1992, The sedimentary evolution of the Red Sea and Gulf of Aden: Journal of Petroleum Geology, v. 15, p. 157–172.

Davidson, I., S. Al-Kadasi, A.K. Al-Subbary, J. Baker, S. Blakey, D. Bosence, C. Dart, R. Heaton, K. McClay, M. Menzies, G. Nichols, L. Owen, and A. Yelland, 1994, Geological evolution of the southeastern Red Sea rift margin, Republic of Yemen: Geological Society of America Bulletin, v. 106, p. 1474–1493.

Davidson, I., D. Bosence, I. Alsop, and M.H. Al-Aawah, 1995, Deformation and sedimentation around active Miocene salt diapirs on the Tihama Plain, NW Yemen, *in* G.I. Alsop, D.J. Blundell, and I. Davidson, eds., Salt tectonics: Geological Society Special Publication 100, p. 1–18.

Davison, A., and D.C. Rex, 1980, Age of volcanism and rifting in southwest Ethiopia: Nature, v. 283, p. 657–658.

Egloff, F., R. Rihm, J. Makris, Y.A. Izzeldin, M. Bobsien, K. Meier, P. Junge, T. Norman, and W. Warsi, 1991, Contrasting structural styles of the eastern and western margins of the southern Red Sea: the 1988 SONNE experiment: Tectonophysics, v. 198, p. 329–353.

Ehrenberg, S.N., 1990, Relationship between diagenesis and reservoir quality in sandstones of the Garn Formation, Haltenbanken, mid-Norwegian continental shelf: AAPG Bulletin, v. 74, p. 1538–1558.

El-Anbaawy, M.I.H., M.A.H. Al-Anwah, K.A. Al-Thour, and M.E. Tucker, 1992, Miocene evaporites of the Red Sea rift, Yemen Republic: sedimentology of the Saly Halite: Sedimentary Geology, v. 81, p. 61–71.

Evans, J., S.L. Bryant, and C.A. Cade, this volume, Modeling the effects of diagenetic cements on sandstone permeability, *in* J.A. Kupecz, J. Gluyas, and S. Bloch, eds., Reservoir quality prediction in sandstones and carbonates: AAPG Memoir 69, p. 91–102.

Girdler, R.W., and P. Styles, 1978, Seafloor spreading in the western Gulf of Aden: Nature, v. 271, p. 615.

Gluyas, J.G., 1985, Reduction and prediction of sandstone reservoir potential, Jurassic North Sea: Philosophical Transactions of the Royal Society, v. A315, p. 187–202.

Gluyas, J.G., and C.A. Cade, this volume, Sand compaction, *in* J.A. Kupecz, J. Gluyas, and S. Bloch, eds., Reservoir quality prediction in sandstones and carbonates: AAPG Memoir 69, p. 19–28.

Gluyas, J.G., and M.L. Coleman, 1992, Material flux and porosity changes during diagenesis: Nature, v. 356, p. 52–53.

Gluyas, J.G., and A.J. Leonard, 1995, Diagenesis of the Rotliegend Sandstone: the answer ain't blowin' in the wind: Marine and Petroleum Geology, v. 12, p. 491–497.

Gluyas, J.G., and N.H. Oxtoby, 1995, Diagenesis: a short (2 million year) story—Miocene sandstones of Central Sumatra, Indonesia: Journal of Sedimentary Research, v. A65, p. 513–521.

Gluyas, J.G., A.G. Robinson, D. Emery, S.M. Grant, and N.H. Oxtoby, 1993, The link between petroleum emplacement and sandstone cementation: Geological Society of London, Petroleum Geology of NW Europe, Proceedings of the 4th Conference, March 29–April 1, 1992, Barbican, London, p. 1395–1402.

Hall, S.A., 1989, Magnetic evidence for the nature of the crust beneath the southern Red Sea: Journal of Geophysical Research, v. 94.

Heaton, R.C., M.P.A. Jackson, M. Bamahmoud, and A.S.O. Nani, 1993, Superimposed Neogene extension, contraction and salt canopy emplacement in the Yemeni Red Sea (abs.): AAPG Hedberg Research Conference on Salt Tectonics, September 13–17, Bath, England, Abstract Volume.

Hughes, G.W., and Z.R. Beydoun, 1992, The Red Sea—Gulf of Aden: biostratigraphy, lithostratigraphy and palaeoenvironments: Journal of Petroleum Geology, v. 15, p. 135–156.

Hughes, G.W., O. Varol, and Z. Beydoun, 1991, Evidence for Middle Eocene rifting of the Gulf of Aden and Late Oligocene rifting in the southern Red Sea: Marine and Petroleum Geology, v. 8, p. 354–358.

Kazmin, V., 1973, Geological map of Ethiopia, Adis Ababa: Ministry of Mines, Geological Survey of Ethiopia, scale 1:2,000,000.

Loucks, R.G., M.M. Dodge, and W.E. Galloway, 1984, Regional controls on diagenesis and reservoir quality in Lower Tertiary sandstones along the Texas Gulf Coast, *in* D.A. McDonald and R.C. Surdam, eds., Clastic diagenesis: AAPG Memoir 37, p. 15–45.

Merla, G., 1979, A geological map of Ethiopia and Somalia: Rome, Consiglio Naz. Reirche, scale 1:2,000,000.

Mitchell, D.J.W., R.B. Allen, W. Salana, and A. Abouzahu, 1992, Tectonostratigraphic framework and hydrocarbon potential of the Red Sea: Journal of Petroleum Geology, v. 15, p. 187–210.

Primmer, T.J., 1993, Regional diagenesis in sandstones and controls on reservoir quality: a petroleum industry perspective (abs.): Abstracts of the 5th Cambridge Clay Diagenesis Meeting of the Clay Minerals Group, March 25–26, 1993.

Primmer, T.J., C.A. Cade, I.J. Evans, J.G. Gluyas, M.S. Hopkins, N.H. Oxtoby, P.C. Smalley, E.A. Warren, and R.H. Worden, this volume, Global patterns in sandstone diagenesis: their application to reservoir quality prediction for petroleum exploration, *in* J.A. Kupecz, J. Gluyas, and S. Bloch, eds., Reservoir quality prediction in sandstones and carbonates: AAPG Memoir 69, p. 61–78.

Robinson, A.G., and J.G. Gluyas, 1992, Model calculations of sandstone porosity loss due to compaction and quartz cementation: Marine and Petroleum Geology, v. 9, p. 319–323.

Rønnevik, H., S. Eggen, and J. Vollset, 1983, Exploration of the Norwegian Shelf, *in* J. Brooks, ed., Petroleum geochemistry and exploration of Europe: Geological Society of London Special Publication 12, p. 71–94.

Schmoker, J.W., and D.L. Gautier, 1988, Sandstone porosity as a function of thermal maturity: Geology, v. 16, p. 1007–1010.

Small, J.S., D.L. Hamilton, and S. Habesch, 1992, Experimental simulation of clay precipitation within reservoir sandstone 1: techniques and examples: Journal of Sedimentary Petrology, v. 62, p. 508–519.

Sudan, 1963, Sudan geological map, Khartoum: Sudan Survey Department, 3d edition, scale 1:4,000,000.

U.S. Geological Society, 1963, Geologic map of the Arabian Peninsula: U.S. Geological Society Misb. Geol. Inves. Map 1-270-A Washington, D.C.: U.S. Geological Survey, scale 1:2,000,000.

Chapter 12

Ramm, M., A.W. Forsberg, and J.S. Jahren, 1997, Porosity–depth trends in deeply buried Upper Jurassic reservoirs in the Norwegian Central Graben: an example of porosity preservation beneath the normal economic basement by grain-coating microquartz, *in* J.A. Kupecz, J. Gluyas, and S. Bloch, eds., Reservoir quality prediction in sandstones and carbonates: AAPG Memoir 69, p. 177–199.

Porosity–Depth Trends in Deeply Buried Upper Jurassic Reservoirs in the Norwegian Central Graben: An Example of Porosity Preservation Beneath the Normal Economic Basement by Grain-Coating Microquartz

Mogens Ramm[1]
Norsk Hydro Research Centre
Bergen, Norway

Arne W. Forsberg
Norsk Hydro Exploration
Stabekk, Norway

Jens S. Jahren
Department of Geology, University of Oslo
Oslo, Norway

ABSTRACT

Successful prior-to-drilling prediction of anomalously good reservoir quality in prospects at deep burial requires an understanding of diagenetic processes and quantitative models on how porosity is related to sandstone composition and to burial history. Quartz cementation and compaction are, in many cases, the most important porosity-reducing processes in quartz- and feldspar-rich arenites, capable of destroying all useful porosity during burial toward 4000 m. Hence, the recognition of factors that may hinder porosity loss by these processes, and thereby preserve good reservoir quality to depths beneath those usually considered as economic basement, is crucial during prospect evaluation of deep structures.

In two deep (>4000 m) oil discoveries in Upper Jurassic sandstones in the Norwegian Central Graben, high porosity (>20%) appears to be preserved due to the presence of a ubiquitous microquartz coating on framework grains, and not due to any burial history-dependent factor such as high pore pressure, low thermal maturity, or early oil emplacement. In these sandstones, the microquartz coating has hindered quartz precipitation and late diagenetic chemical compaction. In interbedded sandstones without

---

[1] Present address: Norsk Hydro Exploration, Stabekk, Norway.

microquartz coating, the porosity is low (<10%) due to extensive quartz cementation. The microquartz coating appears within specific isochronous layers, and its presence is probably caused by input of amorphous silica (volcanic glass and sponge spicules) during deposition. The recognition of the inhibiting effect of this coating on quartz cementation, combined with quantitative models on the relationship between sandstone composition and diagenetic processes such as compaction and quartz cementation, allows confident porosity predictions. Hence, future porosity prediction in deeply buried Upper Jurassic sandstone in this area should focus on establishing sedimentological models addressing prediction of sandstone facies within intervals deposited during periods with high amorphous silica production and deposition.

INTRODUCTION

In the Norwegian sector of the North Sea and in the hydrocarbon provinces off mid-Norway, the porosity in most arenitic sandstone reservoirs appears to follow linear porosity vs. depth trends that can be expressed as $\phi = \phi_0 - G \times Z$, where ϕ_0 is the porosity at the time of deposition, G is the porosity depth gradient, and Z is burial depth (Selley, 1978; Bjørlykke et al., 1986, 1989; Ehrenberg, 1990; Ramm and Bjørlykke, 1994). Furthermore, it appears that much of the deviation from such trends is related to variations in petrographic characteristics, such as variations in total clay content (Ramm and Bjørlykke, 1994) or the presence or absence of clay coatings (Ehrenberg, 1993). However, empirical data from the Norwegian shelf also document that deviations from the general trends may be related to burial history-dependent factors; for example, due to retarded compaction in highly overpressured reservoirs (Ramm, 1992; Ramm and Bjørlykke, 1994) or due to variations in degree of quartz cementation due to differences in the time-temperature–dependent "diagenetic maturity" (Walderhaug, 1994a).

During the evaluation of deeply buried (4000–5000 m) structures in the Cod Terrace area, in the Norwegian Central Graben (Figure 1), it was observed that the reservoir quality in Upper Jurassic sandstones from the Mime field and neighboring structures showed some large, unpredicted porosity variations. Of special interest in this respect was the observation that some sandstone intervals have porosities significantly higher than usually found at depths of 4000 m in the North Sea. Most important, Jurassic reservoir sandstones from the Norwegian shelf follow approximately the same porosity vs. depth trend (Figure 2A). However, only some of the Upper Jurassic sandstones from the Cod Terrace area follow this trend, and it is apparent that one group of sandstones, buried to depths of ~4000 m, has significantly better porosity than the "normal trend." These high-porosity outliers are replotted with other high-porosity outliers from the Norwegian shelf in Figure 2B. In these other high-porosity outliers, the high porosity can either be explained with retarded compaction due to high pore pressures or with retarded quartz cementation due to chlorite coatings on the detrital grains. Apparently, the Upper Jurassic sandstones have porosities of the same magnitude as the other high-porosity outliers, but these sandstones do not contain chlorite coatings, and the reservoirs are not highly overpressured. Hence, other explanations must be sought, and a successful prior-to-drilling porosity prediction in this area requires a good quantitative model on how porosity is related to sandstone composition and burial history; it also requires recognition of the factors or processes that preserve good reservoir quality to depths beneath those usually considered as the economic basement.

REGIONAL SETTING

The Norwegian part of the Central Graben area in the southernmost parts of the Norwegian shelf represents one of the richest hydrocarbon provinces in the North Sea; hydrocarbon discoveries have been made both within Cretaceous chalk reservoirs and within Upper Jurassic and Triassic sandstone reservoirs. The early exploration in the area was concentrated on Cretaceous chalk plays; the important discoveries Cod, Ekofisk, Eldfisk, Edda, Tommeliten, and Tor fields were made during the first decade of hydrocarbon exploration in the Norwegian North Sea. The Upper Jurassic sandstone plays, however, have a shorter history. The first well in block 7/12 tested a Cretaceous chalk play and was abandoned about a hundred meters above the subsequently discovered Upper Jurassic Ula field. The chalk interval was dry, however, and the block was relinquished. A renewed interest in the area during the late 1970s and early 1980s led to the Ula discovery ($70 \times 10^6\,Sm^3$ recoverable oil) in 1976 ($1\,Sm^3$ = standard cubic meter = 6.3 barrel), the Gyda discovery ($32 \times 10^6\,Sm^3$ recoverable oil) in 1980, and the

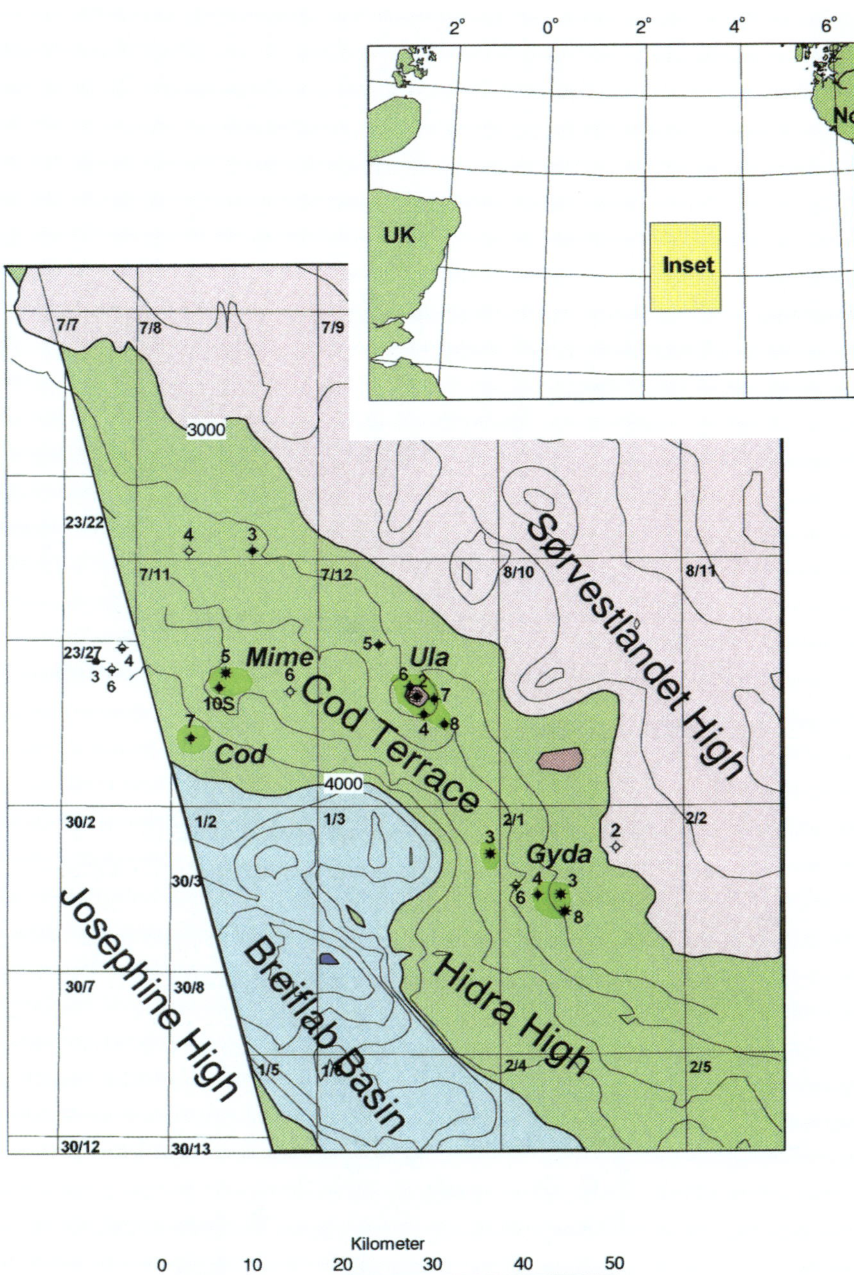

Figure 1. Location map showing depth (two-way traveltime in milliseconds) to the base-Cretaceous unconformity, Cod Terrace area, Norwegian Central Graben.

minor Mime discovery ($0.6 \times 10^6 \, \text{Sm}^3$ recoverable oil) in 1982. These three reservoirs have been developed for production; the Ula field came onstream in 1986, the Gyda Field in 1990, and the Mime Field in 1992. Today, these reservoirs represent the deepest that have been brought into production in the Norwegian sector of the North Sea (Figure 3) and, as such, the good reservoir properties, particularly in the two deepest reservoirs (the Mime and Gyda fields), are very noteworthy.

GEOLOGICAL FRAMEWORK

The Upper Jurassic section in the Central Graben was deposited in an open shelf environment following an Oxfordian relative sea level rise (Home, 1987; Forsberg et al., 1994; Oxtoby et al., 1995). Forsberg et al. (1994) introduced a sequence stratigraphical nomenclature for the Upper Jurassic sections in the area based on a multi-disciplinary genetic sequence stratigraphical approach. They were able to draw timelines through the shallow marine sandstone units from the Cod Terrace area (formerly termed the Ula Formation) and deeper marine muddy sequences (formerly termed the Haugesund and Farsund formations) of the central parts of the Feda Graben. Accordingly, five Upper Jurassic sequences were identified (Figure 4).

The Cod Terrace area was subaerially exposed before the Oxfordian transgression during the Early and Middle Jurassic, and the Upper Jurassic strata are frequently overlying Triassic rocks. Only locally are

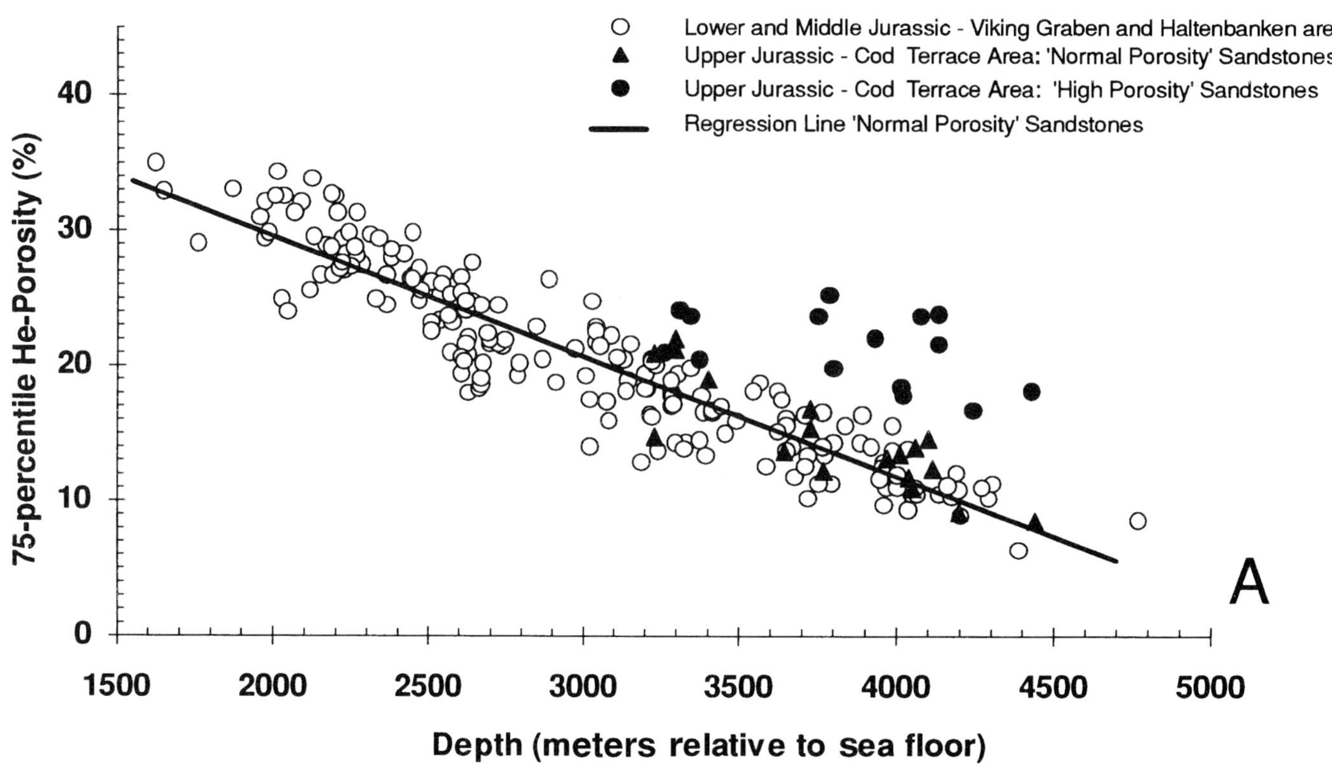

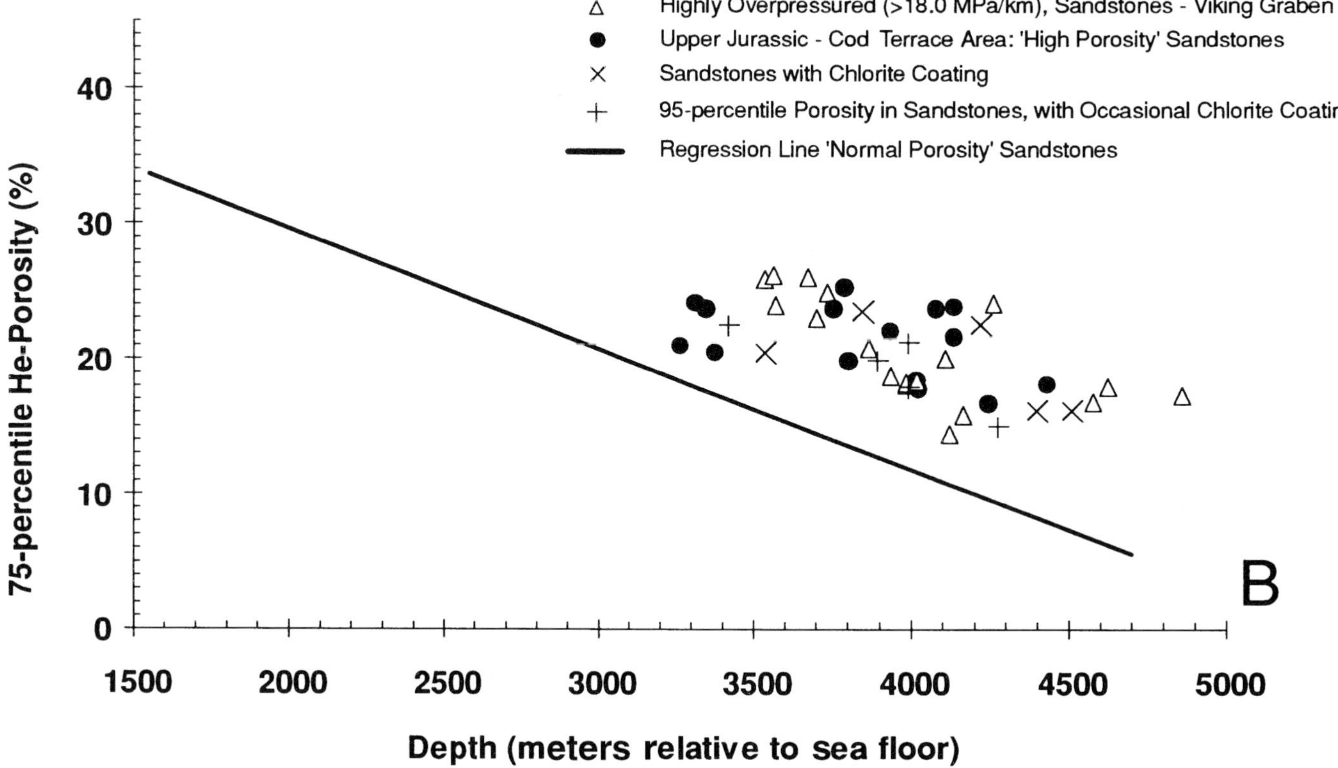

Figure 2. Porosity vs. depth plots showing the 75% value of He-porosity measurements within different Jurassic sandstone units from wells from the North Sea and the Haltenbanken areas, offshore Norway. (A) Upper Jurassic sandstones from the Cod Terrace area compared to normal-porosity reservoir sandstones. The "normal" porosity depth trend is expressed as $\phi = 47.4 - 0.0089 \times Z$, $r^2 = 0.87$, $N = 214$. Only clean arenitic sandstone units not containing chlorite coatings and not representing highly overpressured reservoirs (pressure gradient >18 MPa/km) are shown. (B) Upper Jurassic high-porosity sandstones from the Cod Terrace area compared to high-porosity outliers from other areas on the Norwegian Shelf.

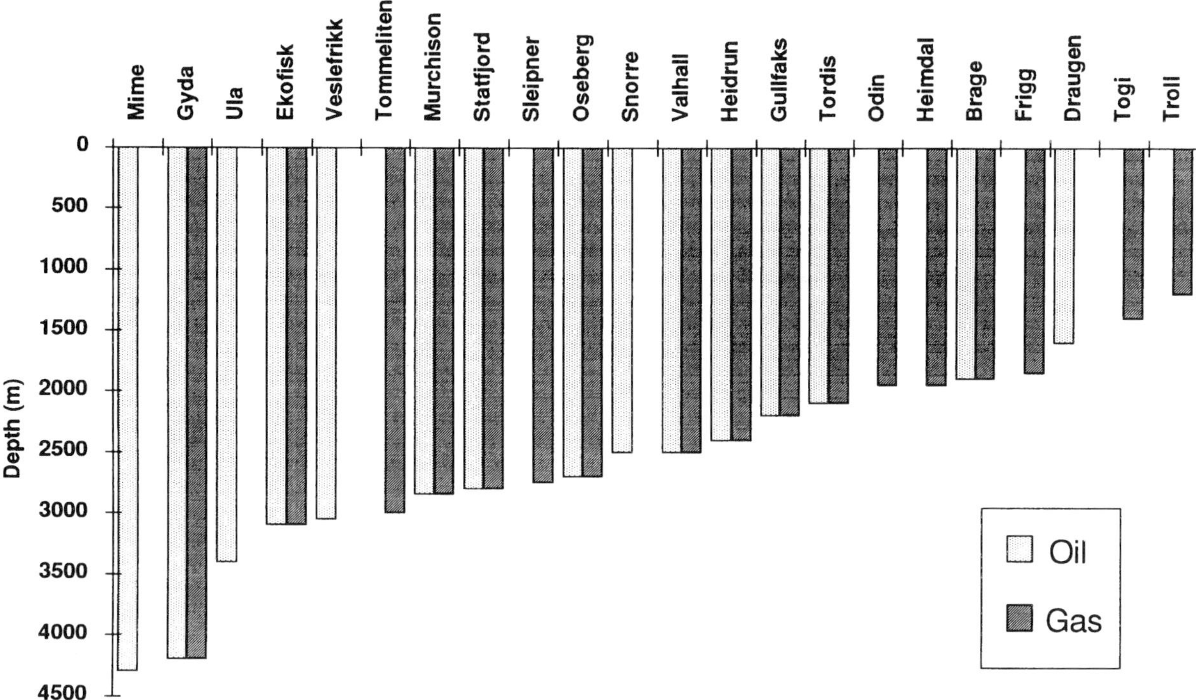

Figure 3. Burial depths for oil and gas fields off Norway.

fluvial deposits of the Bathonian to Bajocian Bryne Formation observed. The reservoir sandstones in the Cod Terrace area accumulated after the Late Oxfordian transgression, mainly in topographic depressions on the downfaulted side of the major Ula-Gyda fault zone (Home, 1987), which separates the Sørvestlandet High from the Cod Terrace (Figure 1).

The Upper Jurassic sequences comprise mudstones to fine- and medium-grained sandstones, deposited in an offshore marine shelf setting. Most sandstones contain appreciable fine-grained clay material; primary sedimentary structures are rare due to extensive bioturbation. The grain size is rather uniform, varying mainly between 100 and 300 µm. Textural cleaning upward trends are pronounced and formed as response to relative sea level changes. Abrupt flooding events followed by slow sediment aggradation, with the most well sorted clean sandstones deposited on top of each sequence, are well illustrated by the gamma-ray logs. A lithostratigraphic subdivision of the Upper Jurassic section, based on the sequence stratigraphical nomenclature of Forsberg et al. (1994), is shown in Figure 5, and a more detailed definition of the units is presented in Appendix A.

POROSITY TRENDS

Porosity vs. Depth

High- and low-porosity zones are stratigraphically correlatable in wells from the Gyda area (Figure 5A). High-quality reservoir zones are seen in units $C1_2$ and B_2, whereas nonporous sandstones are seen in units $C1_4$, $C1_1$, B_1, and A (Figure 5A). The transgressive

sandstone in unit A_1, cored in well 2/1-2, has moderate porosity compared to its burial depth (3340 m RKB [relative to Kelly Bushing]). Low porosities in the muddy intervals of units $C1_4$ and B_1 are probably related to the high clay content (e. g., high gamma-ray signals) of these rocks. The low porosity in the cleaner interval of unit $C1_1$, however, contrasts the good porosity in the apparently analogous sandstones of unit $C1_2$ and the upper parts of unit B.

The porosity is more uniformly distributed throughout the cored sections in the sandstones from the Ula area (Figure 5B). Low porosities are observed within muddy intervals and in carbonate cemented zones. The wells from the central part of the Ula field (i.e., wells 7/12-2, 7/12-4, and 7/12-6) have porosities of ~20% at ~3400 m RKB. The more deeply buried sections of well 7/12-7 and 7/12-5, however, have porosities of only 15% and 12% at 3800 m and 3900 m, respectively. Wells 7/11-5 and 7/11-6 also show large porosity variations within apparently homogeneous sandstones. A reversed relationship between porosity and clay content is indicated in units A and B in well 7/11-5. Here the porosity is higher in unit B than in unit A, although the gamma-ray log signals indicate cleaner sandstones in upper parts of A than in B.

The lithostratigraphical units are divided into three "reservoir-quality facies" with respect to the relation between porosity and depth. Facies 1, the high-porosity outliers, is characterized by high porosity (>20%) at depth >4 km and represents units B_2 and $C1_2$. Facies 2, the normal-porosity sandstones, are those characterized by low gamma-ray signals and low porosity at great burial depth (>4000 m) (i.e., units A_1, A_3, and $C1_1$). Facies 3, the poor reservoir quality sandstones, are those with

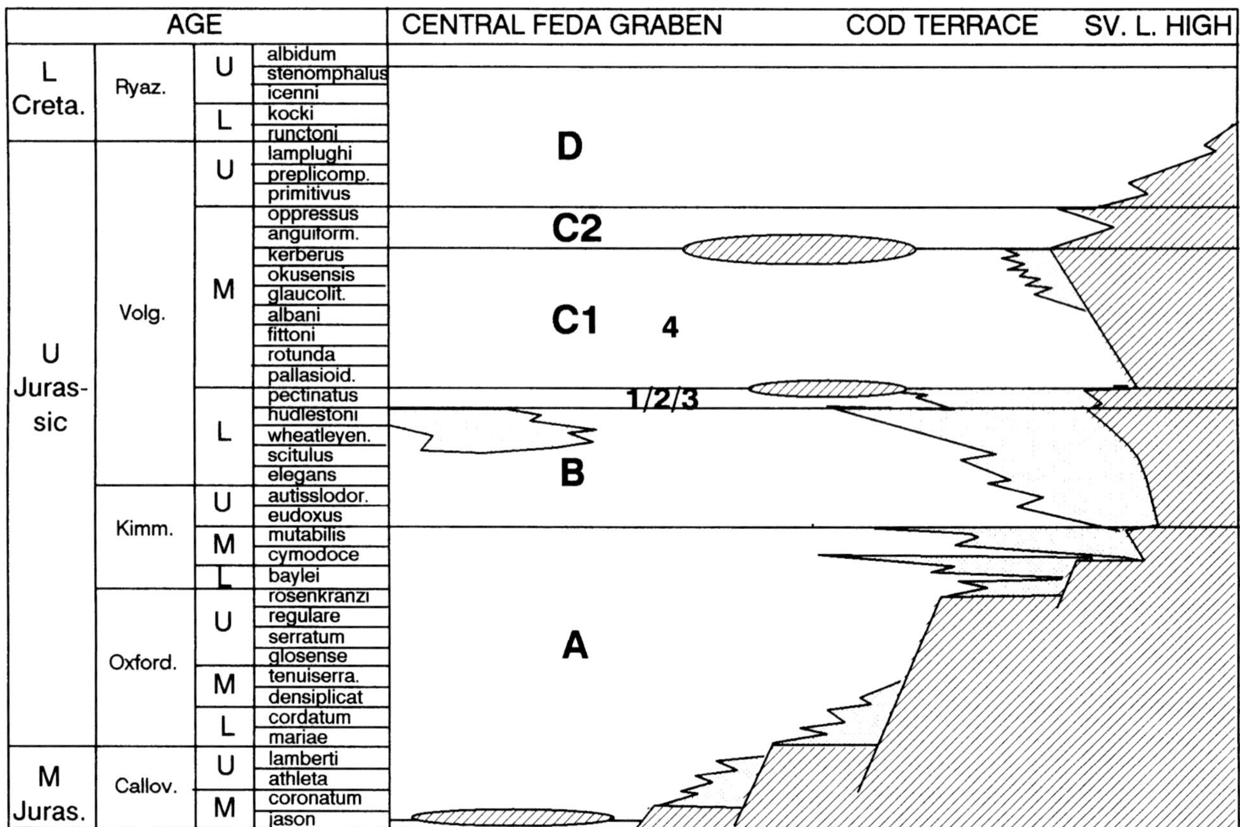

Figure 4. Upper Jurassic stratigraphy in the Cod Terrace area, showing the chronostratigraphic and sequence stratigraphic division of the Upper Jurassic strata and distribution of shales and sandstones in the Feda Graben and Cod Terrace area (modified from Forsberg et al., 1994). Letters refer to sequences discussed in the text.

high gamma-ray signals and poor reservoir quality at all depths (Table 1). Least-squares regression lines based on the porosity vs. depth relationship for the three reservoir-quality facies are shown in Figure 6. Near 3500 m, the porosity is approximately the same in facies 1 and 2, where the porosity in both clean, low gamma-ray sandstone facies is ~25%. Below 3500 m, however, the porosity in the two facies differentiates. In the facies 2 sandstones, the porosity–depth gradient is steep, and at 4500 m these sandstones have porosities <10%. To the contrary, the facies 1 sandstones lose porosity more gently with depth, and porosities near 20% are preserved to 4500 m. The muddy facies 3 sandstones have porosities <15% at all depths, and these units are in essence unprospective at depths below 3000 m.

Porosity vs. Pore Fluid Composition

The porosities of facies 1 and 2 sandstone units from oil zones and from the water legs/dry wells are plotted vs. depth in Figure 7. The figure depicts no significant and systematic difference in porosity between water- and hydrocarbon-saturated reservoirs.

The porosity difference across the oil-water contact is particularly large in well 7/11-5. In the oil-filled unit B and the water-filled unit A, the 75% porosities are 23.6% and 12.3%, respectively. In well 1/3-3, however,

the oil-water contact is located about 10 m below the boundary between the low-porosity unit $C1_3$ and the high-porosity unit $C1_2$. However, a minor decrease in porosity is observed across the oil-water contact at 4221 m RKB. The average porosity between 4213–4221 m is 23.7% (standard deviation 2.0%), whereas between 4221 and 4247.5, it is 22.1% (standard deviation 2.3%). The standard deviations in the two subpopulations are approximately equal, and a simple one-way analysis of variance can be used to reject, at the 95% confidence level, the hypothesis that the porosity difference above and below the water zone (1.6%) is not statistically different (the calculated ϕ value for the data is 9.6, with 1 and 102 degrees of freedom, which is larger than $\phi_{(0.05;1102)} = 3.9$). However, this does not necessarily mean that it is the difference in the pore-fluid composition that causes the porosity difference.

Porosity Related to Lithology

The high-porosity zones in units $C1_2$ and B_2 in well 2/1-6 contain extensively bioturbated, fine-grained sandstones. The low-porosity unit $C1_4$ comprises muddy graywackes and siltstones, whereas the low-porosity sandstone in unit $C1_1$ is a relatively clean, medium-grained sandstone. The coarsening/cleaning-upward sequence in unit B indicates a clear inverse

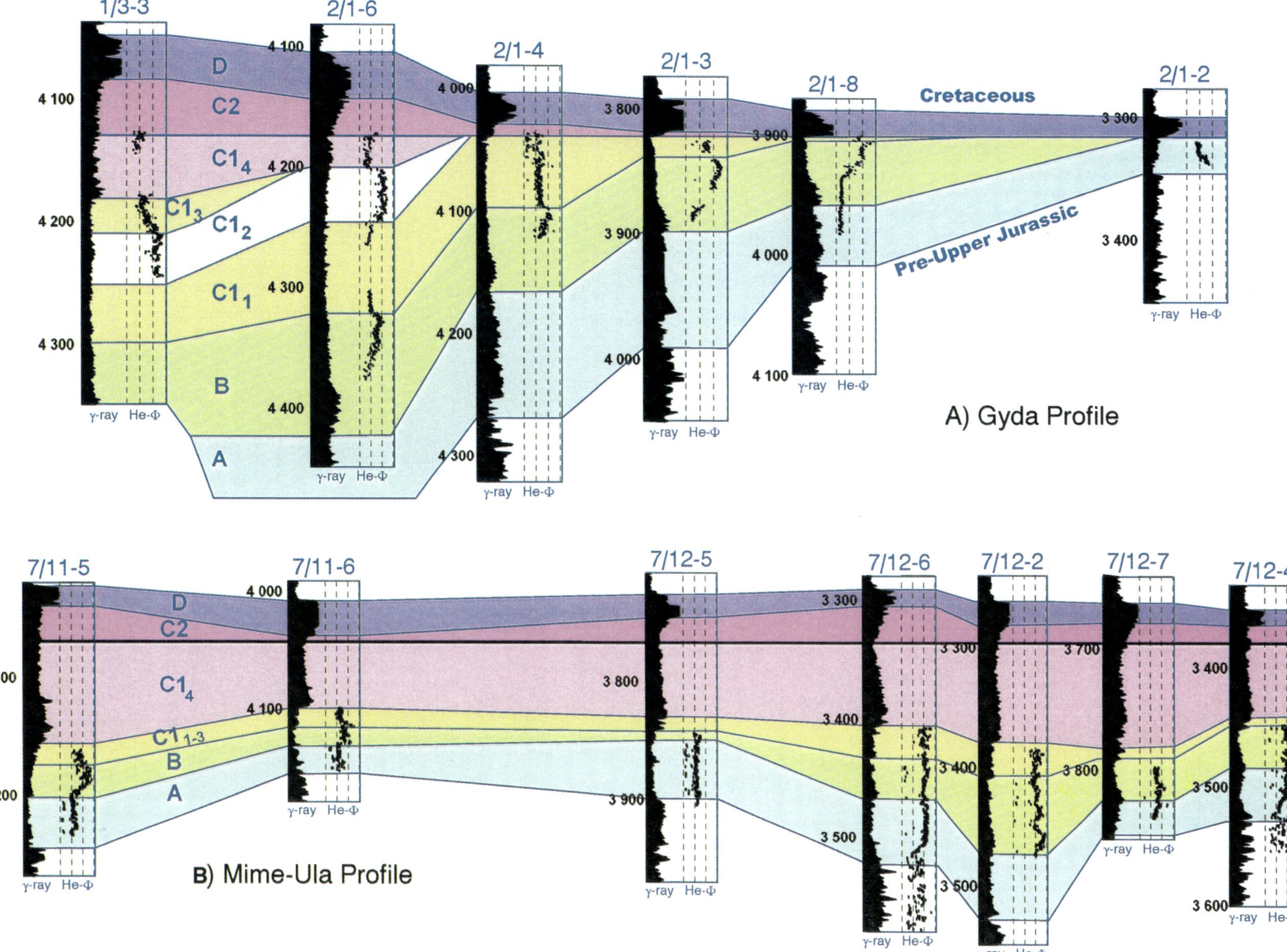

Figure 5. Well correlation between wells from the Gyda area (A) and Mime-Ula area (B) showing gamma-ray log and core-porosity panels for the individual wells. The range in the porosity panels is 0%–30%. Location of wells is shown in Figure 1.

Table 1. Characteristics of Wells.[*]

Well	Unit	Reservoir Quality Facies	Hydro-carbon Present	Depth Interval (m RKB[***])	Thickness (m)	Mean Burial Depth (m RSF[***])	75-Percentile Porosity	Number of Plugs
1/3-3	$C1_4$	3		4138–4148	10	4049	8.2	24
	$C1_3$	2	+	4181–4200	19	4098	14.6	61
	$C1_2$	1	+/-	4210–4248	38	4135	23.8	115
2/1-2	A_2	3		3318–3329	11	3230	9.0	33
	A_1	2	-	3330–3336	9	3239	14.8	19
2/1-3	$C1_1$	2	+	3823–3832	9	3731	15.3	24
	B_2	1	+	3840–3862	22	3755	23.7	53
	B_1	3		3880–3888	8	3788	5.4	18
2/1-4	$C1_1$	2	+	4036–4090	54	3972	13.2	163
	B_2	1	+	4095–4125	30	4021	18.4	79
	B_1	3		4132–4138	6	4044	5.9	20
2/1-6	$C1_4$	3		4171–4200	29	4094	8.9	65
	$C1_2$	1	-	4202–4245	43	4133	21.6	126
	$C1_1$	2	-	4250–4315	65	4201	9.2	49
	B_2	1	-	4320–4350	30	4244	16.7	90
	B_1	3		4361–4376	15	4277	8.5	16
2/1-8	B_2	1	+	3898–3923	25	3807	19.8	60
	B_1	3		3931–3955	24	3840	4.1	63
	A	3		3955–3981	26	3865	3.4	65
7/8-3	Ula Fm[**]	(2)	+	3731–3768	37	3618	13.7	113
7/11-5	$C1_{1-3}$	2	+	4159–4171	22	4060	16.9	26
	B	1	+	4171–4191	20	4083	23.6	67
	A	2	-	4205–4238	33	4117	12.3	112
7/11-7	$C1_{1-3}$	2	-	4100–4111	11	4005	13.5	46
	B	1	-	4110–4131	21	4020	17.8	74
	A	2	-	4131–4145	14	4037	11.7	55
7/11-7	C2	(1)	-	4549–4557	8	4435	18.0	25
	C2	(2)	-	4558–4565	7	4443	8.6	15
7/11-9	A?	(2)	-	4172–4177	5	4068	13.9	16
7/12-2	$C1_{1-3}$	2	+	3385–3410	25	3302	21.6	63
	B	1	+	3410–3476	66	3347	23.7	138
7/12-4	B	1	+	3450–3492	42	3375	20.5	102
	A_3	2	+	3492–3510	18	3406	19.0	48
	A_2	3		3511–3525	14	3423	12.5	30
7/12-5	A	2	+/-	3850–3900	50	3772	12.2	139
7/12-6	$C1_{1-3}$	2	+	3407–3434	27	3327	20.9	62
	B	1	+	3434–3474	40	3361	21.0	106
	A_3	2	+	3474–3507	33	3397	22.0	93
712-7	A-B	2	+/-	3800–3842	42	3737	16.8	112
23/27-3	Ula Fm	(1)	+	4010–4047	37	3920	22.1	96
23/27-4	Ula Fm	(1)	-	3405–3425	20	3309	24.2	63
23/27-6	Ula Fm	(1)	-	3869–3909	40	3778	25.2	126

[*]Measured depth, burial depth, presence of hydrocarbons, and porosity of individual sandstone units.
[**]Ula Fm = Ula Formation.
[***]m RKB = depth relative to Kelly Bushing; m RSF = depth relative to sea floor.

correlation between clay content and porosity. However, this correlation is disrupted by the clean, low-porosity sandstones in unit $C1_1$.

Inverse correlation between clay content and porosity is also observed in units B and $C1_{1-3}$ of well 7/11-5. The gamma-ray logs from this well indicate a slightly higher clay content in the low-porosity unit $C1_{1-3}$ than in the high-porosity unit B (Figure 5). The upper part of unit A, however, resembles unit $C1_1$ in well 2/1-6 and has very low porosity, in spite of a very low clay content, as indicated by the gamma-ray signal.

In unit A in well 7/12-4, a strong negative correlation between clay content and porosity is indicated and, as such, the unit resembles unit B of wells 7/11-5

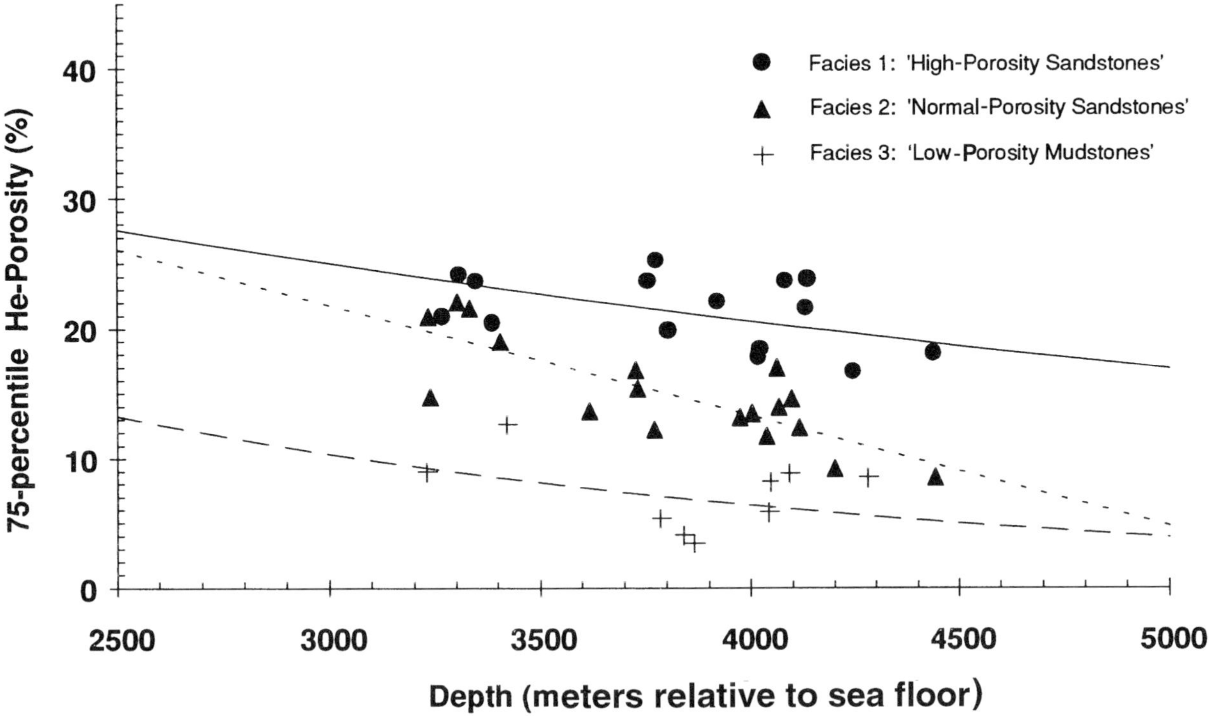

Figure 6. Porosity in individual sandstone units vs. depth. The sandstone units are divided into three reservoir-quality facies showing different porosity depth relationships (Table 1). Facies 1 (high-porosity sandstones): $\phi = 45 \times e^{(-0.196 \times Z/1000)}$; Facies 2 (normal-porosity sandstones): $\phi = 47.3 - 0.0085 \times Z$; Facies 3 (poor reservoir quality mudstone): $\phi = 45 \times e^{(-0.490 \times Z/1000)}$. The exponential regression lines for the high- and poor-porosity facies are obtained by locking the pre-exponential factor to 45%.

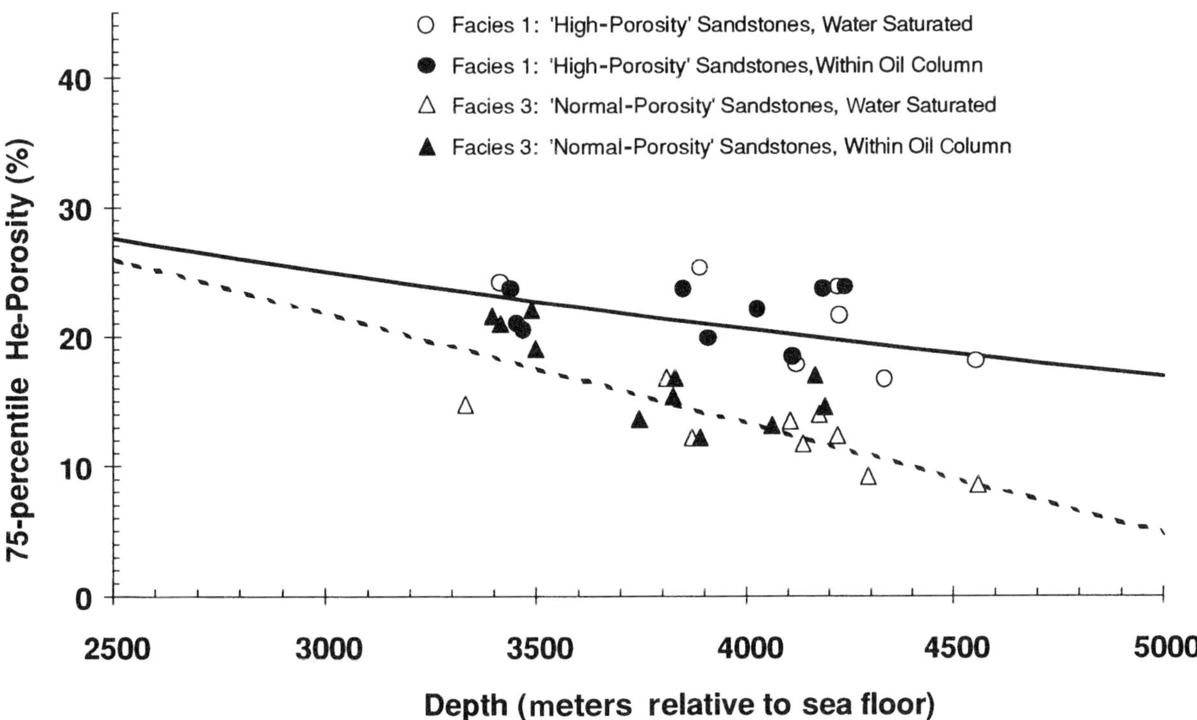

Figure 7. Porosity in facies 1 (high-porosity) sandstones and 2 (normal-porosity) sandstones replotted with respect to the presence or absence of hydrocarbons.

and 2/1-6. Within unit B, however, no such clear correlation between porosity and clay content is indicated. Except for the low-porosity levels corresponding to carbonate cemented layers and the reduced porosity in the lower part of unit A, the porosity is rather homogeneously distributed throughout the cored section in this well.

Porosity vs. Bulk Mineralogy

Eighty-eight samples from wells 7/11-5, 7/11-6, 7/12-4, and 2/1-6 have been analyzed for their bulk mineralogy composition by quantitative XRD (X-ray diffraction) measurements (methods used are documented by Ramm, 1991). To test the correlation between bulk mineralogy and porosity, correlation coefficients between porosity and mineral content from the complete XRD data set and a number of sample subsets were examined (Table 2).

When all samples were considered (Table 2, column 1), there is a significant positive correlation between porosity, quartz, and feldspar content, but there is a negative correlation between porosity, clay, and calcite content. The positive correlation between quartz content and porosity is influenced by the low quartz contents in seven carbonate cemented samples and in four mudstones samples, all having low porosity.

When the mudstones and the carbonate cemented samples (>10% total carbonate) are excluded, porosity is negatively correlated to depth and clay content but positively correlated with feldspar content. In this sample subset, there is no positive correlation between quartz content and porosity (Table 2, column 2).

The correlation between porosity and depth is mainly caused by differences in porosity between samples from well 7/12-4 (3450–3525 m) and the samples from the three other wells (4100–4355 m). Hence, considering the data from the three deeper wells gives a possibility to assess the mineralogical influence on the porosity variations at deep burial (Table 2, column 3). Although significant negative correlation between porosity, clay, and calcite content, and positive correlations between porosity and feldspar content, are indicated, the correlation coefficients are low, and few statistically significant values are found.

Figure 8 depicts the variations in the Clay Index (the ratio of total clays to quartz plus feldspar content) and quartz content vs. porosity. It is observed that most samples follow a trend of reducing porosity with increasing Clay Index. Some samples with little clay have very low porosity, however. All of these samples are extensively cemented either with carbonate or quartz cement. When these samples are excluded, systematic and highly significant relationships between porosity bulk mineralogy are observed (Table 2, column 4). Significant and negative correlation between porosity and clay content is particularly apparent.

The quartz cemented intervals in unit $C1_1$ in well 2/1-6, and upper parts of unit A in well 7/11-5 do not follow the same trend between porosity and clay content as do the other samples. The samples from the low gamma-ray interval 4201–4345 m in well 2/1-6, representing units B_2, $C1_1$, and $C1_2$, show distinctly different relations between porosity and bulk mineralogy (Table 2, column 5). These samples all contain little clay material, but the porosity is very variable. Within this group of samples, the quartz content shows a strong *negative* correlation with porosity. This relation is illustrated in Figure 8B, where the quartz cemented samples from unit $C1_1$ in well 2/1-6 and those from the upper part of unit A in well 7/11-5 cluster in the lower right corner of the diagram and are characterized by having low porosity and very high contents of quartz plus feldspar.

Petrographic Observations from Thin Sections

Thin sections from wells 7/11-5, 7/11-6, 7/11-10S, 7/12-4, and 2/1-6 have been studied and point counted. The point counting was done with emphasis on estimating the amount of intergranular (primary) and intragranular (secondary) porosity and the amount of intergranular cements (mainly quartz and carbonate). The petrographical characteristics of the different stratigraphical units are indicated through a brief description of samples from well 2/1-6 in Appendix B.

Intergranular volume (IGV) vs. cement diagrams including data from well 2/1-6 are shown in Figure 9. The compactional and cementational porosity loss (*COPL* and *CEPL*, respectively) are estimated by equations 1 and 2, which are modified versions of those presented by Ehrenberg (1989)

$$COPL = \phi_0 - \frac{(100 - \phi_0) \times (IP + TC)}{100 - (IP + TC)} \tag{1}$$

$$CEPL = (\phi_0 - COPL) \times \frac{TC}{IP + TC} \tag{2}$$

where ϕ_0 is the original porosity (which here is assumed to be 45%), TC is the total cement, and IP is the intergranular porosity. All parameters are expressed as volume percentages of total rock volume.

Most samples (e.g., those from units B, $C1_2$, and $C1_4$) have little quartz cement and have lost most of their intergranular porosity by compaction. Their present intergranular porosity is inversely correlated with the matrix content. Substantial amounts of quartz cement are observed in the samples from unit $C1_1$. The samples from well 2/1-6 may be divided into three groups. The well-sorted arenites from units $C1_2$ and B_2 are characterized by low contents of matrix and quartz cement and by high porosity. These sandstones contain 25%–30% intergranular porosity plus cement and ~7% cement. According to equations 1 and 2, they have lost ~21%–27% porosity (~50% of the original porosity) by compaction and 5%–6% (~10% of the original porosity) by cementation. The fine-grained graywackes from units $C1_4$ and B_1 are characterized by low porosity and low contents of quartz cement. On the average, they contain 17% intergranular porosity plus cement and 5%

Table 2. Correlation Coefficients Between Porosity and Mineral Content.

	All Samples (n = 88)	All Samples— Mudstones and Carbonate Cemented (n = 77)	Only Wells 2/1-6 + 7/11-5 + 7/11-6 (n = 51)	Quartz-Cemented Samples from 7/11-5 + 2/1-6 Excluded (n = 41)	Only Interval 4201-4345 mKB in Well 2/1-6 (n = 11)
Depth	−0.3031*	−0.3914[†]	−0.2069	−0.0018	−0.4115
Chlorite/Quartz	−0.3504[†]	−0.3668*	−0.3041[‡]	−0.5321*	−0.1920
Chlorite	−0.3787[†]	−0.3819[†]	−0.3107[‡]	−0.5521[†]	−0.2095
Illite/Quartz	−0.3813[†]	−0.3143*	−0.1950	−0.4826*	0.2682
Illite	−0.3881[†]	−0.3394*	−0.2027	−0.5274[†]	0.2572
Clay Index	−0.3894[†]	−0.4891[†]	−0.3925[‡]	−0.8490[†]	0.3224
Quartz	0.3259	−0.1650	−0.1524	0.0765	−0.6529[‡]
K-Feldspar/Quartz	0.0835	0.3380[‡]	0.2458[‡]	0.2795[‡]	0.6568[‡]
K-Feldspar	0.3951[†]	0.3361*	0.2135	0.3980	0.1021
Albite/Quartz	0.0121	0.2525[‡]	0.2645[‡]	0.0944	0.4392
Albite	0.4083[†]	0.3234*	0.3717*	0.2849[‡]	0.3306
Calcite/Quartz	−0.4371[†]	−0.1327	−0.2808[‡]	−0.2733[‡]	−0.2900
Calcite	−0.4537[†]	−0.1443	−0.3541	−0.2890[‡]	−0.5252[‡]
Ankerite/Quartz	−0.0699	−0.1297	0.0434	−0.1410	0.6065[‡]
Ankerite	−0.0160	−0.1665	0.0023	−0.1608	0.5382[‡]
Siderite/Quartz	−0.1313	−0.0450	−0.0469	−0.0688	0.2321
Siderite	−0.0242	−0.0466	−0.0896	−0.0656	−0.1048
Pyrite/Quartz	−0.0785	−0.1195	−0.0361	−0.1655	−0.0937
Pyrite	−0.0392	−0.1284	−0.0467	−0.1894	−0.1689

[†] Significance level > 99.9%.
* Significance level > 99%.
[‡] Significance level > 90%.

cement and have, according to equations 1 and 2, lost ~34% porosity (75% of their original porosity) by compaction and 3% (~7% of the original) by cementation. The quartz cemented arenites from unit $C1_1$ are characterized by low porosity and matrix content and high content of quartz cement. On the average, they contain ~20% intergranular porosity and 15% cement and have, according to equations 1 and 2, lost ~30% (67% of the original) porosity by compaction and 10% (25% of the original) by cementation.

Petrographic Observations Using Scanning Electron Microscopy

The petrographic observations from bulk mineralogy analyses by XRD and from thin sections revealed that much porosity variation can be related to variations in the clay content. However, the samples from unit $C1_1$ in well 2/1-6 and the uppermost part of unit A in well 7/11-5 have low porosity in spite of low clay content, and this is due to the extensive quartz cementation. Sample chips from wells 2/1-6 and 7/11-5 have been examined in SEM in order to describe textures that might explain why some of the clean samples are extensively quartz cemented, whereas others are not. Secondary electron images of characteristic samples from high-porosity zones in well 2/1-6 are presented in Figure 10. The samples are characterized by little pore-occluding cement and high preserved primary intergranular porosity. It is observed, however, that clean quartz grain surfaces are not present. All grains are coated with clay minerals, and more commonly by small (1 mm) microquartz crystals. Occasionally 10- to 50-mm-large euhedral overgrowths stand out from the coated framework grains.

Larger euhedral quartz overgrowths are extremely rare in the high-porosity samples, and it appears that there is a close relationship between the occurrence of the microquartz coating and amount of euhedral quartz cement. Hence, minor porosity loss caused by quartz cementation may be due to efficient inhibition of late diagenetic growth of quartz cement by the clay and microcrystalline quartz coating on framework grains. Thus, the high porosity in the clean sandstones of unit B in well 7/11-5 and units $C1_2$ and B_2 of well 2/1-6 appears to be related to inhibited quartz cementation by the coating. In unit $C1_1$ of well 2/1-6 and the upper part of unit A in well 7/11-5, quartz precipitation has not been inhibited; much porosity is destroyed by chemical compaction and quartz cementation in these sandstones.

Constraints on the Quartz Precipitation from Fluid Inclusion Homogenization Temperatures

Quartz cement is the volumetrically most important cement in the deeply buried, clean sandstones with low porosity, but nearly absent in the good-porosity sandstones. Quartz cement accounts for ~15% in unit $C1_1$ in well 2/1-6, 20% in unit A in well 7/11-5 (Walderhaug, 1994b), and 5% in well 7/12-6 (Nedkvitne et al., 1993).

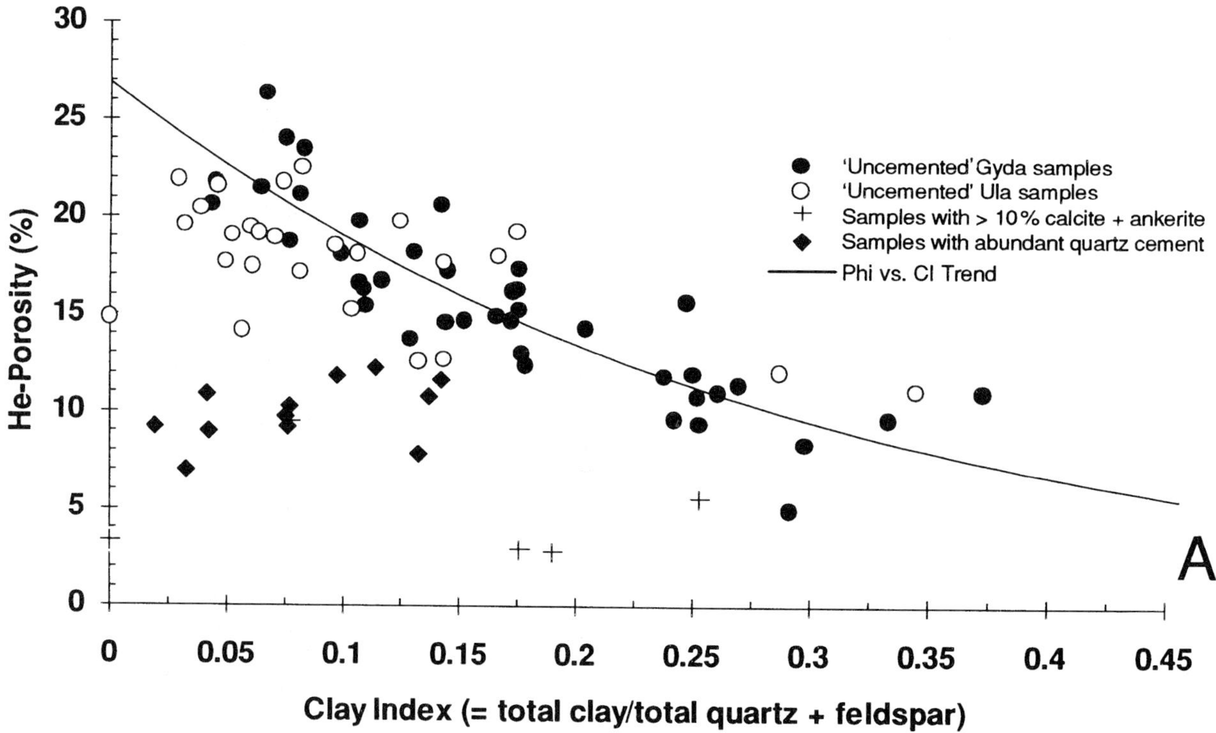

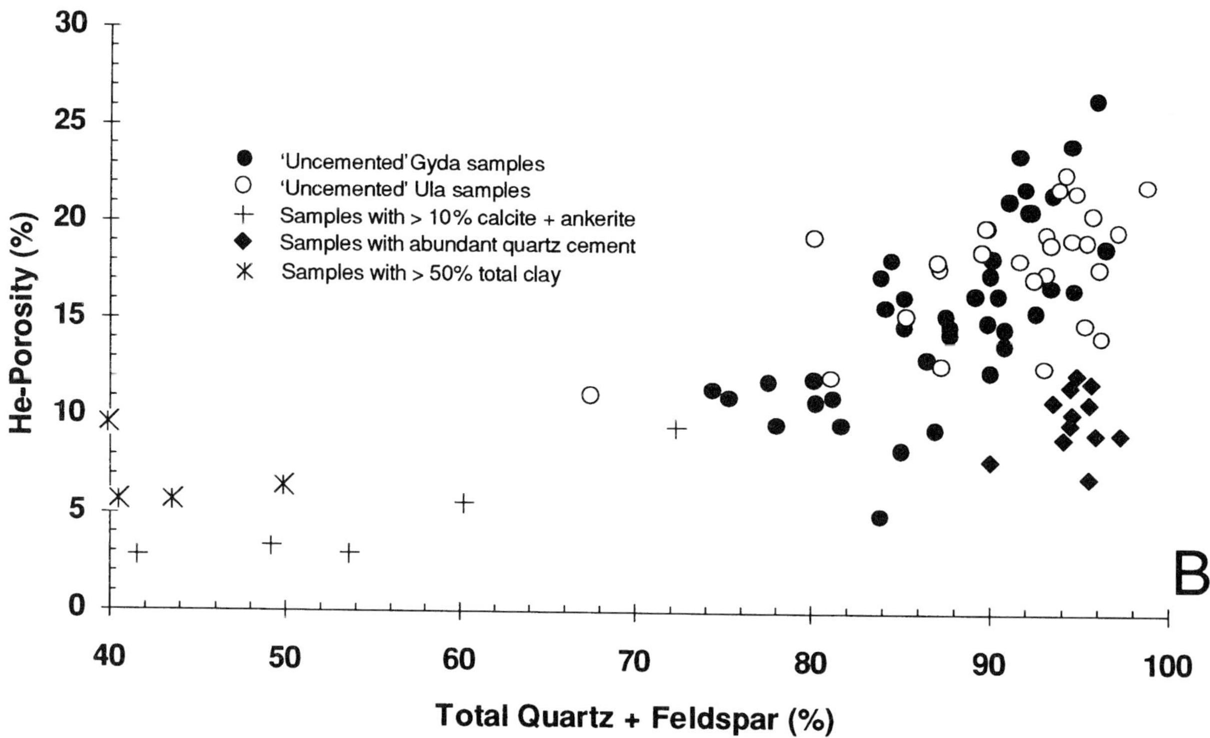

Figure 8. Bulk mineralogy vs. porosity. (A) Clay Index vs. He-porosity (Clay Index = sum of all phyllosilicates divided by the sum of quartz and feldspar, here the normalized ratio of the 19.8° 20 XRD-peak to the normalized quartz and feldspar XRD-peaks) (B) Quartz content from bulk XRD vs. He-porosity. The regression line for the porosity vs. Clay Index trend ($\phi = 26.9\ e^{(-3.46 \times CI)}$, $r^2 = 0.75$) is based on samples without abundant quartz and/or carbonate cement.

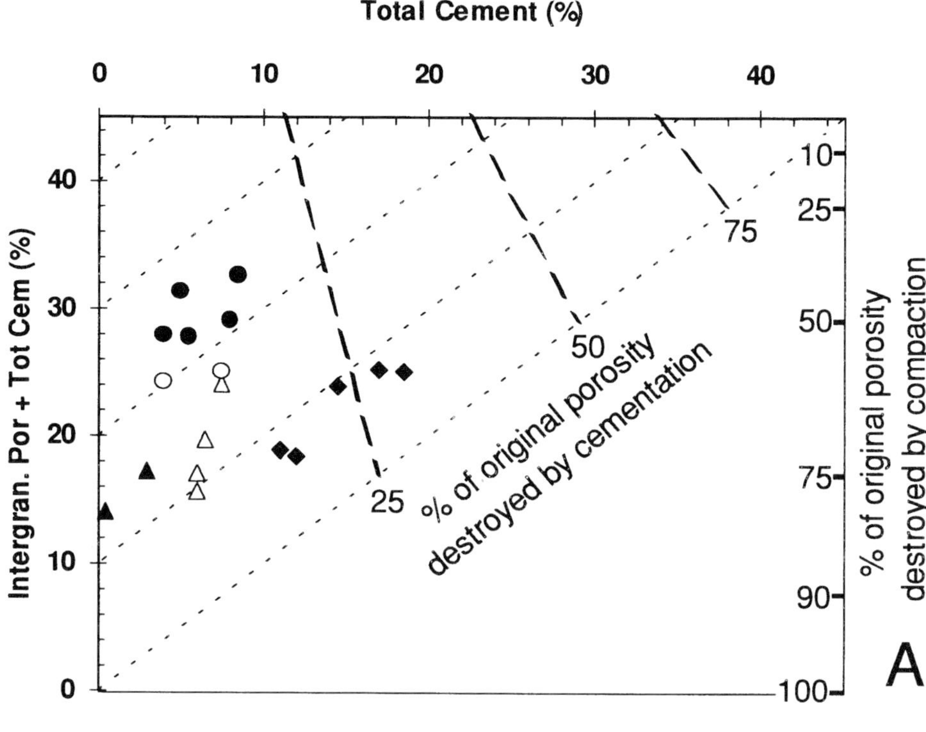

Figure 9. Diagrams modified from Houseknecht (1987) showing (A) the total cement vs. total intergranular cement and porosity, and (B) quartz cement vs. intergranular porosity plus quartz cement in samples from well 2/1-6. These diagrams illustrate the relative effect of compaction and cementation on porosity. The most porous sandstones in units $C1_2$ and B_2 have lost about half their initial porosity by compaction, whereas the muddy sandstones from units $C1_4$ and B_1 have lost more than 75% of their initial porosity by compaction. Neither of these sandstones have lost much porosity by quartz cementation. The quartz-cemented samples from unit $C1_1$ have also lost significantly more porosity by compaction than by cementation.

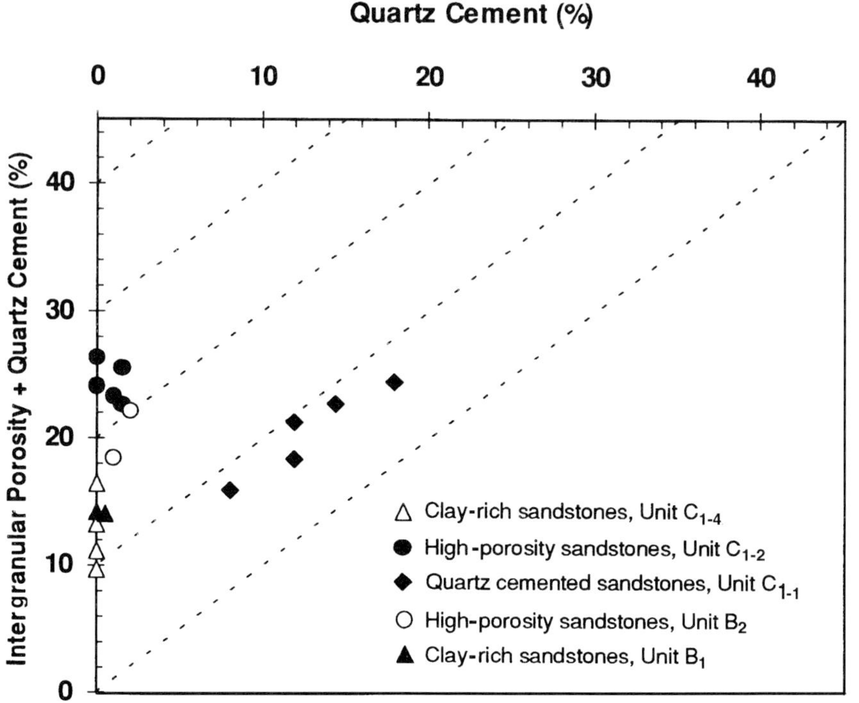

The homogenization temperatures in petroleum inclusions in wells 7/12-6 and 7/11-5 are 30–50°C lower than in the aqueous inclusions (Figure 11), which probably implies that the petroleum inclusions comprise undersaturated oils (with respect to gas), while the aqueous inclusions are saturated. The validity of fluid inclusion homogenization temperatures as a tool to pinpoint the quartz precipitation temperatures has been questioned by Osborn and Hazeldine (1993), who suggested common stretching and resetting of the homogenization temperatures toward equilibration with the bottom-hole temperature. A close correlation between maximum temperatures and homogenization temperatures may not necessarily reflect resetting, but merely reflect the burial and temperature history of sandstones as suggested by a model by Walderhaug

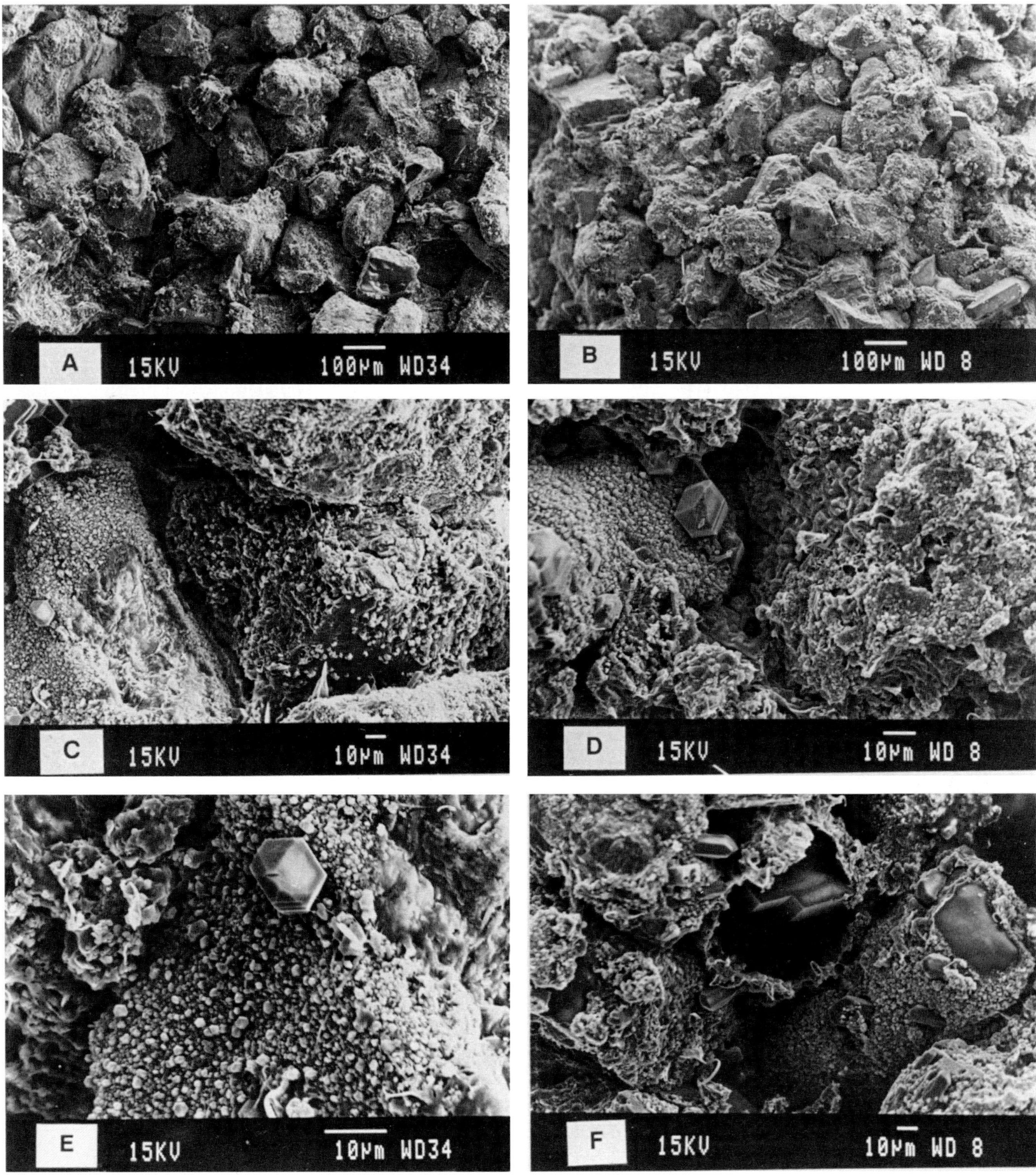

Figure 10. Secondary electron images of rock chips from high-porosity sandstones in well 2/1-6. (A, C, and E) 4209.00 m RKB; (B, D, and F) 4219 m RKB. Note absence of large euhedral quartz overgrowths and abundant microcrystalline quartz coatings on framework grains. Some larger (10–50 mm) overgrowths are observed. These larger crystals occur exclusively on quartz grains; where more than one occur on the same grain, they are parallel, indicating growth in optical continuity with the parent grain.

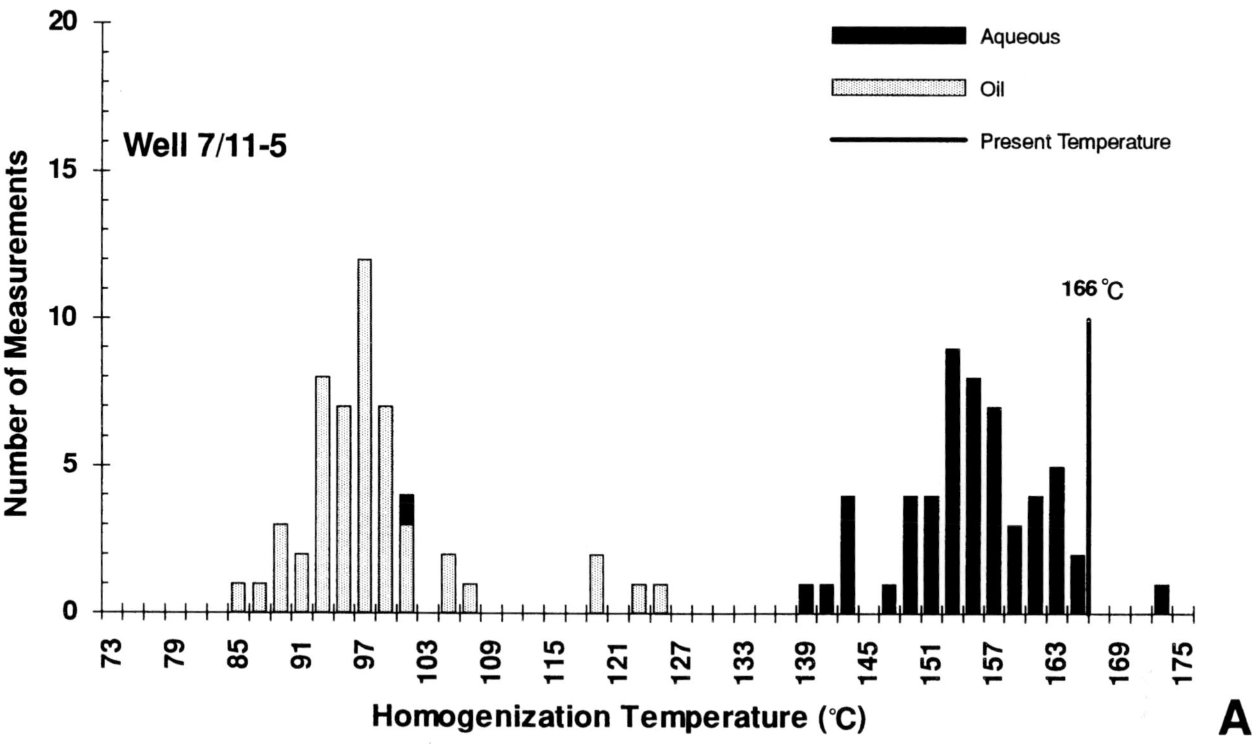

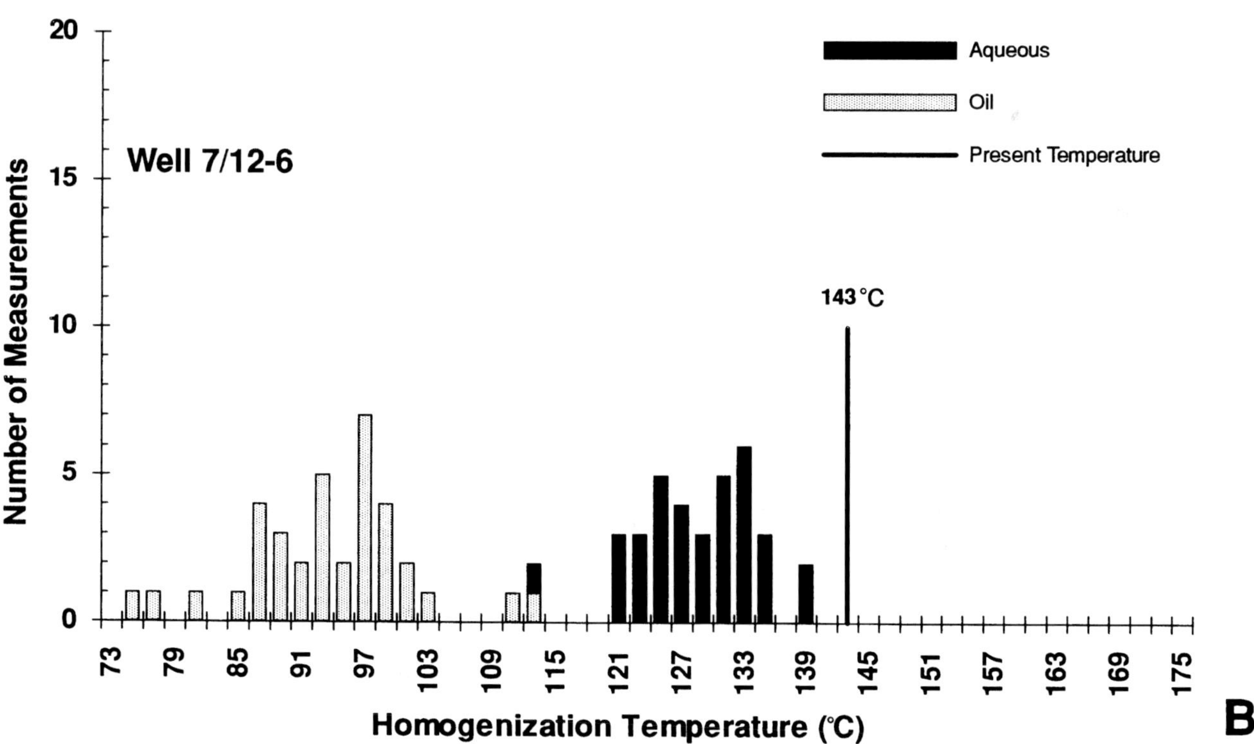

Figure 11. Fluid inclusion homogenization temperatures from quartz overgrowths. Data from Nedkvitne et al. (1993) and Walderhaug (1994b).

(1994a). According to this model, correlation between the "lowest" homogenization temperature in a sandstone and the present reservoir temperature may follow from the heating history rather than from resetting of the inclusions. If sandstones of similar age are buried at different depths in neighboring structures, then the sandstones in the most deeply buried structure may have passed into deeper parts of the "quartz cementation window" more rapidly than the sandstones in the more shallowly buried reservoir. Hence, the homogenization temperatures in the aqueous inclusions shown in Figure 11 may reflect the true trapping temperature, whereas a significant pressure correction is required for the petroleum inclusions.

According to the above assumptions, it follows that the fluid inclusion homogenization temperatures from wells 7/12-6 and 7/11-5 indicate that most of the quartz cement precipitation occurred at temperatures >120–130°C, which translates to burial depths >3000 m, and that precipitation has continued until the present (Figure 11). The overall high "lowest" homogenization temperatures in the two wells (and the higher "lowest" values in well 7/11-5 than in well 7/12-6) probably reflect rapid and differential subsidence during the Late Oligocene to early Miocene. The area received ~1000 m of overburden during this period, and the Upper Jurassic sandstones in wells 7/12-6 and 7/11-5 were buried to ~2500 and 3000 m, respectively, during this short period. Hence, it is likely that little quartz precipitated prior to or during this period of rapid burial; most of the cement precipitated, at slightly different burial depth and temperature in the two wells, after the period of rapid subsidence (Walderhaug, 1994a).

DISCUSSION

Burial Control on Porosity Variations

Theoretical models and empirical data from the Norwegian Shelf indicate that porosity–depth trends are affected by pore pressure gradients and by time-temperature history (Ramm, 1992; Ramm and Bjørlykke, 1994; Walderhaug, 1994a). Within the Jurassic sections on the Cod Terrace, the pore pressures are moderately high and show overall small variations (Figure 12). Furthermore, sandstones within continuous sandstone intervals, representing identical pressure compartments, show large porosity variation. Hence, pore pressure variations may not explain the varying porosity within the arenitic sandstone units. Similarly, differences in thermal maturation cannot explain the large porosity variations observed within some of the continuous sandstone sequences.

Hydrocarbon emplacement has been claimed to halt or retard diagenetic processes and preserve porosity during subsequent burial (Hancock and Taylor, 1978; Selley, 1978; Sommer, 1978; Gluyas et al., 1990, 1993). Many of the Upper Jurassic sandstone sequences described in this study contain hydrocarbon-saturated pore spaces, and oil emplacement may accordingly be suggested to play a controlling role on the porosity variations. The timing of the oil emplacement is critical

if an effect of such emplacement is expected. If the oil entered the sandstone after burial to approximately the present burial depth, porosity–depth trends would not be expected. In the Ula field, the most significant period of oil emplacement has been the last 3–5 m.y., and petroleum migration is probably still continuing (Nedkvitne et al., 1993). However, the Upper Jurassic sandstones in well 7/11-5 have been buried to 700–800 m, and probably have been heated from <150°C to >160°C during the last 3–5 m.y.; the tightly quartz cemented sandstone in upper parts of unit A have, as indicated by the fluid inclusion data (Figure 11), probably lost substantial porosity during this period.

The presence of petroleum inclusion within quartz overgrowth in the reservoir sandstones in the Ula and Mime fields (Figure 11) implies that silica to some extent is mobile after oil emplacement, at least within the zone of immovable oil (the transition zone) (Walderhaug, 1990; Oxtoby et al., 1995). As long as pore throats and grain surfaces are water wet, dissolution along grain contacts and stylolites may continue, and the silica may diffuse through water films on grain contacts into open pores and precipitate as cement. Furthermore, as soon as hydrocarbons enter the pore space, the relative permeability for water is severely reduced, and the hydrocarbon traps represent dead ends for any basinwide advective pore water flow. To the contrary, isochemical cementational processes within the irreducible water may still be active. Hence, if quartz cementation is mainly supported by mass transfer by diffusion rather than by flowing pore waters, chemical compaction and quartz cementation are probably not halted by hydrocarbon emplacement.

The most convincing observation, indicating that timing of oil emplacement into the sandstones is not the principal cause of the porosity variations, is the fact that high-porosity sections occur in dry wells or within the water leg beneath the oil-water contacts (e.g., high-porosity zones in wells 2/1-6, 1/3-3, 7/11-6, and 7/11-7). The only case in which a dramatic shift in porosity is observed associated with the oil-water contact is in well 7/11-5. However, it is possible that the presence of oil in the highly porous sandstones of unit B and absence of oil in unit A is due to differences in porosity and permeability formed prior to oil emplacement, and that oil was able to enter into the porous sandstones of unit B, but not into the tightly cemented sandstones of unit A. Hence, the porosity variation across the oil-water contact in this well may merely reflect a "filling down to situation" than the effect of pore fluid on diagenesis.

From the discussion above, it appears that external factors such as pore pressure variations, thermal maturity, and oil emplacement are not the main factors controlling the porosity variation within the Upper Jurassic sandstone reservoirs on the Cod Terrace. Oil emplacement may have retarded the porosity loss slightly, which may explain the 1%–2% higher porosities within the oil zones (e.g., the difference across the oil-water contact in well 1/3-3). Original or pre-deep

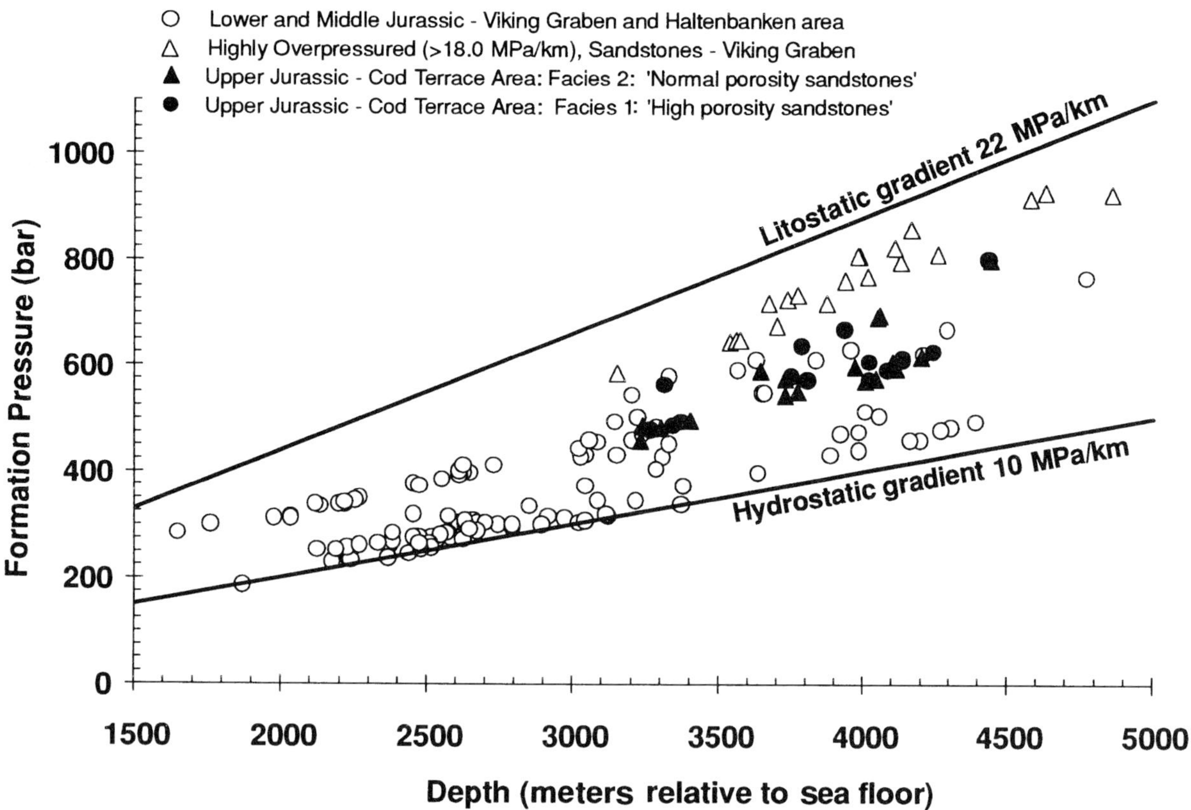

Figure 12. Pore pressure data from the Upper Jurassic reservoir sandstones from the Cod Terrace area compared to sandstones plotting along the normal porosity vs. depth trend shown in Figure 2A and to sandstones, where high porosities appear to be related to extremely high pore pressures.

burial rock characteristics are therefore more likely to control the diagenetic evolution and porosity distribution of the sandstones. Hence, the porosity variation must be linked to the lithological and petrographical characteristics of the different units.

Facies Control on Porosity

Two different, mutually contradictory, tendencies of porosity variation vs. mineralogical composition have been identified. The observation from the well logs that porosity appears to correlate inversely with clay content is verified by petrographical data. High-porosity intervals are in general cleaner and have lower contents of chlorite and illite and lower Clay Index than the less-porous samples. However, it is also observed that the low-porosity samples from unit $C1_1$ in well 2/1-6 and from the upper part of unit A in well 7/11-5 are quartz rich and contain little clay.

Three major diagenetic processes contribute to the general loss of pore space with depth: mechanical compaction, chemical compaction, and cementation. It is probably the varying effect of these processes on the different facies that mainly causes the observed porosity trends. *Mechanical compaction* is driven by the overburden stress (i.e., the net stress = geostatic stress – pore pressure). The bulk volume reduction is due to reorientation, cleavage, and fracturing of brittle grains and pseudoplastic deformation of ductile grain clay

matrix. *Chemical compaction* is the compaction related to dissolution of framework grains within stylolites or within stressed grain contacts and is frequently associated with reprecipitation of solids within adjacent open pores. The net result is reduced pore space due to volume reduction and cementation. *Cementation* is thus frequently related to chemical compaction, but may occur following precipitation of authigenic phases following import of ions from outside. In this case, the porosity is decreased without any associated bulk volume reduction, and the IGV remains constant.

Cementation

Calcite cement is present in most samples, but constitutes larger proportions only in thin zones. In these zones, the intergranular volumes are large (35–45%), indicating that the precipitation occurred at relatively shallow burial, before severe compaction. Furthermore, all extensively cemented zones are found in association with accumulations of bivalve fossils, whereas remnants of leached calcite-rich plagioclases are rare. Hence, a biogenic origin of the calcium is indicated. In most samples, ankerite constitutes minor proportions of the bulk volume (1%–5%). This cement has probably been formed at the expense of calcite at relatively deep burial, following release of magnesium and iron from illitization of smectites and mixed-layer smectites. In the Tertiary sandstones of the Gulf Coast, Boles and

Franks (1979) suggested a similar origin for ankerite cement and indicated that ankerite was formed mainly at temperatures >125°C. Note that in well 2/1-6, calcite is more abundant in the very clean sandstones of unit $C1_1$ than in the slightly more clay rich units $C1_2$ and B_2, where ankerite is more abundant.

Euhedral quartz cement is the volumetrically most important cement in the clean, low-porosity sandstones, but it is nearly absent in the high-porosity sandstones. Quartz cement is normally not common in Jurassic reservoirs from the Norwegian Shelf at depths <2000 m, but frequently becomes very abundant beneath 2500–3000 m (Bjørlykke et al., 1989, 1992; Ehrenberg, 1990; Ramm and Ryseth, 1996). Fluid inclusion homogenization temperatures also indicate that the quartz precipitation is normally initiated at burial depths beneath 2500–3000 m (Walderhaug, 1994a). Hence, during burial toward ~2500–3000 m, the "high-porosity outliers" and the "normal-porosity sandstones" may have followed similar porosity–depth trends. At about this burial depth, the "normal-porosity sandstones" started to lose more porosity by quartz cementation, whereas the "high-porosity sandstones" were not cemented and remained highly porous.

Mechanical Compaction

In samples in which quartz and carbonate cement constitute minor proportions of the rock volume, mechanical compaction seems to be the dominant mechanism of porosity destruction. Mechanical compaction of sandstones will normally cause an exponential porosity reduction with depth (Wood, 1989; Ramm, 1992), and the porosity–depth trend may be expressed with equations on the form

$$\phi = \phi_0 \times e^{(-\alpha \times Z)} \qquad (3)$$

where Z is depth and ϕ is a rate factor depending on the framework strength of the rock. Equation 3 has the same form as the nonlinear regression lines used to fit porosity–depth trends shown in Figure 6. In the resulting porosity–depth profiles the initial, zero-depth porosity equals the pre-exponential factor, ϕ_0, and α describes the rate of porosity loss with depth. The nonlinear regression lines plotted in Figure 6 indicate that equation 3 can be used to model the porosity evolution of the two end-member groups, the facies 1 (high-porosity) sandstones and the facies 3 (muddy) sandstones, by assuming $\phi_0 = 45\%$ and $\alpha = 0.20$ and 0.49 km^{-1}, respectively. The XRD data from samples that are not extensively quartz or carbonate cemented indicate a systematic trend of decreasing porosity with increasing clay-to-framework grain ratio (Figure 8). A regression line, $\phi = 26.9 \times e^{-3.46 \times CI}$, was found to describe this relationship. Combining the regression line of porosity vs. Clay Index at 4100–4350 m (approximated to 4.2 km) with equation 3 and assuming $\phi_o = 45\%$ yields:

$$\phi = 26.9 \times e^{-3.46 \times CI} = 45 \times e^{-\alpha \times 4.2} \qquad (4)$$

and

$$\alpha = \left(ln\frac{45}{26.9} + 3.46 \times CI \right) / 4.2 = 0.12 + 0.82 \times CI \qquad (5)$$

Hence, a framework stability factor of 0.12 km^{-1} is expected to fit the data with zero clay content. This gives a better porosity vs. depth trend than indicated by the facies 1 (high-porosity) sandstones shown in Figure 6. Except for one sample, the maximum porosity among the analyzed samples from the three deeper wells is <25%. Furthermore, in Figure 6, it is observed that porosities >25% are rare in all sandstones buried beneath 4 km. Theoretical considerations on mechanical compaction of sandstones containing varying amounts of ductile and stable framework grains suggest that when the content of ductile grains is low, bridging of nonductile grains may prevent deformation of the ductile grains (Rittenhouse, 1971). Thus, with an ideal distribution of soft spherical grains, one in 21 grains could be present without notable reduction in framework stability. If the clay matrix in the studied sandstones behaves similarly, it may be suggested that variations in the Clay Index between 0 and 0.05 do not affect the rate of mechanical compaction. Accordingly, a framework stability factor of 0.16 ($0.12 + 0.05 \times 0.82$) should fit the porosity data of the clean sandstones as long as they are unaffected by chemical compaction and cementation. Furthermore, the porosity–depth trend of the "high-porosity sandstones" and the "poor-porosity sandstones" shown in Figure 6 corresponds to the expected trend for rocks having a Clay Index equal to 0.10 ($0.12 + 0.82 \times 0.10 = 0.20$) and 0.45 ($0.12 + 0.82 \times 0.45 = 0.49$), respectively.

Chemical Compaction

Theoretical modeling of pressure solution (Ramm, 1992), fluid inclusion data from varying reservoir sandstones (Walderhaug, 1994a), and the distribution of quartz cement in the Garn Formation off mid-Norway (Ehrenberg, 1990) and in the Brent Group and Statfjord Formation in the Northern Viking Graben (Bjørlykke et al., 1992; Ramm and Ryseth, 1996) suggest that porosity loss by chemical compaction and quartz cementation is important only at burial depths beneath 2500–3000 m. The quartz precipitation temperatures deduced from the fluid inclusion data from the Ula and Mime fields (Nedkvitne et al., 1993; Walderhaug, 1994b) indicate that quartz cementation and chemical compaction are important only beneath 2500–3000 m in the Cod Terrace area as well. Dissolution along grain contacts and within stylolites is the obvious source of the quartz cement, but the amplitude of individual stylolites (1–3 mm) and the spacing between stylolites in the intensively quartz cemented zones (10 cm) indicate that solution along stylolites is insufficient in accounting for the observed amount of cement. The IGVs in the quartz-cemented samples is 21%–27%, and the sum of intergranular porosity and quartz cement is 15%–20% (Figure 9), which is lower than the densest packing of spherical grains. Furthermore, the IGVs are lower in the quartz-cemented sandstones than the IGV in the most porous samples from units $C1_2$ and B_2. These observations indicate that intergranular pressure solution has contributed, together with dissolution along stylolites, as a principal source of quartz cement. This is supported by the observation that straight, concave-convex, and suturated contacts between detrital quartz grains are more common than tangential grain contacts.

In the Ula field samples, the amounts of quartz overgrowths vary 2%–8% in clean sandstone samples, but they are rare or absent in muddy samples (Nedkvitne et al., 1993). Similar relations have previously been observed by Tada and Siever (1989), who reported efficient inhibition of quartz precipitation in sandstones containing more than 4% clay matrix and by Ramm and Ryseth (1996), who reported particularly small amounts of quartz cement (<3%) in samples with more than 8% detrital clay. The samples from this unit have a lower Clay Index than the adjacent porous samples from units C1$_2$ and B$_2$. Similar relationships have been reported in the literature: Tada and Siever (1989) documented efficient inhibition of quartz precipitation in sandstones containing >4% clay matrix. There is, however, no clear relationship between clay content and quartz cement in the clean samples from wells 2/1-6 and 7/11-5; several of the quartz-cemented samples from unit C1$_1$ in well 2/1-6 and from unit A in well 7/11-5 have higher clay contents than some of the porous samples from adjacent sandstones.

MICROCRYSTALLINE QUARTZ COATING

Pervasive microcrystalline quartz coatings on framework grains have been observed in samples from wells 7/11-5, 7/11-6, 7/11-10S (unit B), and 2/1-6 (units C1$_2$ and B2). The samples containing this coating always contain little euhedral macroquartz, and it appears that the coating has prevented normal quartz cementation. Similar "tiny, double-ended quartz crystals" have been observed in a study of Upper Jurassic sandstones in the Claymore oil field, in the British sector of the North Sea (Spark and Trewin, 1986). Those researchers noted that "The early deposition of small quartz crystals on grain surfaces was an important factor in the porosity preservation. These crystals provided cement when primary porosity was high, but only occupied a small proportion of the porosity. With further burial the crystals inhibited the deposition of larger pore-filling quartz and feldspar overgrowths." If this interpretation is correct, similar relationships can also be inferred for the Gyda and Mime reservoirs, implying that the good reservoir quality is a direct consequence of the microquartz coating. To use this observation in a predictive manner, fundamental questions related to its occurrence and to its ability to preserve porosity must be answered.

Occurrence

Calcedonic quartz cement is observed in samples from unit B in wells 7/11-5 and 7/11-6. Precipitation of micro- and cryptocrystalline quartz requires initial elevated silica activities due to presence of amorphous silica and frequently follows the generalized diagenetic evolution: Opal-A—Opal-CT—cryptocrystalline—microcrystalline quartz (Williams and Crerar, 1985; Williams et al., 1985; Hendry and Trewin, 1995). The microcrystalline quartz coating on framework grains in the studied well probably formed following a similar evolution path; it is likely that the large number of small crystals formed by self-nucleation from a pore water that was highly supersaturated with respect to quartz. This self-nucleation probably occurred at a time when an earlier opaline phase dissolved and kinetic reaction barriers were overcome due to increased temperatures.

Similar features (microcrystalline quartz and calcedonic cement) have also been observed at approximately 1700 m burial depth in sandstones within the volcanic dominated Balder Formation (Eocene) in wells from block 25/11 and 25/8 in the Viking Graben (Ramm et al., 1992, unpublished Norsk Hydro Reports). Furthermore, recrystallized volcanic glass shards (≤1.5 cm) have been found in samples from unit B in well 7/11-5; altered sponge spicules are rather common in the sections containing grain-rimming microquartz. Hence, the microcrystalline coatings are most probably a consequence of initial deposition of volcanic or biogenic amorphous silica.

Porosity Preservation

The small microcrystalline quartz crystals are less stable (~3%; Jahren, in press) than the normal macroquartz, because of a larger surface energy. Hence, that these small crystals are preserved implies that the pore water has been supersaturated with respect to quartz ever since the formation of the microquartz, probably since burial to <1700 m (by analogy with the Balder Formation).

Precipitation of euhedral quartz overgrowths normally occurs, at low degrees of supersaturation (5%–20%), by screw dislocation growth (or spiral growth). If physical hindrances like coatings or poisoning atoms fixed on the growth sites are present, a higher degree of supersaturation is required before growth can commence. The crystallographic axes of the small (0.5–5 mm) quartz crystals are randomly oriented, and only occasionally will the microcrystal be oriented parallel to the underlying detrital grain. Hence, during spiral growth, precipitation at outcropping dislocations will commence against the microquartz crystals. The alternatively two-dimensional nucleation on flat surfaces probably does not occur because of the required higher degree (~35%) of quartz supersaturation, which is not achieved during normal clay-induced quartz dissolution or pressure solution between grains and in stylolites (Jahren, in press).

QUANTITATIVE POROSITY VS. DEPTH MODEL

Figure 13 illustrates a generalized quantitative model to explain the porosity variations in the Upper Jurassic sandstones in the Cod Terrace area. During evaluation of new prospects, the model is applicable when combined with depositional models predicting the distribution of sandstone facies (e.g., shale content) and potential occurrence of zones with microquartz coatings.

During shallow to intermediate burial, most porosity reduction is due to mechanical compaction. The porosity–depth gradient at a particular depth is

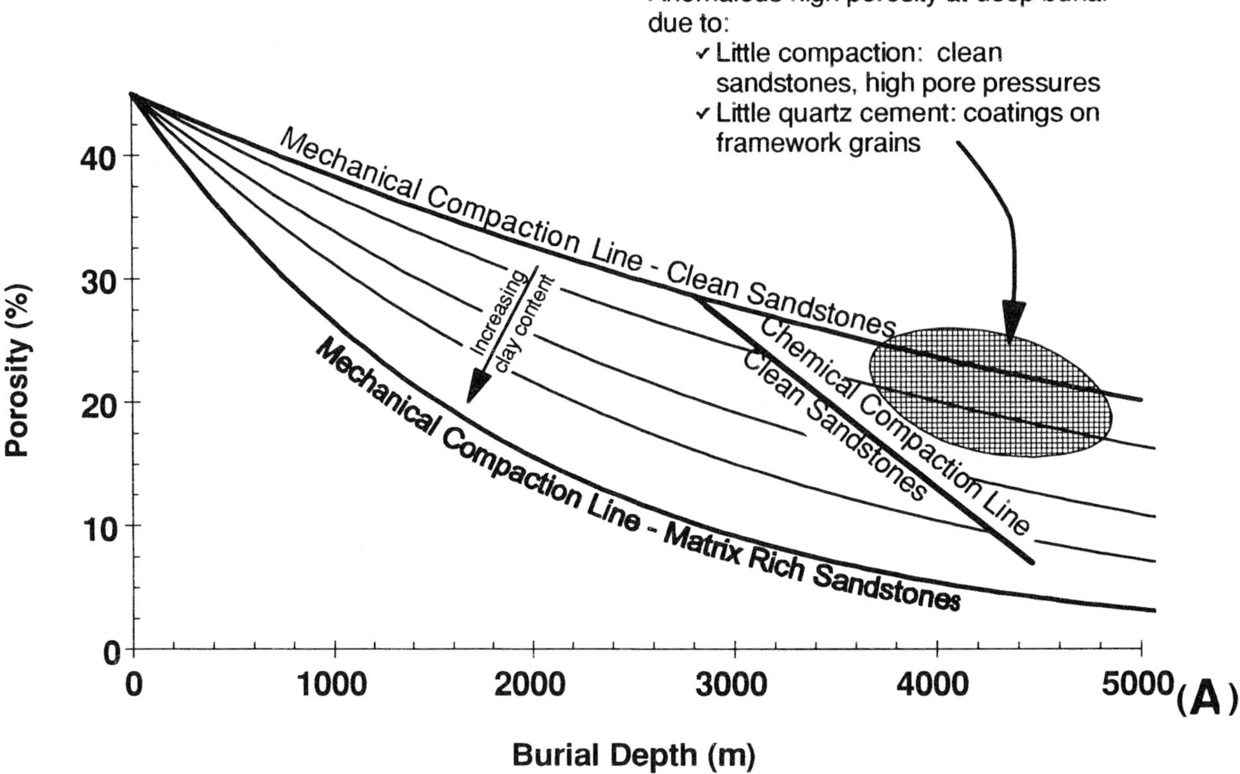

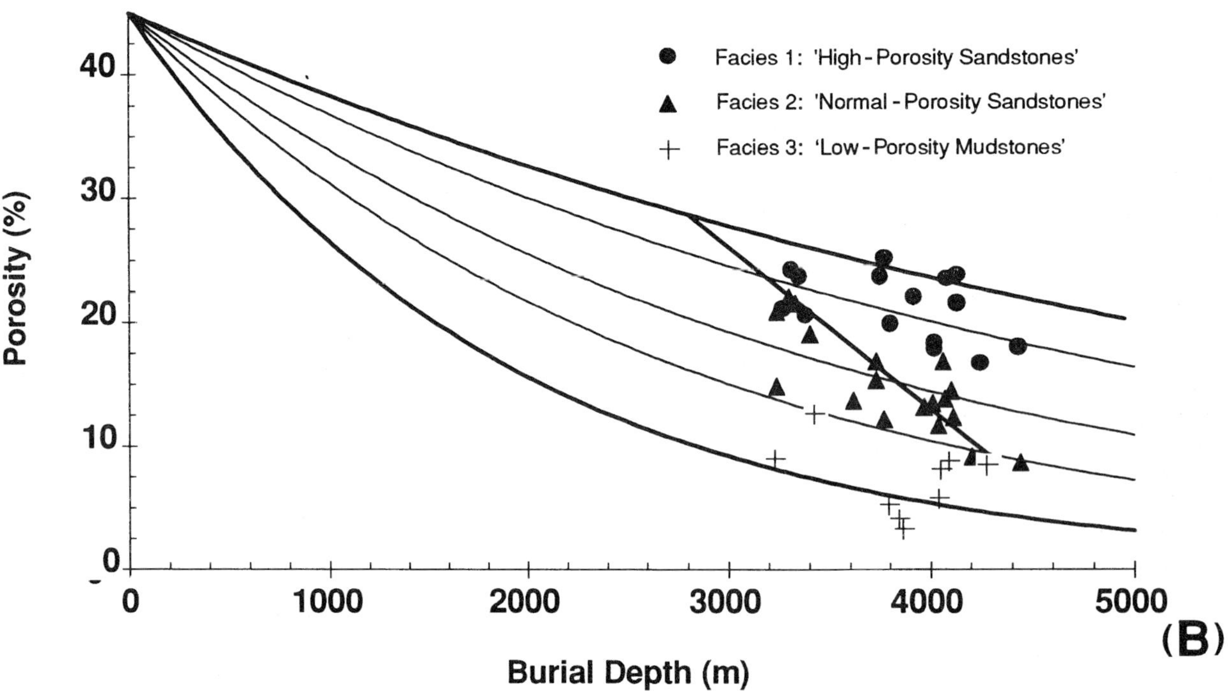

Figure 13. (A) Porosity depth model. (B) Model compared to the 75% He-porosity data. According to the generalized model, the illustrated empirical porosity–depth trends reproduce the porosity variations in the data set. The porosity in the sandstones that are not prone to chemical compaction is expressed as $\phi = 45\,e^{(-0.16 \times Z)}$ when CI < 0.05, and $\phi = 45\,e^{(-(0.12 + 0.82 \times \text{CI}) \times Z)}$ when CI > 0.05, while the porosity in the sandstones prone to chemical compaction is expressed as $\phi = 45\,e^{(-0.16 \times Z)}$ when Z < 2.8 km, and $\phi = 29 - 13 \times (Z - 2.8)$, when Z > 2.8 km.

proportional to the existing porosity and a rate factor, depending on the framework stability. The framework stability is mainly a function of the clay-to-framework grain (quartz plus feldspar) ratio. Sandstones with low clay contents have high framework stability, and at 2500–3000 m, they still have porosities >25%. Clay-rich sandstones and mudstones, however, have low framework stability and, at 3000 m, porosities <10% may be encountered.

At 2500 m, chemical compaction and quartz cementation are initiated in clean sandstones without coatings on framework grains. Below ~2800 m, the rate of porosity loss by chemical compaction becomes faster than the rate of porosity loss by mechanical compaction. Empirically, these sandstones are indicated to lose ~13% porosity per kilometer burial during burial from ~2800 to ~4300 m. Thus, occurrences of good reservoir quality in sandstones buried beneath 4000 m require high framework stability and retarded chemical compaction. Inhibition of quartz cementation due to microcrystalline quartz coatings on framework grains seems to have been efficient in retarding the chemical compaction in deeply buried reservoir sandstones in the Gyda field and in block 7/11.

CONCLUSIONS

1. The porosity variations within the Upper Jurassic sandstones from the Cod Terrace area reflect initial and pre-deep burial rock composition and texture, and are little affected by pore pressure and time/temperature variations or by timing of hydrocarbon emplacement.
2. At shallow to intermediate burial, porosity reduction is mostly by mechanical compaction. The porosity–depth gradient at a particular depth is a function of present porosity and framework stability. The factor found to be most influential on the framework stability of the sandstone is the clay-to-framework grain (quartz plus feldspar) ratio.
3. Below ~2800 m, two groups of relatively clean sandstones diverge rapidly with respect to porosity vs. depth trends. Those sandstones prone to chemical compaction and quartz cementation are severely affected and lose ~13% porosity per kilometer burial between 2800–4300 m. Other sandstones are less affected by chemical compaction and experience a slower rate of porosity loss. At 4200 to 4500 km burial depth, these sandstones still have porosities between 20% and 25%.
4. Retarded chemical compaction in relatively clean arenites is mainly due to inhibited quartz precipitation and dissolution because of microcrystalline quartz on framework grains.
5. Prior-to-drilling prediction of porosity in these deeply buried sandstones will be successful only if the inhibiting effect of the microquartz coating on quartz cementation is recognized. Furthermore, the geological evaluation should emphasize the prediction of the distribution of different sandstone facies and the presence of sandstones within the intervals deposited during periods with high amorphous silica deposition.

ACKNOWLEDGMENTS

This paper is mainly based on research and development projects funded by Norsk Hydro, but early parts were conducted when M. Ramm was employed at the University of Oslo. M. Ramm would like to thank Knut Bjørlykke for criticism and suggestions during this period. Norsk Hydro is acknowledged for permission to publish this study. The authors would also like to acknowledge constructive reviews by Charles Curtis and Pete Turner.

REFERENCES CITED

Bjørlykke, K., P. Aagaard, H. Dypvik, D.S. Hastings, and A.S. Harper, 1986, Diagenesis and reservoir properties of Jurassic sandstones from the Haltenbanken area, offshore mid-Norway, *in* A.M. Spencer et al., eds., Habitat of hydrocarbons on the Norwegian continental shelf: Norwegian Petroleum Society, London, Graham & Trotman, p. 275–386.

Bjørlykke, K., M. Ramm, and G.C. Saigal, 1989, Sandstone diagenesis and porosity modification during basin evolution: Geologische Rundschau, v. 78, p. 243–268.

Bjørlykke, K., T. Nedkvitne, M. Ramm, and G.C. Saigal, 1992, Diagenetic processes in the Brent Group (Middle Jurassic) reservoirs of the North Sea—an overview, *in* A.C. Morton, R.S. Haszeldine, M.R. Giles, and S. Brown, eds., Geology of the Brent Group: Geological Society of London Special Publication 61, p. 265–289.

Boles, J.R., and S.G. Franks, 1979, Clay diagenesis in Wilcox Sandstones of Southwest Texas: implications of smectite diagenesis on sandstone cementation: Journal of Sedimentary Petrology, v. 49, p. 55–70.

Ehrenberg, S.N., 1989, Compaction and porosity evolution of Pliocene sandstones, Ventura Basin, California: discussion: AAPG Bulletin, v. 73, p. 1274–1276.

Ehrenberg, S.N., 1990, Relationship between diagenesis and reservoir quality in sandstones of the Garn Formation, Haltenbanken, mid-Norwegian continental shelf: AAPG Bulletin, v. 74, p. 1538–1558.

Ehrenberg, S.N., 1993, Preservation of anomalous high porosity in deeply buried sandstones by grain-coating chlorite: examples from the Norwegian Shelf: AAPG Bulletin, v. 77, p. 1260–1286.

Forsberg, A.W., M.B. Gowers, and E. Holtar, 1994, Multidiscipline stratigraphic analysis of the Upper Jurassic strata of the Norwegian Central Trough, *in* A.M. Spencer, ed., Generation, accumulation and production of Europe's hydrocarbons III: Proceedings of the European Association of Petroleum Geologists Special Publication, New York, Springer-Verlag, v. 3, p. 45–58.

Gluyas, J.G., A.J. Leonard, and N.H. Oxtoby, 1990, Diagenesis and petroleum emplacement: the race for space—Ula Trend, North Sea (abs.): Utrecht, International Association of Sedimentologists, 13th International Sedimentologist Congress Abstracts, p. 193.

Gluyas, J.G., A.G. Robinson, D. Emry, S.M. Grant, and N.H. Oxtoby, 1993, The link between petroleum emplacement and sandstone cementation, *in* J.R. Parker, ed., Petroleum geology of Northwest Europe: Geological Society of London, p. 1395–1402.

Hancock, N.J., and A.M. Taylor, 1978, Clay mineral diagenesis and oil migration in the Middle Jurassic Brent Sand Formation: Journal of the Geological Society of London, v. 135, p. 69–72.

Hendry, J.P., and N.H. Trewin, 1995, Authigenic quartz microfabrics in Cretaceous turbidites: evidences for silica transformation processes in sandstones: Journal of Sedimentary Research, v. 65, p. 380–392.

Home, P.C., 1987, Ula, *in* A.M. Spencer et al., eds., Geology of Norwegian oil and gas fields: London, Graham and Trotman, p. 143–151.

Houseknecht, D.W., 1987, Assessing the relative importance of compactional processes and cementation to reduction of porosity in sandstones: AAPG Bulletin, v. 71, p. 633–642.

Jahren, J.S., 1993, Microcrystalline quartz coatings in sandstones: a scanning electron microscopy study *in* Karlson, ed., Extended abstracts of the 45th annual meeting of the Scandinavian Society for Electron Microscopy (abs.): SCANDEM 93.

Jahren, J.S., and M. Ramm, in press, The porosity preserving effects of microcrystalline quartz coatings in arenitic sandstones; examples from the Norwegian Continental Shelf, *in* R.H. Worden and S. Morad, eds., Quartz cementation in oil field sandstones: International Association of Sedimentologists Special Publication.

Nedkvitne, T., D.A. Karlsen, K. Bjørlykke, and S.R. Larter, 1993, Relationship between reservoir diagenetic evolution and petroleum emplacement in the Ula field, North Sea: Marine and Petroleum Geology, v. 10, p. 255–270.

Osborn, M., and S. Hazeldine, 1993, Evidence for resetting of fluid inclusion temperatures from quartz cements in oilfields: Marine and Petroleum Geology, v. 10, p. 271–278.

Oxtoby, N.H., A.W. Mitchell, and J.G. Gluyas, 1995, The filling and emptying of the Ula Oilfield: fluid inclusion constraints, *in* J.M. Cubitt and W.A. England, eds., The geochemistry of reservoirs: Geological Society of London Special Publication 86, p. 141–157.

Ramm, M., 1991, On quantitative mineral analysis of sandstone using XRD: Department of Geology, Oslo, Internal Papers, v. 63, 23 p.

Ramm, M., 1992, Porosity–depth trends in reservoir sandstones: theoretical models related to Jurassic sandstones, offshore Norway: Marine and Petroleum Geology, v. 9, p. 553–567.

Ramm, M., and K. Bjørlykke, 1994, Porosity/depth trends in reservoir sandstones: assessing the quantitative effects of varying pore-pressure, temperature history and mineralogy, Norwegian Shelf data: Clay Minerals, v. 29, p. 475–490.

Ramm, M., and A.E. Ryseth, 1996, Reservoir quality and burial diagenesis in the Statfjord Formation, North Sea: Petroleum Geosciences, v. 2, p. 313–324.

Rittenhouse, G., 1971, Mechanical compaction of sands containing different percentages of ductile grains: a theoretical approach: AAPG Bulletin, v. 55, p. 92–96.

Selley, R.C., 1978, Porosity gradients in the North Sea oil-bearing sandstones: Journal of the Geological Society of London, v. 135, p. 119–132.

Sommer, F., 1978, Diagenesis of Jurassic sandstones in the Viking Graben: Journal of the Geological Society of London, v. 135, p. 63–67.

Spark, I.S.C., and N.H. Trewin, 1986, Facies related diagenesis in the main Claymore oilfield sandstones: Clay Minerals, v. 21, p. 479–496.

Stewart, I.J., and K. Scherverud, 1993, Structural controls on the Late Jurassic age shelf system, Ula Trend, Norwegian Sea, *in* J.R. Parker, ed., Petroleum geology of Northwest Europe: Geological Society of London, p. 469–483.

Tada, R., and R. Siever, 1989, Pressure solution during diagenesis: Annual Review of the Earth Planetary Science, v. 17, p. 89–118.

Walderhaug, O., 1990, A fluid inclusion study of quartz cemented sandstones from offshore mid-Norway—possible evidence for continued quartz cementation during oil emplacement: Journal of Sedimentary Petrology, v. 60, p. 203–210.

Walderhaug, O., 1994a, Precipitation rates of quartz cement in sandstones determined by fluid inclusion microthermometry and temperature-history modelling: Journal of Sedimentary Research, v. 64, p. 324–333.

Walderhaug, O., 1994b, Temperatures of quartz cementation in Jurassic sandstones from the Norwegian continental shelf—evidence from fluid inclusions: Journal of Sedimentary Research, v. 64, p. 311–324.

Williams, L.A., and D.A. Crerar, 1985, Silica diagenesis, II. General mechanisms: Journal of Sedimentary Petrology, v. 55, p. 312–321.

Williams, L.A., G.A. Parks, and D.A. Crerar, 1985, Silica diagenesis, I. Solubility controls: Journal of Sedimentary Petrology, v. 55, p. 301–311.

Wood, J.R., 1989, Modelling the effect of compaction and precipitation/dissolution on porosity, *in* I.E. Hutcheon, ed., Short Course in Burial Diagenesis: Mineralogical Association of Canada, v. 15, p. 311–362.

APPENDIX A. DEFINITION OF UPPER JURASSIC LITHOSTRATIGRAPHICAL UNITS ON THE COD TERRACE

In the Cod Terrace area, sequence A is represented by a transgressive basal sandstone (unit A_1), an open marine upward-coarsening shale to muddy sandstone (A_2), and an upper, more well sorted sandstone (A_3). Sequence B is represented by a lower, fine-grained muddy unit (B_1) and an upper, more well sorted sandstone unit (B_2). In well 1/3-3, sequence C1 is represented by four units: $C1_1$ comprises clean low-porosity sandstones; $C1_2$ comprises sandstones with slightly higher clay contents having very good reservoir quality; $C1_3$ entails clean, low-porosity sandstones equivalent to those in unit $C1_1$; $C1_4$ comprises dark mud and fine-grained muddy sandstones. Unit $C1_3$ is present in well 1/3-3 only, whereas unit $C1_2$ occurs in both wells 1/3-3 and 2/1-6. Successively more of this sequence is absent from wells 2/1-4, 2/1-3, and 2/1-8 due to erosion during the Middle Volgian unconformity. The lithostratigraphic units $C1_1$, $C1_2$, and $C1_3$, observed in the wells from the Gyda area, cannot be distinguished as individual units in wells from blocks 7/11 and 7/12. In those wells, unit B is overlain by a finer grained sandstone termed unit $C1_{1-3}$. Unit $C1_4$, however, is recognized in all wells.

The total thickness of the Upper Jurassic section varies considerably within the Cod Terrace area. Pre-Middle Volgian sections >300 m are encountered in wells 2/1-6 and 1/3-3, while sections thinner than 10–20 m occur in the crestal part of the Gyda structure. This thinning across the Gyda field is mainly due to Middle to Late Volgian erosion; only the lowermost units A, B, and $C1_1$ are present in wells 2/1-4, 2/1-3, and 2/1-8. Initial thickness variations due to differential subsidence are, however, also apparent. Units A and B are thinner in the wells from the crestal area of the Gyda structure than in the downflank wells 2/1-6 and 1/3-3.

The total thickness of the pre-Middle Volgian–Upper Jurassic section is thinner in blocks 7/11 and 7/12 than in the Gyda field area; in this area, it is practical to define units A and B as the lithostratigraphical equivalents to sequences A and B. Unit A is of about uniform thickness in the wells from block 7/12, whereas unit B is thicker in the central part of the Ula field (i.e., in well 7/12-2) than in the downflank wells. This relation probably reflects syndepositional differential subsidence, which probably was triggered by early salt movements (Home, 1987; Stewart and Scherverud, 1993). Well 7/12-5, representing a smaller structure to the northwest of the Ula field, contains a condensed unit B, indicating very slow subsidence during the Late Kimmeridgian to Early Volgian. The condensed unit $C1_{1-3}$ reflects slow Early Volgian subsidence in block 7/11 and 7/12 areas compared to the Gyda field area.

Along the fault edge toward the Breiflab Basin and in the northeastern part of the Cod Terrace, erosion rather than deposition was the case during the Late Jurassic. Hence, Upper Jurassic sandstones are either absent or much thinner than in the wells shown in Figure 5 from the footwall upland (e.g., in wells 7/11-8 and 7/8-4). In well 7/11-7, a 30- to 40-m-thick sandstone is encountered between unit D and Triassic strata. Palynological dating of this sandstone indicates Late Middle Volgian age corresponding to sequence C2. The sandstones probably cannot be correlated to the other wells. An ~10-m-thick sandstone is encountered in well 7/11-9 above Triassic strata; this sandstone can probably be correlated to unit A. The Upper Jurassic section is several hundred meters thick in the downfaulted area in UK-block 23/27.

APPENDIX B. SHORT DESCRIPTION OF PETROGRAPHIC CHARACTERISTICS OF SAMPLES FROM WELL 2/1-6

Unit $C1_4$ comprises dark, bioturbated mudstones and fine-grained feldspathic graywackes, with occasional belemnite fossils. Framework grains account for 50%–60%, and quartz is much more abundant than feldspar. Albite is more abundant than K-feldspar. The porosity varies between 5% and 12%, mostly comprising microporosity within dispersed intergranular clays and matrix. Quartz overgrowths are rare, while carbonate cements (mainly ankerite) account for 5%–10%. The matrix content (15%–30% by volume) is dominated by illite, with minor amounts of chlorite.

Unit $C1_2$ comprises fine-grained, bioturbated arkoses. The content of framework grains is mostly close to 60%. Feldspars are more abundant than in the unit above and account for a fifth to a third of the framework grains. Potassium-feldspars are more important than albite. The porosity varies between 18% and 25%, mostly comprising intergranular macroporosity. Quartz overgrowths are rare, while carbonate cements (mainly ankerite) normally account for 5%–10% (volume). The clay matrix (5% by volume) is dominated by illite, with minor amounts of chlorite.

Unit $C1_1$ comprises clean, very well sorted, fine- to medium-grained arkoses. The content of framework grains is mostly close to 70% (volume). The feldspar content is about the same as in unit $C1_2$. The porosity varies between 7% and 10%, mostly comprising intergranular macroporosity. Quartz overgrowths are very abundant, normally accounting for 10%–20%. Carbonate cements (calcite and ankerite) are less abundant than in the other units (2%–3% by volume). The content of clay matrix is low (2%–4% by volume) and dominated by illite.

Units B_1/B_2 comprise fine-grained, bioturbated, feldspathic graywackes (B_1) coarsening upward into cleaner arkoses (B_2). The content of framework grains increases from ~50% in the lower parts to ~60% (volume) in the cleaner parts. The quartz and feldspar distributions are approximately the same as in unit $C1_2$. The porosity increases from <10% near the base of the cored interval to ~20% near the top. Most of the porosity in unit B_1 consists of intergranular microporosity, but macroporosity dominates in unit B_2. Quartz overgrowths are rare, while carbonate cements (mainly ankerite) normally account for 5%–10% (volume). The clay matrix (5% to >15% by volume) is dominated by illite, with minor amounts of chlorite.

Gluyas, J.G., 1997, Poroperm prediction for reserves growth exploration: Ula Trend, Norwegian North Sea, *in* J.A. Kupecz, J. Gluyas, and S. Bloch, eds., Reservoir quality prediction in sandstones and carbonates: AAPG Memoir 69, p. 201–210.

Chapter 13

Poroperm Prediction for Reserves Growth Exploration: Ula Trend, Norwegian North Sea

Jon G. Gluyas[1]
*BP Exploración de Venezuela S.A., Edificio Centro
Seguros de Sud America
Caracas, Venezuela*

ABSTRACT

Much of the remaining prospectivity in the Ula trend (Norwegian Central Graben) is deep (>3.5 km). A major risk to successful petroleum exploration in the trend is reservoir effectiveness. A few oil discoveries are not yet commercial because they occur in low-permeability sandstone. No simple porosity–depth relationship exists for the whole of the Ula trend. As such, mapping of economic basement is difficult. There are, however, simple porosity vs. depth relationships within the two main producing fields: Ula and Gyda.

The porosity–depth relationships in the fields are due to downflank cementation by quartz. Quartz cementation was synchronous with oil emplacement, and evidence from petroleum-filled fluid inclusions has led to the conclusion that cementing fluids and petroleum competed in a "race for space." The Ula trend displays evidence of all three outcomes of such a race: petroleum emplacement ahead of cementation, synchronous processes, and cementation ahead of petroleum emplacement.

Porosity prediction for undrilled prospects and prospect segments was made by risking the three possible outcomes of such a race for space. The reservoir in prospect 7/12-JU4 was predicted to be oil bearing and have a mean porosity of about 16.4%: a function of synchronous petroleum emplacement and cementation. The well, however, was dry. It had a mean porosity of 14%; this compares well with the predicted porosity (13.9%) at the well location for a system in which cementation was completed before oil emplacement (equivalent to a porosity estimate for a dry hole).

---

[1]Present address: Monument Oil and Gas plc, London, United Kingdom.

INTRODUCTION

The Ula trend (Figure 1) of the Norwegian Central Graben contains three producing oil fields. In decreasing size order they are: Ula (reserves 435 mmstb [million stock tank barrels]; Home, 1987; Brown et al., 1992), which is now in decline; Gyda (200 mmstb; Gluyas et al., 1992), which is on plateau; Mime (a few tens of mmstb), which ceased production in 1994. Much of the exploration activity in the Ula trend today concentrates on reserves growth; field extensions can be tied back to existing production facilities. Bjørnseth and Gluyas (1995) report, "remaining prospectivity in the Ula Trend is subtle, but a large number of small to medium sized prospects have been defined (reserves generally less than 100 mmstb)."

In such a mature province, the perceived exploration risk drives the economic viability of prospects. Even the smallest prospects can look attractive if risk is low and tieback costs are acceptable. An analysis of drilling statistics for the trend showed that, although risks on trap presence and charging exist, they are low relative to those associated with reservoir presence and effectiveness (Bjørnseth and Gluyas, 1995). The Ula trend contains enough examples of light petroleum trapped within low-permeability rock for reservoir effectiveness to be a major concern.

This chapter examines the way an attempt was made to evaluate reservoir effectiveness (porosity and permeability) of a prospect ahead of drilling.

GEOLOGICAL BACKGROUND

Well 7/12-2 was the discovery well for the Ula field (Figures 1, 2), a giant field that produces light oil from a high-quality Upper Jurassic, shallow marine sandstone reservoir (Home, 1987; Oxtoby et al., 1995). The trap is a well-defined four-way-dipping dome, with >500 m of vertical closure. There are at least six oil-water contacts; the shallowest in the west is ~300 m shallower than the deepest in the east. All of the differences in oil-water contacts are due to faulting rather than to stratigraphic effects.

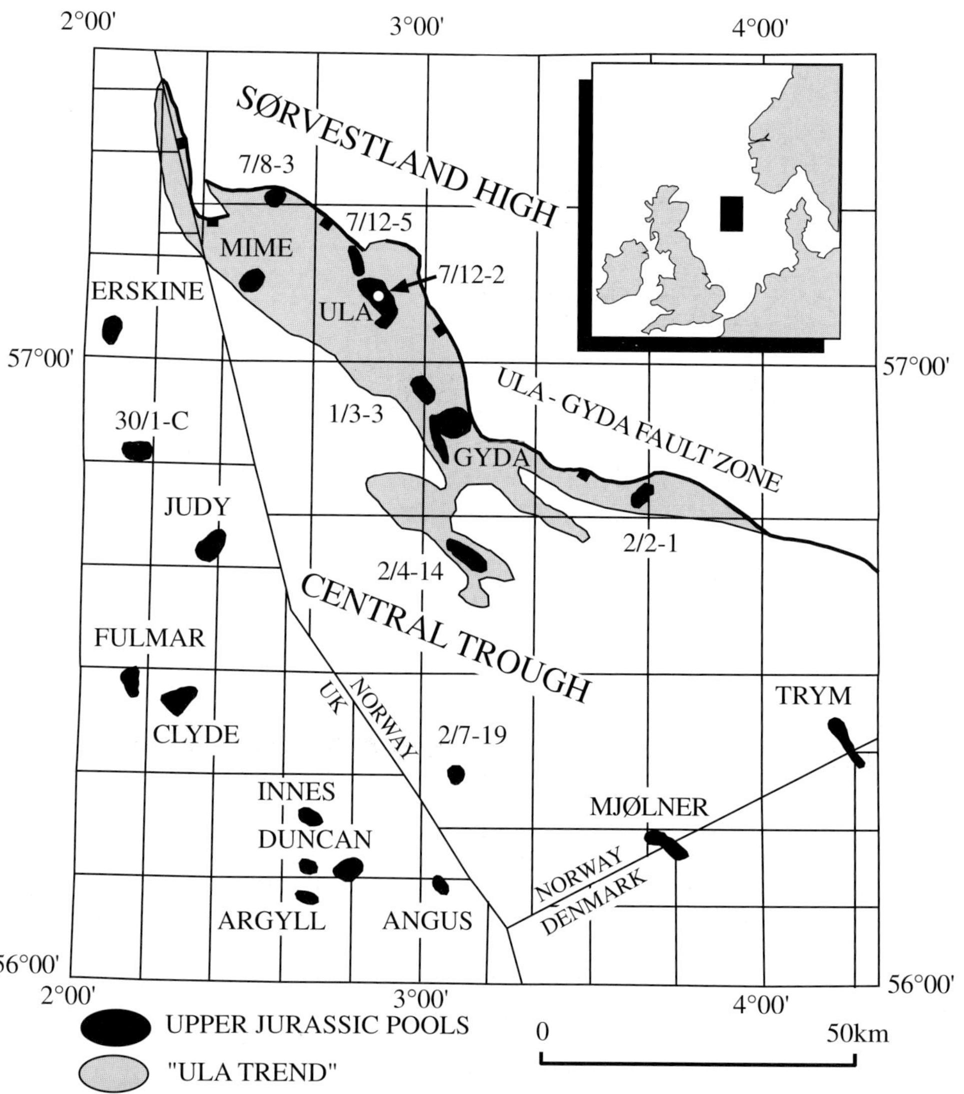

Figure 1. Location map for the Ula trend (from Bjørnseth and Gluyas, 1995). The 7/12-5 and 7/8-3 discoveries remain unnamed.

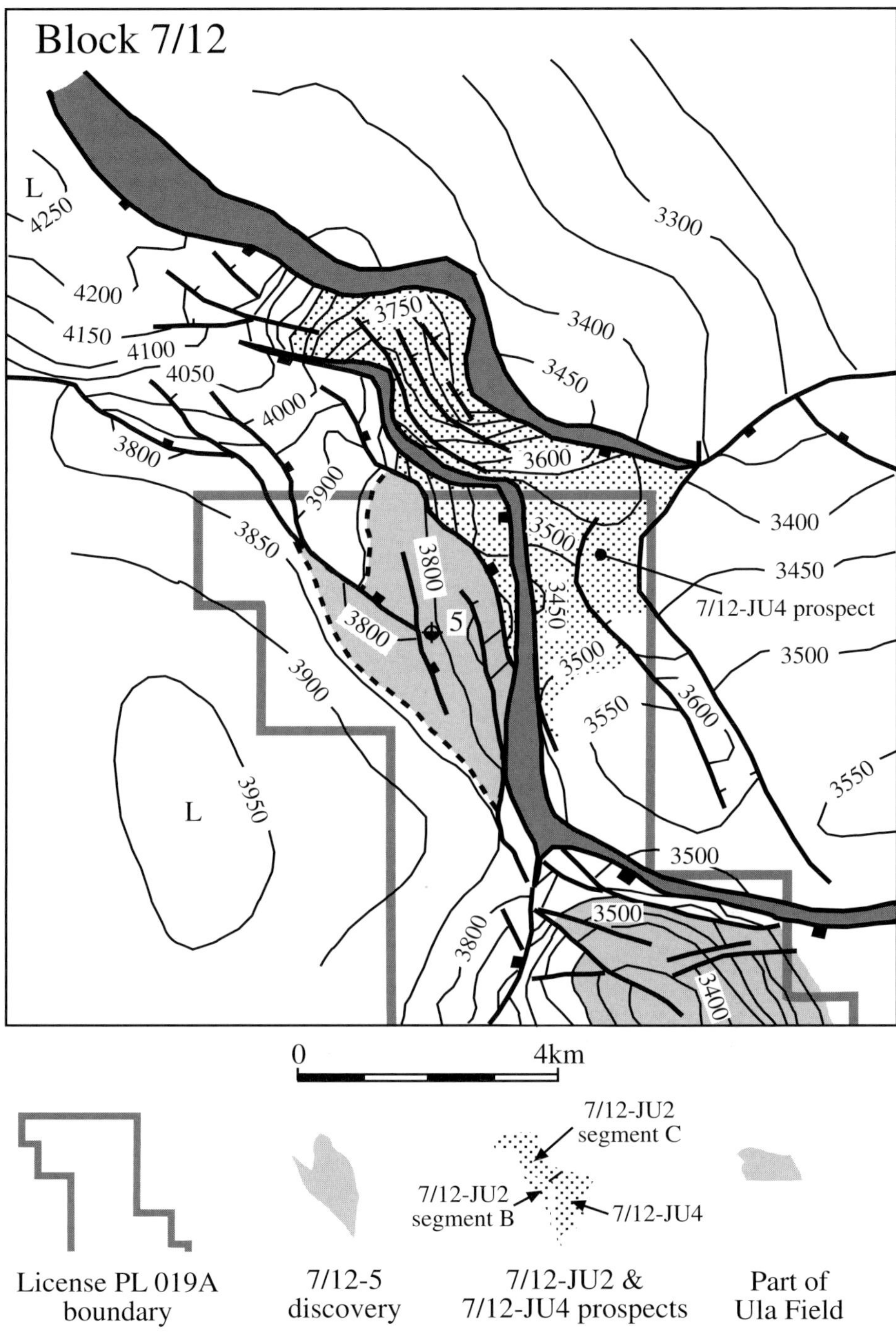

Figure 2. Structural map for the 7/12 block showing both the Ula field and the prospects and prospect segments around the 7/12-5 discovery.

Well 7/12-5 was drilled on a similar but smaller structure in the same license a few years after 7/12-2. Depth to crest for 7/12-5 is about 400 m deeper than in the Ula field. It too had light oil. The reservoir is the same Upper Jurassic shallow marine sandstone present in the Ula field. However, the average permeability of the reservoir in 7/12-5 (2 md) is 2 orders of magnitude lower than that for Ula (192 md). The discovery has not been developed or named. Deeper still is the bulk of the Gyda field (Block 2/1; Gluyas et al., 1992), which lies in the adjacent license. The quality of the same reservoir in the Gyda field (40 md) is midway between that of Ula and that of 7/12-5 (Figures 3, 4).

Following completion of 2-D and 3-D seismic surveys across the licenses in the late 1980s and early 1990s, a large number of leads and prospects were identified within the Upper Jurassic play fairway. The key risks associated with exploration of these prospects were believed to be reservoir presence and permeability (Bjørnseth and Gluyas, 1995). The aim of this chapter is to illustrate an attempt to quantify one of those key risks—reservoir effectiveness—and to predict the reservoir quality in prospects and prospect segments (JU2 and JU4) around the 7/12-5 discovery (Figure 2). There was clearly no wish in the License Group to discover oil in the reservoir similar to that encountered by 7/12-5.

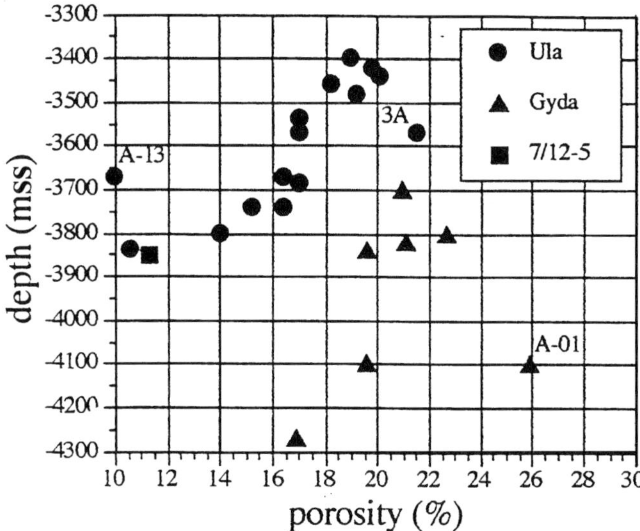

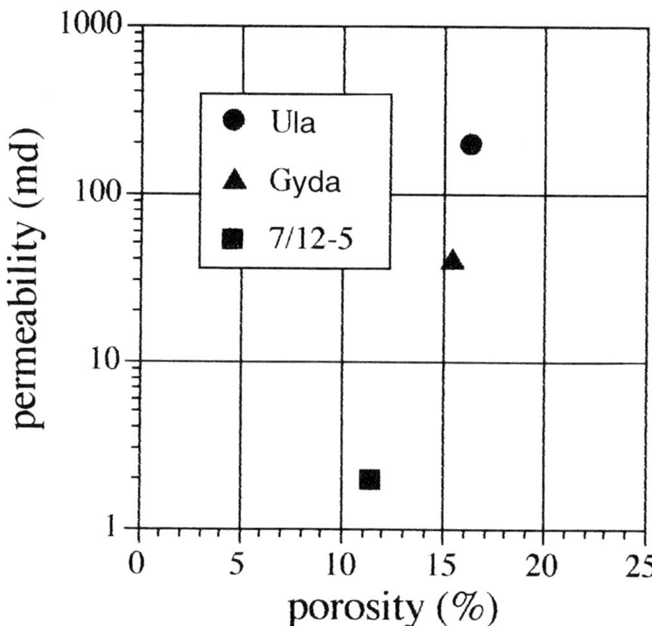

Figure 3. Porosity–depth plot for wells in Ula, Gyda, and 7/12-5. Data are well averages for clean grainstones (most of the reservoir was cored in the wells on this figure). Three anomalous points are highlighted. The relatively high porosity in 3A on the western flank has not been satisfactorily explained. The A-13 well on Ula lies at the north end of the field where the sandstones shale out. The low porosity may be due to increased compaction associated with the shaly sandstone in A-13. The A-01 well on Gyda has high porosity; the origin has been attributed to the retardation of cementation in A-01 by oil on its migration from flank to crest of Gyda. mss = meters subsea level.

Figure 4. Mean arithmetic porosity and permeability data for the Ula, Gyda, and 7/12-5 wells.

RESERVOIR SEDIMENTOLOGY

The Ula trend sandstones are Upper Jurassic shallow marine deposits (Figure 5). Most of the sandstones accumulated below fair-weather wave base as storm deposits. Accommodation space was created by a combination of active rifting and associated movement of Permian evaporites in the underlying section (Stewart, 1993). The sands accumulated, base to top, as a series of progradational, aggradational, and finally retrogradational packages (Oxtoby et al., 1995) (Figure 5). The largely fine- to medium-grained arkosic sand was probably second cycle, shed from emergent Triassic "pods" (Bjørnseth and Gluyas, 1995). As a result of the heterogeneous development of accommodation space associated with salt withdrawal, reservoir thickness can vary dramatically over short distances. For example, over a distance of ~5 km in the Ula field, the reservoir thickness changes from 200 m to just a few tens of meters.

RESERVOIR QUALITY

Porosity, permeability, and depth data for the Ula trend reservoir sandstone are presented in Tables 1 and 2, and are plotted in Figure 3. Data are well averages for the prime grainstone-texture reservoir sandstones. Argillaceous packstone, wackestone-texture sandstones are of very low reservoir quality everywhere, and have not been included in the plots. The most important feature of Figure 3 is that the porosity of the reservoir prime reservoir (grainstone-texture) sandstones declines dramatically from crest to flank in both the Ula and Gyda fields. The porosity gradients are about the same but the intercepts are different. The fieldwide porosity–permeability relationships are plotted in Figure 4. Prospects with average permeability <10 md are probably not economically viable.

RESERVOIR DIAGENESIS

The paragenetic sequence for the reservoir sandstones in the Ula trend is displayed in Figure 6. The timing of the events shown in Figure 6 is based on a combination of microthermometric determinations on fluid inclusions, radiometric age dating of illitic clay, and stable isotope analysis of carbonate cements linked to burial history and thermal history calculations. Supporting evidence for both the relative and absolute timing of diagenetic events has been published by Gluyas et al. (1990), Gluyas and Coleman (1992), Gluyas et al. (1993), Nedkvitne et al. (1993), Oxtoby et al. (1995), and Ramm et al. (this volume).

Most of the precompaction processes have little effect on the quality of the sandstones, although local but pervasive calcite cementation has reduced the net-gross rate in part of Gyda. The main processes to have affected the quality of the extant reservoirs are compaction and quartz cementation (Gluyas et al., 1990; Nedkvitne et al., 1993; Oxtoby et al., 1995; Ramm et al., this volume).

Of compaction and quartz cementation, quartz cementation is the key variable within individual oil fields. Figure 7 is a crossplot of quartz cement content

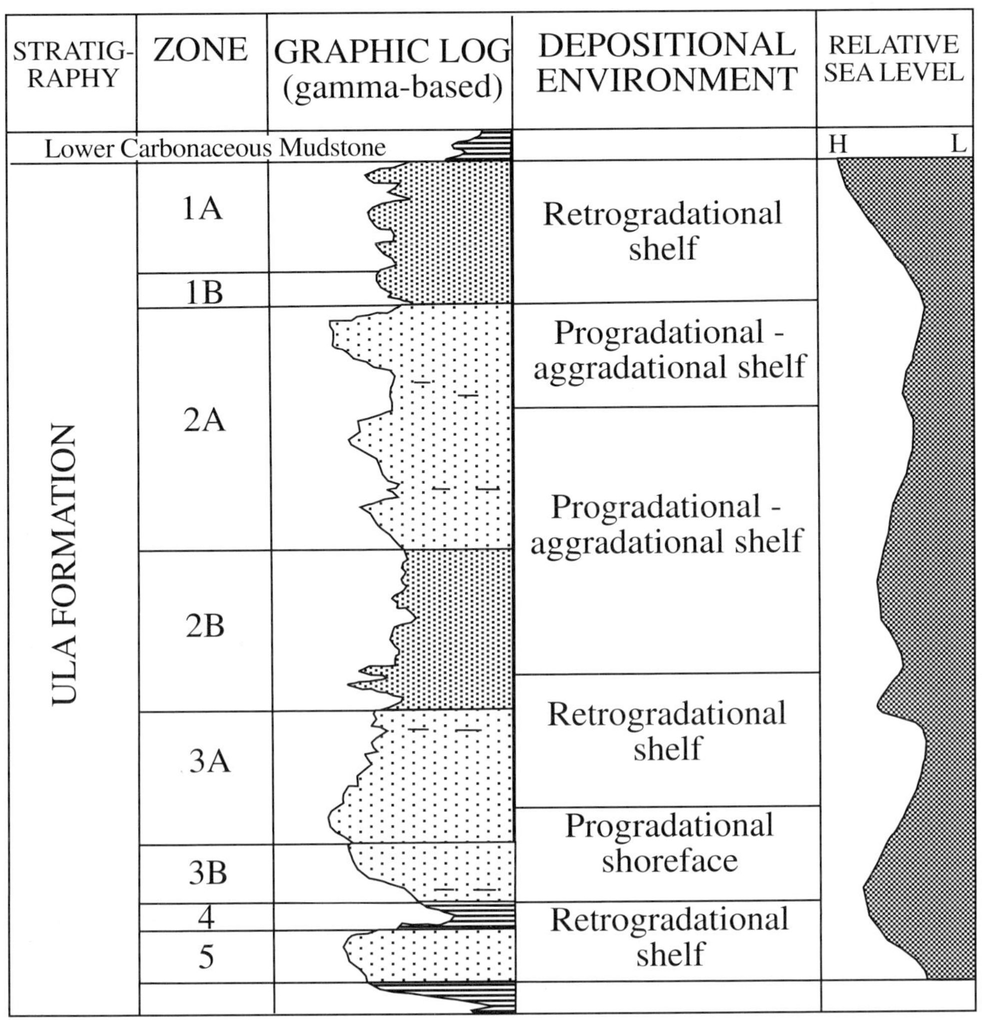

Figure 5. Graphic log of the Ula reservoir sandstone (from Oxtoby et al., 1995).

against porosity for the Ula field wells. Highly porous (and permeable) sandstones have little cement, while tight sandstones have much cement. The relationship between well-cemented sandstones of poor reservoir quality and little cemented sandstones of much better reservoir quality is manifest in two ways. For Ula trend fields, reservoir quality diminishes dramatically as a function of burial depth as quartz cement content increases (Figure 8). The rate of porosity decline is commonly double the regional gradient. However, high-porosity anomalies exist. These are either the crestal parts of oil accumulations or short intervals of medium-grained sandstones on the field flanks.

Two quite different explanations for the origin of the anomalously porous sandstones have been published. These explanations have also sought to explain the steep porosity gradients.

Hypothesis 1: Cementation Retarded by Petroleum Emplacement

The steep cementation gradients and high-porosity anomalies within fields are postulated to be a product of the way in which cementation and petroleum emplacement interacted during filling of the trap with

oil (Gluyas et al., 1990, 1993). The hypothesis states that the presence of petroleum retarded cementation by quartz. The highly porous field crests were filled with oil before much cementation. The steep cementation gradients record simultaneous cementation and oil emplacement, creating a "race for space," during which downward filling of the trap with petroleum progressively slowed cementation (Oxtoby et al., 1995).

Supporting evidence for this hypothesis is provided by the abundance and distribution of petroleum-filled and aqueous inclusions trapped within the quartz cements. The crestal parts of Ula contain only a few percentiles of quartz cement. However, this cement is quite literally full of petroleum-filled fluid inclusions (Oxtoby et al., 1995). The abundance of such inclusions declines down-flank and in the water leg of the field, and petroleum-filled fluid inclusions are rare. The reservoir in 7/12-5 contains only aqueous inclusions. This is taken to imply that cementation was complete in the area around 7/12-5 before oil emplacement.

Further supporting evidence for the hypothesis comes from anomalies within the distribution of petroleum-filled inclusions and the distribution of high-quality reservoir rock. On the eastern flank of

Table 1. Average Porosities for Ula Trend Wells.

Well	Middepth (mss)	Porosity (%)
2/1-3	−3800	22.7
2/1-4	−4100	19.6
2/1-6	−4270	9.8
2/1-A-01	−4100	17.9
2/1-A-02	−3700	21.0
2/1-A-04	−3840	19.6
2/1-A-07	−3820	21.1
7/8-3	−3740	13.7
7/12-5	−3850	11.3
7/12-2	−3420	19.8
7/12-3A	−3570	21.5
7/12-4	−3460	18.2
7/12-6	−3440	20.1
7/12-7	−3800	14.0
7/12-8	−3740	15.2
7/12-9	−3740	16.4
7/12-A-01	−3670	16.4
7/12-A-03	−3570	17.0
7/12-A-03A	−3685	17.0
7/12-A-08	−3840	10.6
7/12-A-12A	−3509	19.7
7/12-A-13	−3672	10.0
7/12-A-13A	−3540	17.0
7/12-A-15	−3400	19.0
7/12-A-18	−3480	19.2

Table 2. Field Average Porosity and Permeability.

Field	Porosity (%)	Permeability (md)
Ula	16.3	192
Gyda	15.5	40
7/12-5	11.3	2
7/8-3	13.7	42

the field, well 7/12-7 encountered the reservoir ~400 m below the field crest. The average reservoir properties of the reservoir in this well fall on the field-wide porosity and permeability trends. However, the well contains ~2 m of medium-grained sandstone at the top of the principal reservoir interval. It is both highly porous (27%–28%) and permeable (>1 darcy), having similar reservoir quality to the field crest. This sand has very little quartz cement, but what it does contain has abundant petroleum-filled fluid inclusions. The inference at the time was that this particular 2-m interval represented one of the oil migration routes into the Ula trap. Subsequent geochemical modeling of the source rock confirmed the easterly filling direction for the field. A similar pattern exists in the Gyda field, where short intervals of anomalously porous and permeable sandstones are present in western downflank well 2/1-A-01.

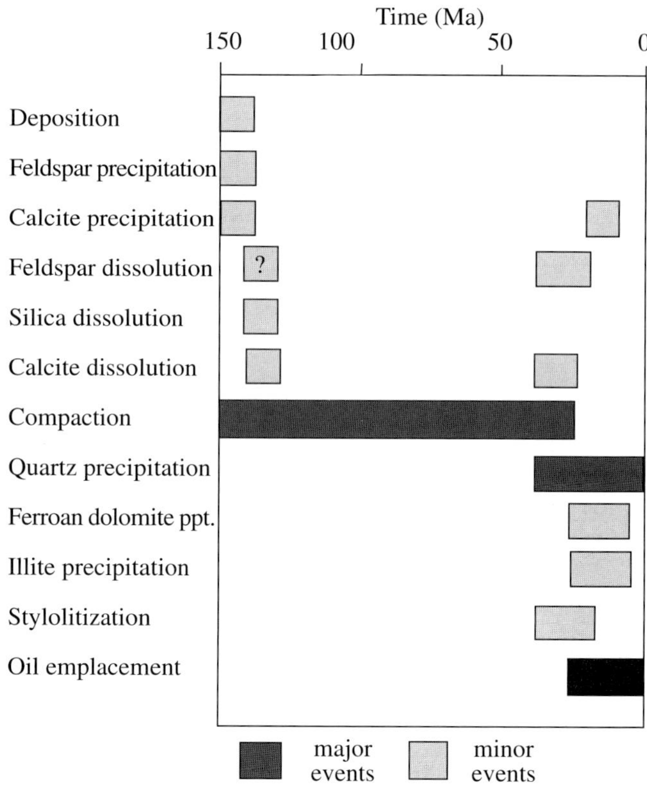

Figure 6. Summarized diagenetic history for the Ula sandstone in the Ula field. The Gyda field reservoir has a similar history, although in Gyda calcite is locally an important cement, forming impermeable layers. The figure is modified from Bjørnseth and Gluyas (1995). Quantitative diagenetic data for the Ula and Gyda wells have been published in Gluyas et al. (1990), Gluyas and Coleman (1992), Gluyas et al. (1993), Nedkvitne et al. (1993), Oxtoby et al. (1995), and Ramm et al. (this volume).

Hypothesis 2: Cementation Retarded by Early Diagenetic Precipitation of Microquartz

In this hypothesis, the presence of highly porous sandstones at depth is attributed to retardation of quartz cementation in sandstones that have grains coated by microcrystalline quartz. This hypothesis is fully described by Ramm et al. (this volume). The occurrence of microcrystalline quartz cement has been linked to the distribution of relic remains of siliceous sponge spicules and/or volcanic glass. The spiculitic sands accumulated as shoals, and as such the presence of high porosity at depth is seen to be a function of the lithofacies distribution at the time of sand deposition.

The two hypotheses are clearly different, and their use in methodologies for prediction of porosity would give very different results. However, at the time that the Ula trend prospects were under evaluation, the exploration team working the issue accepted the hypothesis that cementation could be retarded by oil emplacement. This chapter reports their work.

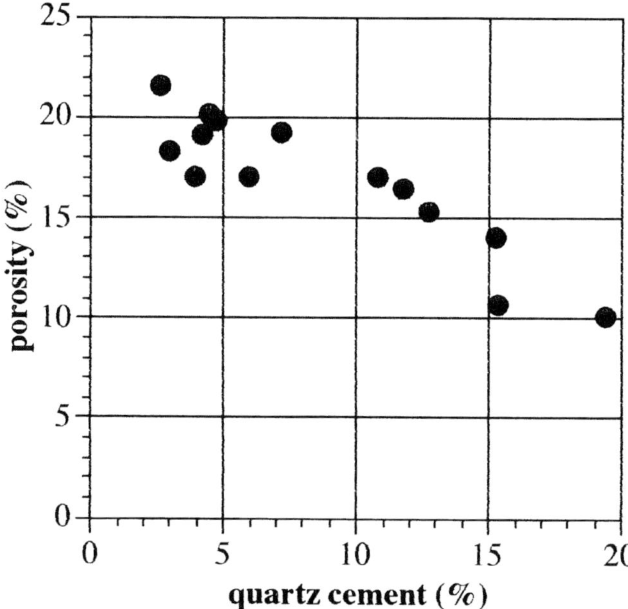

Figure 7. Relationship between quartz cement content and porosity for the reservoir sandstones of the Ula field. Data for both porosity and quartz cement content are average values for each well. The number of samples for each porosity point is approximately 100; for each quartz cement point, about 10.

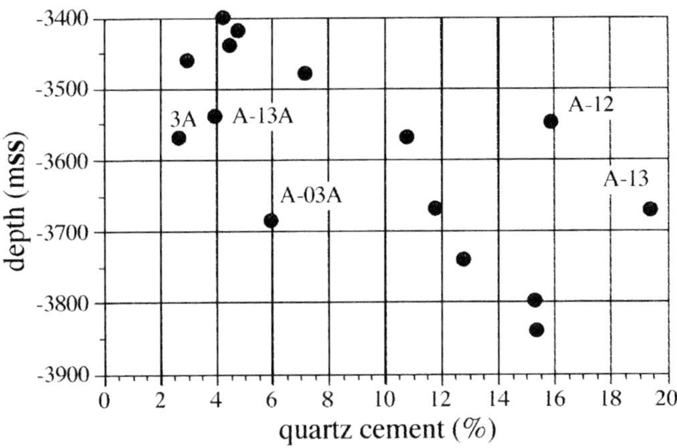

Figure 8. Quartz cement vs. depth [meters subsea (mss) level] for Ula field. The two main anomalies are at –3550 m, well 7/12-3A (Figure 3), and –3690 m, well 7/12-A12. The A12 anomaly may be due to point-count error since it lies on the porosity–depth plot for the field.

RESERVOIR PRESSURE

Each of Ula, Gyda, and 7/12-5 wells was overpressured at discovery by ~12–14 MPa at the reservoir midpoint. Overpressure in the Ula trend reservoirs is probably a function of burial disequilibrium caused by rapid burial in the Neogene. The whole of the area of the Ula trend is covered by ~2 km of Neogene mud-rich sediments. For lack of better data, the 12–14 MPa overpressure estimate was used in the porosity calculation for the prospects.

ESTIMATION OF POROSITY AND PERMEABILITY

Prediction of porosity is commonly treated as simply an estimation of uncertainty; that is, the spread around a most likely value. The porosity of the sandstone could be controlled by one of three largely distinct processes: (1) compaction alone, (2) compaction with cementation, or (3) compaction with a degree of cementation inversely controlled by petroleum emplacement. The possibilities were captured by risking three models. The outcomes for each of these models have varying degrees of uncertainty.

The three porosity evolution models (Figure 9) are:

- Model 1—Quartz cementation was complete before petroleum emplacement (regional porosity decline).
- Model 2—Quartz cementation and petroleum emplacement occurred at the same time (Ula trend porosity decline).
- Model 3—Petroleum emplacement occurred before quartz cementation. The sandstones remain largely uncemented (no cementation).

Model 1—Quartz Cementation Complete Before Oil Emplacement

Many North Sea Jurassic sandstones have similar porosity–depth gradients (8% ±1% km^{-1}) (Selley, 1978; Gluyas, 1985). Emery et al. (1993) have shown that many of these sandstones have only water-bearing fluid inclusions in quartz cement. They concluded that for these sandstones, cementation by quartz was completed in the absence of petroleum. The sandstones in well 7/12-5 contain only aqueous inclusions in quartz cement.

For model 1, the prospect porosities are calculated using 7/12-5 data and a regional porosity gradient of –8% km^{-1}. There are too few data to develop a water leg gradient specifically for the Ula trend.

Model 2—Quartz Cementation Synchronous with Oil Emplacement

The Ula trend sandstones exhibit very steep porosity declines with depth that are associated with equally rapid increases in quartz cement with depth. The porosity–depth gradients in the Ula and Gyda fields are similar, but the intercepts differ. The distribution of petroleum-filled fluid inclusions in the quartz cement mimics that of the porosity decline. The highly porous field crests with little cement contain an abundance of petroleum-filled fluid inclusions in that cement. The well-cemented flanks of the fields show few or no petroleum-filled fluid inclusions in their quartz cement (Oxtoby et al., 1995).

Porosities are calculated using the 7/12-5 data for the intercept and the Gyda/Ula data for the slope. The

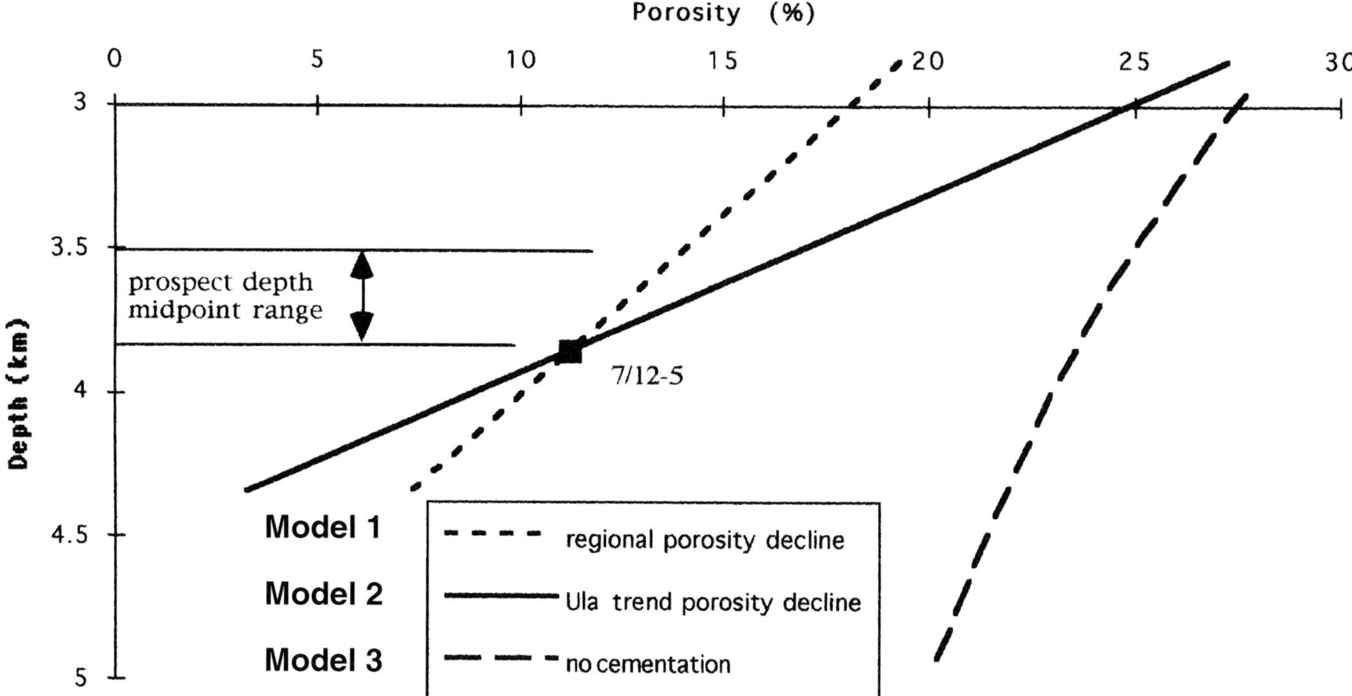

Figure 9. Modeled porosity–depth relationships. Model 1: Oil emplacement before cementation uses a regional porosity gradient of 8% porosity loss for each additional kilometer of burial. The line passes through the porosity–depth point for 7/12-5. Model 2: Oil emplacement and simultaneous quartz cementation uses a porosity gradient of 16% km^{-1} derived from the local Ula trend data. It too passes through the porosity–depth point for well 7/12-5. Model 3: Oil emplacement without (before) cementation is based on the compaction curve of Gluyas and Cade (this volume), with an overpressure correction of 14 MPa. The curve is conditioned to the high-porosity, uncemented sandstones at the crest of the Ula field and downflank to similarly high-porosity sandstones in 7/12-7.

7/12-5 data are considered a reasonable intercept because the prospects are clustered around this discovery. The application of a steep Ula trend gradient through the data of 7/12-5 is analogous to the Ula or Gyda fields in which quartz–cemented sandstones without petroleum-filled inclusions occur just above the deepest oil-water contacts (Oxtoby et al., 1995).

Model 3—Oil Emplacement Before Quartz Cementation

The crestal parts of both Ula and Gyda contain sandstones with very little or no quartz cement. Such sandstones are simply compacted according to the effective burial stress (lithostatic load minus fluid overpressure).

An estimate of porosity prior to drilling is calculated using the compaction equation:

$$\phi = 50 \exp\left(\frac{-10^{-3}z}{2.4 + 5 \times 10^{-4}z}\right) \tag{1}$$

(Robinson and Gluyas, 1992), in which ϕ = porosity (%) and z = depth (m), with appropriate adjustment for an overpressure correction. An overpressure of 14 MPa is equal to an effective burial depth ~1120 m less than the real burial depth using the equation of Gluyas and Cade (this volume):

$$z' = z - \left(\frac{u}{(\rho_r - \rho_w)g(1 - \phi_\Sigma)}\right) \tag{2}$$

in which z' = effective burial depth (in meters), u – overpressure (in megapascals), ρ_r and ρ_w are density of rock and water (kgm^{-3}), g = acceleration due to gravity (ms^{-2}), and ϕ_Σ = bulk fractional porosity of overlying sand and mud sediment column. The porosity–depth relationships associated with each of these models are illustrated by Figure 9.

RISKED POROSITY MODELS

The following risks were assigned on the basis of empirical observations.

- About 1 in 20 of the Ula and Gyda wells have significant portions of their reservoir interval free of quartz cement.
- Two fields—Ula and Gyda—had synchronous oil emplacement and cementation.
- Only 7/12-5 was cemented before oil emplacement.

There was, therefore, a nominal risk of 2:1 in favor of synchronous cementation and oil emplacement. However, given that a significant portion of the 7/12-JU4

Table 3. Parameters for Ula Trend Prospects.

Depth (m)	JU4	JU2 Segment C	JU2 Segment A	JU2 Segment B
Depth to crest	3440	3440	3720	3540
Depth to closure	3525	3900	3846	3846
Mid-volume depth	3510	3675	3820	3675

Table 4. Predicted Porosities for Ula Trend Prospects.

Porosity (%)	JU4	JU2 Segment C	JU2 Segment A	JU2 Segment B
Model 1	13.9	13.1	Proven 11.5	12.7
Model 2	16.5	14.6	—	14.0
Model 3	25.7	25.2	—	25.0

prospect was updip of 7/12-5, the chance of simultaneous cementation and oil emplacement was estimated to be higher. The consequent estimated risks were: model 1 = 0.2, model 2 = 0.75, and model 3 = 0.05.

The porosity calculations were based on the trap configuration in Figure 2 and data in Table 3. The resultant risked porosities were: JU4 = 16.4%, JU2 segment C = 14.8%, JU2 segment A = 11.5%, and JU2 segment B = 14.5%.

PERMEABILITY CALCULATION

Permeabilities were calculated from the empirical relationship between porosity and permeability using field average data (Ula and Gyda, 7/12-5) (Table 4, Figure 4).

$$\log_{10} k = 0.36\phi - 3.80 \qquad (3)$$

where k = permeability (in millidarcys) and ϕ = porosity (in percent).

The resultant risked permeabilities were: JU4 = 120 md, JU2 segment C = 50 md, JU2 segment A = 4 md, and JU2 segment B = 20 md.

UNCERTAINTY CALCULATIONS

The uncertainty surrounding the porosity and permeability data predictions was calculated using the spread of Ula field data in Figures 3 and 4. The porosity range for the Ula field at a given depth is 6% at 95% confidence limits. Hence, 2σ on the porosity quoted above is ±3%; this porosity variation corresponds to a permeability variation of ~0.75 magnitude (Figure 4).

WELL RESULTS

Well 7/12-10 was drilled on prospect 7/12-JU4. The reservoir was present, but the oil was missing and the well was dry. The fault system lying to the west of the prospect is now believed to be sealing, having stopped access of petroleum to the prospect.

In consequence, the appropriate model should have been model 3, cementation complete before oil emplacement. The porosity predicted by this model was 13.9%, and that from core analysis in 7/12-10 was 14.0%; this was within the confidence limits, and therefore a perfect prediction.

DISCUSSION AND CONCLUSIONS

This approach to reservoir quality prediction may seem sophisticated. Moreover, because this approach uses, as support, a hypothesis that is disputed, it is tempting to conclude that the approach is not worthwhile. However, the hypothesis is used only to explain the porosity-to-depth relationships and not to generate a methodology for porosity prediction. The model curves in Figure 9 are based wholly on empirical observation.

The "no cementation" curve is founded on the observation that some of the sandstones in the Ula trend are not cemented. The shape of the curve is based upon experimental and empirical data (Gluyas and Cade, this volume).

The "regional porosity decline" curve of Figure 9 is based on empirical data from the Central and Northern North Sea. The sandstones in this data set are like those of 7/12-5 because their quartz cements do not contain petroleum-filled fluid inclusions (Emery et al., 1993).

The "Ula trend porosity decline" is the local curve. The porosity gradient is steep when compared with regional data; the component sandstones that make up the Ula trend porosity decline are distinct insofar as they contain petroleum trapped in inclusions in quartz cement. Thus, the three models for porosity prediction can be used without reference to a hypothesis to explain the models.

In conclusion, the lack of a simple porosity-to-depth relationship for the Ula trend as a whole drove investigations to reveal how porosity was destroyed. This in turn delivered a methodology that allowed better use of the empirical porosity–depth data for reservoir quality prediction.

ACKNOWLEDGMENTS

I wish to thank Per Svela, Grete Block-Valge, and Per Christian Mjelde for helping improve this manuscript. I also thank BP Norge and partners for giving me permission to publish this work.

REFERENCES CITED

Bjørnseth, H.M., and J.G. Gluyas, 1995, Petroleum exploration in the Ula trend, *in* S. Hanselein, ed., Petroleum exploration in Norway: Norsk Petroleumsforening/NPF, Special Publication 4, Proceedings of the Norwegian Petroleum Conference, December 9–11, 1991, Stavanger, Norway, Elsevier, Amsterdam, p. 85–96.

Brown, A., A.W. Mitchell, I.R. Nilssen, I.J. Stewart, and P.T. Svela, 1992, Ula field: relationship between structure and hydrocarbon distribution, *in* B.T. Larsen and R.M. Larsen, eds., Structural and tectonic modelling and its application to petroleum geology: Norsk Petroleumsforening/NPF, Special Publication 1, Elsevier, Amsterdam.

Emery, D., P.C. Smalley, and N.H. Oxtoby, 1993, Synchronous oil migration and cementation in sandstone reservoirs demonstrated by quantitative description of diagenesis: Philosophical Transactions of the Royal Society of London, v. 344, p. 115–125.

Gluyas, J.G., 1985, Reduction and prediction of sandstone reservoir potential, Jurassic North Sea: Philosophical Transactions of the Royal Society of London, v. A315, p. 187–202.

Gluyas, J.G., K. Byskov, and N. Rothwell, 1992, A year in the life of Gyda production: IBC, Advances in Reservoir Technology, London—Conference Proceedings, p. 187–202.

Gluyas, J., and C.A. Cade, this volume, Prediction of porosity in compacted sands, *in* J.A. Kupecz, J. Gluyas, and S. Bloch, eds., Reservoir quality prediction in sandstones and carbonates: AAPG Memoir 69, p. 19–28.

Gluyas, J.G., and M.L. Coleman, 1992, Material flux and porosity changes during diagenesis: Nature, v. 356, p. 52–53.

Gluyas, J.G., A.J. Leonard, and N.H. Oxtoby, 1990, Diagenesis and petroleum emplacement: the race for space—Ula Trend, North Sea (extended abs.): Nottingham, England, 13th International Sedimentological Congress, International Association of Sedimentologists, Utrecht, p. 193.

Gluyas, J.G., A.G. Robinson, D. Emery, S.M. Grant, and N.H. Oxtoby, 1993, The link between petroleum emplacement and sandstone cementation, *in* J.R. Parker, ed., Petroleum geology of Northwest Europe: Barbican, London, Geological Society of London Proceedings of the 4th Conference, p. 1395–1402.

Home, P.C., 1987, The Ula oilfield block 7/12, Norway, *in* A.M. Spencer et al., eds., Geology of the Norwegian oil and gas fields: Norwegian Petroleum Society, London, Graham & Trotman, p. 143–152.

Nedkvitne, T., D.A. Karlsen, and K. Bjørlykke, 1993, Relationship between diagenetic evolution and petroleum emplacement: Marine and Petroleum Geology, v. 10, p. 225–270.

Oxtoby, N.H., A.W. Mitchell, and J.G. Gluyas, 1995, The filling and emptying of the Ula oilfield (Norwegian North Sea), *in* J.M. Cubitt and W.A. England, eds., The geochemistry of reservoirs: Geological Society Special Publication 86, p. 141–158.

Ramm, M., A.W. Forsberg, and J.S. Jahren, this volume, Porosity-depth trends in deeply buried Upper Jurassic reservoirs in the Norwegian Central Graben: an example of porosity preservation beneath the normal economic basement by grain-coating microquartz, *in* J.A. Kupecz, J. Gluyas, and S. Bloch, eds., Reservoir quality prediction in sandstones and carbonates: AAPG Memoir 69, p. 177–199.

Robinson, A.G., and J.G. Gluyas, 1992, Model calculations of sandstone porosity loss due to compaction and quartz cementation: Marine and Petroleum Geology, v. 9, p. 319–323.

Selley, R.C., 1978, Porosity gradients in North Sea oil-bearing sandstones: Journal of the Geological Society of London, v. 135, p. 119–132.

Stewart, I.J., 1993, Structural controls on the Late Jurassic age shelf system, Ula trend, Norwegian North Sea, *in* J.R. Parker, ed., Petroleum geology of Northwest Europe: Barbican, London, Geological Society of London Proceedings of the 4th Conference, p. 469–484.

Cavallo, L.J., and R. Smosna, 1997, Predicting porosity distribution within oolitic tidal bars, *in* J.A. Kupecz, J. Gluyas, and S. Bloch, eds., Reservoir quality prediction in sandstones and carbonates: AAPG Memoir 69, p. 211–229.

Chapter 14

◆

Predicting Porosity Distribution Within Oolitic Tidal Bars

Larry J. Cavallo
Stonewall Gas Company
Jane Lew, West Virginia, U.S.A.

Richard Smosna
Department of Geology and Geography, West Virginia University
Morgantown, West Virginia, U.S.A.

◆

ABSTRACT

The Mississippian Greenbrier Limestone is a major gas reservoir in the Appalachian basin, but its complex porosity patterns often deter active exploration. In southern West Virginia, the reservoir consists of oolitic tidal bars that are composites of smaller shoals. Porosity trends closely follow the ooid-grainstone facies that occupied shoal crests where coarse-grained, well-sorted ooid sand was generated with either unidirectional or bidirectional cross-beds. Nonporous packstone occurred in adjacent tidal channels, and a transitional grainstone/packstone facies of marginal porosity was situated along the flanks of the shoals. The key to drilling successful wells is in understanding the complex internal geometry of Greenbrier ooid shoals. A well penetrating the oolite with good porosity and bimodal cross-beds should be offset perpendicular to the dip directions; that is, parallel to the shoal axis. However, a well penetrating thin, porous limestone with one dominant cross-bed azimuth should be offset opposite to that dip direction; that is, up the flank of the ooid shoal. Shaly interbeds characterize the edges of the shoals and mark the limit of productive wells. Schlumberger's Formation MicroScanner log, which provides data on both lithology and cross-bedding, has proven to be a useful tool in predicting the distribution of oolite porosity.

INTRODUCTION

Oolitic reservoirs in the Mississippian Greenbrier Limestone (Union Member; Figure 1) have historically produced significant quantities of natural gas across the central Appalachian basin. Exceptional wells in West Virginia, for example, will ultimately produce 1–2 billion ft^3 (3–6 × 107 m^3) and occasionally ≤9 billion ft^3 (27 × 107 m^3) of gas. Nevertheless, a complex geological setting with seemingly random porosity patterns, rapid facies changes, and involved diagenetic histories has deterred active exploration. Gas production varies widely, depending on pay thickness (15–97 ft; 4.6–30 m), porosity (3%–28%),

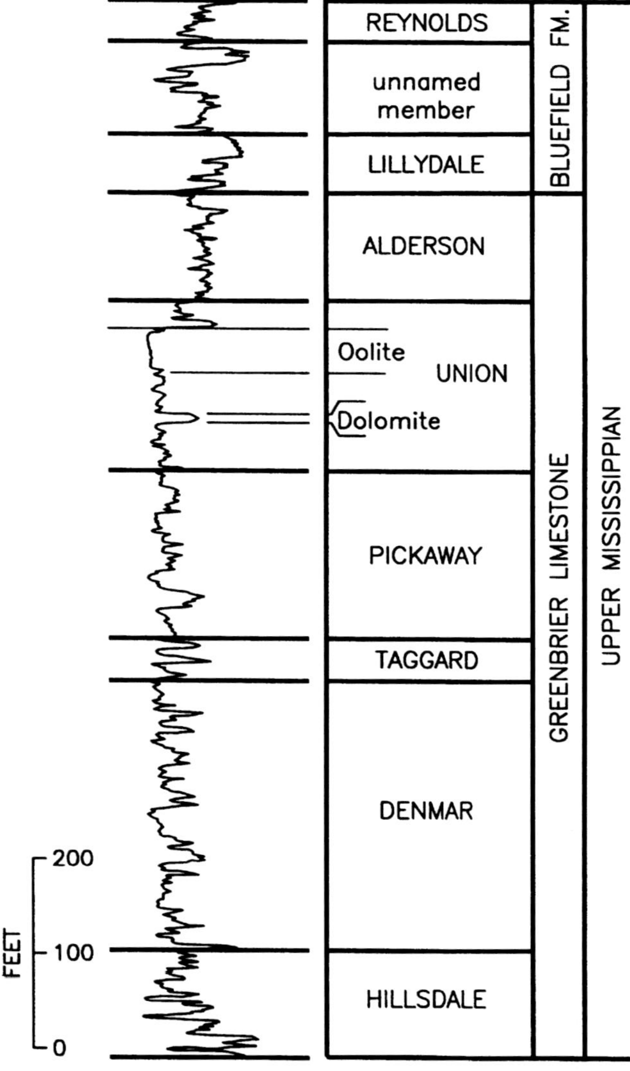

Figure 1. Stratigraphic members of the Greenbrier Limestone and lowest Bluefield Formation picked on the gamma-ray log of well 118, Mercer County, West Virginia.

and permeability (0.01–15 md). Well spacing in producing fields, although not restricted, averages 2400 ft (730 m), and typical treatment of the pay zone consists simply of acidization with 15% HCl.

Only recently have productive zones within the Union Member been fit into a regional depositional model (Kelleher and Smosna, 1993). Union oolites of southern West Virginia were deposited as a belt of tidal bars positioned along a northeast–trending hinge line in the Greenbrier gulf (Figure 2). Along this hinge line (or series of hinge lines), a shallow-marine shelf to the northwest dropped off into a somewhat deeper

basin to the southeast (Donaldson, 1974; Kelleher and Smosna, 1993). Strong tidal currents striking the shelf edge are thought to have generated north-west–trending oolitic bars and intervening channels, similar to those that border Tongue of the Ocean and Exuma Sound, Bahamas (Ball, 1967; Halley et al., 1983). Kelleher and Smosna (1993) delineated by isopach maps eight tidal bars in McDowell, Wyoming, Raleigh, and Mercer counties (Figure 3); four central bars were well defined by previous drilling, but the existence of the outer bars was at that time somewhat speculative. More recent drilling illustrates that the belt of tidal bars does in fact continue along trend to the northeast and southwest.

This study concentrates primarily on the drilling results of Stonewall Gas Company in the northeasternmost oolitic bar, named Blue Jay from the lease name of its discovery well (Figure 3). Ten successful wells (out of 13) have been drilled on the southern terminus of that tidal bar. The geometry and makeup of the Blue Jay bar, however, is more complicated than originally believed. Our purpose in this chapter is to refine Kelleher and Smosna's (1993) model so that it accurately reflects the highly variable nature of porosity development within the reservoir. In this way, porosity trends can be better predicted, leading to new discoveries and an effective exploitation of these oolitic bars. To test our refined model and illustrate its usefulness, we apply concepts developed for the Blue Jay bar to another bar where drilling is ongoing. At present, eight wells have been drilled on the southern terminus of the Poca Land bar, situated 10 km to the west (Figure 3), and an additional five wells are scheduled for the near future. Locations for new wells in the Poca Land bar will be based on geological predictions from the more detailed depositional model.

Our analyses make full use of Schlumberger's Formation MicroScanner (FMS) log, a relatively new technology that accurately measures minute differences in rock resistivity (Serra, 1989). Computer processing of FMS resistivities produces a color image of the inside of the well bore (resembling a core photograph); darker hues represent more conductive elements or beds, and lighter hues more resistive elements or beds. More than a simple dipmeter, the FMS tool allows continuous observation of detailed lateral and vertical changes in rock properties. From the shapes and patterns revealed on the FMS image, an experienced geologist can interpret the rock texture (in this study, oolites, shale partings, mottling, and interbedding of lithologies), sedimentary structures (cross-beds, bed thickness, the nature of bedding contacts, and stylolites), and structural features (regional dip, fractures). Lithologic identifications based on FMS logs are confirmed by petrographic analysis of 12 sidewall cores recovered from the Union Member.

GEOLOGICAL SETTING

The Mississippian Greenbrier Limestone accumulated during a major transgression of an epeiric sea into the Appalachian foreland basin. Deposition occurred in a broad gulf that extended across parts of six Appalachian states (Figure 2). The gulf was bordered by

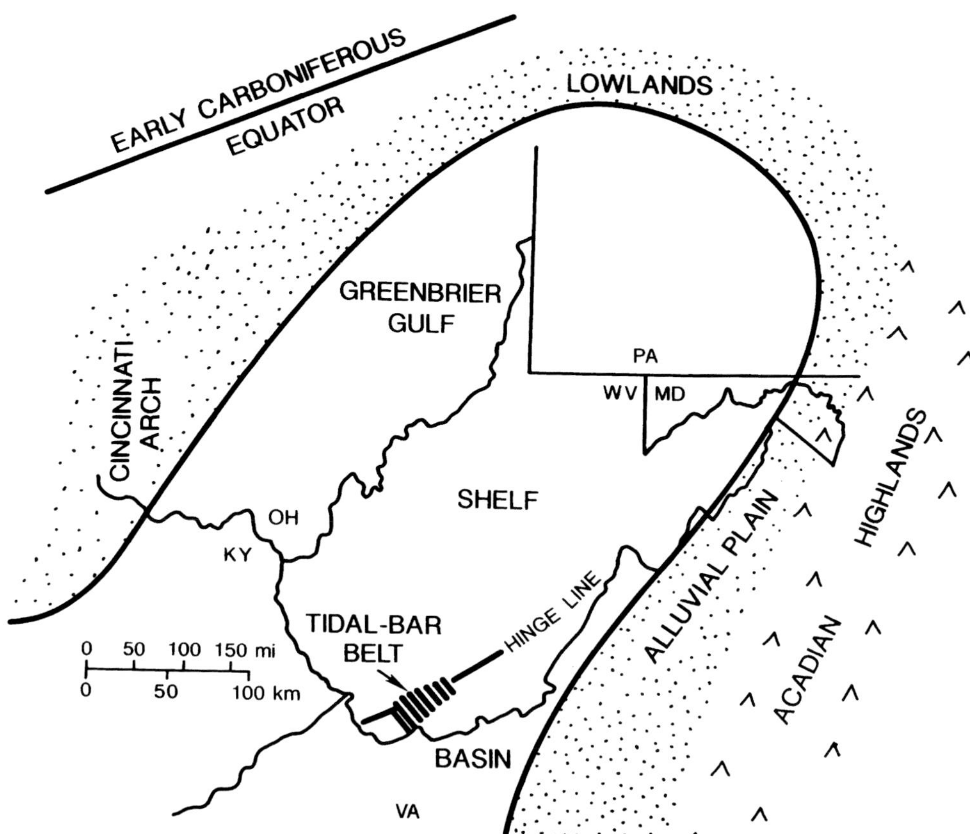

Figure 2. Paleogeography of the Greenbrier gulf in the central Appalachian basin. A belt of tidal bars in the Union oolite member was positioned along a hinge line that separated the basin on the south from the broad shelf to the north.

the Acadian highlands to the east, exposed lowlands to the north, and the Cincinnati Arch to the west (deWitt and McGrew, 1979). Northern lowlands shed small volumes of terrigenous sediment into the basin as paralic sandstones that interfinger with sandy marine limestones, while minor red shales represent an eastern alluvial plain that rimmed the Acadian highlands (Adams, 1970; Brezinski, 1989; Carney and Smosna, 1989; Smosna and Koehler, 1993). Paleomagnetic data place the basin ~10° south of the Equator (Scotese, 1984). The climate may have been fairly arid due to a rain-shadow effect behind the Acadian highlands, leading to reduced runoff and a low level of terrigenous input from surrounding landmasses (Cecil, 1990).

During rapid basin subsidence, cherty skeletal wackestones, packstones, and mudstones accumulated in relatively deep-water environments of the southeastern basin (basal Hillsdale and Denmar members; Figure 1). Red silty shales (overlying Taggard Member) mark a brief progradation of the eastern alluvial plain. Subsidence then slowed, the sea transgressed to its greatest limits, and ooid grainstones formed on the shallow shelf (Pickaway and Union members). These latter oolites are the focus of our study. Finally, an increase in terrigenous sediments (Alderson Member) signifies the close of carbonate deposition. The overlying Mauch Chunk Group (including the Lillydale Shale and Reynolds Limestone members of the Bluefield Formation) consists of marine and fluvial-deltaic shales and sandstones with only minor limestones.

A number of Mississippian structural/stratigraphic hinge lines have been identified across West Virginia and Kentucky on the basis of regional and local isopach mapping (Flowers, 1956; Donaldson, 1974; MacQuown and Pear, 1983; Carney and Smosna, 1989; Kelleher and Smosna, 1993). In each case, the formation thickens markedly over a short distance. In the area of this study, for instance, the Greenbrier Limestone thickens to the southeast at a rate of 6.5 m/km north of the hinge line and 8.9 m/km south of the line. The formation attains its maximum thickness of ~900 m in neighboring Virginia, and it is thought that basin subsidence may have taken place along deeply seated normal faults beneath these down-to-the-south hinge lines (Donaldson, 1974; MacQuown and Pear, 1983). Rapid deposition was able to keep pace with the differential subsidence, so even the basin center remained relatively shallow.

COMPOSITE BARS

A stratigraphic cross section ~3 km long has been constructed along the axis of the Blue Jay bar using gamma-ray and bulk-density well logs (Figure 4). The stratigraphic interval extends from the Reynolds Limestone Member, a marker bed across the entire state, down to a thin dolomite near the middle of the Union Member that serves as a local datum. Thin and shaly in the south (well 80), the Union oolite becomes thicker and less argillaceous to the north (well 145). Furthermore, three distinct units within the oolite can

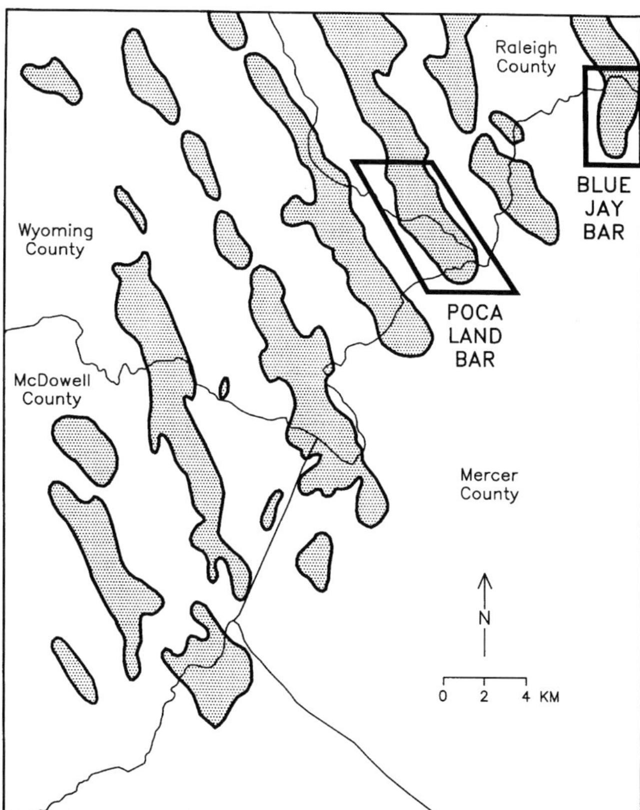

Figure 3. Outline of north-west–trending tidal bars and intervening channels in the Union Member, as marked by the 9-m isopach contours (from Kelleher and Smosna, 1993). The two detailed study areas of this chapter include the southern ends of the Blue Jay bar in Mercer County and the Poca Land bar in Wyoming and Raleigh counties.

be identified on gamma-ray well logs in the south, although they merge northward into a single blocky signature. Interbeds of shale also occur on the eastern and western margins of the Blue Jay bar. Presence or absence of shale interbeds thus provides some indication of a well's position on the tidal bar.

The isopach map of the Union oolite, ranging in thickness from 9 m on the bar's eastern edge to 19 m along its axis, delineates a north-south orientation (Figure 5). The bar is shown to terminate south of the existing wells for two reasons. First, the gamma-ray log for well 80 (Figure 4) shows the oolite to be rather shaly and to have lost its blocky signature. It is predicted, therefore, that its thickness decreases markedly toward the south. Second, this southern termination of the bar closely coincides with the southern termination of other bars to the west (Figure 3) (Kelleher and Smosna, 1993). A line connecting these southern terminations parallels the trend of the tidal-bar belt and the postulated hinge line; presumably, water depth was too great and tidal currents too weak for substantial ooid formation basinward of this line.

Quite apparent on the cross section of Figure 4 is the variable nature of porosity development along the bar,

where porosities ≥6% (density <2.60 g/cm^3) are noted on the bulk-density logs. Porosity is present in the lowest unit of wells 80 and 151 only, in the lower two units of well 159, in the upper two units of well 133, in the upper unit of well 145 only, and in all three units of well 118.

An isopach map of the lower unit is depicted in Figure 6A. Shaly deflections in the gamma-ray well logs determine the base and top of this unit, although the top becomes more difficult to distinguish in the central area of the bar. Varying in thickness from 2 to 8 m, the unit appears as two laterally linked shoals. The shoals are ~1500 m wide and 3200 m long; they are of equal size and shape, both reveal a parallel northwest-southeast orientation, and a thin intershoal area separates the two. This pattern may continue along the entire length of the Blue Jay bar, perhaps 30 km to the north.

An isopach map of the middle unit indicates a thickness range between 3 and 9 m (Figure 6B). Two shoals are similar to those of the lower unit in terms of size, shape, orientation, and intershoal thin. Shoals of the middle unit are situated immediately above the intershoal areas of the lower unit, and shoals of the lower unit are capped by the intershoal area of the middle unit. The isopach map of the upper unit (Figure 6C) illustrates a single curvilinear shoal. This shoal is 1200 m wide, 3600 m long, and 2–7 m thick. A smaller shoal of probable limited extent and based on only one well may be located to the east. The curvilinear shoal lies directly above the combined crests of shoals identified in the lower and middle units.

Figure 7 shows porosity-isopach maps constructed for the lower, middle, and upper units. These maps depict stratigraphic thicknesses with a porosity ≥6%. Trends in porosity-thickness correspond closely to the total thickness of each unit (Figures 6, 7). Porosity in Union grainstones consists of intercrystalline micropores between calcitic microrhombs that make up the individual ooids; primary interparticle porosity has been occluded by calcite cement. Kelleher and Smosna (1993) suggested that growth of the bars above sea level or periodic lowstands enabled meteoric water to enter and diagenetically alter the ooids to calcitic microrhombs, thereby creating secondary microporosity within the grains. Comparable interpretations for the recrystallization of porous ooids in other limestone reservoirs have been offered by Keith and Pittman (1983) and Ahr (1989). Close agreement between stratigraphic thickness of the Union oolitic units and their porosity thickness supports this hypothesis: crests of the shoals where sedimentation was greatest would have stood higher and been exposed longer to the infiltration of meteoric water and would now possess the greatest porosity.

LITHOFACIES

The FMS image logs provide lithologic information for the Blue Jay bar. On the basis of rock texture (in particular, oolites, shale partings, mottling, and interbedding of lithologies) and sedimentary structures (cross-beds, thickness of bedding or bed sets, and

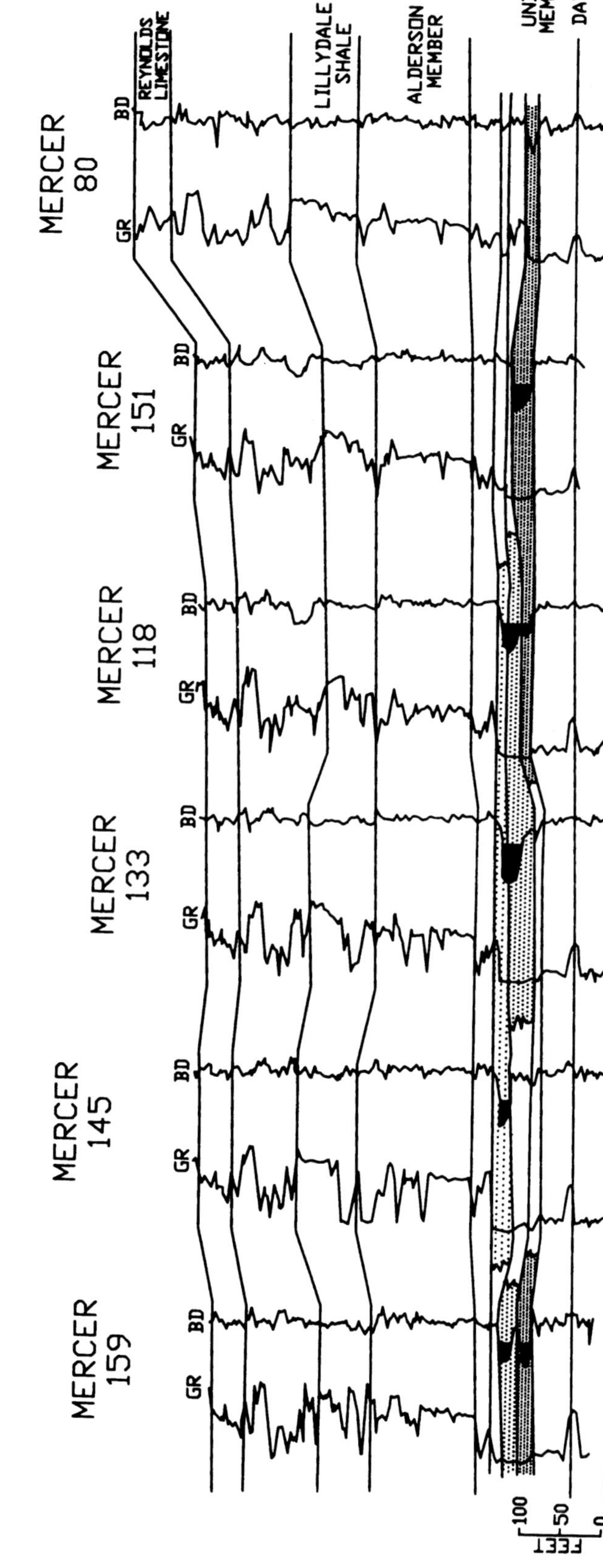

Figure 4. Cross section AA' along the axis of the Blue Jay tidal bar. Gamma-ray (GR) and bulk-density (BD) well logs show three distinct porous units within the oolite of the Union Member, and stratigraphic intervals with >6% porosity are marked in black on the bulk-density logs. The lowest unit (dark stippling) is present in wells 159, 118, 151, and 80; the middle unit (medium stippling) in wells 159, 133, and 118; and the upper unit (light stippling) in wells 145, 133, and 118. Line of section is marked on Figure 5.

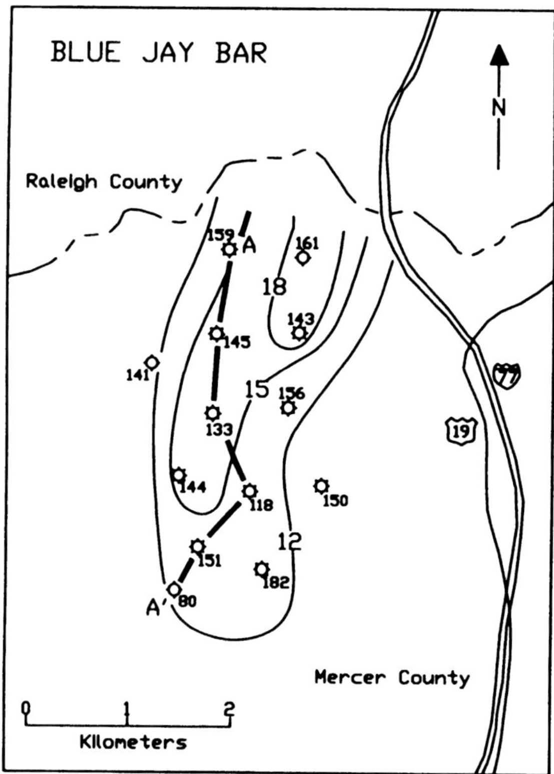

Figure 5. Isopach map of the Union oolite delineates a north-south orientation for the southern termination of the Blue Jay tidal bar. Contour interval equals 3 m.

reactivation surfaces), three lithofacies have been defined: cross-bedded ooid grainstone, bioturbated packstone, and a transitional grainstone/packstone.

Cross-Bedded Ooid Grainstone

Evident on FMS images for the lower unit of well 151 (Figure 8) is a grainy, pockmarked texture, a characteristic feature of this lithofacies. Although resolution of the FMS tool is not fine enough to observe individual ooids, the range of resistivities present among the ooids accounts for the diagnostic graininess. Equally characteristic of the facies is an abundance of well-defined cross-bed sets. Images reveal these sets as a series of stacked sinusoidal curves with the same dip azimuth and magnitude. Dip angles range from 2° to 29°, averaging 16° (regional structural dip is <2°). Frequently associated with the cross-beds are reactivation surfaces, relatively flat-lying surfaces that separate cross-bed sets (Klein, 1977; Smosna and Koehler, 1993). Cross-beds above and below these erosional surfaces often display vastly different dip azimuths and magnitudes. These bed sets, reaching a maximum thickness of 60 to 70 cm, represent sand waves or megaripples that migrated across the surface of the ooid shoals.

The amount of shale partings is low in the ooid-grainstone facies. Shale, less resistive than surrounding limestone, appears darker on the images. Where present, shales are thinner than 5 cm. They are usually at the top or bottom of the unit and outside the porous pay zones.

Six sidewall cores recovered from the Union oolite in well 145 (Figure 5) are of this lithofacies. Thin-section analysis establishes these rocks as ooid grainstones. Ooids account for >80% of the total grains and have a mean grain size of 0.65 mm (coarse sand). Most ooids possess a thick cortex around a nucleus of fossils, peloids, or small intraclasts, although several of the largest ooids have a relatively thin coating. These six sidewall cores have a mean porosity of 9.3% and mean permeability of 0.145 md.

Bioturbated Packstone

This lithofacies displays a mottled texture on the FMS logs (Figure 9), a texture vastly different from that of the ooid grainstone. Unfortunately, no samples have been retrieved for thin-section analysis. A packstone lithology, however, is considered most likely, because in modern carbonate settings, channels between oolitic tidal bars are floored by burrowed, muddy pelletal sand (Ball, 1967; Harris, 1979). In an analogous manner, mottled patches with varying resistivity resulted from churning and mixing of the Greenbrier sediment by burrowing infauna.

Cross-bedding and other sedimentary structures are not observed in FMS images. Instead, drilling-induced fractures are a diagnostic feature of this lithofacies, an indication of the packstone's lower physical strength. Where not extensively bioturbated, bedding is on the order of a few centimeters. Stylolites are a common feature, appearing on the image as thin, highly irregular traces of less-resistive minerals. Also, the number of interbedded shales is relatively high. Rocks of this facies have no appreciable porosity and are not of reservoir quality.

Transitional Grainstone/Packstone

The third lithofacies is a combination of the previous two, containing characteristics of both the cross-bedded ooid grainstone and the bioturbated packstone (Figure 10), but subtle differences justify designating these rocks as a separate facies. Its grainy texture is less extensive than that of the ooid grainstone, due to significant interbeds of mottled packstone and shale. Cross-beds are common but not pervasive, and seldom are they truncated by reactivation surfaces. Drilling-induced fractures are absent in the burrowed-mottled sections of this facies. Six sidewall cores from well 145 show the rock type to be an ooid grainstone. In contrast to ooids of the pure grainstone lithofacies, these have a finer grain size (0.45 mm, or medium sand) and relatively thin coatings. Moreover, ooids account for only 50%–80% of the total grains; uncoated fossils, peloids, and intraclasts are abundant.

Porosity in the transitional facies is typically <4% and of marginal reservoir quality. Cored samples have a mean porosity of 2.7% and permeability of <0.01 md.

Lithofacies Distribution

Figure 11 depicts the geographical extent of the three Blue Jay lithofacies during deposition of the

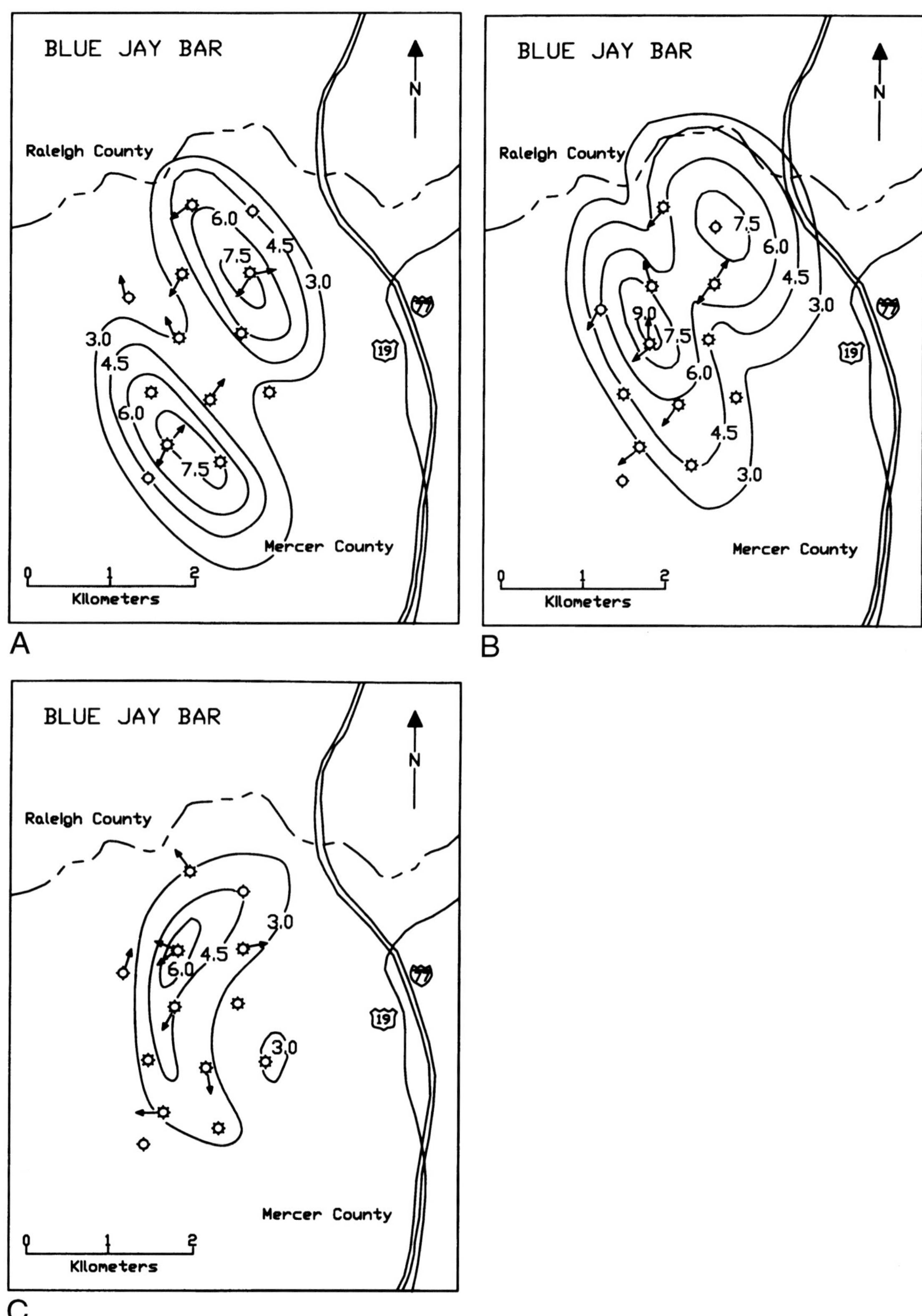

Figure 6. Isopach maps of the lower unit (A), middle unit (B), and upper unit (C) illustrate the individual ooid shoals that make up the Blue Jay tidal bar. Arrows indicate main dip azimuths of cross beds (regional structural dip is negligible). Contour interval equals 1.5 m.

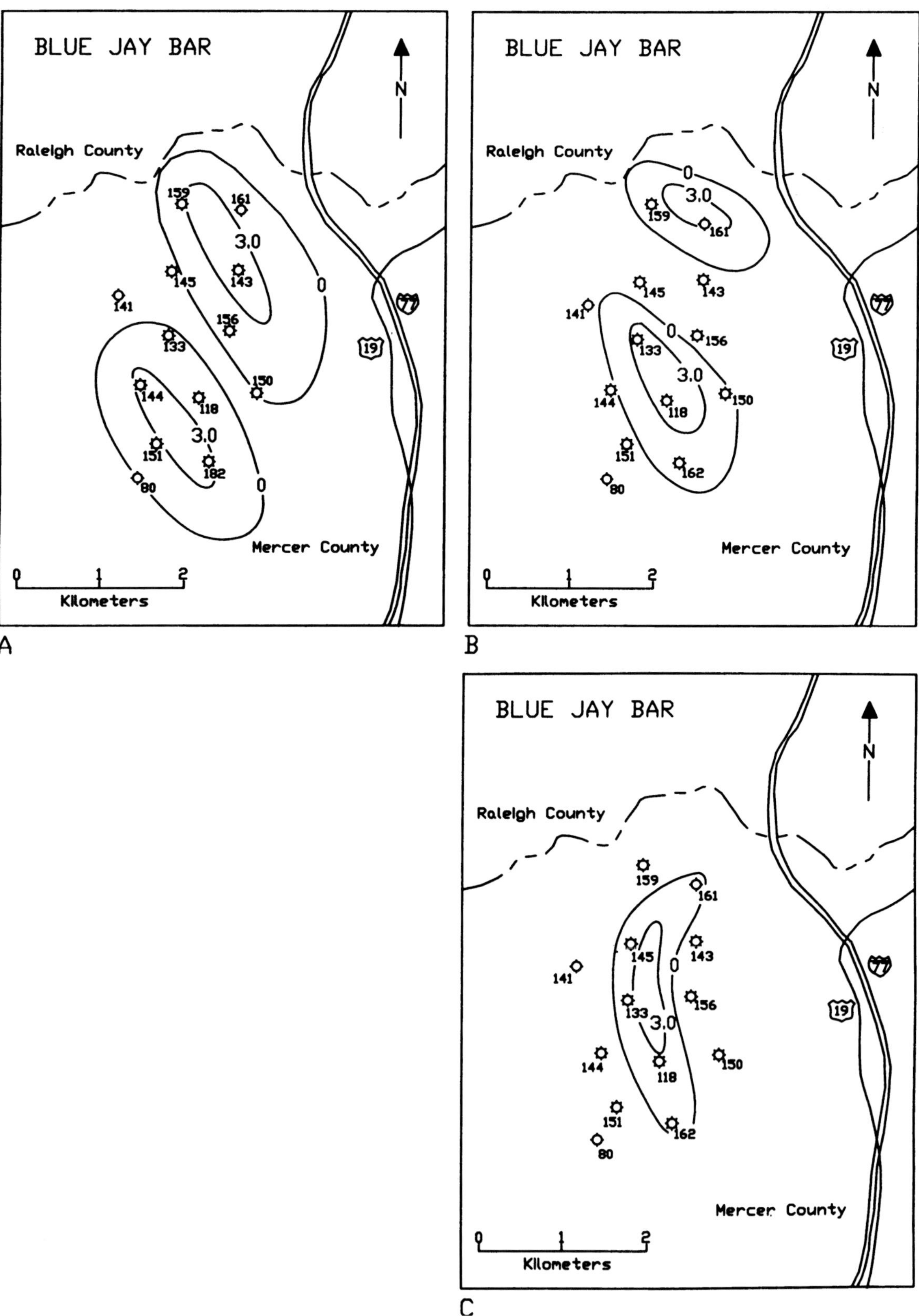

Figure 7. Porosity-isopach maps (stratigraphic thickness with ≥6% porosity) of the lower unit (A), middle unit (B), and upper unit (C) correspond closely to the total thickness of each unit. Compare with Figure 6. Contour interval equals 3 m.

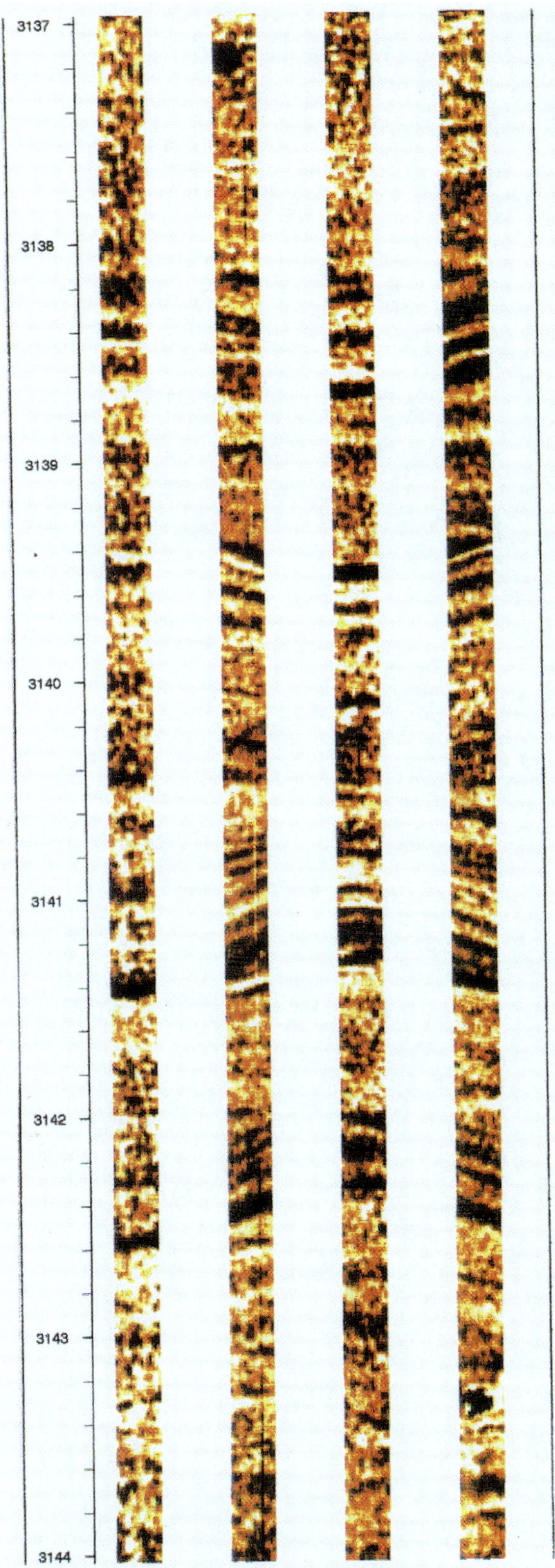

Figure 8. An FMS image log of an ooid grainstone, demonstrating its grainy texture as well as cross-bed sets (stacked sinusoidal curves with the same dip azimuth and magnitude). Reactivation surfaces at depths 3139.4, 3140.5, and 3142.0 ft abruptly terminate the cross bedding. Lower unit of Union oolite, well 151, Mercer County.

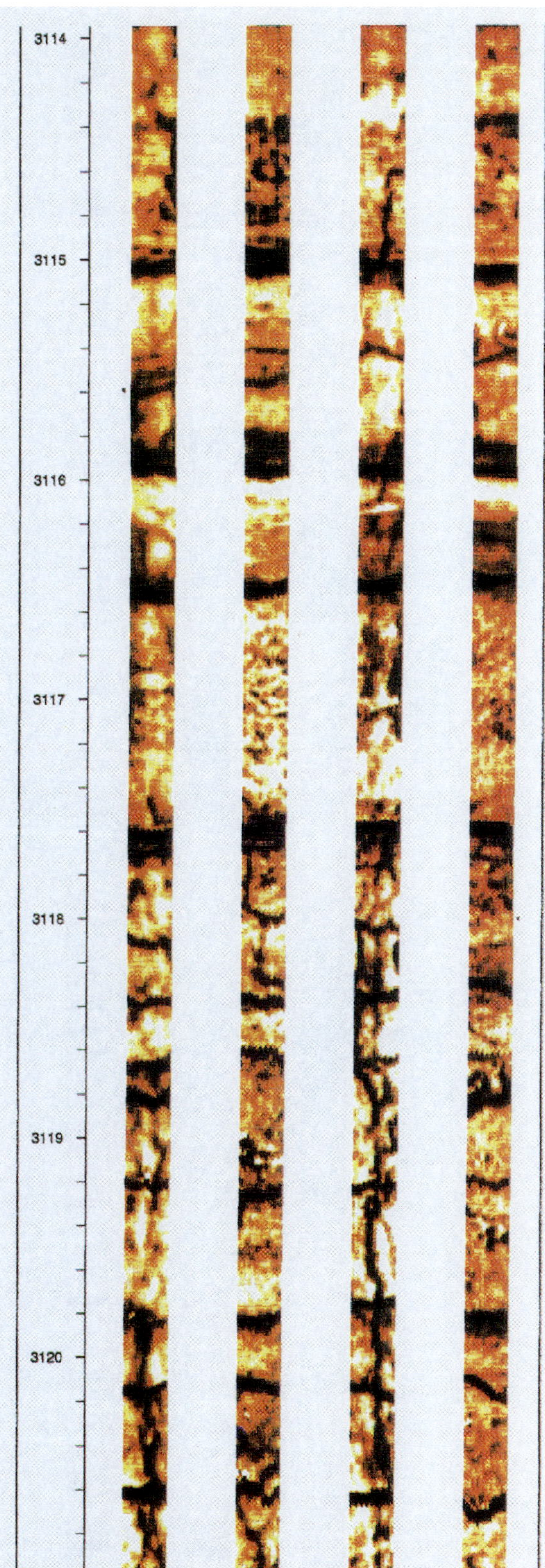

Figure 9. An FMS image log of a bioturbated packstone, demonstrating its mottled texture as well as shale interbeds (thick dark horizontal bands), stylolites (thin dark horizontal bands at depths 3115.5, 3118.1, 3120.2, and 3120.6 ft), and vertical fractures. Upper unit of Union oolite, well 151, Mercer County.

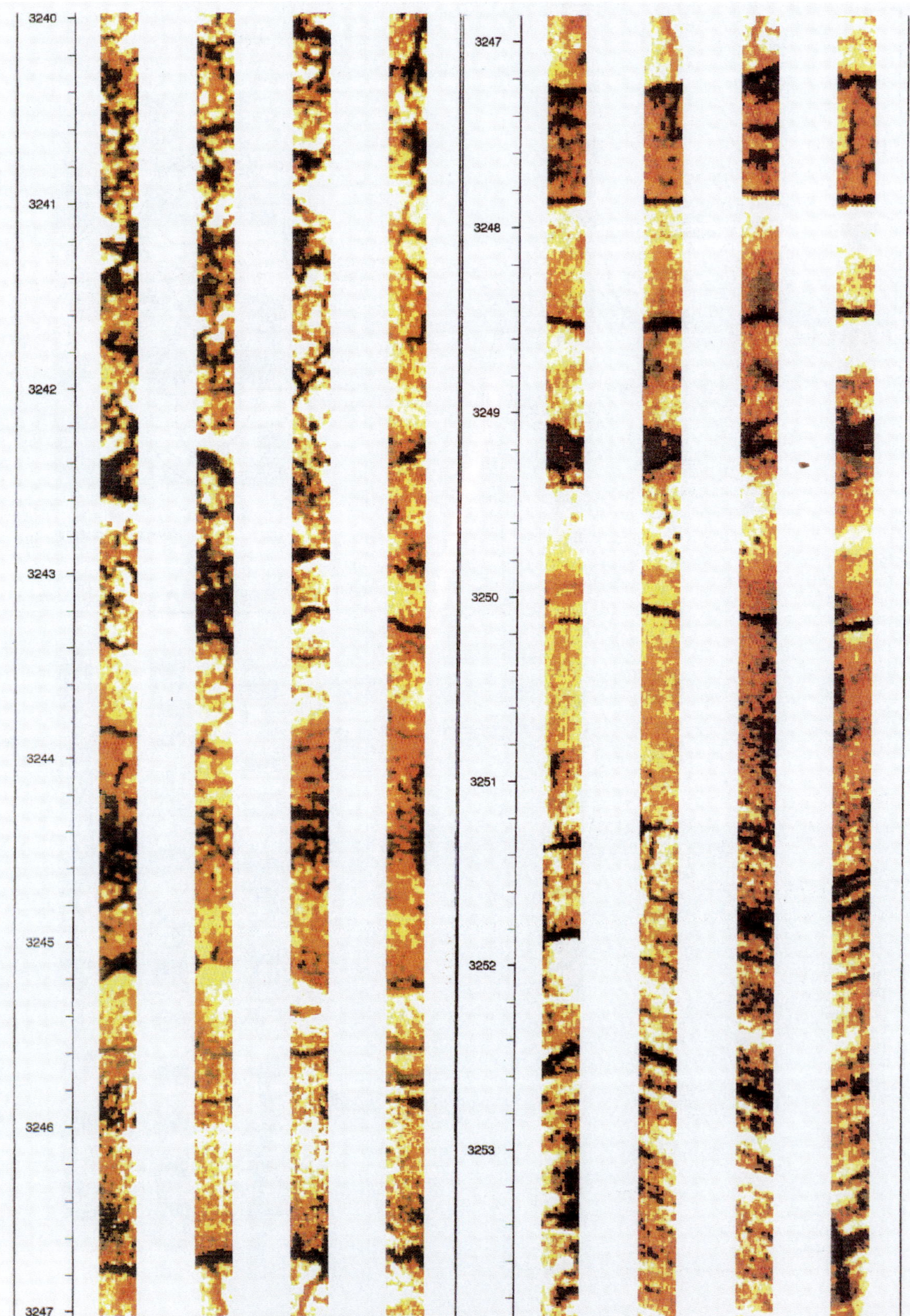

Figure 10. An FMS image log of the transitional grainstone/packstone, demonstrating a mottled texture [depths 3240–3243 ft and 3247–3249 ft] and a grainy texture [depths 3243–3247 ft and 3249–3254 ft]. Middle unit of Union oolite, well 143, Mercer County.

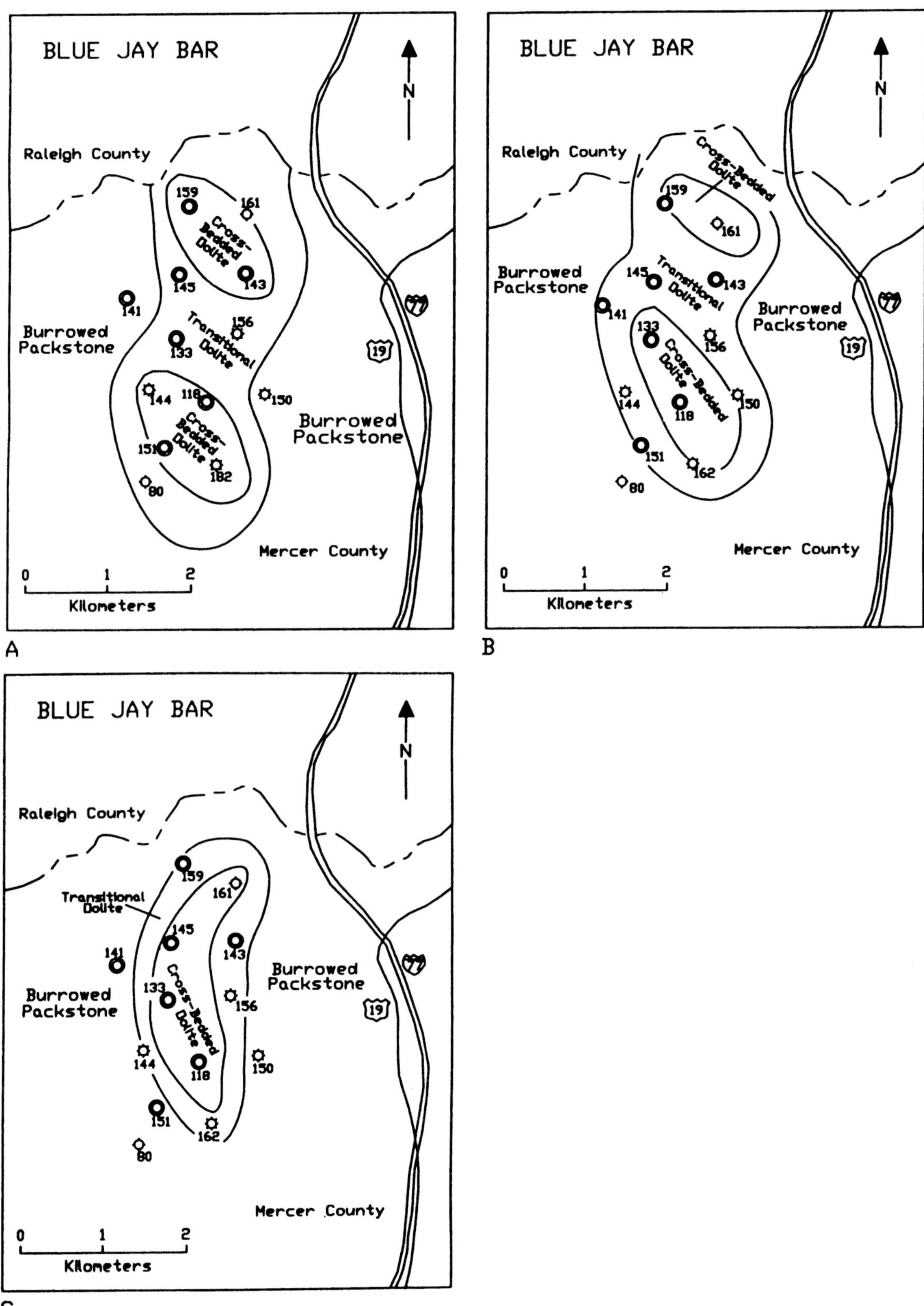

Figure 11. Facies maps of lower unit (A), middle unit (B), and upper unit (C) of the Union oolite at Blue Jay bar. Each shoal consists of a central cross-bedded oolite-grainstone facies surrounded by a transitional grainstone/packstone facies. Burrowed packstone occupies the adjacent tidal channels. Circled wells are those with FMS logs.

lower, middle, and upper units. Comparisons of these maps to unit isopach maps (Figure 6) and porosity-isopach maps (Figure 7) illustrate a clear relationship among lithofacies, shoal thickness, and porosity.

Cross-bedded ooid grainstone occurs where unit thickness and porosity are greatest, corresponding to the crest of shoals atop the Blue Jay tidal bar. The well-washed, well-sorted, coarse-grained ooids represent a high-energy setting in which water depth did not exceed 2 m (Newell et al., 1960; Ball, 1967; Harris, 1979). The thickness of cross-bed sets generally increases upward within each stratigraphic unit. Bed sets are thickest (maximum ~70 cm) at the base, reflecting the greatest water depth, and decrease to ~20 cm near the top as sediments aggraded into shallower water (Ball, 1967). At the very top of the unit, bedding again thickens to ~60 cm; perhaps these thicker cross-bed sets represent beach ridges or dunes that capped the shallowing-upward sequence (Halley et al., 1983).

Bioturbated packstone is positioned on the lower flanks of the shoals and in adjacent tidal channels, where the unit is thin and nonporous. Water depth exceeded 4–5 m, and the energy level was drastically lower.

Transitional grainstone/packstone is situated around and between ooid shoals, and inside the boundary of the tidal bar. In this position, tidal-current intensity fluctuated between the high energy of ooid grainstone (shoal crest) and the low energy of bioturbated packstone (adjacent channel), and the two end-member lithologies became interbedded. The greater number of uncoated grains, finer grain size, thinner oolitic coating, and shale/packstone interbeds collectively suggest a shoal environment of slightly deeper water and less favorable for ooid formation. On Andros Island and Joulters Cays, in the Bahamas, modern and Pleistocene ooid shoals can similarly be divided into two subenvironments: a central, very oolitic area surrounded by an area of lower ooid concentration (Harris, 1979; Boardman et al., 1993). In the Mississippian Ste. Genevieve Limestone of Indiana, Zuppann (1993) described a very oolitic lithofacies in the central portion of an ooid shoal, with a decreasing percent of ooids down the flanks.

PALEOCURRENT ANALYSIS

The FMS image logs also depict cross-bed dip directions for the Union oolite as a whole and for each of the three units. These data, displayed by arrows on the several isopach maps, are then used to infer paleocurrent directions at the time of deposition. Throughout the oolite, cross-beds have a consistent bimodal dip azimuth: northeast-southwest and separated by ~180°. This bimodal pattern confirms a tidal influence during construction of the bar (Klein, 1977; Smosna and Koehler, 1993).

Cross-bed dip directions in the unit subdivisions (Figure 6) define three scenarios: (1) wells drilled on or near the shoal axis exhibit bimodal cross-beds with 180° separation of dips, (2) wells on the flanks of the ooid shoals exhibit cross-beds that generally dip away from the shoal axis, and (3) wells located outside a shoal exhibit a dip pattern that may or may not indicate the position of any nearby shoal.

Figure 12, a schematic cross section, portrays the three scenarios and their relationship to shoal thickness. Recognition of these relationships becomes vital when using the FMS image log as a predictive tool to aid in the placement of offset wells. A well penetrating the Union oolite with good porosity and 180° bimodal cross-dips should be offset perpendicular to the cross-bed dip directions; that is, parallel to the shoal axis. A well penetrating a thin, porous limestone section with one dominant cross-bed dip direction should be offset in the direction opposite to that dip direction; that is, up the flank of the ooid shoal. And a well penetrating limestone with no porosity (that is, outside the shoal) may exhibit a dip pattern unrelated to the shoal's position.

SEDIMENTARY MODEL

The origin of Union tidal bars can be explained following Ball's (1967) interpretation of modern bars in the Bahamas. Greenbrier tidal currents may have repeatedly swept over the hinge line separating the deeper basin to the southeast from the shallow shelf to the northwest. Ocean water was perhaps rapidly displaced with each tidal change, and the interface between shelfal and basinal water masses became unstable. Splitting into a number of equally spaced digits, these fingerlike flows were then responsible for the oolite-belt geometry (Ball, 1967). In shallow-marine environments with strong reversing tidal currents, sand accumulates in zones of shear where, for part of the daily cycle, adjacent currents flow in opposite directions (Swift and Niedoroda, 1985). Union bars probably began their development in such a manner, forming in areas between the digitate currents. Ooid sand deposited in the shear zones retarded the tidal flows, resulting in additional sand deposition, which further retarded the tidal flows. Due to this continuing feedback between sea-floor topography and diminishing current strength, the bars grew upward through time (Swift and Niedoroda, 1985). These processes ultimately produced a broad belt of alternating bars and channels. The bars have a somewhat sinuous axis, and the southern termination of the Blue Jay bar, in particular, trends north-south (Figures 3, 5). Tidal currents here are inferred to have flowed north-south through the intervening channels.

Flow over the Blue Jay bar itself was, however, oblique to its axis (that is, northeast-southwest), as indicated by the position of individual shoals (Figure 13). This oblique flow direction resulted from refraction of tidal currents where they approached the bar: the currents turned through 45° as they passed over the flank of the bar and upward toward the crest. The ooid shoals, the most active portions of the tidal bar, were aligned crosswise to this flow. Furthermore, the currents must have accelerated as they moved up the shoals, due to the rapidly decreasing cross-sectional area of the water column (Swift, 1985). Flow along the shoal crest was

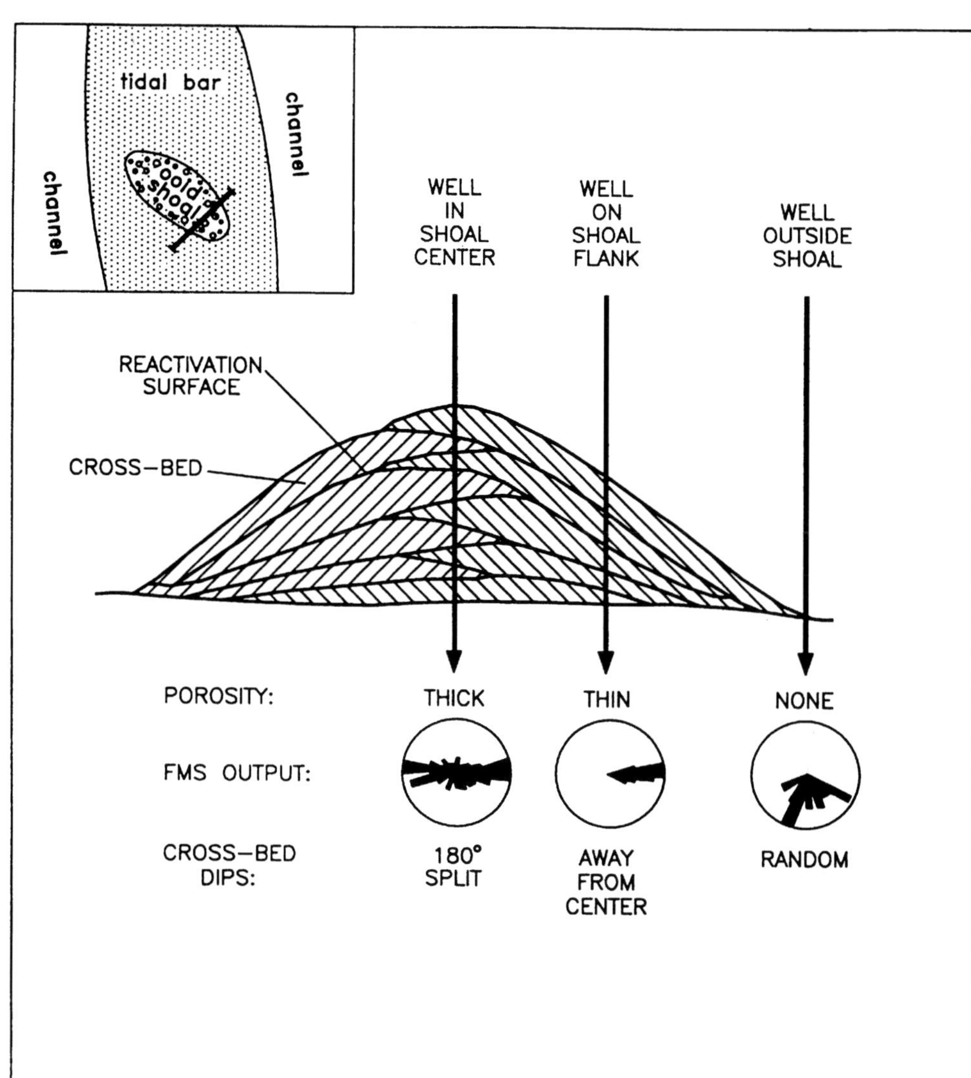

Figure 12. Schematic cross section of an ooid shoal, indicating internal sedimentary structures, thickness of pay zone, and FMS cross-bed dip directions. Inset depicts position of this cross section relative to shoal's axis.

reduced by friction in shallow water, and when passing over the crest, tidal flow decelerated further, where the water column expanded on the downcurrent flank. The orientation of individual shoals reveals that flood currents dominated the western margin of the Blue Jay bar, and ebb currents, its eastern margin.

At the beginning of a tidal cycle, ooids were eroded from the upcurrent (southwestern) flanks of the shoals by accelerating, north-directed flood currents (Figure 13). Erosion and entrainment of sand produced the numerous reactivation surfaces observed on FMS image logs. In contrast, ooids were deposited on the crest and downcurrent (northeastern) flank from the decelerating flood currents. Sand here accumulated in the form of sand waves or megaripples, with cross-beds that dip northeastward. Halley et al. (1983) documented the presence of sand waves, similarly oriented 45° to the long axis, that adorn the crests of modern tidal bars at Tongue of the Ocean, Bahamas. With a reversal of the tide, south-directed ebb currents—refracted 45° to a more southwesterly direction—eroded ooids from the (now) upcurrent northeastern

flank, producing reactivation surfaces. Grains were transported to the crest and to the southwestern downcurrent flank, and deposited in cross-beds dipping to the south-southwest.

Frequently reversing tides were therefore responsible for the characteristic dip patterns of the FMS logs. Cross-beds on the crest exhibit a bidirectional dip with 180° separation; on the flanks, cross-beds are unidirectional and dip away from the shoal's crest. As an example, the FMS image in Figure 8 depicts typical sedimentary structures that formed near the axis of an ooid shoal. This 7-ft (2-m) stratigraphic interval occurs within the lower oolite unit of well 151, Mercer County (see Figure 11A for location). A set of cross-beds may be observed at depth 3140.5–3141.4 ft, with a dip direction of N35E produced by the flood current. The set has been truncated above by a reactivation surface (depth 3140.5 ft), in turn, is overlain by another set of cross-beds (depth 3139.4–3140.5 ft) S25W and produced by the ebb current. Other sets of northeast- and southwest-dipping cross-beds, as well as additional reactivation surfaces, are visible in Figure 8.

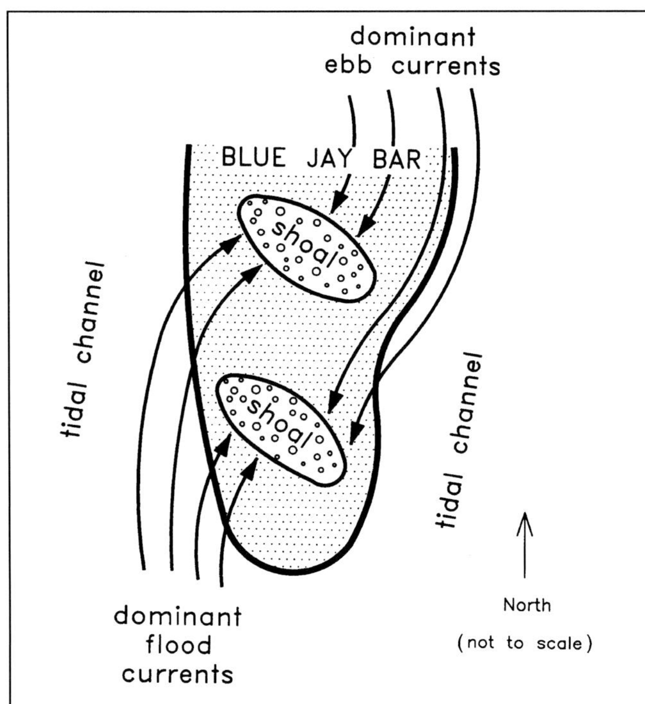

Figure 13. Interpreted regime of dominant tidal currents responsible for constructing ooid shoals of the Blue Jay bar.

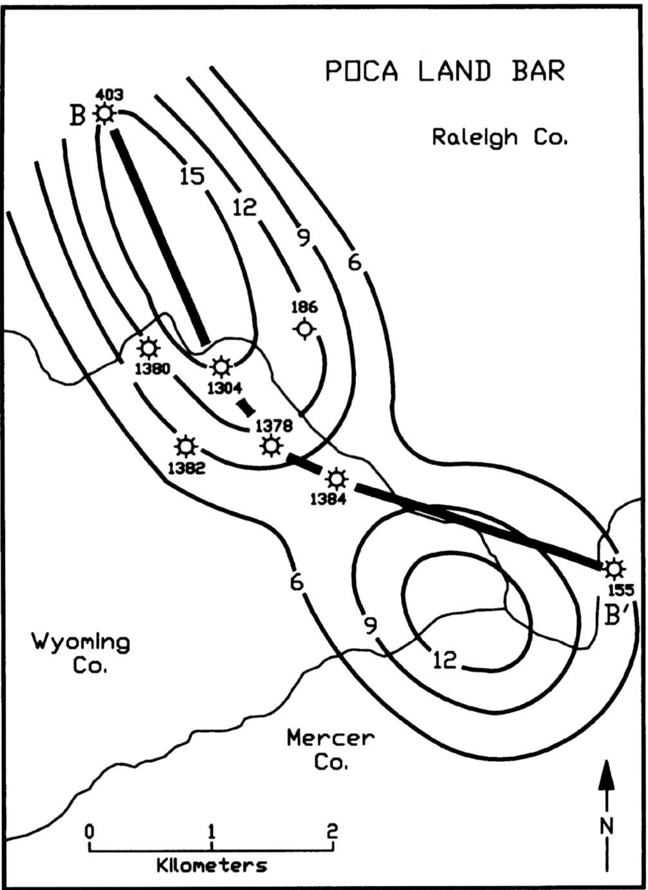

Figure 14. Isopach map of the total Union oolite delineates a northwest-southeast orientation for the Poca Land tidal bar. Contour interval equals 3 m. Due to mechanical problems, well 186 was never completed.

The Blue Jay bar is not a single, homogeneous deposit; it is composed of three vertically stacked shoals. These shoals have an average thickness of 8 m, width of 1.5 km, length of 3.5 km, and spacing of 2.1 km. Similar to composite bars described by Evans (1970), Klein (1977), and Zuppann (1993), we assume that migrating ooid shoals of the Union Member merged through time into the larger bars. Each generation of shoals, however, is offset somewhat from those beneath (Figures 6, 11). Those of the lower unit built upward to sea level by rapid sand deposition, at which time sea-floor topography caused a lateral shift in the tidal currents. Sites of ooid deposition moved to the intershoal areas. Shoals of the middle unit then grew upward to sea level, forming a continuous bar composed of two generations of smaller shoals. With a slight relative rise of sea level, additional accommodation space was created for the shoal in the upper unit. This upper shoal lies across the crests of those underlying it, demonstrating that its position was governed by remnant topography of the composite lower/middle bar. In addition, the curvilinear extension of the upper shoal toward the northeast (Figure 6C) resembles the Holocene spitlike buildup that evolved around the open northern margin of Joulters ooid shoal, where ooids are swept along by longshore transport (Harris, 1979; Boardman et al., 1993). At the close of Union time, a major sea level rise brought ooid deposition to an end across the region (Carney, 1993).

POROSITY PREDICTIONS

The Poca Land study area lies two tidal bars away from Blue Jay. Development drilling using FMS image logs has just begun, and a bit of artistic license or geological intuition enters into the construction of the several maps. However, the refined depositional model developed for the Blue Jay bar serves as a valuable guide in predicting the distribution of porosity at Poca Land and in locating additional wells.

The Union oolite thins to the southeast, reflecting a southward termination of the tidal bar (Figure 14). Shale interbeds in the oolite of the southernmost four wells suggest that they are situated near the bar's margin. Well 403, in contrast, centrally located along the axis, exhibits a blocky gamma-ray signature. Termination of the Poca Land bar near well 155 matches the southern termination of other bars in the Greenbrier tidal-bar belt (Figure 3). As discussed above, water depth north of this line proved ideal for the formation of ooids, but was presumably too great south of the line. Thickness variations along the bar axis (~15 m in wells 403 and 1304, 6.7 m in well 1384, and 7.0 m in well 155) result from differential stacking of the constituent ooid shoals.

Gamma-ray and bulk-density well logs allow the oolite to be subdivided into three informal units (Figure 15). An isopach map of the lower unit (Figure 16A) depicts two laterally linked, north-south shoals of equal shape and size (2–8 m thick, 1.6 km wide, and 2.4 km

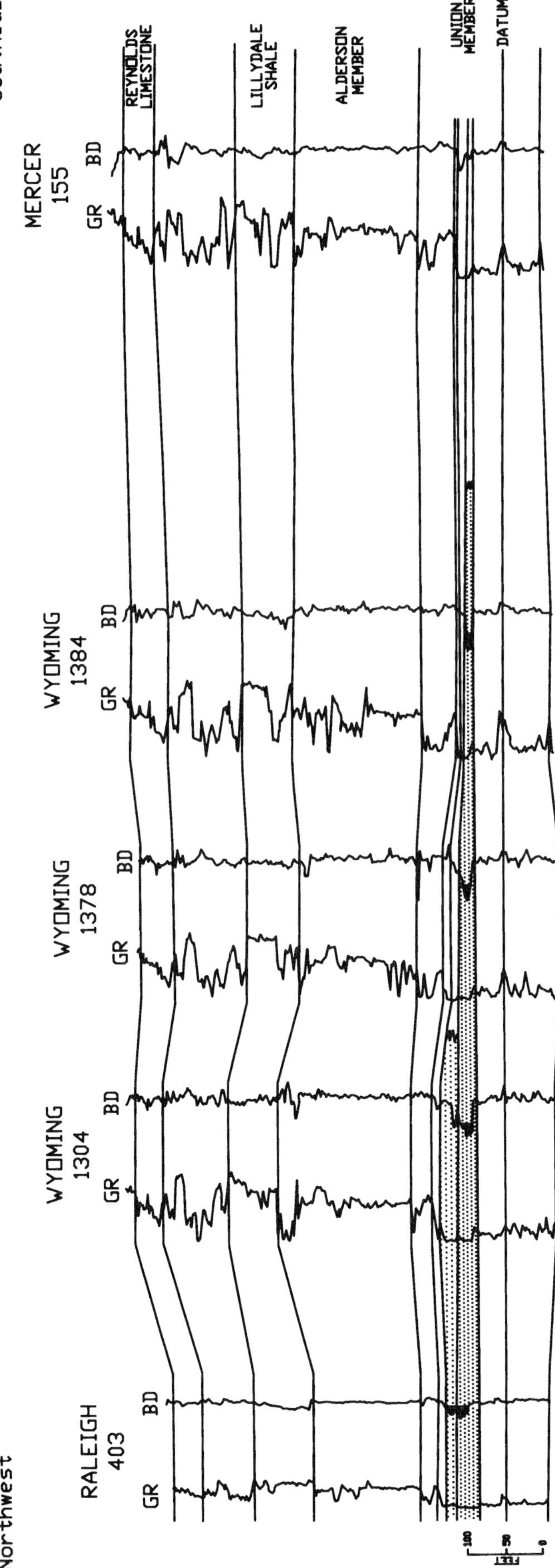

Figure 15. Cross section BB' along the axis of the Poca Land bar. Gamma-ray (GR) and bulk-density (BD) well logs show three distinct units within the oolite of the Union Member; stratigraphic intervals with >6% porosity are marked in black on the BD logs. The lowest unit is porous (medium stippling) in wells 403, 1304, 1378, and 1384; the middle unit is porous (light stippling) in wells 403 and 1304; the upper unit has no porosity. Line of section is marked on Figure 14.

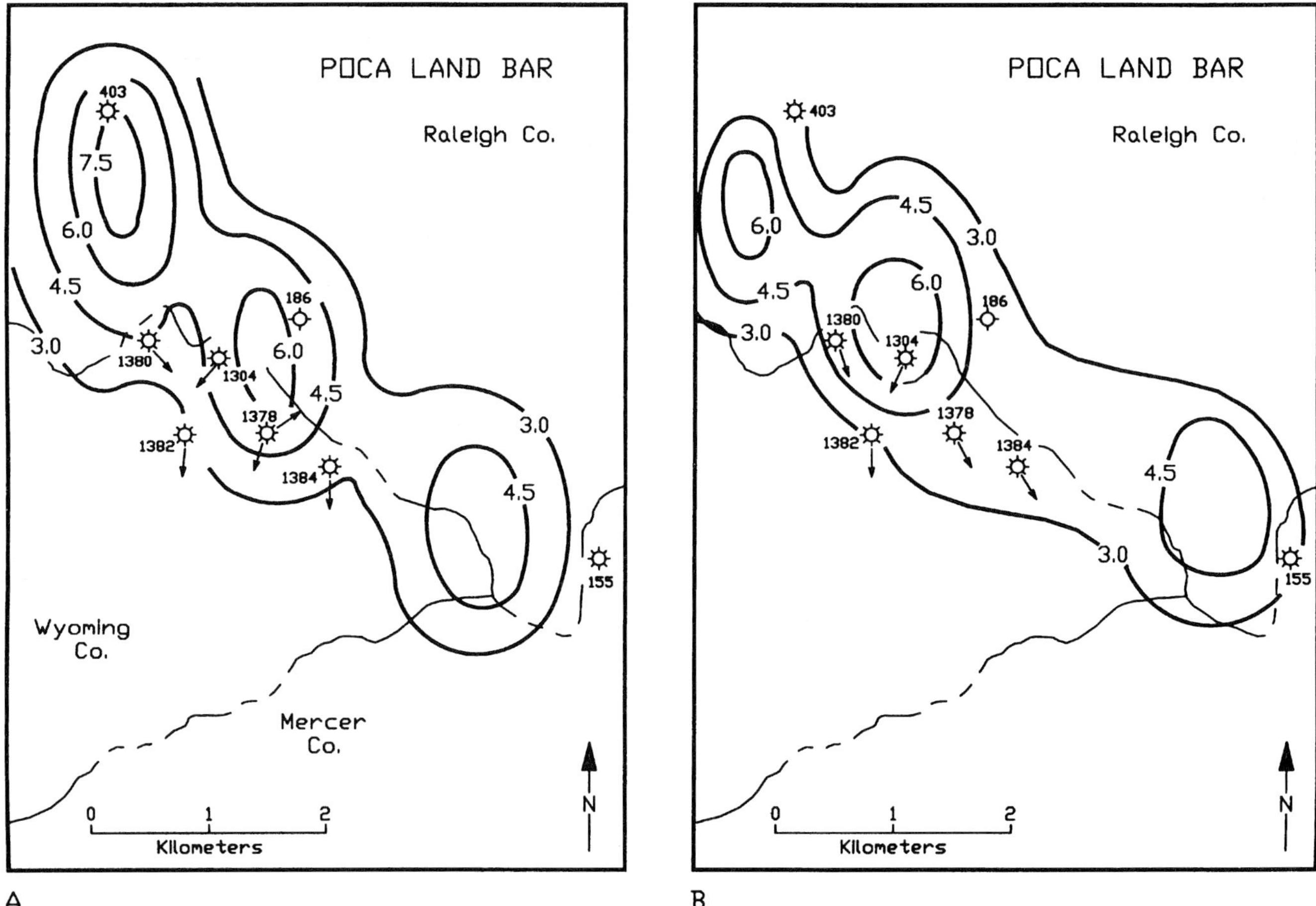

Figure 16. Isopach maps of the lower unit (A) and middle unit (B) illustrate the individual ooid shoals that make up the Poca Land tidal bar. Arrows indicate main dip azimuths of cross beds. Contour interval equals 1.5 m.

long), separated by an intershoal thin. A third shoal is projected to the south—in the wide gap between wells 1384 and 155. As with the Blue Jay bar, these mapped shoals may persist to the northwest along the entire 30-km length of the Poca Land tidal bar. The middle unit (Figure 16B) also consists of two shoals with similar orientation, size, shape, and intershoal thin. Placement of a third shoal near well 155 is somewhat awkward, because it is unusually far from the two known shoals. Perhaps two smaller shoals could be projected here instead of one. As observed on isopach maps for the Blue Jay bar, constituent shoals in the Poca Land bar appear to be offset: shoals of the middle unit are situated immediately above the intershoal areas of the lower unit, and shoals of the lower unit are capped by the intershoal area of the middle unit. The upper unit of the Poca Land bar occurs as a thin blanket of 1–3 m (isopach map not included) without any ooid shoals.

Porosity-isopach maps for the lower and middle units (Figure 17) correspond closely to the unit isopach maps (Figure 16). The trends of greatest porosity parallel crests of the ooid shoals. The interpretation is the same as for the Blue Jay bar: shoal crests stood higher above the surrounding sea floor; they were subsequently exposed longer when relative sea level

dropped, infiltrating meteoric water created the microporosity within ooids; consequently, areas of greatest porosity match stratigraphic thicks (Kelleher and Smosna, 1993). Ooid shoals of the lower unit have the best developed porosity; limestone thickness with porosity >6% ranges ≤4.6 m. Porosity in shoals of the middle unit occurs only in two northern wells, where the maximum thickness of porous limestone is 3.7 m. No porosity has been encountered in the upper unit.

Lithofacies of the Poca Land bar have been interpreted based on five FMS image logs. A lithofacies map of the lower unit (Figure 18A) illustrates that cross-bedded ooid grainstones occupy the crest of the central shoal; this facies is projected to the other two shoals where the unit is porous and thick. The transitional grainstone/packstone facies surrounds the shoal crests and extends into slightly deeper water. No wells actually penetrated the burrowed packstone of the adjacent tidal channels; existence of this lithofacies is postulated from the Blue Jay sedimentary model.

The cross-bedded grainstone facies has yet to be found in the middle unit (Figure 18B). However, the fact that all five logged wells fall within the transitional facies strongly suggests that the grainstone must be nearby. Furthermore, well 1304 contains 0.6 m of

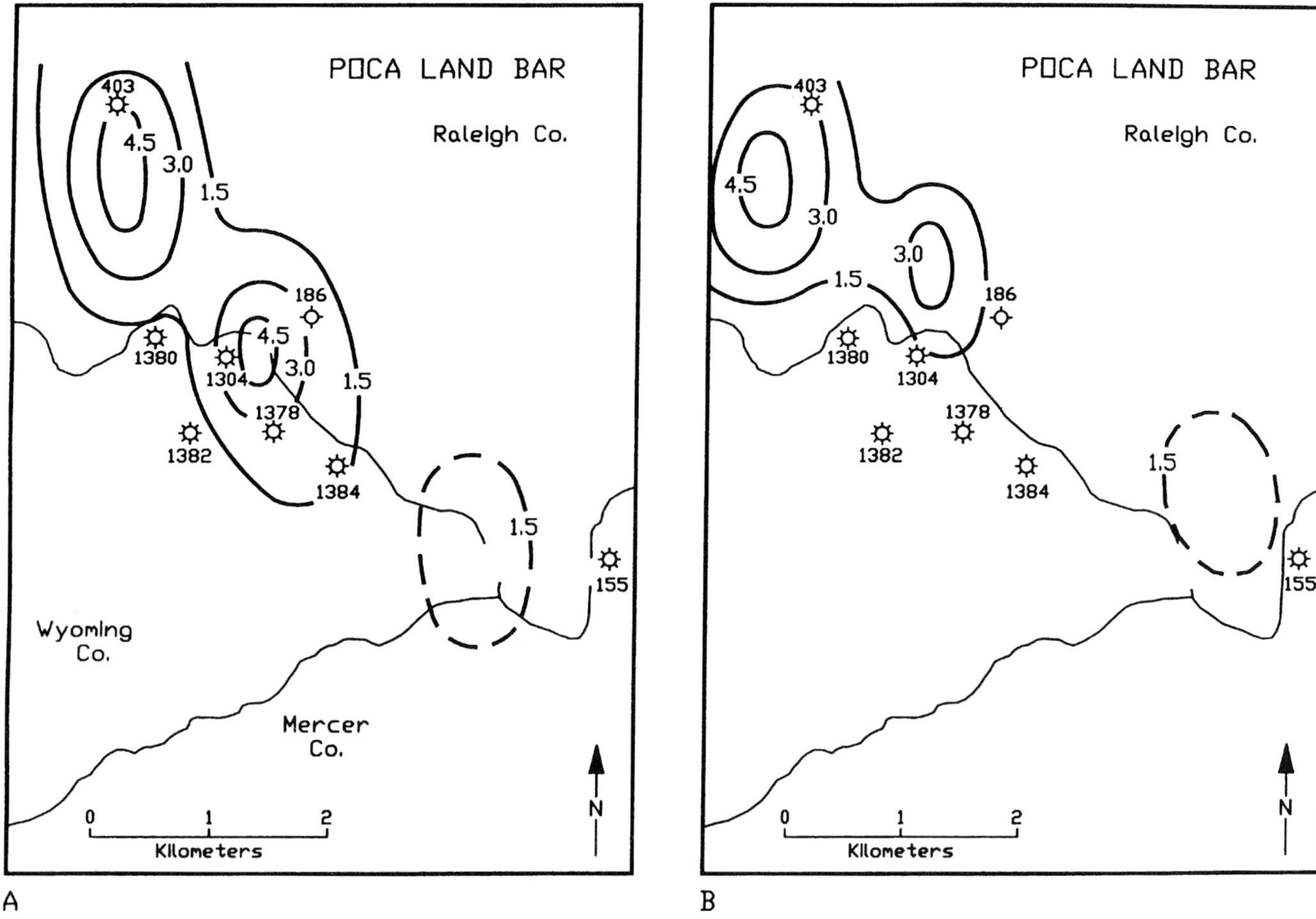

Figure 17. Porosity-isopach maps (stratigraphic thickness with >6% porosity) of the lower unit (A) and middle unit (B), Poca Land tidal bar. Contour interval equals 1.5 m.

limestone with 6% porosity, hinting that the ooid grainstone facies must lie immediately to the north. In a similar manner, the grainstone facies is projected to well 403 on the northernmost shoal (not logged with FMS), based on the greater unit thickness and higher porosity.

The upper unit consists solely of burrowed packstone (lithofacies map not included). In contrast to the Blue Jay bar, water depth at Poca Land during deposition of the upper unit was presumably too great for the maintenance of ooid shoals because (1) the Greenbrier shelf may have had a slight tilt, deepening somewhat to the southwest from the Blue Jay bar (Kelleher and Smosna, 1993) with (2) a rise of sea level near the end of Union deposition (Carney, 1993). No ooid grainstones or transitional grainstone/packstones are expected in the upper unit of the Poca Land study area.

Cross-bed dip directions for the total Union oolite of the Poca Land bar (Figure 16) reveal a curious pattern when compared to those of Blue Jay. Instead of displaying bimodal north-south dip azimuths, dips are almost consistently to the south (southwest to southeast). The northern component does not generally exist. A lack of bimodal dips does not invalidate the idea of tidal construction of the bars; the five logged wells are interpreted to be situated on the

southwest side of the tidal bar, where tidal currents would have been south-directed.

Cross-bed dips in the lower unit (Figure 16A) show all three scenarios discussed for the Blue Jay bar. Well 1378, situated on the axis of the central shoal, exhibits a 180° separation (northeast-southwest) in dip directions. Well 1304 shows southwest dips indicative of a position on the western flank of the same shoal, and well 1384 shows southern dips indicative of a position on its southern nose. Wells 1380 and 1382, positioned off the shoals, have dip patterns not influenced by the nearby shoal. Cross-bed dips in the middle unit (Figure 16B) lie outside of the main shoal bodies, but their consistent southward dips give evidence of an ooid shoal north of well 1304. The FMS image logs illustrate random dip patterns in the nonporous packstones of the upper unit (map not included).

Using the information for the Poca Land study area, we can predict the locations for future successful wells. Three new wells in Raleigh County (indicated by the stars in Figure 18A) are anticipated to penetrate the central ooid shoal of the lower stratigraphic unit. This prediction is based on a combination of (1) the thick illustrated on our isopach map, (2) a local maximum in porosity thickness, (3) the proximity of other wells that

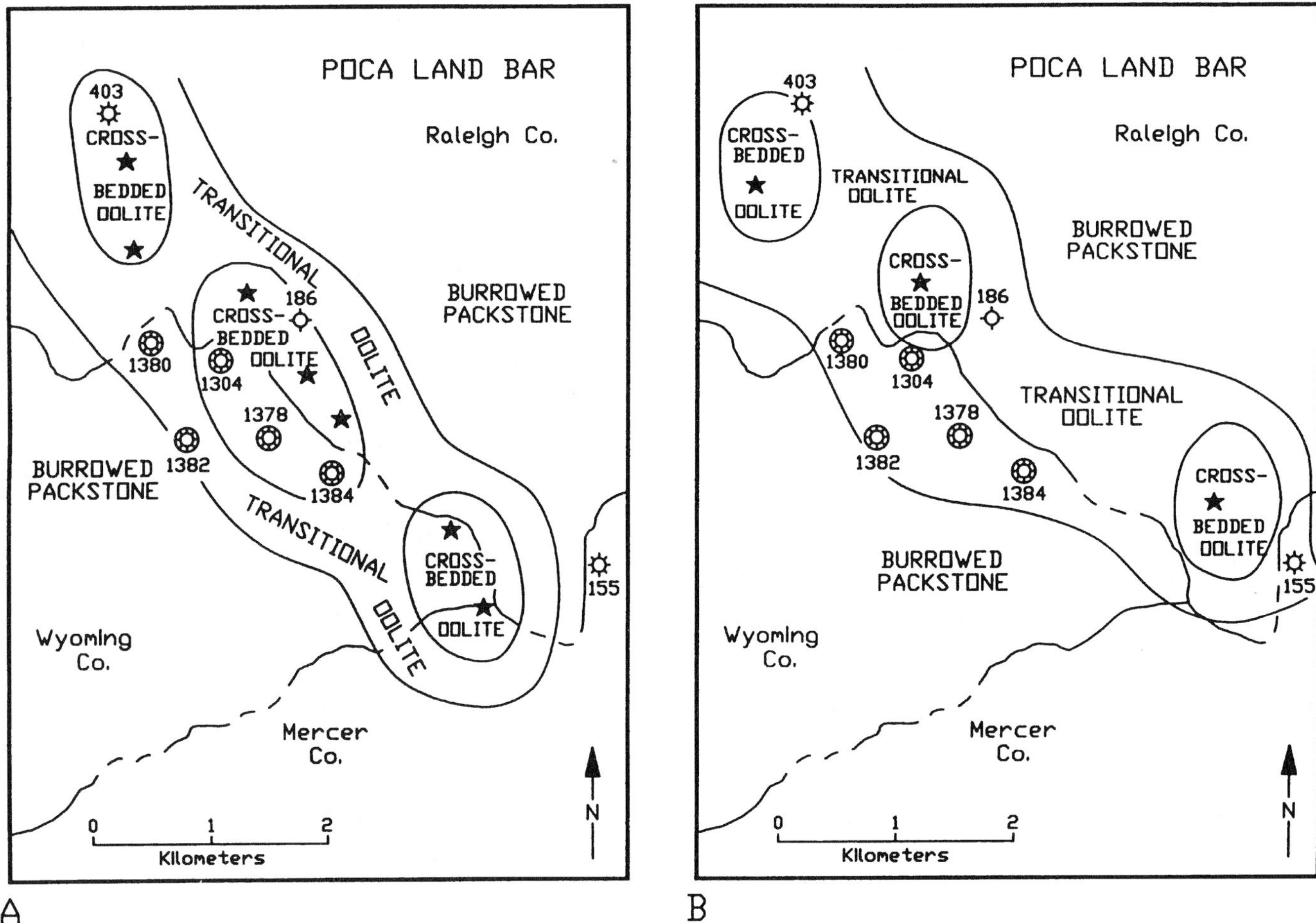

Figure 18. Facies maps of lower unit (A) and the middle unit (B) of the Union oolite at Poca Land bar. Circled wells are those with FMS logs; stars are the selected locations for future wells.

penetrated ooid grainstone, and (4) cross-beds in adjacent wells indicative of a flank position. Two additional wells in Raleigh County are anticipated to encounter the northern ooid shoal, but this prediction is less certain. We chose these locations based on the isopach thick and a local maximum in porosity thickness. Lastly, two future wells in Wyoming County may discover a southern shoal in the lower Union unit. This prediction is made solely by extending the trend of ooid shoals into the area between wells 1384 and 155 with the same spacing established to the north.

Applying the same reasoning, we predict the locations for three successful wells in the middle Union oolite (indicated by the stars in Figure 18B). One well should penetrate the central ooid shoal, based again on a combination of (1) the thick illustrated on our isopach map, (2) a local maximum in porosity thickness, (3) the proximity of other wells that penetrated ooid grainstone, and (4) cross-beds in adjacent wells indicative of a flank position. A second well will encounter the northern ooid shoal, a prediction supported by the isopach thick and a local maximum in porosity thickness. A third well may discover a southern shoal, but this final prediction is the least hopeful. The location is proposed solely by continuing the trend of ooid shoals from the northwest; however, spacing between these shoals remains uncertain.

CONCLUSIONS

The oolitic tidal-bar belt in the Union Member of the Greenbrier Limestone is not a continuously porous body. Rather, it is composed of three stratigraphic units that may contain highly porous shoal facies, marginally porous transitional facies, and nonporous channel facies. The three-dimensional geometry of these natural-gas reservoirs is quite complex: contemporaneous shoals within a single tidal bar were laterally linked; moreover, they stacked with a vertical offset as they grew through time. Their orientation is oblique to the general trend of the bar, a result of refraction of the tidal currents responsible for their development. Shoals, measuring 8 m thick, 1500 m wide, and 3500 m long, consist of ooid grainstone. On FMS logs, this lithofacies displays a grainy texture, abundant cross-bed sets, and reactivation surfaces. Cross-beds along the crest have a bimodal dip direction perpendicular to the shoals' axes, whereas those on the flanks are unimodal and directed away from the shoals' axes. The thickest pay zones (≤3 m of grainstone with >6% porosity) occur along the shoals' crests. Shale partings become common around the shoal margins where the ooid grainstone passes through a transitional lithofacies into nonporous bioturbated packstone of the adjacent tidal channel.

ACKNOWLEDGMENTS

The authors acknowledge the reviews of Julie Kupecz, Laura S. Foulk, and Neil Hurley, whose comments and suggestions improved the manuscript. Stonewall Gas Company provided the data and gave permission to publish the results. Alison Hanham and Debbie Benson drafted the illustrations.

REFERENCES CITED

Adams, R.W., 1970, Loyalhanna Limestone—crossbedding and provenance, *in* G.W. Fisher, F.J. Pettijohn, J.C. Reed, and K.N. Weaver, eds., Studies of Appalachian geology—central and southern: New York, Interscience Publishers, p. 83–100.

Ahr, W.M., 1989, Early diagenetic microporosity in the Cotton Valley Limestone of East Texas: Sedimentary Geology, v. 63, p. 275–292.

Ball, M.M., 1967, Carbonate sand bodies of Florida and the Bahamas: Journal of Sedimentary Petrology, v. 37, p. 556–591.

Boardman, M.R., C. Carney, and P.M. Bergstrand, 1993, A Quaternary analog for interpretation of Mississippian oolites, *in* B.D. Keith and C.W. Zuppann, eds., Mississippian oolites and modern analogs: AAPG Studies in Geology 35, p. 227–241.

Brezinski, D.K., 1989, Late Mississippian depositional patterns in the north-central Appalachian basin, and their implications to Chesterian hierarchal stratigraphy: Southeastern Geology, v. 30, p. 1–23.

Carney, C., 1993, The drowning of ooid shoals: Mississippian Greenbrier Limestone near the West Virginia dome, *in* B.D. Keith and C.W. Zuppann, eds., Mississippian oolites and modern analogs: AAPG Studies in Geology 35, p. 141–148.

Carney, C., and R. Smosna, 1989, Carbonate deposition in a shallow marine gulf, the Mississippian Greenbrier Limestone of the central Appalachian Basin: Southeastern Geology, v. 30, p. 25–48.

Cecil, C.B., 1990, Paleoclimate controls on stratigraphic repetition of chemical and siliciclastic rocks: Geology, v. 18, p. 533–536.

deWitt, W., and L.W. McGrew, 1979, The Appalachian basin region, *in* L.C. Craig and C.W. Connor, eds., Paleotectonic investigations of the Mississippian System in the United States: U.S. Geological Survey Professional Paper 1010, p. 13–48.

Donaldson, A.C., 1974, Pennsylvanian sedimentation of the central Appalachians, *in* G. Briggs, ed., Carboniferous of the southeastern United States: Geological Society of America Special Paper 148, p. 47–78.

Evans, W.E., 1970, Imbricate linear sandstone bodies of Viking Formation in Dodsland-Hoosier area of southwestern Saskatchewan, Canada: AAPG Bulletin, v. 54, p. 469–486.

Flowers, R.R., 1956, A subsurface study of the Greenbrier Limestone in West Virginia: West Virginia Geological & Economic Survey, Report of Investigation No. 15, 17 p.

Halley, R.B., P.M. Harris, and A.C. Hine, 1983, Bank margin, *in* P.A. Scholle, D.G. Bebout, and C.H. Moore, eds., Carbonate depositional environments: AAPG Memoir 33, p. 463–506.

Harris, P.M., 1979, Facies anatomy and diagenesis of a Bahamian ooid shoal: Sedimenta VII, University of Miami, Florida, 163 p.

Keith, B.D., and E.D. Pittman, 1983, Bimodal porosity in oolitic reservoir—effect on productivity and log response, Rodessa Limestone (Lower Cretaceous), East Texas basin: AAPG Bulletin, v. 67, p. 1391–1399.

Kelleher, G.T., and R. Smosna, 1993, Oolitic tidal-bar reservoirs in the Mississippian Greenbrier Group of West Virginia, *in* B.D. Keith and C.W. Zuppann, eds., Mississippian oolites and modern analogs: AAPG Studies in Geology 35, p. 163–173.

Klein, G.D., 1977, Clastic tidal facies: Champaign, Illinois, Continuing Education Publication Co., 149 p.

MacQuown, W.C., and J.L. Pear, 1983, Regional and local geologic factors control Big Lime stratigraphy and exploration for petroleum in eastern Kentucky: Kentucky Geological Survey, Series XI, Special Publication 9, p. 1–20.

Newell, N.D., E.G. Purdy, and J. Imbrie, 1960, Bahamian oolitic sand: Journal of Geology, v. 68, p. 481–497.

Scotese, C.R., 1984, Paleozoic paleomagnetism and the assembly of Pangea, *in* R. Van der Voo, C.R. Scotese, and N. Bonhommet, eds., Plate reconstruction from Paleozoic paleomagnetism: American Geophysical Union, Geodynamic Series, v. 12, p. 1–10.

Serra, O., 1989, Formation MicroScanner image interpretation: Houston, Schlumberger Educational Services, 117 p.

Smosna, R., and B. Koehler, 1993, Tidal origin of a Mississippian oolite on the West Virginia Dome, *in* B.D. Keith and C.W. Zuppann, eds., Mississippian oolites and modern analogs: AAPG Studies in Geology 35, p. 149–162.

Swift, D.J.P., 1985, Response of the shelf floor to flow, *in* R.W. Tillman, D.J.P. Swift, and R.G. Walker, eds., Shelf sands and sandstones: SEPM Short Course Notes 13, p. 135–241.

Swift, D.J.P., and A.W. Niedoroda, 1985, Fluid and sediment dynamics on continental shelves, *in* R.W. Tillman, D.J.P. Swift, and R.G. Walker, eds., Shelf sands and sandstones: SEPM Short Course Notes 13, p. 47–133.

Zuppann, C.W., 1993, Complex oolite reservoirs in the Ste. Genevieve Limestone (Mississippian) at Folsomville field, Warrick County, Indiana, *in* B.D. Keith and C.W. Zuppann, eds., Mississippian oolites and modern analogs: AAPG Studies in Geology 35, p. 73–89.

Chapter 15

Major, R.P., and M.H. Holtz, 1997, Predicting reservoir quality at the development scale: methods for quantifying remaining hydrocarbon resource in diagenetically complex carbonate reservoirs, *in* J.A. Kupecz, J. Gluyas, and S. Bloch, eds., Reservoir quality prediction in sandstones and carbonates: AAPG Memoir 69, p. 231–248.

Predicting Reservoir Quality at the Development Scale: Methods for Quantifying Remaining Hydrocarbon Resource in Diagenetically Complex Carbonate Reservoirs

R.P. Major
Mark H. Holtz
The University of Texas at Austin, Bureau of Economic Geology
Austin, Texas, U.S.A.

ABSTRACT

The Jordan (San Andres) reservoir comprises ~400 ft (120 m) of upward-shoaling subtidal to peritidal carbonate strata, which is now thoroughly dolomitized and partly cemented by sulfates. Subtidal facies include dominant pellet packstone/grainstone, with local bryozoans, algae, and coral bioherms and associated skeletal grainstone flanking beds. The lower part of the subtidal section is characterized by stratigraphically distinct zones in which permeability has been enhanced by a postburial carbonate-leaching event. These diagenetically altered (leached) zones crosscut subtidal depositional facies. Peritidal facies are nonporous mudstone and generally nonporous pisolite packstone characterized by abundant sulfate cement. The pisolitic rocks are locally porous and permeable where sulfate cement is either leached or absent from fenestrae.

Cumulative production is 68 million stock tank barrels (MMSTB) of 218 MMSTB original oil in place, which is a recovery efficiency of 31%. A total of 47 MMSTB of remaining mobile oil occurs as bypassed oil in the contacted upper part of the reservoir, which has been penetrated by well bores; 12 MMSTB of mobile oil is in the uncontacted lower part, which has not been penetrated by well bores. The most prospective areas for increased production by waterflood profile modification in the contacted part of the reservoir are the southwest corner of the field, where low-permeability, diagenetically unaltered subtidal rocks are incompletely swept, and the eastern central part of the field, where heterogeneous permeability in peritidal rocks has resulted in an incomplete sweep. The most prospective areas for increased production through well-bore deepening into the uncontacted part of the reservoir are the southeast corner of the field, where high-permeability, diagenetically altered subtidal rocks are uncontacted, and the central part of the field, where high-permeability, diagenetically altered subtidal rocks are uncontacted. An understanding of diagenetically controlled reservoir properties can be used to predict the locus of remaining resource and to design recovery strategies.

INTRODUCTION

A major challenge for predicting reservoir quality at the development scale is having a sufficiently detailed understanding of the geometry and extent of individual flow units (sensu Ebanks, 1987) within a reservoir. Our knowledge of depositional facies patterns and geometries in sedimentary rocks is based on extensive documentation of modern sediments, sedimentary processes, and ancient rocks exposed in outcrops. Our ability to predict depositional facies relationships at a scale that is meaningful in maturely developed reservoirs is fairly advanced; our ability to predict diagenetic patterns that control reservoir quality is, however, at a much more primitive stage. We present here a case study in which diagenetic alteration of a carbonate reservoir controls flow-unit geometry and, based on subsurface mapping, the geometry of these diagenetically controlled flow units can be used to predict reservoir quality and to quantify remaining resource. The Guadalupian (Upper Permian) San Andres Formation of the Permian Basin, West Texas and southeastern New Mexico, provides an opportunity to test new reservoir characterization and resource assessment techniques.

The general depositional facies tracts of San Andres reservoirs are divided into four categories: inner ramp, ramp crest, outer ramp, and slope/basin. Flow units in outer ramp facies tract reservoirs may be controlled in large part by postdepositional diagenetic alteration of relatively homogeneous depositional facies (Ruppel et al., 1995). In this chapter, we review the geologic and engineering parameters that control reservoir quality, and the volume and distribution of remaining oil, in a mature, outer ramp San Andres reservoir—the Jordan San Andres reservoir on University of Texas Lands (University Lands) in Ector and Crane counties, Texas.

GEOLOGIC SETTING AND PRODUCTION HISTORY

The paleogeography of the Permian Basin was controlled by Pennsylvanian tectonism that deformed Precambrian basement and pre-Pennsylvanian sedimentary rocks (Galley, 1958; Ward et al., 1986). During the Permian, sedimentation in the region occurred in two basins, the Delaware Basin on the west and the Midland Basin on the east, separated by the south-southeast–trending Central Basin Platform (Figure 1). The Central Basin Platform was the site of shallow-water ramp carbonate sedimentation, whereas the central portions of the Delaware and Midland basins were the sites of siliciclastic deposition (Galley, 1958; Ward et al., 1986).

The Permian stratigraphic section on the Central Basin Platform contains Wolfcampian, Leonardian, and Guadalupian shallow-water carbonate strata, many now thoroughly dolomitized, and includes relatively minor zones of siliciclastic-rich carbonates. Guadalupian carbonates are in conformable and gradational contact with overlying Ochoan evaporites and siliciclastic red beds deposited during increasingly restricted marine conditions in the Permian Basin.

Jordan field is one of a complex of five fields—Penwell, Jordan, Waddell, Dune, and McElroy, termed the PJWDM field complex (Major et al., 1988)—that produce from both San Andres and Grayburg reservoirs (Longacre, 1980, 1983; Harris et al., 1984; Bebout et al., 1987; Major et al., 1988; Harris and Walker, 1990). Jordan field produces from a San Andres reservoir located on a low-relief, broad anticlinal structure with a northwest–trending axis (Figure 1). The structure was created by drape of Permian sediments over buried Pennsylvanian faults that trend oblique to the approximate eastern margin of the Central Basin Platform (Ward et al., 1986).

The San Andres reservoir at Jordan field is composed of dolomitized rocks exhibiting textures indicative of sediments deposited in subtidal, open-marine environments that shoaled upward to tidal-flat environments. These facies prograded from west to east across the platform, and the tidal-flat section thickens westward. This westward thickening of low-porosity and low-permeability tidal-flat facies provides an updip seal, and oil production is mainly from the eastern flank of the broad anticline.

San Andres reservoirs in the Permian Basin can be categorized into four facies tracts: (1) inner ramp, (2) ramp crest, (3) outer ramp, and (4) slope/basin (Kerans et al., 1994; Ruppel et al., 1995). The relatively distal setting of outer ramp reservoirs, such as Jordan field, results in relatively low depositional facies diversity. In this setting, minor fluctuations in relative sea level did not result in exposure to shoaling, higher energy environments, or subaerial exposure. Thus, although the upper part of the Jordan San Andres reservoir represents shoaling to tidal-flat depositional environments, much of the reservoir is composed of subtidal, open-marine facies that have a

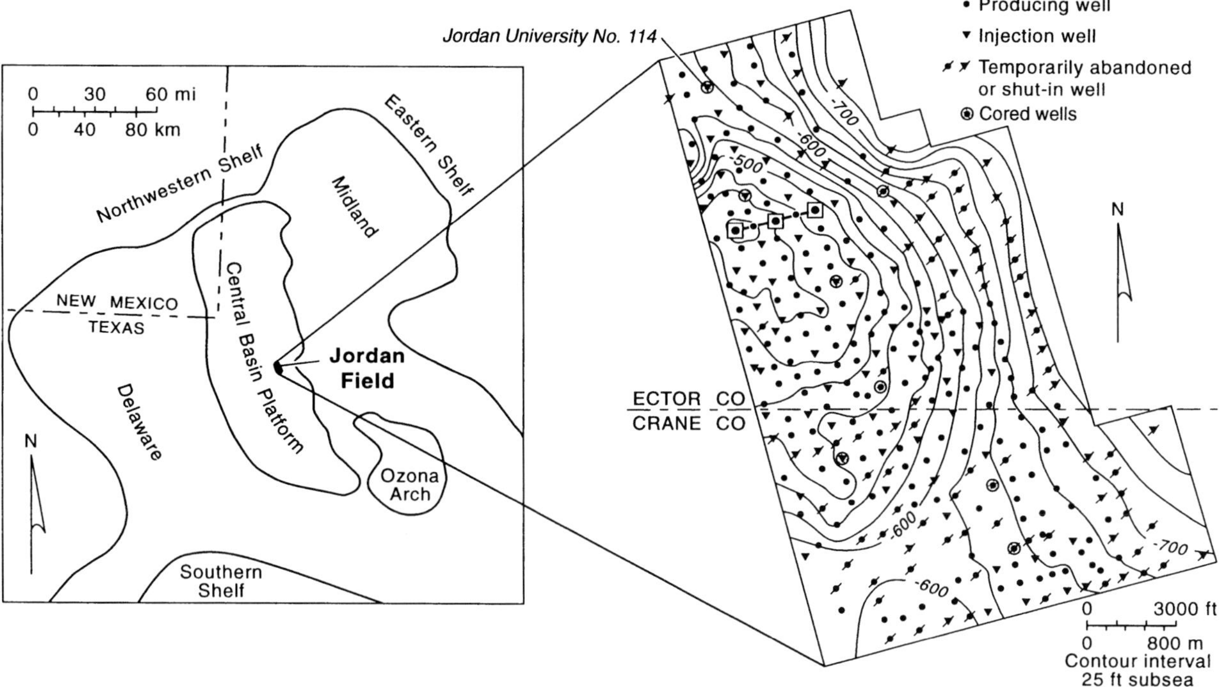

Figure 1. Permian Basin paleogeography during San Andres (middle Guadalupian) time and location of Jordan field. The inset is a structure map of the University Lands part of the Jordan (San Andres) reservoir.

low depositional texture diversity. In this depositional setting, the influences of postdepositional diagenetic alteration can play a major role in control of reservoir flow units (Ruppel et al., 1995).

Jordan field is, in terms of discovery and development history, typical of San Andres reservoirs of the Central Basin Platform. It is a very mature field, having been discovered in 1937. Typical early wells had initial potential flow rates of several hundred to 1000 bbl of oil per day. In the late 1940s, annual production reached 1.9 million stock tank barrels (MMSTB). Annual production declined through the 1950s and 1960s to a low of 1.0 MMSTB. A program of infill drilling, well deepening, and conversion of producing wells to water injection wells began in 1969, following peripheral waterflooding in 1968. By 1971, a modified five-spot waterflood was in place, and annual production peaked in 1975 at 2.2 MMSTB. Annual production steadily declined through the late 1970s and early 1980s and is now ~650 thousand stock tank barrels (MSTB). The present well spacing is ~20 acres per well, and the two Jordan field units on University Lands have a cumulative production of 68 MMSTB.

STUDY AREA AND AVAILABLE DATA

The study area is The University of Texas Lands part of Jordan field, which comprises ~4500 acres and is 66% of the field. Although some form of wireline-log data are available for nearly all wells in the study area, the majority of logs are neutron or density-neutron logs. As is discussed below, the most useful porosity tool in this reservoir is the acoustic log, and most of these were run in the 1970s during infill drilling for conversion to waterflood. Virtually all resistivity logs are post-1970. Thus, all resistivity data are postwaterflood and, because flooding is assumed to have substantially changed the resistivity of interstitial pore waters, these resistivity logs cannot be used to reliably calculate fluid saturations.

Seven conventional cores, generally 300–400 ft (90–120 m) long, are available from within the study area (Figure 1). These were augmented by two Jordan field cores immediately west of the University Lands boundary and 14 cores in the East Penwell San Andres Unit, which offsets Jordan field to the north (Major et al., 1990). All but two of the cores from Jordan field have been analyzed for porosity and permeability using high-temperature analytic techniques. As is discussed in detail below, the presence of gypsum in this reservoir requires more expensive, more time-consuming, low-temperature core analysis for accurate porosity and permeability measurements. Thus, although there are numerous cores for lithologic description, there are relatively few reliable core-derived porosity and permeability data.

LITHOLOGIC RESERVOIR DESCRIPTION

The Jordan San Andres reservoir is interpreted to have been deposited in the outer ramp facies tract. As

Figure 2. Core photograph of pellet packstone/grainstone, which is a lithology in the open-marine facies. The pellets originated as fecal pellets in a carbonate mud depositional environment. This rock type is commonly porous and permeable; this particular sample also contains fusulinids [East Penwell San Andres Unit No. 431, 3510 ft (1070 m), scale in centimeters].

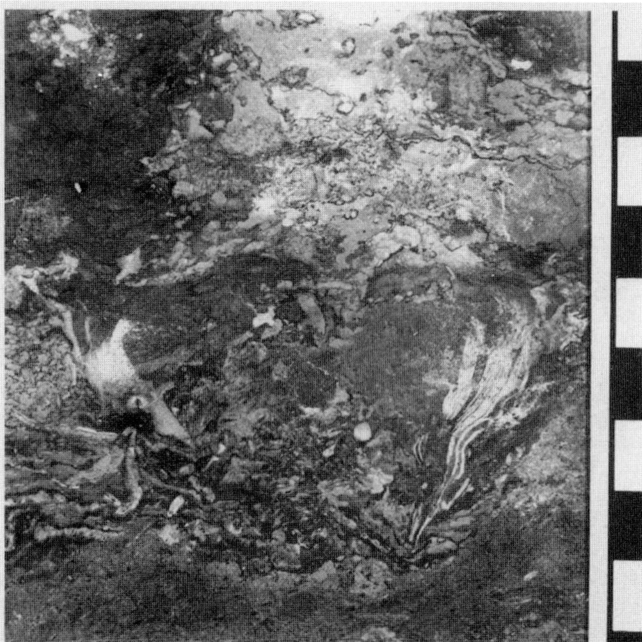

Figure 3. Core photograph of a bioherm, which is a lithology in the open-marine facies. Note the prominent bryozoan in growth position. Bioherms contain internal mud sediment and generally have low porosity and permeability [Jordan University No. 431, 3626 ft (1105 m), scale in centimeters].

discussed above, this location on the ramp margin of the Central Basin Platform results in relatively low depositional facies diversity (Ruppel et al., 1995), and postdepositional diagenetic alteration can play an important role in variations in petrophysical properties. Lithologic description of the reservoir is divided into depositional facies and diagenetic overprint; the goal of lithologic description is to divide the reservoir into flow units (sensu Ebanks, 1987), which is a critical first step for predicting reservoir quality.

Depositional Facies

The San Andres reservoir at Jordan field is assigned to the outer ramp facies tract of Ruppel et al. (1995) because it is dominantly composed of rocks deposited at or below fair-weather wave base. These open-marine rocks are overlain by rocks deposited in a tidal-flat setting during a period of relative sea level lowstand. Accordingly, the depositional facies described here are divided into two parts.

Open-Marine Depositional Facies

Open-marine facies are pellet packstone/grainstone and bioherms composed of bryozoans, algae, and corals, with associated flanking facies of skeletal grainstone. Calcium sulfate cements are common. The pellet packstone/grainstone facies, which is the volumetrically dominant reservoir facies, is composed of variable amounts of mud matrix and spherical to ovoid fecal pellets ~0.2–0.5 mm in diameter. Fossils of open-marine invertebrates are common, especially fusulinids and

bivalves. Burrow structures are rare, and there is a general lack of laminations due to thorough bioturbation. Fecal pellets were deposited as soft carbonate mud and exhibit a wide range in degree of preservation, as is characteristic of many modern low-energy settings (Wanless et al., 1981). The pellets in this thoroughly dolomitized rock are commonly not visible on slabbed core surfaces. Thus, these rocks may be incorrectly described as mudstone or, where skeletal grains are abundant, as wackestone. Where pellets are well preserved, the rock has interparticle porosity; where pellets have been destroyed by compaction, porosity is low and is generally intercrystalline, moldic, or both. Extensive bioturbation and presence of abundant fossils of open-marine invertebrates within pelleted mud (Figure 2) indicate that this sediment was deposited in a shallow subtidal setting in an environment similar to Holocene carbonate shelf and ramp settings.

Thin, generally less than 15 ft (4.5 m) thick, bioherms composed of sponges, algae, corals, and bryozoans occur locally and are laterally discontinuous (cannot be correlated between wells) in the lower part of the open-marine section. Crinoid fragments are a common accessory grain in this facies. Bioherms (Figure 3), which are generally nonporous, contain abundant internal mud sediment that displays geopetal structures. Skeletal grainstone, composed principally of bryozoan and crinoid fragments and, less abundantly, fusulinid and mollusk fragments, is closely associated with bioherms. The presence of abundant

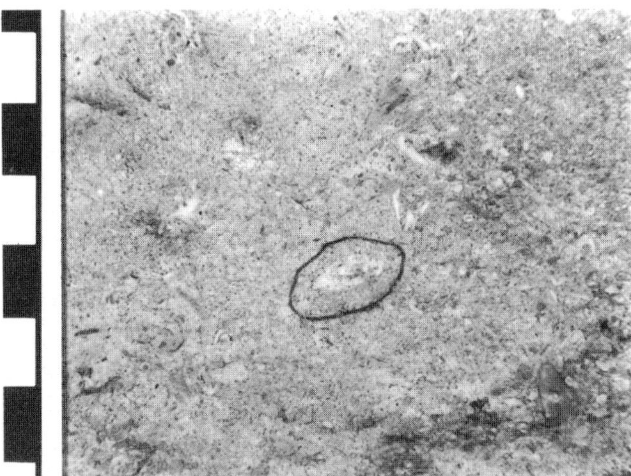

Figure 4. Core photograph of skeletal grainstone, part of the open-marine facies. This facies is closely associated with bioherms and is commonly porous and permeable. The circled feature is a well-preserved bryozoan [Jordan University No. 431, 3656 ft (1114 m), scale in centimeters].

fossils of open-marine organisms (Figure 4), lack of desiccation features, and stratigraphic proximity to pellet packstone/grainstone indicate that bioherms and skeletal grainstone were deposited in a subtidal, open-marine environment.

Tidal-Flat Depositional Facies

Tidal-flat facies are pisolite packstone and mudstone. Pisolite packstone is composed of poorly sorted symmetrical and asymmetrical pisolites having diameters generally in the range of 0.2 to 4 mm and fine-grained carbonate mud matrix. Pisolites commonly have a fitted fabric. This facies is characterized by abundant caliche, fenestrae (Figure 5), desiccation cracks, tepee structures, and sheet cracks. Locally, minor karst dissolution is indicated by severe brecciation and infilling by greenish-gray siltstone. The karsted intervals are generally <3 ft (<1 m) thick. This facies is commonly cemented with anhydrite and gypsum cement, and generally has very low porosity and permeability. Locally, however, cementation with calcium sulfates is incomplete, and this facies may be porous and permeable. The abundant evidence for syndepositional desiccation and the presence of minor karst dissolution indicate that pisolite packstone formed in a tidal-flat environment that was frequently subaerially exposed.

Mudstone is composed of cream-colored, generally massive dolomite, although some mudstone is faintly laminated (Figure 6). Stromatolitic laminae are present but rare. Mudstone is composed of dolomite crystals generally smaller than 0.02 mm. With the exception of stromatolites, this facies is barren of fossils, suggesting that it was deposited in a hypersaline environment in which stromatolites could survive but marine invertebrates were excluded. The

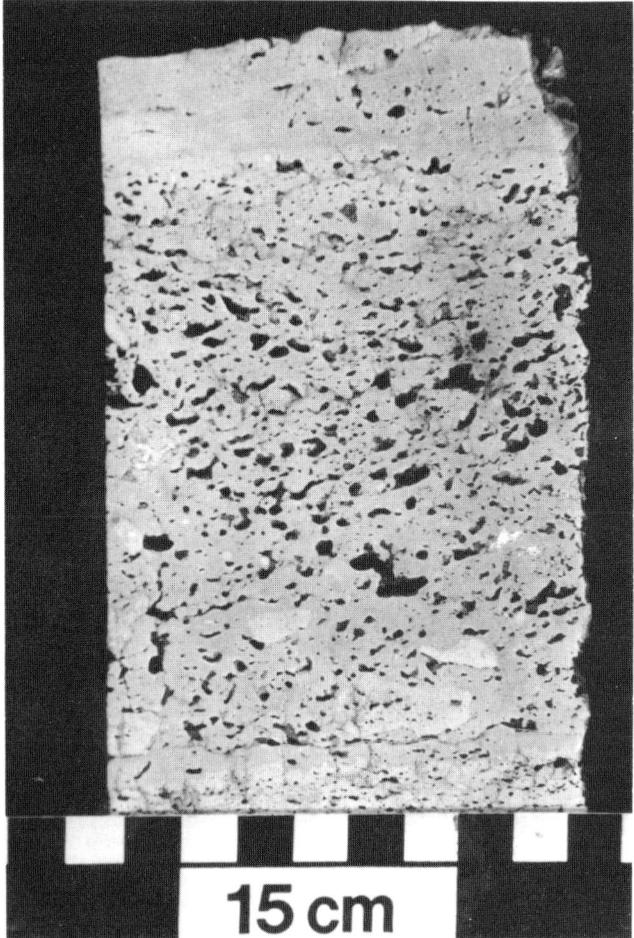

Figure 5. Core photograph of pisolite packstone, which is a lithology in the tidal-flat facies. This sample is porous because fenestrae are incompletely filled with sulfate cements [Jordan University No. 431, 3355 ft (1023 m), scale in centimeters].

absence of fossils and the close association with the pisolite packstone facies indicate deposition in hypersaline ponds on a tidal flat, isolated and probably landward of an open-marine depositional environment.

Pisolite packstone and mudstone are interbedded with three intervals of siliciclastic silt that may be correlated regionally using gamma-ray logs. Most siltstone is massive, although locally it is finely laminated. Siltstone is commonly calcareous and in transitional contact with pisolite packstone and mudstone. The presence of siltstone interbedded with rocks containing evidence of subaerial exposure, and the lack of any regional sources for siliciclastic detritus, suggest that these sediments were transported to the tidal-flat environment by eolian processes. Some reworking in shallow water subsequent to eolian transport is indicated by the laminations.

Tidal-flat facies are separated from subjacent open-marine facies by an interval of greenish-gray organic-rich shale that may be correlated throughout Jordan field using gamma-ray logs.

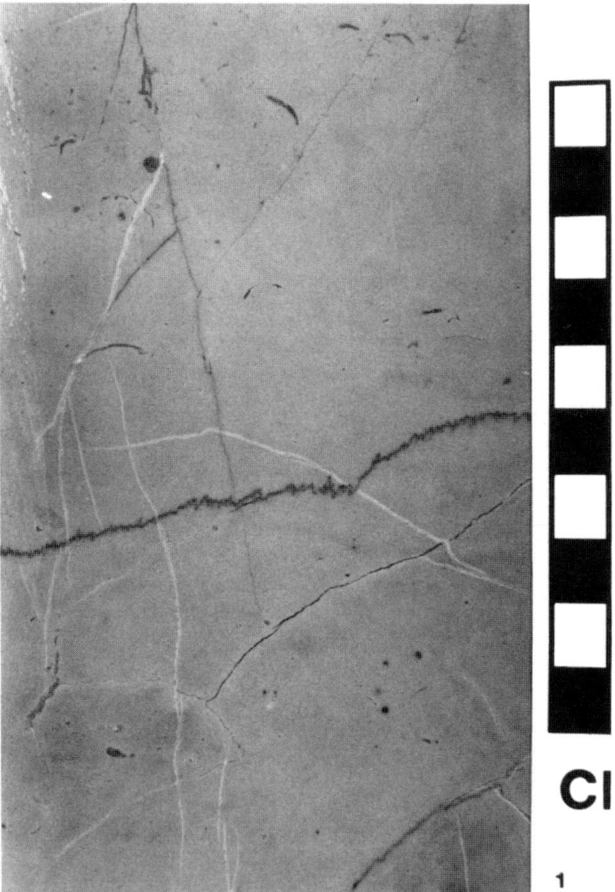

Figure 6. Core photograph of mudstone, which is a lithology in the tidal-flat facies. This lithology is nearly completely nonporous [East Penwell San Andres Unit No. 431, 3329 ft (1015 m), scale in centimeters].

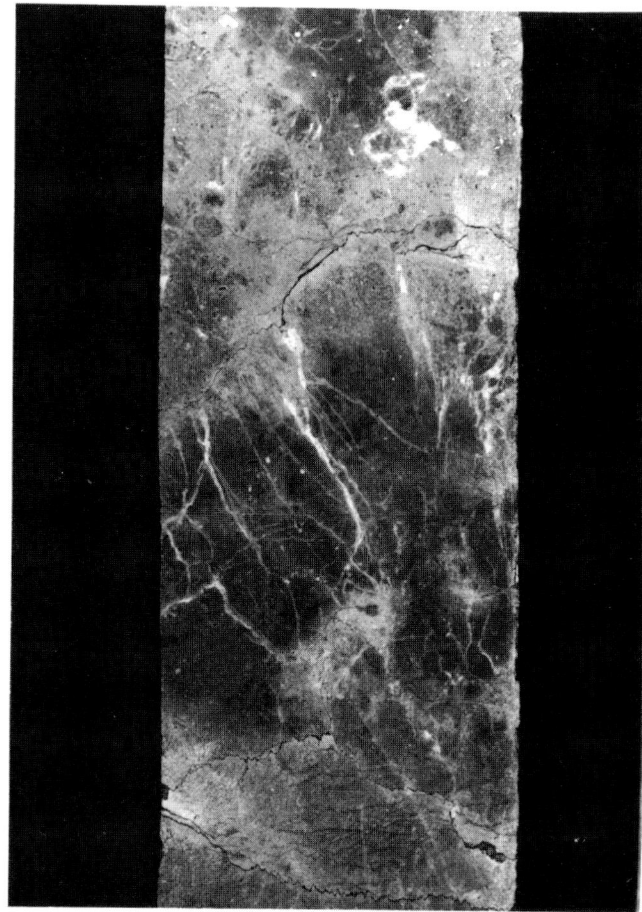

Figure 7. Pellet packstone/grainstone partially altered by postburial leaching (area of lighter color). Note that the leached parts of this sample are associated with stylolites, suggesting that undersaturated fluids moved along stylolites [Jordan University No. 638W, 3478 ft (1060 m), core is ~3 in. (7.7 cm) wide].

Diagenetic Overprint

Tidal-flat pisolite packstone generally has low porosity because fenestrae and sheet cracks are cemented with calcium sulfates. Locally, calcium sulfate cementation was either incomplete, did not occur, or, more likely, sulfate cements were leached. Where little or no cement occurs in pisolite packstone, this facies is porous and permeable (Figure 5). The volumetrically dominant pore type is fenestral. This diagenetically controlled porous texture is important because, where porous, the pisolite packstone facies can be oil productive.

Open-marine facies have been partly to completely altered by a postburial leaching event. This diagenetically altered dolomite can be identified on slabbed core surfaces as tan to brown rock that contrasts with the dark-gray color of unaltered dolomite. Altered dolomite in some cases mimics the geometry of burrows, whereas in other cases it forms aureoles around stylolites (Figure 7), suggesting that the fluids causing this alteration preferentially flowed along stylolites (Carozzi and Von Bergen, 1987; Von Bergen and Carozzi, 1990). This association demonstrates that diagenetic alteration was a postburial, postcompaction event.

Diagenetically altered dolomite is more permeable than unaltered dolomite, as indicated by minipermeameter data illustrated in Figure 8 (for description of this instrument, see Eijpe and Weber, 1971; Kittridge, 1988; Chandler et al., 1989). The mottled geometry of this diagenetic alteration results in such close association of these two rock types that the order-of-magnitude difference in permeability illustrated in Figure 8 is commonly below the sampling resolution of conventional core-plug or whole-rock permeability analyses.

This permeability-enhancing diagenetic alteration is apparently the result of leaching and partial dissolution of dolomite crystals. Hollow and corroded dolomite crystals are visible at the light microscope and scanning electron microscope level of resolution (Figure 9). Apparently this late-stage diagenetic event widened intercrystalline pore throats, resulting in increased permeability. This diagenetic process may also partly cause alteration of nodules of anhydrite to gypsum. Anhydrite nodules with outer edges altered to gypsum are commonly surrounded by a "halo" of

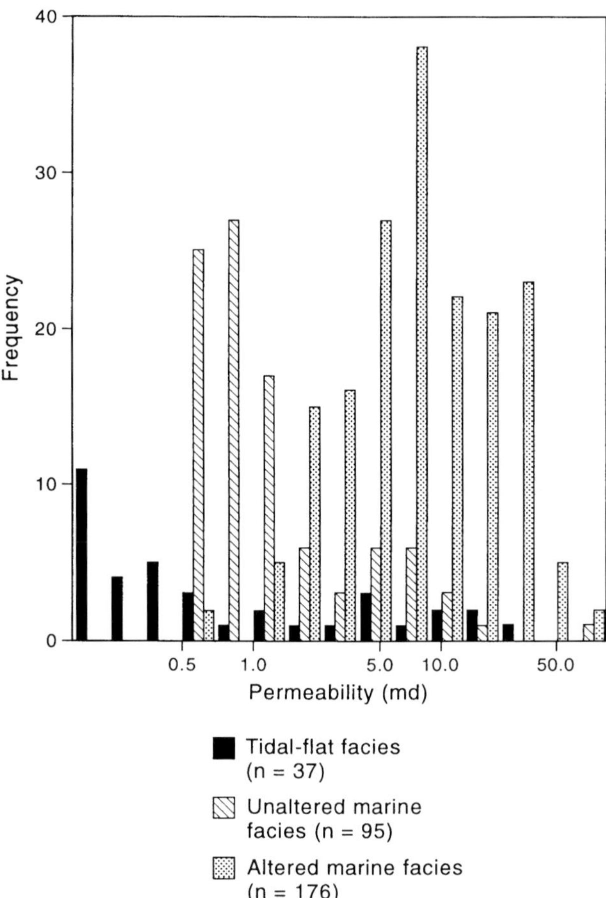

Figure 8. Permeability histograms for tidal-flat facies (pisolite packstone), unaltered marine facies (pellet packstone/grainstone), and altered marine facies (pellet packstone/grainstone). The highest permeabilities are in diagenetically altered pellet packstone/grainstone. Data are from West Jordan Unit No. 12-4, East Penwell San Andres Unit No. 207, and East Penwell San Andres Unit No. 1914. All data were collected from cores that were not subjected to high-temperature, gypsum-destructive handling.

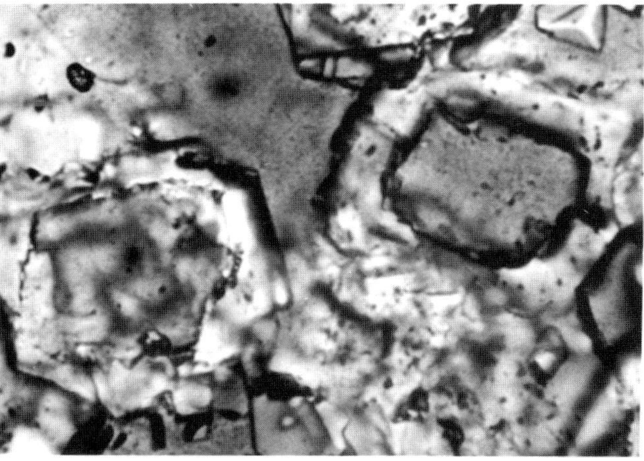

Figure 9. Thin-section photomicrograph illustrating leached, hollow dolomite rhombs in diagenetically altered pellet packstone/grainstone. [East Penwell San Andres Unit No. 1313, 3922 ft (1495 m), dolomite crystals are ~50 mm in width].

altered dolomite (Figure 10), and some samples of diagenetically altered rock contain ≤20% gypsum. It can be inferred from this relationship that the fluids leaching the dolomite also altered (hydrated) some of the anhydrite nodules and cements to gypsum. Inasmuch as gypsum in San Andres Formation and Grayburg Formation reservoirs is restricted to the central and southern parts of the Central Basin Platform, and formation water resistivities in this area increase in a southerly direction (M.H. Holtz and R.P. Major, 1995, unpublished data), we infer that the fluids that created the high-permeability diagenetically altered dolomite at Jordan field originated from the south.

The unaltered and altered dolomite textures have similar carbon isotope compositions, but they may be distinguished by different ranges of oxygen isotope composition (Figure 11). Carbon isotopic compositions are generally in the range of 4.5 to 6‰. The unaltered

dolomites have oxygen isotope compositions that range from 3 to 5.5‰, whereas the altered dolomite has more depleted oxygen isotope compositions of 1–4‰.

A similar range of isotopic compositions in San Andres dolomites from the Central Basin Platform was interpreted by Leary and Vogt (1990) as indicating that the altered dolomite was recrystallized either at elevated temperatures or in the presence of water with a depleted oxygen isotope composition. As outlined above, however, the textures of the altered dolomite suggest an episode of leaching. For example, textural evidence suggests that diagenetic alteration included removal of some cloudy, inclusion-rich cores of dolomite rhombs.

Dolomitization of the San Andres Formation has been attributed to hypersaline fluids on the basis of stratigraphic proximity to overlying evaporites and dolomite geochemistry (for example, Bein and Land, 1983). If the cloudy dolomite cores in the Jordan San Andres reservoir were formed from hypersaline brines early in the diagenetic history of these rocks, it can be inferred that the inclusion-rich cores of these crystals are enriched in ^{18}O relative to the limpid rims of the crystals. Thus, textural as well as geochemical data suggest that the diagenetically altered texture is the result of a permeability-increasing leaching event that preferentially removed ^{18}O-enriched, presumably less stoichiometric, dolomite. This resulted in a bulk rock with relatively depleted ^{18}O composition and enhanced permeability. Alternatively, the light oxygen isotope signature of the altered texture may be the result of recrystallization that preceded leaching; the limpid rims of altered dolomite rhombs have a different oxygen isotope composition than do the rims of unaltered dolomite rhombs. This cannot be determined with currently available sampling technology for isotope analysis.

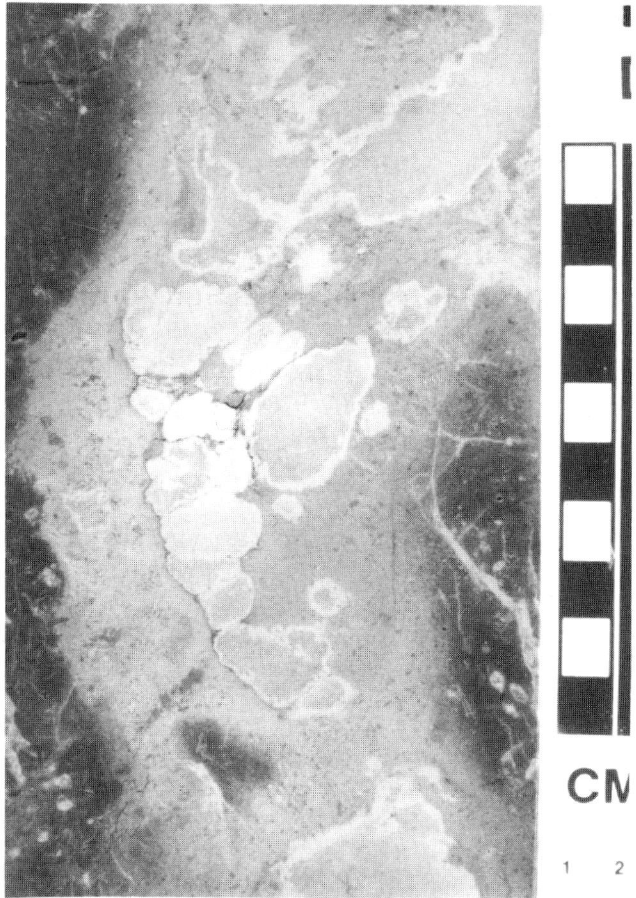

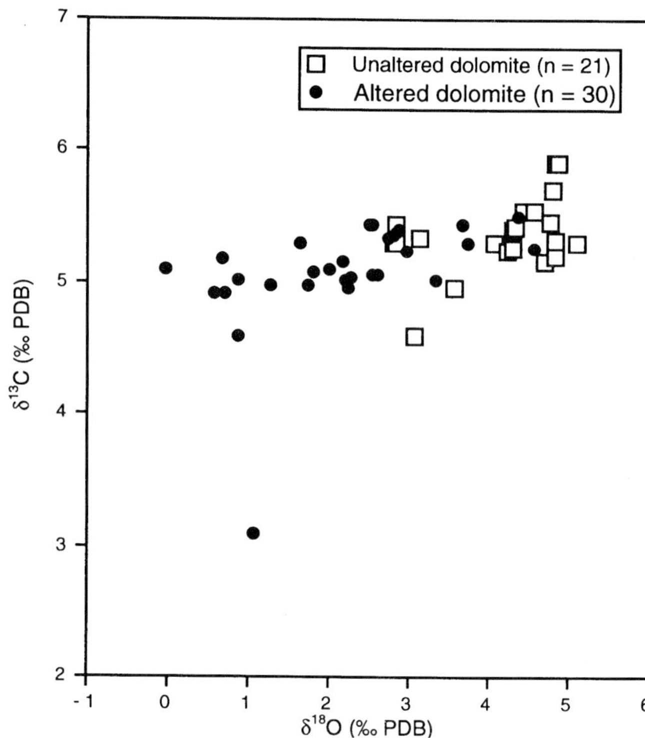

Figure 11. Carbon and oxygen isotope cross plot of diagenetically altered and unaltered pellet packstone/grainstone. The unaltered rock has oxygen isotope compositions of 3‰–5.5‰, and the altered rock has compositions of 1‰–4‰. All stable isotope data reported relative to the PDB standard.

Figure 10. Core photograph of anhydrite nodules partly altered to gypsum in diagenetically altered pellet packstone/grainstone. Note that the diagenetically altered (lighter colored) pellet packstone/grainstone is associated with alteration of gypsum to anhydrite [East Penwell San Andres Unit No. 1914, 3918 ft (1194 m)].

Flow Units

Because responses of neutron and acoustic logs to gypsum-bearing rocks differ, these two logs can be used to identify diagenetically altered rock textures in wells that are not cored. As indicated previously, the high-permeability diagenetically altered rock is associated with higher gypsum content than unaltered rock. Thus, altered reservoir rock containing abundant gypsum may be identified on wireline logs where dolomitic neutron log porosity exceeds acoustic porosity normalized to a dolomite matrix. The relationship of acoustic log porosity, neutron porosity, and percent of altered texture observed in slabbed core demonstrates the use of wireline logs to identify the diagenetically altered facies (Figure 12).

Jordan San Andres reservoir is divided into four flow units (sensu Ebanks, 1987) on the basis of both depositional facies and diagenetic overprint. Open-marine rocks are divided into three flow units defined by the stratigraphic patterns of diagenetically altered

facies as identified using wireline logs. The lowermost flow unit A is 100%, or nearly 100%, altered-texture rock and is characterized by a neutron log–acoustic log porosity-curve separation. The overlying flow unit B is composed of diagenetically unaltered rock characterized by a normalized neutron log that is in good agreement with a normalized acoustic log. The overlying flow unit C is composed of a mottled mixture of diagenetically altered and unaltered rock and is characterized by a neutron log–acoustic log separation (Figures 12, 13).

The uppermost flow unit D is composed of tidal-flat rocks that occur above the organic-rich shale identified by a gamma-ray marker (Figure 13). This marker can be correlated across the field. Porosity in this section occurs in pisolite packstone in which fenestrae and sheet cracks are not plugged with calcium sulfate cements.

Open-marine rocks contain thin zones of siliciclastic silt concentrated along stylolite swarms and shaly partings, many of which can be correlated throughout the field using gamma-ray logs. These intervals are interpreted to document low sedimentation rates associated with rapid rise of relative sea level. As such, these beds define cycle boundaries even where overlying and underlying depositional textures are nearly indistinguishable. Indeed, correlation of these features westward (updip) indicates that in more landward depositional environments these surfaces divide cycles defined by upward-shallowing depositional

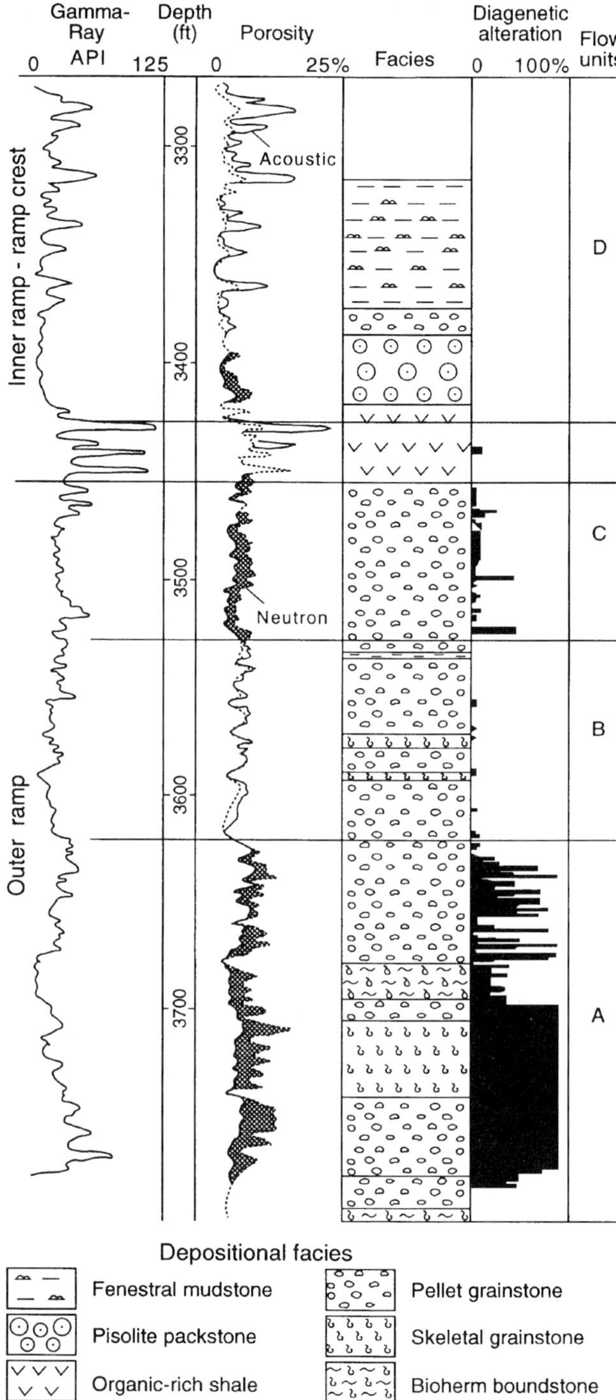

Figure 12. Calibration of acoustic log porosity, neutron log porosity, and percent of diagenetically altered facies observed in core. Note that the flow units and depositional rock types do not correlate. Flow units are dominantly controlled by diagenetic alteration. Data from Jordan University No. 114.

facies (S.C. Ruppel, 1995, personal communication). Thus, these gamma-ray correlations are approximate time lines (Ruppel et al., 1995).

The boundaries of between flow units A and B and flow units B and C are approximately parallel to time lines, which suggests that the shaly partings and stylolite swarms acted as aquatards to flow of the fluids that leached the open-marine parts of this reservoir. Note, however, that flow unit C pinches out in a downdip direction. Thus, despite the influence of depositional textures on flow-unit geometry, diagenetic overprint crosscuts depositional facies and is the major control on flow-unit geometry (Figure 13).

PETROPHYSICAL RESERVOIR DESCRIPTION

The second step in predicting reservoir quality is to make a petrophysical description of the reservoir. In this step, we take the geologic reservoir model, interpreted in terms of flow units, and describe porosity, permeability, and estimates of initial water and oil saturation within that geologic context.

Calibration of Porosity Between Logs and Cores

Although acoustic, neutron, and density porosity logs are available at Jordan field, the presence of abundant gypsum in this reservoir precludes the use of neutron and density logs for reliable porosity measurements. Neutron logs measure the bound water of hydration in gypsum as porosity, and the low density of gypsum (2.35 g/cm^3) relative to that of dolomite (2.88 g/cm^3) and anhydrite (2.98 g/cm^3) results in significant uncertainty concerning matrix density used in density log calculations (Tilly et al., 1982; Bebout et al., 1987; Holtz and Major, 1994). Furthermore, calibration of logs with core data necessitates the use of core porosity data collected using low-temperature nongypsum-destructive analytic techniques. Conventional high-temperature analysis volatilizes the bound water of hydration in gypsum crystals and yields incorrect values.

The most reliable log-derived porosities were made from a calibration of acoustic transit time with core porosity measured by low-temperature, nondestructive techniques. Use of all three open-hole porosity tools resulted in a poorer fit than the use of acoustic logs alone. The Jordan field porosity-acoustic transit time relationship is

$$\phi = -44.159 + 1.006\Delta t \qquad (1)$$

where ϕ = porosity (%) and Δt = acoustic transit time or two-way traveltime (μsec/ft).

Permeability Character

Division of the reservoir into flow units is based on identification of three rock types with distinctly different permeabilities: tidal-flat facies, unaltered open-marine facies, and altered open-marine facies. Minipermeameter data are used to evaluate open-marine facies rocks because available low-temperature whole-core data do not have sufficient sampling resolution to discriminate between unaltered and altered open-marine facies (there are no low-temperature core analyses from flow unit A). The geometric mean of

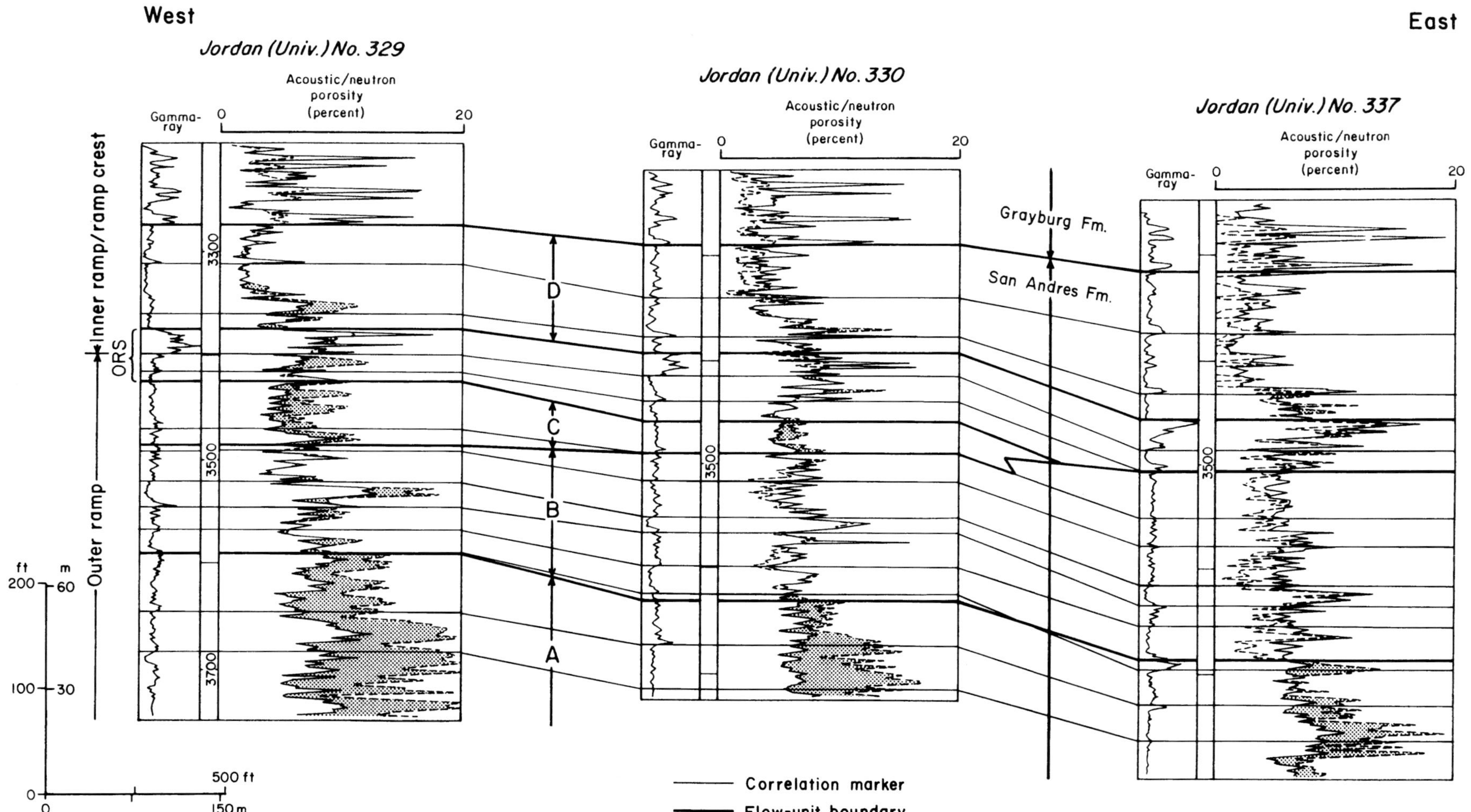

Figure 13. West-east cross section through Jordan (San Andres) field illustrating flow units. Diagenetically altered, high-permeability zones (shaded) are readily distinguished by comparing neutron and acoustic log porosity, because the latter is affected by the presence of gypsum. ORS = organic-rich shale.

minipermeameter data is 1.3 md for the unaltered open-marine facies and 8.4 md for altered open-marine facies (Figure 8). A relatively small amount of low-temperature, conventional core permeability data is available from tidal-flat facies cored in wells offsetting University Lands. The geometric mean of these data is 2.1 md (Figure 8), although the variance of these data is much greater than that of other rock types due to the large percentage of vuggy porosity in this rock type.

Core plugs carefully chosen to sample flow unit rock types were analyzed for porosity and permeability. Porosity-permeability relationships for the three rock types demonstrate that, for any porosity >10%, tidal-flat facies rocks have the highest permeabilities, followed by altered open-marine rocks and unaltered open-marine rocks. For porosities <10%, tidal-flat rocks have only slightly higher permeabilities than unaltered open-marine rocks (Figure 14).

Empirical equations that describe the relationship between porosity and permeability were derived for each flow unit. Because the relationship between porosity and permeability for tidal-flat rocks is bimodal, there are separate equations for flow unit D rocks having <10% porosity and >10% porosity. The 10% porosity threshold corresponds to the point at which vuggy porosity becomes interconnected, causing higher permeability per incremental increase in porosity.

$$\text{Flow unit A}: k = 3.5 \times 10^{-3} * 10^{0.234\phi} \tag{2}$$

$$\text{Flow unit B}: k = 2.5 \times 10^{-3} * 10^{0.188\phi} \tag{3}$$

$$\text{Flow unit C}: k = 3.08 \times 10^{-3} * 10^{0.209\phi} \tag{4}$$

$$\text{Flow unit D}: (<10\% \text{ porosity}): \tag{5}$$
$$k = 0.0271 * 10^{0.125\phi}$$

$$\text{Flow unit D}: (>10\% \text{ porosity}): \tag{6}$$
$$k = 4.096 \times 10^{-3} * 10^{0.329\phi}$$

where k = permeability (md) and ϕ = porosity (%).

The geometric mean of vertical permeability in West Jordan Unit Well No. 17-2 is 0.15 md and that of horizontal permeability is 0.3 md [similar to the relationship for the San Andres reservoir at Slaughter field (Ebanks, 1990)]. This suggests that vertical flow can occur on a small scale but not on a large scale. However, the Dystra Parsons coefficient value of 0.93 in well 17-2 documents a high degree of vertical permeability variation. This suggests that, on a 1-ft (0.3-m) scale, oil can flow vertically if it can flow horizontally. On a larger scale of tens of feet, permeability varies widely. Indeed, Ford and Kelldorf (1976) demonstrated that zones of bypassed oil and high-permeability "thief zones" occur in the Jordan (San Andres) reservoir because of the lack of vertical permeability continuity.

Capillary Pressure Curves

Nearly all of the resistivity logs in the Jordan (San Andres) reservoir postdate initiation of the waterflood. Thus, the formation waters at the time these logs were made may have been contaminated with floodwaters, and some zones may have been more thoroughly flushed than others. Under these circumstances, water

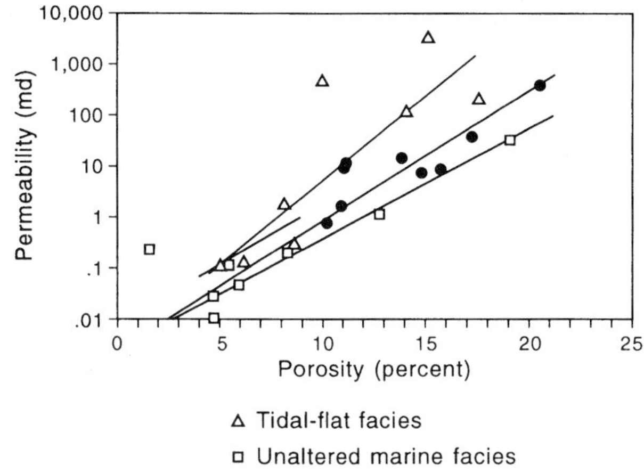

Figure 14. Relationship between porosity and permeability for the unaltered open-marine facies, altered open-marine facies, and tidal-flat facies.

resistivity is unknown; therefore, resistivity data cannot be used to calculate original saturations. An alternate approach is to use capillary-pressure data to estimate hydrocarbon saturation, following the procedure of Amyx et al. (1960). Nine plug samples were selected from undamaged parts of cores that had been previously sampled for high-temperature core analysis. These were analyzed by brine-injection capillary-pressure tests, using procedures that do not dehydrate gypsum. Plugs were chosen to represent the three principal rock types: diagenetically unaltered open-marine facies, diagenetically altered open-marine facies, and porous tidal-flat facies.

The shapes of capillary-pressure curves indicate the nature of the pore structure (Murray, 1960; Keith and Pittman, 1983). Both altered and unaltered open-marine rocks (Figure 15a) have capillary-pressure curve shapes that suggest unimodal pore-throat size distribution, thus substantiating the observation that either interparticle or intercrystalline pores dominate. This indicates that vugs do not make a major contribution to porosity, and explains why the core porosity–acoustic transit time function is a reliable method of calculating log-derived porosity. The shape of tidal-flat facies capillary-pressure curves (Figure 15b) also indicates a unimodal pore-throat-size distribution. However, the variability in pore-throat size is less for tidal-flat facies than for altered and unaltered open-marine facies. This suggests that the fenestral pores are not connected as touching vugs but, rather, are connected by sheet cracks of similar pore-throat size.

Capillary-pressure curves may be used to estimate original water and hydrocarbon saturations in various rock types provided that permeability and height above the free-water table are known (Amyx et al., 1960). The height of the free-water table in the Jordan (San Andres) reservoir is estimated to be 950 ft (290 m) subsea, as will be discussed in the next section. Analysis of capillary-pressure data by applying multiple nonlinear regression resulted in two equations, one

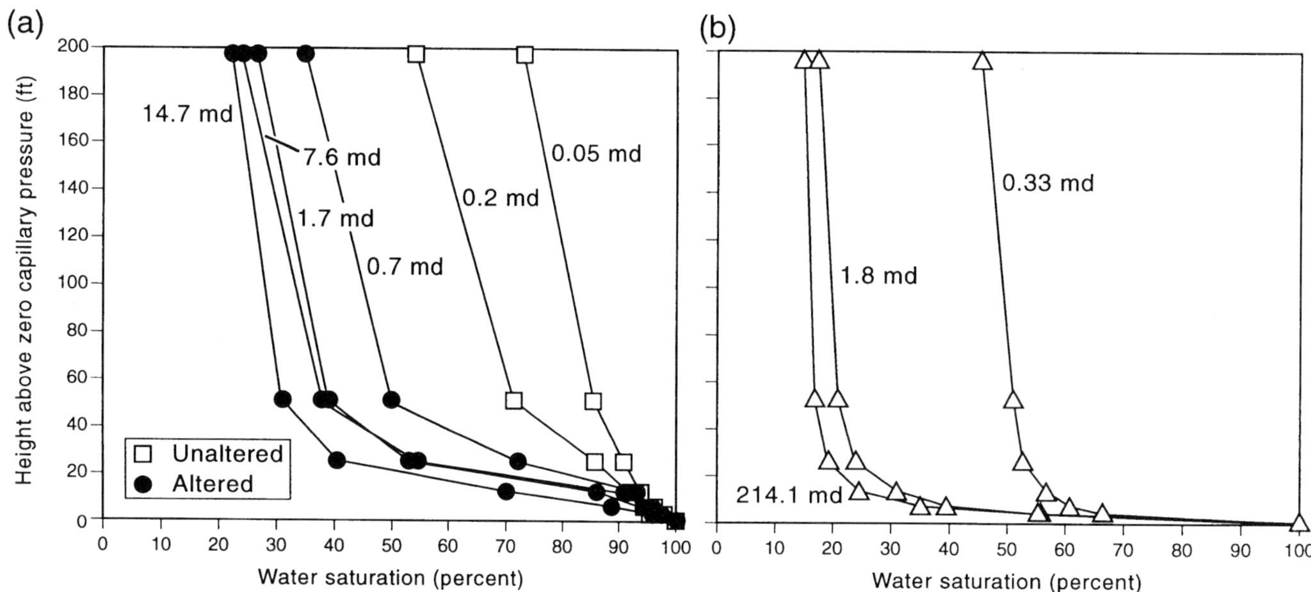

Figure 15. Brine capillary-pressure curves for the unaltered and altered open-marine facies (a) and tidal-flat facies (b).

that calculates water saturation for altered and unaltered open-marine facies (flow units A, B, and C), and one that calculates water saturation for tidal-flat facies (flow unit D).

Flow units A, B, and C:

$$S_{wi} = 120.22 - 36.016 * \log(h_{aw}) - 11.968 * \log(k) \qquad (7)$$

Flow unit D:

$$S_{wi} = 68.5 - 19.7 * \log(h_{aw}) - 6.59 * \log(k) \qquad (8)$$

where S_{wi} = initial water saturation (%) and h_{aw} = height above oil-water contact (ft).

Equation 7 is statistically significant, having an F value (analysis of variance) of 80.4 (F critical = 3.15) at a 95% confidence level and an r^2 of 0.84. Equation 8 is also statistically significant at a 95% confidence level with an F value of 60.69 (F critical = 3.49) and an r^2 of 0.87.

RESOURCE EVALUATION

Interpretation of reservoir volumes is the last step in a comprehensive prediction of reservoir quality. The goal of this part of the study is to quantitatively estimate the amount of remaining oil and to map its distribution in the reservoir.

Production Patterns

Per-well production data are available only for the postwaterflood time period, 1969 to the present. Production data before 1969 are available on a per-lease basis (the two Jordan San Andres units on University Lands were composed of several separate leases during different periods). Flow tests were performed periodically on individual wells, and these test data were used to apportion annual lease production to each well within the lease.

The updip part of the reservoir has been more densely drilled than other parts (Figure 1). To compensate for variations in well density, the drainage area of each well was approximated within a grid of square 40-ac cells. The fraction of well drainage areas within each 40-ac cell was used to apportion production. Thus, a single data point for each 40-ac cell expresses production in units of million stock tank barrels per acre. These data were contoured to produce a cumulative production map (Figure 16). This map exhibits a trend of high production extending from the updip central-western margin of the field to the downdip southeastern corner. The southwestern corner of the field is an area of low production.

Original Oil in Place

Calculation of original oil in place (OOIP) requires knowledge of the height above free-water level. Well logs cannot be used to calculate water saturation, and thus identify the depth at which water saturations are 100%, because all resistivity logs in the Jordan (San Andres) reservoir postdate waterflooding, and original water resistivities are unknown. Two observations were used to estimate the elevation of free water. First, core analysis data indicate no hydrocarbon saturation below 950 ft (290 m) subsea. A core from University Well No. 638W contained no residual oil saturation below 950 ft (290 m) subsea; a core from University Well No. 114 had residual oil saturation to 930 ft (283 m) subsea, at which depth the rock has less than 0.1 md permeability; and a core from University Well No. 219W had no residual oil saturation below 945 ft (288 m) subsea. Second, shallow-resistivity values recorded by microresistivity logs were compared with medium- and deep-resistivity values in porous zones. These resistivities indicate that, below 950 ft (290 m) subsea, there was

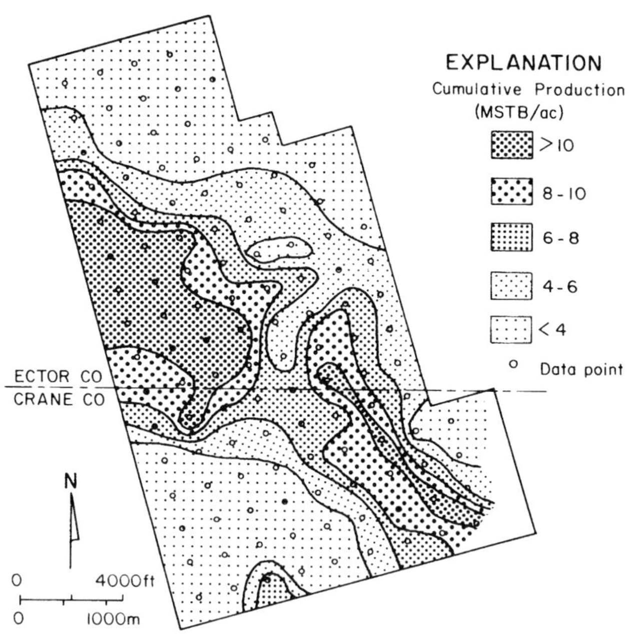

Figure 16. Cumulative production map based on reconstructed per-well production, 1937–1988, normalized on a 40-ac grid. Note that the axis of highest cumulative production has a northwest-southeast orientation. MSTB/ac = million stock tank barrels per acre.

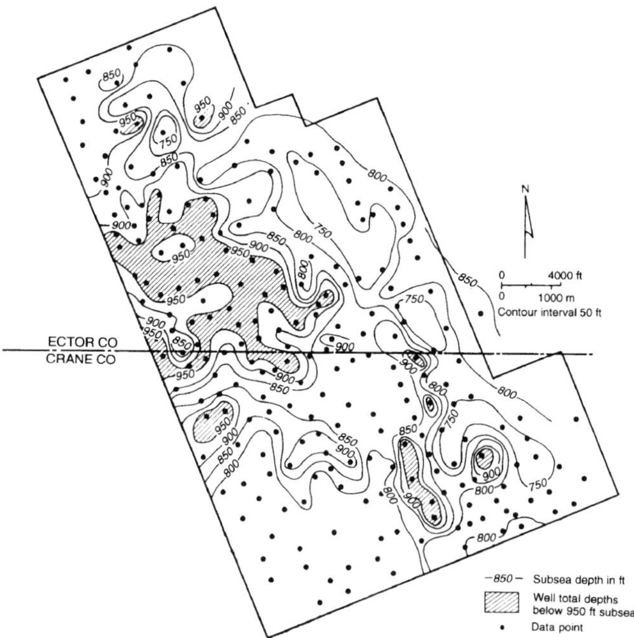

Figure 17. Contour map of total depths of wells. Note that most wells do not reach 950 ft (290 m) subsea, which is the interpreted free-water level.

no flushing of movable oil adjacent to well bores. Thus, we interpret the free-water level to be 950 ft (290 m) subsea. This interpretation is supported by the results of well-deepening associated with initiation of waterflooding in the 1970s. Wells that did not previously reach the 9-ft (2.7-m) free-water level and were deepened [generally to depths ≤950 ft (290 m)] began producing oil at greater rates, indicating that much of the lower part of the reservoir had not been drained.

Many wells in the Jordan (San Andres) reservoir do not penetrate the lower part of the reservoir, even though some wells were deepened in the 1970s. For this reason, we have divided the reservoir into two zones for the purpose of evaluating remaining oil. A map of total depths of wells (Figure 17) illustrates the geometry of the part of the reservoir between total depths of wells and the free-water level of 950 ft (290 m) subsea. Remaining oil in the reservoir was calculated separately for the contacted and uncontacted parts of the reservoir using an oil formation volume factor of 1.28 (Holtz et al., 1991).

Contacted Reservoir

Original oil in place was mapped in units of oil saturation × porosity × thickness ($S_o\phi h$) for each flow unit. Saturations were calculated using equations 7 and 8 with porosity values from acoustic logs calculated using equation 1. Volumes of original oil in place were calculated by gridding $S_o\phi h$ values over their representative areas and multiplying $S_o\phi h$ by area to yield volume. This yielded original oil-in-place values of 36.3

MMSTB for flow unit A, 53.5 MMSTB for flow unit B, 30.9 MMSTB for flow unit C, and 96.8 MMSTB for flow unit D (Figure 18). The sum of original oil in place calculated for all four flow units is 218 MMSTB, which yields a 31.2% recovery efficiency.

Maps of $S_o\phi h$ illustrate the spatial distribution of original oil in place for each flow unit. The flow unit A map (Figure 18a) indicates that the highest oil volumes were contained in the updip central part of the field. Flow unit A is close to the free-water level in the downdip (east) part of the field; the zero contour marks the point at which the top of flow unit A meets the free-water level. The flow unit B $S_o\phi h$ map (Figure 18b) indicates highest oil volumes in the updip central part of the field and in the southwest part of the field. The volume of oil generally decreases downdip. Flow unit B is absent in the west-central and southeast parts of the field; therefore, flow unit C is immediately superjacent to flow unit A in these areas. The flow unit C map (Figure 18c) illustrates an area of highest oil volume crosscutting structure from the downdip southeastern part of the field to the updip central part of the field. Flow unit C is absent in the northeast and southwest areas. The flow unit D map (Figure 18d) indicates generally increasing oil volumes in the downdip (east) part of the field and in the updip north-central part of the field.

Uncontacted Reservoir

Estimating the amount of original oil in place in uncontacted parts of the reservoir clearly requires some assumptions. We start with the conservative assumption that those parts of the reservoir below total depths of wells will, in most instances, be close to

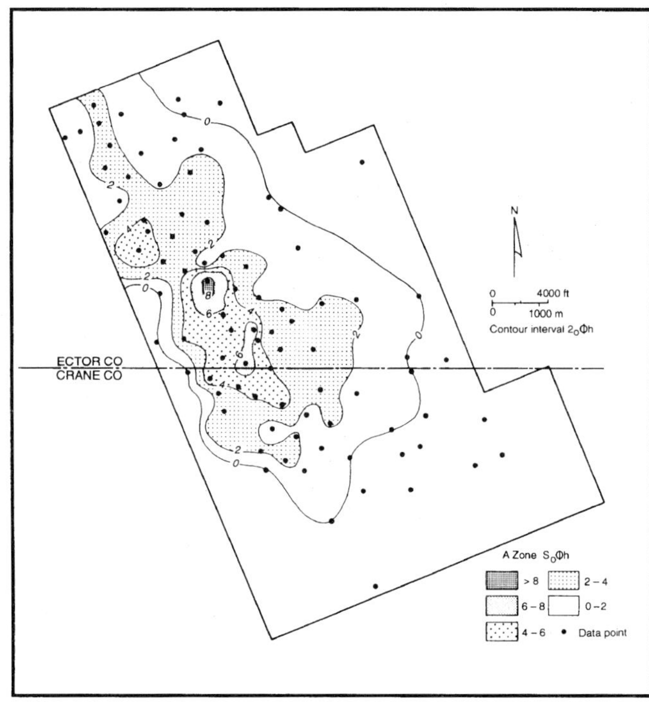

a.

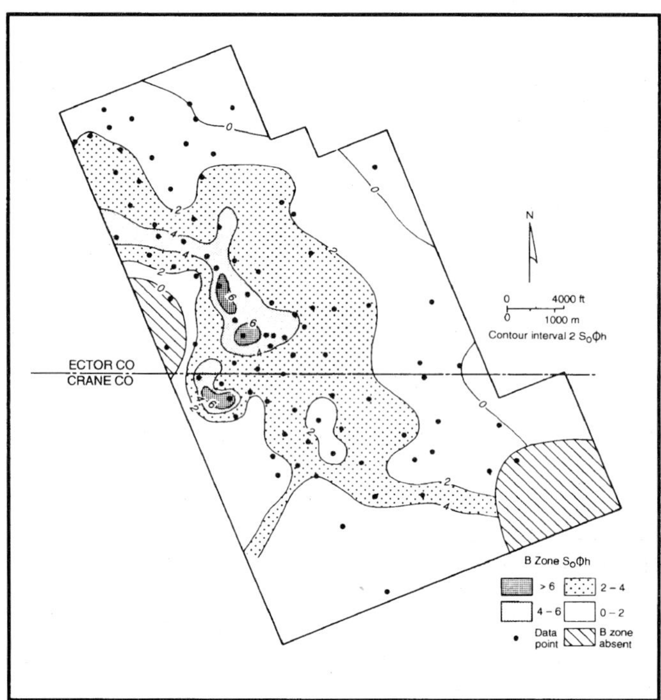

b.

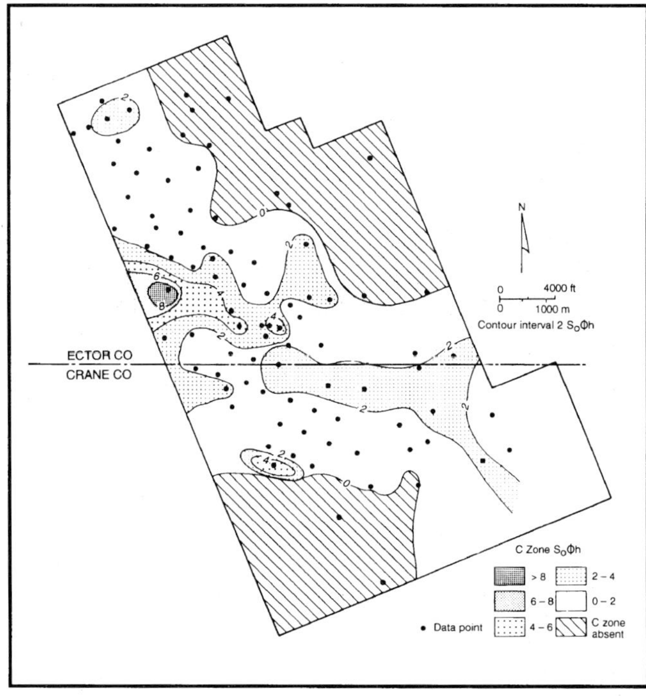

c.

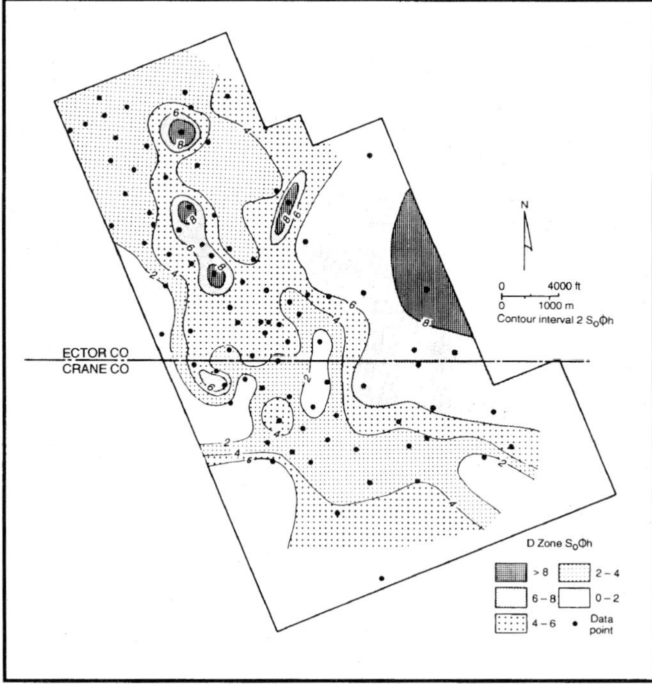

d.

Figure 18. (a) Flow unit A oil saturation × porosity × thickness $S_o\phi h$ map. The zero contour on the western (downdip) side of the field indicates where flow unit A dips below the free-water level. (b) Flow unit B $S_o\phi h$ map. (c) Flow unit C $S_o\phi h$ map. The zero contours indicate that this flow unit pinches out to the northeast and southwest. (d) Flow unit D $S_o\phi h$ map.

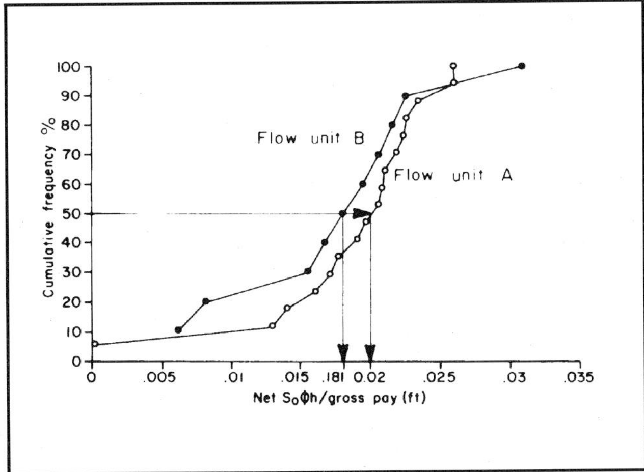

Figure 19. Cumulative frequency of net $S_o\phi h$/gross pay ratio for flow units A and B for wells in which these flow units are in contact with free water.

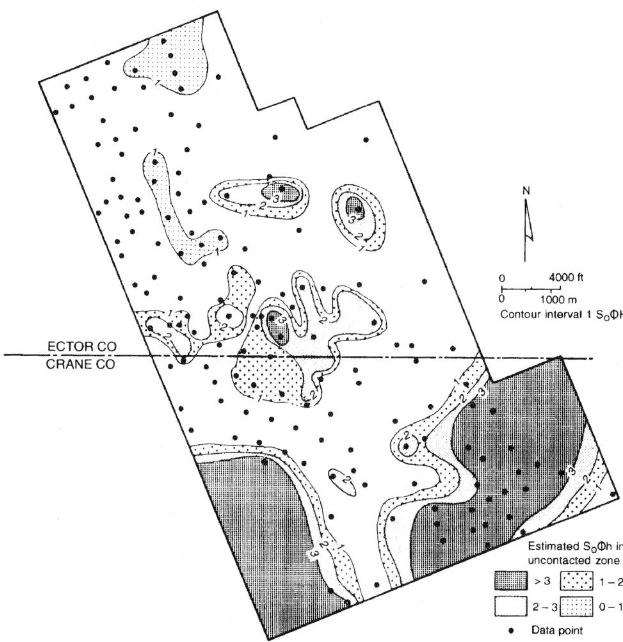

Figure 20. Contour map of $S_o\phi h$ for the uncontacted part of the reservoir.

the free-water level and, therefore, have relatively low saturations because of buoyancy considerations (Amyx et al., 1960). In the absence of log data, we estimated original oil in place by calculating the average $S_o\phi h$/gross pay ratio, by flow unit, where well-log data are available and, importantly, where the flow unit is near the free-water level. This ensures that the database we use for estimating the average $S_o\phi h$/gross pay ratio contains only those parts of each flow unit that have relatively low oil saturations. The average $S_o\phi h$/gross pay ratios for flow units A and B are illustrated in Figure 19. There were only two logged wells that contained flow unit C and one that contained flow unit D in contact with free water; we have made calculations for these flow units using the flow unit B average from Figure 19 because this is the lowest value and we seek to make conservative estimates.

The uncontacted part of the reservoir contains an OOIP of 23 MMSTB: 6.6 MMSTB in flow unit A, 4.7 MMSTB in flow unit B, 1.0 MMSTB in flow unit C, and 10.7 MMSTB in flow unit D. The highest volumes of uncontacted oil are in the southwest and southeast corners of the field. Some local high values are present in the central part of the field (Figure 20).

Distribution of Remaining Oil

Our calculation of total OOIP for both contacted and uncontacted parts of the reservoir is 240.5 MMSTB, which indicates a recovery efficiency of 28%. Thus, 173 MMSTB remain in this reservoir. By applying an average saturation of oil residual to waterflood for the San Andres/Grayburg reservoirs (Finley et al., 1990), we estimate that 113 MMSTB are residual oil and 59.5 MMSTB of remaining oil are mobile to waterflood.

Contacted Reservoir

Within the contacted part of the reservoir there remain 47.3 MMSTB of mobile oil and 102.2 MMSTB of oil residual to waterflood. To obtain a conservative

remaining oil distribution, all oil production was assumed to have come from the contacted part of the reservoir. Cumulative production (Figure 16) was subtracted from OOIP, calculated by summing OOIP in all flow units (Figure 18), resulting in a remaining-oil map for the contacted zone (Figure 21).

Four areas of high remaining oil occur in the contacted part of the reservoir (Figure 21). The north-trending area of high remaining oil in the eastern central part of the field is coincident with a relative high OOIP in flow unit D (Figure 18d), suggesting that most of this oil is in flow unit D. The high remaining oil in the updip western-central part of the field is approximately coincident with high OOIP values in all flow units (Figure 18). The area of high remaining oil in the northern part of the field is approximately coincident with high OOIP in flow unit D and, to a lesser extent, flow unit B, and the high remaining oil in the southwest corner of the field is also approximately coincident with high OOIP in flow units B and D (Figure 18b, d). This suggests that this oil is primarily in flow units B and D. Most of the remaining oil in the contacted part of this reservoir is in low-permeability, diagenetically unaltered marine facies of flow unit B and tidal-flat facies of flow unit D.

Because the remaining-oil map (Figure 21) illustrates volumes of remaining mobile oil and residual oil, it is difficult to evaluate the extent to which this remaining resource can be produced by increased sweep efficiency. Unfortunately, there are no irreducible oil saturation values for each of the flow units identified in the Jordan (San Andres) reservoir. However, using an average irreducible oil saturation value of 32% for dolomitized San Andres and Grayburg reservoirs in the Central Basin Platform (Finley et al.,

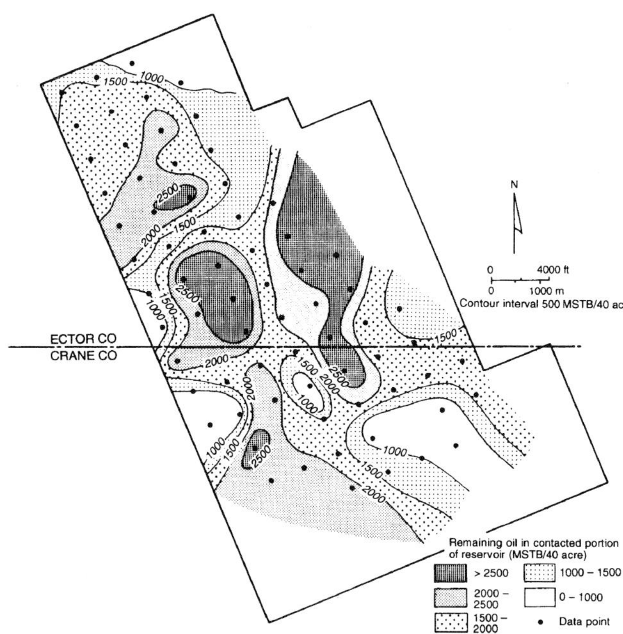

Figure 21. Contour map of remaining oil in the contacted part of the reservoir, constructed by subtracting cumulative production from original oil in place. MSTB = thousand stock tank barrels.

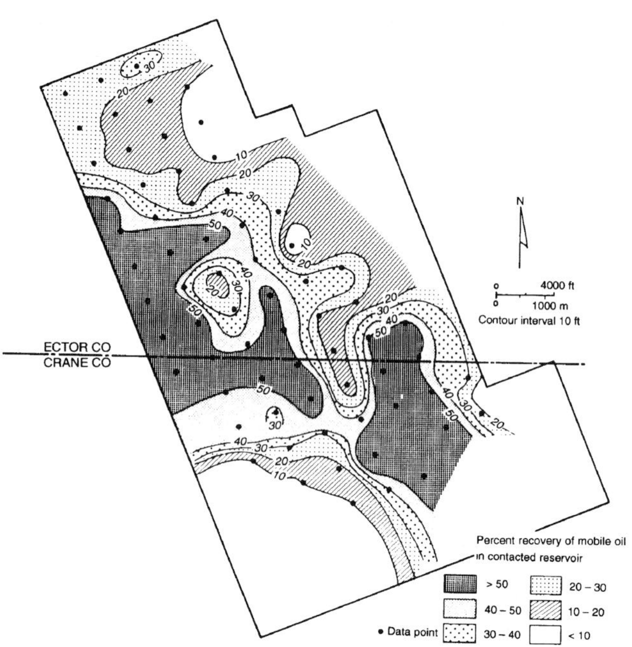

Figure 22. Contour map of sweep efficiency defined as the percent of original mobile oil in place that has been produced.

1990), we can calculate the approximate sweep efficiency based on cumulative production. A map of current recovery efficiency, calculated as percent of original mobile oil in place that has been produced, indicates that the north, east, and southwest parts of the field have low recovery efficiencies (Figure 22).

Uncontacted Reservoir

About 12.2 MMSTB of mobile oil and 10.8 MMSTB of residual oil remain within the uncontacted part of the reservoir, based on the assumption that cumulative production from the field has come exclusively from the contacted part of the reservoir. A large area of high remaining oil occurs in the southeast corner of the field (Figure 20), where flow unit A is below free-water level (Figure 18a) and flow unit B is absent (Figure 18b), indicating that most of the oil in this area of the uncontacted part of the reservoir is in flow units C and D. A large area of high remaining oil also occurs in the southwest corner of the field, where most wells do not penetrate below 800 ft (244 m) subsea. In this area, flow unit A is below the oil-water contact (Figure 18a) and flow unit C is absent (Figure 18c), indicating that most of the oil in this area is in flow units B and D.

APPLICATION OF RESERVOIR QUALITY PREDICTION

This prediction of reservoir quality in the Jordan (San Andres) field provides three avenues for increased production: (1) focusing the waterflood in selected areas of the contacted part of the reservoir, (2) deepening wells to reach the uncontacted part of the reservoir, and (3) initiation of a carbon dioxide flood.

The parts of the contacted reservoir that are most prospective for waterflood profile modification occur in flow unit B, which is characterized by low permeability, and flow unit D, which is characterized by highly heterogeneous permeability. The largest area of unswept remaining mobile oil in flow unit B is in the southwest corner of the field; the largest area of unswept mobile oil in flow unit D is in the east-central part of the field (Figures 21, 22). In these areas, selective well-bore plugging and perforation squeezing could focus the waterflood and increase ultimate oil recovery.

Large parts of the reservoir are below total depths of wells and, therefore, have not been drained. Consider the high cumulative production trend that extends from the updip central-western part of the field to the downdip southeastern corner (Figure 16) and the somewhat similar trend of areas in which well bores have penetrated to the free-water depth of 950 ft (290 m) subsea (Figure 17). The areas of highest cumulative production are due, at least in part, to well bores having been drilled through the entire pay zone. The most prospective areas for deepening wells are those in which high-permeability flow units A and C are uncontacted. The largest area of uncontacted flow unit C is in the southeast corner of the field. Most of flow unit A that is above the free-water level is at least partially penetrated by well bores, but there are deepening opportunities in this flow unit in the central part of the field and near the central-western margin of the field. There are large volumes of remaining oil in flow units B and D in the southwest corner, and flow unit D in the southeast corner, although efficient production from these areas will probably require focusing the waterflood. Because porosity and permeability in tidal-flat facies are extremely heterogeneous and prediction is unreliable, we assign a low priority to well deepening in the southeast corner of the field.

The Jordan (San Andres) reservoir is a good candidate for increased ultimate recovery by miscible-gas enhanced oil recovery operations. The reservoir depth, oil gravity, current oil saturation, and oil viscosity suggest that this reservoir is an excellent candidate for carbon dioxide flooding, according to the criteria of Stalkup (1983). Taber and Martin (1992) predicted that carbon dioxide flooding would increase average recovery for West Texas San Andres reservoirs by 11% of OOIP. Application of this prediction to the Jordan (San Andres) reservoir indicates a potential for an additional 24 MMSTB of oil production from the contacted part of the reservoir.

CONCLUSIONS

Reservoir quality in Jordan (San Andres) field is predicted for four flow units identified on the basis of both depositional facies and subsequent diagenetic alteration. Volumetric calculations and cumulative production patterns indicate that of an original 240 MMSTB of OOIP, 113 MMSTB of which are residual and 127.5 MMSTB of which are mobile, ~59.5 MMSTB of mobile oil remain as both bypassed and uncontacted oil. The largest volumes of bypassed oil occur in low-permeability flow unit B and heterogeneous-permeability flow unit D in areas of low cumulative production. Waterflood profile modification by selective perforation squeezing may focus injection water into the flow units in these areas and contact bypassed oil that would otherwise remain unrecovered. Many of the wells in this reservoir do not penetrate to the free-water level; parts of the reservoir in which the high-permeability flow units A and C are uncontacted by well bores are the principal targets for increased production by well-bore deepening. The Jordan (San Andres) reservoir has physical characteristics that make it an excellent candidate for enhanced oil recovery by carbon dioxide flood.

ACKNOWLEDGMENTS

Funding for this study was provided by The University of Texas System as part of a larger study of reservoirs on The University of Texas Lands. Shell Oil Company, Hondo Oil Company, The University of Texas Lands Office, and the Railroad Commission of Texas provided access to data. We thank M.G. Kittridge for minipermeameter data that were collected in the laboratories of the Department of Petroleum and Geosystems Engineering, The University of Texas at Austin. J.E. Nicol and Mohammed Sattar provided technical support. Brine capillary-pressure analyses were conducted by Bell Laboratories (Midland), and Radian Corporation donated mapping software. We are grateful for the review comments of W.A. Ambrose, P.M. Harris, T.F. Hentz, J.A. Kupecz, and W.G. Zempolich. Parts of this paper have been published by the Society of Petroleum Engineers in the *Journal of Petroleum Technology* (1990, v. 42, no. 10, p. 1304–1309) and *Permian Basin Oil and Gas Recovery Conference Proceedings* (1994, p. 565–576). This material is used here with permission of the society. Publication authorized by the director of the Bureau of Economic Geology, The University of Texas at Austin.

REFERENCES CITED

Amyx, J.W., D.M. Bass, Jr., and R.L. Whiting, 1960, Petroleum Reservoir Engineering: New York, McGraw Hill Book Co., 610 p.

Bebout, D.G., F.J. Lucia, C.R. Hocott, G.E. Fogg, and G.W. Vander Stoep, 1987, Characterization of the Grayburg reservoir, University Lands Dune Field, Crane County, Texas: The University of Texas at Austin, Bureau of Economic Geology Report of Investigations No. 168, 98 p.

Bein, A., and L.S. Land, 1983, Carbonate sedimentation and diagenesis associated with Mg-Ca-Chloride brines: the Permian San Andres Formation in the Texas Panhandle: Journal of Sedimentary Petrology, v. 53, p. 243–260.

Carozzi, A.V., and D. Von Bergen, 1987, Stylolitic porosity in carbonates: a critical factor for deep hydrocarbon production: Journal of Petroleum Geology, v. 10, p. 267–282

Chandler, M.A., D.J. Goggin, and L.W. Lake, 1989, A mechanical field permeameter for making rapid, nondestructive permeability measurements: Journal of Sedimentary Petrology, v. 59, p. 613–615.

Ebanks, W.J., Jr., 1987, Flow unit concept—integrated approach to reservoir description for engineering projects (abs.): AAPG Bulletin, v. 71, p. 551–552.

Ebanks, W.J., Jr., 1990, Geology of the San Andres reservoir, Mallet Lease, Slaughter field, Hockley County, Texas: implications for reservoir engineering projects, *in* D.G. Bebout and P.M. Harris, eds., Geologic and engineering approaches in evaluation of San Andres/Grayburg hydrocarbon reservoirs— Permian Basin: The University of Texas at Austin, Bureau of Economic Geology, p. 75–85.

Eijpe, R., and K.J. Weber, 1971, Mini-permeameters for consolidated rock and unconsolidated sand: AAPG Bulletin, v. 55, p. 307–309.

Finley, R.J., S.E. Laubach, N. Tyler, and M.H. Holtz, 1990, Opportunities for horizontal drilling in Texas: The University of Texas at Austin, Bureau of Economic Geology Geological Circular 90-2, 32 p.

Ford, W.O., Jr., and W.F.N. Kelldorf, 1976, Field results of a short-setting-time polymer placement technique: Journal of Petroleum Technology, v. 28, p. 749–756.

Galley, J.E., 1958, Oil and geology in the Permian Basin of Texas and New Mexico, *in* L.G. Weeks, ed., Habitat of oil: AAPG, Tulsa, OK, p. 395–446.

Harris, P.M., C.A. Dodman, and D.M. Bliefnick, 1984, Permian (Guadalupian) reservoir facies, McElroy field, West Texas, *in* P.M. Harris, ed., Carbonate sands—a core workshop: SEPM Core Workshop 5, p. 136–174.

Harris, P.M., and S.D. Walker, 1990, McElroy field: Development geology of a dolostone reservoir, Permian Basin, West Texas, *in* D.G. Bebout and

P.M. Harris, Geologic and engineering approaches in evaluation of San Andres/Grayburg hydrocarbon reservoirs—Permian Basin: The University of Texas at Austin, Bureau of Economic Geology, p. 275–296.

Holtz, M.H., and R.P. Major, 1994, Effects of depositional facies and diagenesis on calculating petrophysical properties from wireline logs in Permian carbonate reservoirs of West Texas (abs.): AAPG Bulletin, v. 78, p. 494–495.

Holtz, M.H., N. Tyler, C.M. Garrett, Jr., W.G. White, and N.S. Banta, 1991, Atlas of major Texas oil reservoirs database: The University of Texas at Austin, Bureau of Economic Geology, 1 disk.

Keith, B.D., and E.D. Pittman, 1983, Bimodal porosity in oolitic reservoirs—effect on productivity and log response, Rhodessa Limestone (Lower Cretaceous), East Texas Basin: AAPG Bulletin, v. 67, p. 1391–1399.

Kerans, C., F.J. Lucia, and R.K. Senger, 1994, Integrated characterization of carbonate ramp reservoirs using Permian San Andres Formation outcrop analogs: AAPG Bulletin, v. 78, p. 191–216.

Kittridge, M.G., 1988, Analysis of areal permeability variations—San Andres Formation (Guadalupian): Algerita escarpment, Otero County, New Mexico: M.S. thesis, The University of Texas, Austin, Texas, 361 p.

Leary, D.A., and J.N. Vogt, 1990, Diagenesis of the San Andres Formation (Guadalupian), Central Basin Platform, West Texas, in D.G. Bebout and P.M. Harris, eds., Geological and engineering approaches in evaluation of San Andres/Grayburg hydrocarbon reservoirs—Permian Basin: The University of Texas at Austin, Bureau of Economic Geology, p. 21–47.

Longacre, S.A., 1980, Dolomite reservoirs from Permian biomicrites, in R.B. Halley and R.G. Loucks, eds., Carbonate reservoir rocks: SEPM Core Workshop 1, p. 105–117.

Longacre, S.A., 1983, A subsurface example of a dolomitized middle Guadalupian (Permian) reef from West Texas, in P.M. Harris, ed., Carbonate buildups—a core workshop: SEPM Core Workshop 4, p. 304–326.

Major, R.P., D.G. Bebout, and F.J. Lucia, 1988, Depositional facies and porosity distribution, Permian (Guadalupian) San Andres and Grayburg formations, PJWDM field complex, Central Basin Platform, West Texas, in A.J. Lomando and P.M. Harris, eds., Giant oil and gas fields: SEPM Core Workshop 12, p. 615–648.

Major, R.P., G.W. Vander Stoep, and M.H. Holtz, 1990, Delineation of unrecovered mobile oil in a mature dolomite reservoir: East Penwell San Andres Unit, University Lands, West Texas: The University of Texas at Austin, Bureau of Economic Geology Report of Investigations No. 194, 56 p.

Murray, R.C., 1960, Origin of porosity in carbonate rocks: Journal of Sedimentary Petrology, v. 30, p. 59–84.

Ruppel, S.C., C. Kerans, R.P. Major, and M.H. Holtz, 1995, Controls on reservoir heterogeneity in Permian shallow-water platform carbonate reservoirs, Permian Basin: implications for improved recovery: The University of Texas at Austin, Bureau of Economic Geology Geologic Circular 95-2, 30 p.

Stalkup, F.I., Jr., 1983, Miscible displacement: Society of Petroleum Engineers Monograph Series 8, 204 p.

Taber, J.J., and D.F. Martin, 1992, Carbon dioxide flooding: Journal of Petroleum Technology, v. 44, p. 396–400.

Tilly, H.P., B.J. Gallagher, and T.D. Taylor, 1982, Methods for correcting porosity data in a gypsum-bearing carbonate reservoir: Journal of Petroleum Technology, v. 34, p. 2449–2454.

Von Bergen, D., and A.V. Carozzi, 1990, Experimentally-simulated stylolitic porosity in carbonate rocks: Journal of Petroleum Geology, v. 13, p. 179–192.

Wanless, H.R., E.A. Burton, and J.J. Dravis, 1981, Hydrodynamics of carbonate fecal pellets: Journal of Sedimentary Petrology, v. 51, p. 27–36.

Ward, R.F., C.G.St.C. Kendall, and P.M. Harris, 1986, Upper Permian (Guadalupian) facies and their association with hydrocarbons—Permian Basin, West Texas and New Mexico: AAPG Bulletin, v. 70, p. 239–262.

Smosna, R., and K.R. Bruner, 1997, Depositional controls over porosity development in lithic sandstones of the Appalachian Basin: reducing exploration risk, *in* J.A. Kupecz, J. Gluyas, and S. Bloch, eds., Reservoir quality prediction in sandstones and carbonates: AAPG Memoir 69, p. 249–265.

Chapter 16

Depositional Controls Over Porosity Development in Lithic Sandstones of the Appalachian Basin: Reducing Exploration Risk

Richard Smosna
Kathy R. Bruner
Department of Geology and Geography, West Virginia University
Morgantown, West Virginia, U.S.A.

ABSTRACT

Litharenites and sublitharenites of the Devonian Lock Haven Formation contain abundant rock fragments of shale and phyllite. These labile grains suffered varying degrees of destruction in several depositional environments; hence, sedimentary processes largely controlled the sandstones' mineral composition. Fluvial sandstones have a high lithic content, distributary mouth-bar and offshore-shelf sandstones have an intermediate content, and barrier-island sandstones have a low content.

Primary porosity relates inversely to compaction of the lithic grains, decreasing from a maximum minus-cement porosity of $\phi_{mc} = 33\%$ down to zero as lithics increase. The majority of primary porosity, however, has been occluded by cementation. Secondary porosity, created chiefly by dissolution of the chemically unstable rock fragments, is greatest ($\phi_{rf} = 13\%$) for sandstones of a moderate lithic content.

Because of these relationships among depositional processes, lithology, and porosity, we predict that sandstones of different sedimentary environments should exhibit distinct porosity volumes and vary in their reservoir potential. Mouth-bar sandstones will have good total porosity, good secondary porosity, and offer the best reservoir quality. Shelf sandstones will have fair total porosity, most of which is secondary, whereas beach sandstones will have low total porosity, most of which is primary. Fluvial sandstones will be the poorest reservoirs.

INTRODUCTION

The majority of secondary porosity in sandstones is attributed to the dissolution of feldspars and carbonate minerals (Heald and Larese, 1973; Pittman, 1979; Schmidt and McDonald, 1979a, b; Björlykke, 1983, 1984; Shanmugam, 1985, 1990). In contrast, substantial secondary porosity in Upper Devonian reservoirs of the Appalachian basin has been generated by the leaching of metamorphic and sedimentary rock fragments (Bruner and Smosna, 1994). Moreover, we have observed lithmoldic porosity in sandstones throughout the Appalachian stratigraphic section from the Lower Devonian to the Pennsylvanian. Other workers have noted comparable dissolution porosity in sandstones, including major Tertiary reservoirs of the Gulf Coast such as the Wilcox Group and Frio Formation (Moncure et al., 1984; Siebert et al., 1984; Parnell, 1987; Stonecipher and May, 1990).

We suspect that lithmoldic porosity may be even more widespread, although perhaps overlooked or misidentified. Sediments derived from collision-suture belts or foreland fold and thrust belts contain an abundance of chemically unstable metamorphic and sedimentary rock fragments (Dickinson and Suczek, 1979; Bird and Molenaar, 1992; Dickinson, 1988; Potocki and Hutcheon, 1992). Likewise, sands of large river systems that drain passive, Atlantic-type continental margins are enriched in metamorphic and sedimentary rock fragments (Potter, 1978; Loucks et al., 1984). Sandstones of these geologic settings, in particular, may be most susceptible to the creation of similar secondary porosity.

In Upper Devonian reservoirs of the Appalachian basin, creation of secondary porosity was controlled by the number of chemically unstable grains, the amount of primary porosity, and the rocks' early permeability. Primary porosity, in turn, had been influenced by compaction and the volume of ductile grains, quartz overgrowths, dolomite cement, authigenic clays, and the introduction of solid bitumen (Bruner and Smosna, 1994). Figure 1 traces the changing petrographic composition of these reservoir sandstones through time: from deposition, through compaction, to grain dissolution. Mechanical compaction and cementation reduced primary porosity to ~1%, whereas chemical leaching later generated 5% secondary porosity.

Knowledge of the exact relationship between porosity and petrographic composition may allow for the prediction of porosity distribution in development wells. But such correlations generally offer little aid for exploratory wells. In sparsely tested areas, geologists may have few data concerning the content of soluble minerals, ductile minerals, cement volumes, percent clay, or presence of bitumen in a target sandstone. Instead, predicting porosity in advance of the drill must be linked to the kinds of information available at an early stage of exploration, such as depositional environment (Pryor, 1973; Loucks et al., 1984; Stonecipher et al., 1984; Stonecipher and May, 1990), hydrogeological regime (Björlykke, 1983; Galloway, 1984; Shanmugam, 1990), or burial depth (Sclater and Christie, 1980; Chilingarian, 1983; Baldwin and Butler, 1985).

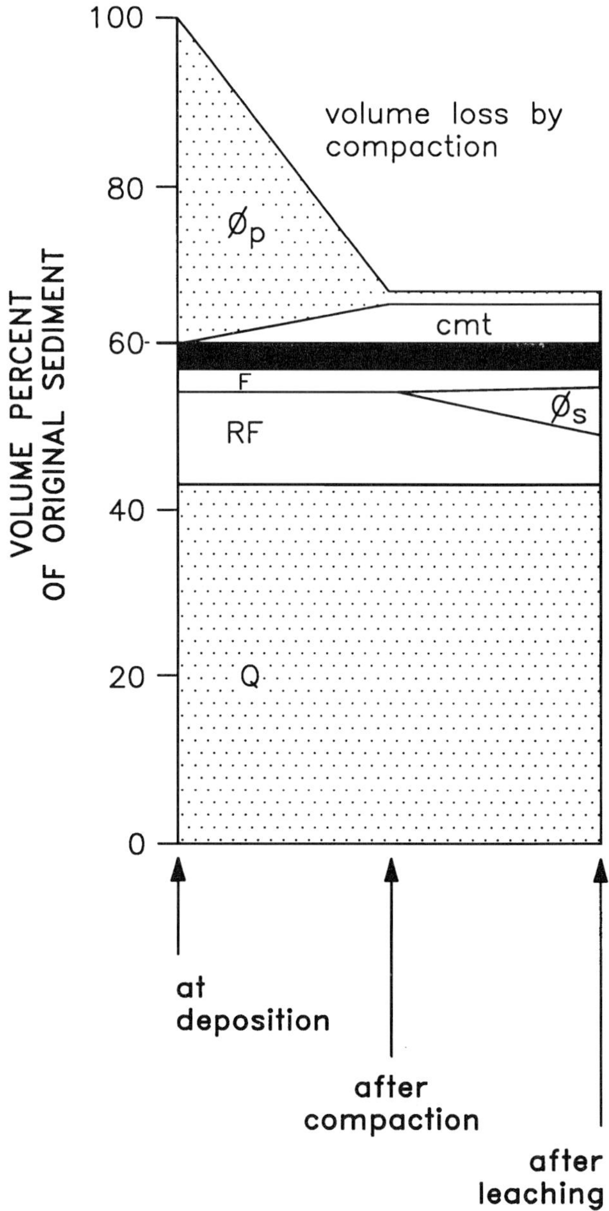

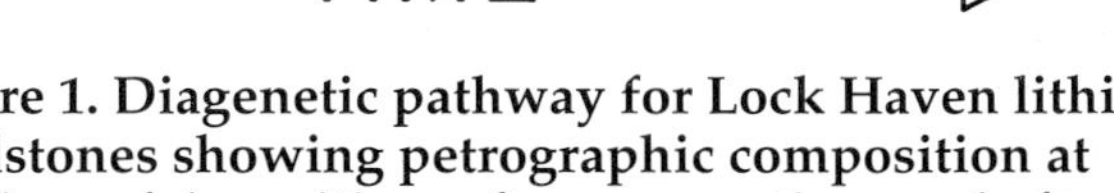

Figure 1. Diagenetic pathway for Lock Haven lithic sandstones showing petrographic composition at the time of deposition, after compaction, and after chemical leaching. Black = other minerals, cmt = cement, ϕ_p = primary porosity, ϕ_s = secondary porosity from the dissolution of rocks fragments and feldspars, Q = sedimentary quartz, RF = rock fragments, F = feldspars.

Upper Devonian sandstones of the Catskill deltaic complex constitute a major exploration play for natural gas in Pennsylvania. Recent activity has focused on the Lock Haven Formation of Centre and Clinton

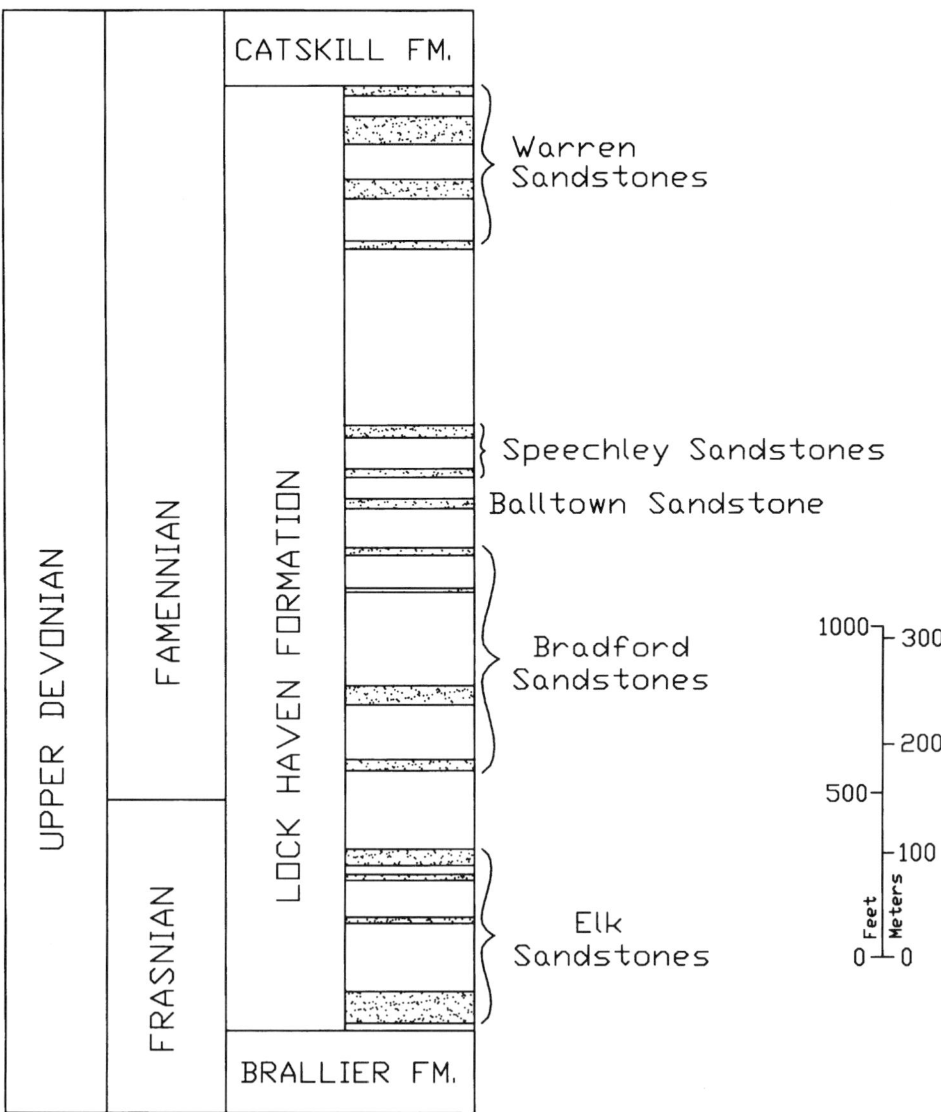

Figure 2. Upper Devonian stratigraphy of Clinton County, Pennsylvania, includes the 900-m-thick Lock Haven Formation. This formation consists of several packages of sandstones (four of which were sampled for this study: Warren, Speechley, Bradford, and Elk) disseminated through a thick interval of shale and siltstone.

counties, ~900 m thick and consisting of interbedded shale, sandstone, and siltstone (Figure 2). Many of the sandstones, however, have a low porosity and permeability; they possess only modest storage capacity and are marginally profitable reservoirs. It is especially important, therefore, under such economic conditions, that new gas prospects be evaluated and appraised judiciously. In this chapter, we document a semiquantitative relationship between observed porosity and inferred depositional facies, a relationship that allows first-order prediction of primary and secondary pore volumes before drilling.

Lock Haven sandstones of the present study go by the drillers' informal member names of Warren, Speechley, Bradford, and Elk. Fifty-one samples were investigated by thin-section microscopy [five more than described in Bruner and Smosna (1994)]; in addition, several of these were investigated by scanning electron microscopy and the X-ray diffraction method. The samples, taken from sidewall and full-bore cores, came from five wells in Clinton County, two in Centre County, and one in Somerset County (Figure 3). The paleoshoreline is thought to have passed through central Pennsylvania at this time (Dennison, 1985), and these sandstones represent a mix of terrestrial, transitional, and shallow-marine environments. Porosities range from 0% to 20%, and horizontal air permeabilities from <0.001 to 6 md.

SEDIMENTARY ENVIRONMENTS

Three full-bore cores of Lock Haven sandstones were recovered in Clinton County and analyzed petrographically. Well numbers, exact locations, and member names, however, are proprietary information and cannot be released. We described these cores in detail, noting lithologies, textures, structures, and vertical trends in order to identify and interpret the several facies. Facies interpretations have been confirmed by comparison with standard, well-recognized sedimentary models (Elliott, 1978a, b; Johnson, 1978; Bouma et al., 1982; Coleman and Prior, 1982; McCubbin, 1982; Miall, 1984; Reinson, 1984; Walker, 1984). The cored

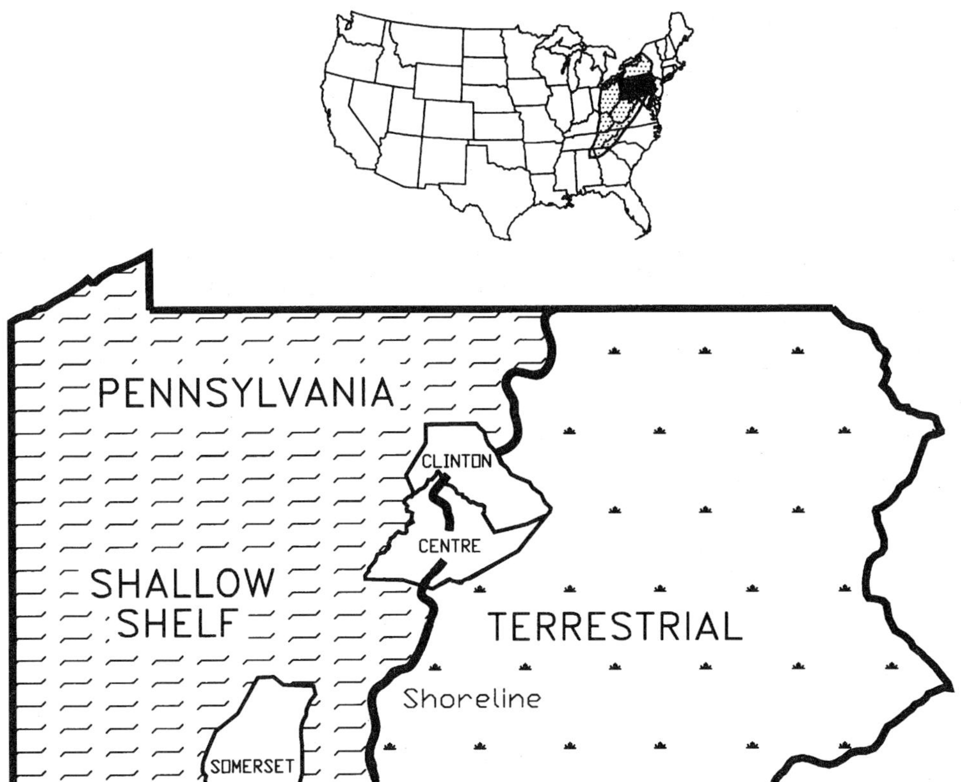

Figure 3. Generalized paleo-geography of Pennsylvania at the end of the Frasnian (after Dennison, 1985). The shoreline passed through or near Centre and Clinton counties, separating eastern rivers and deltas of the Catskill complex from a shallow western sea. Cores of this study were taken from eight wells in Clinton, Centre, and Somerset counties. The inset map shows the location of Pennsylvania (black) and the Appalachian basin (stippled) in the eastern United States.

sandstones compose a systems tract of distributary mouth bars, offshore sand ridges, and barrier islands. A fourth, fluvial facies is present in the systems tract, but only sidewall cores are available. Facies were compared to geophysical well logs, particularly the gamma-ray signatures, and our depositional interpretations could then be extended throughout the region, based on an integration of core analyses, gamma-ray correlations, and sand-body geometries.

Distributary Mouth Bar

The mouth-bar sandstone, 8.5 m thick, constitutes a coarsening-upward sequence, and its funnel-shaped gamma-ray signature suggests a corresponding decrease in clay content (Figure 4). Rocks of the lower part consist of alternating very fine sandstone and silty shale, and the base grades downward into an underlying shale. One sandstone at the base contains abundant shale clasts and carbonate-cemented nodules. Toward the top, the section is made up of fine and very fine sandstone with no shale. Wavy to lenticular bedding and current-ripple bedding are the prevalent sedimentary structures in the lower part, whereas parallel laminae become predominant above. Body fossils (pyritized brachiopods) are rare throughout the unit, although sand- and clay-filled burrows are common in the shales. Ripple cross-laminae, parallel laminae, shale partings, rip-up clasts, and burrows occur at the very top.

This distributary mouth-bar sandstone accumulated during the westward progradation of a Lock Haven delta front across the shelf. The lower section was deposited on distal reaches of the mouth bar near normal wave base. At that depth, deposition frequently alternated between mud and sand. Sedimentation rate of the mud was slow, allowing extensive burrowing by infauna, and shale clasts testify to occasional periods when currents eroded the muddy sea floor. The distal mouth-bar sandstone passes upward in the core to a proximal facies. The upper coarser grained section formed in a somewhat higher energy setting when the bar built itself up to sea level. An increasing energy level is also denoted by the vertical trend in sedimentary structures. Fine sandstone with parallel laminae reflects deposition within the swash zone along the crest of the bar. Very fine, rippled, and burrowed sandstone formed on the back of the mouth bar, where wave action was slightly reduced from that of the crest.

Beneath the mouth-bar sandstone lies a prodeltaic shale with minor siltstone and very fine sandstone. Shales are silty, organic rich, micaceous, and well laminated. The few burrows are mostly small and horizontal; they are filled with silt and sand, and contain pyrite. Several large sand-filled *Rhizocorallium* traces are also identified. The siltstones and sandstones occur in thin beds, isolated laminae, or starved ripples, and the main structure is small-scale cross-laminae. Thicker beds may have a scour base and be parallel laminated. Deformation structures developed where the underlying muds were soupy and weak: load structures along the base of sandstones, distortion of bedding around large burrows and sandstone lenses, and slump structures with small-scale folding and brecciation. Some sandstones contain vertical escape burrows; others are extensively bioturbated.

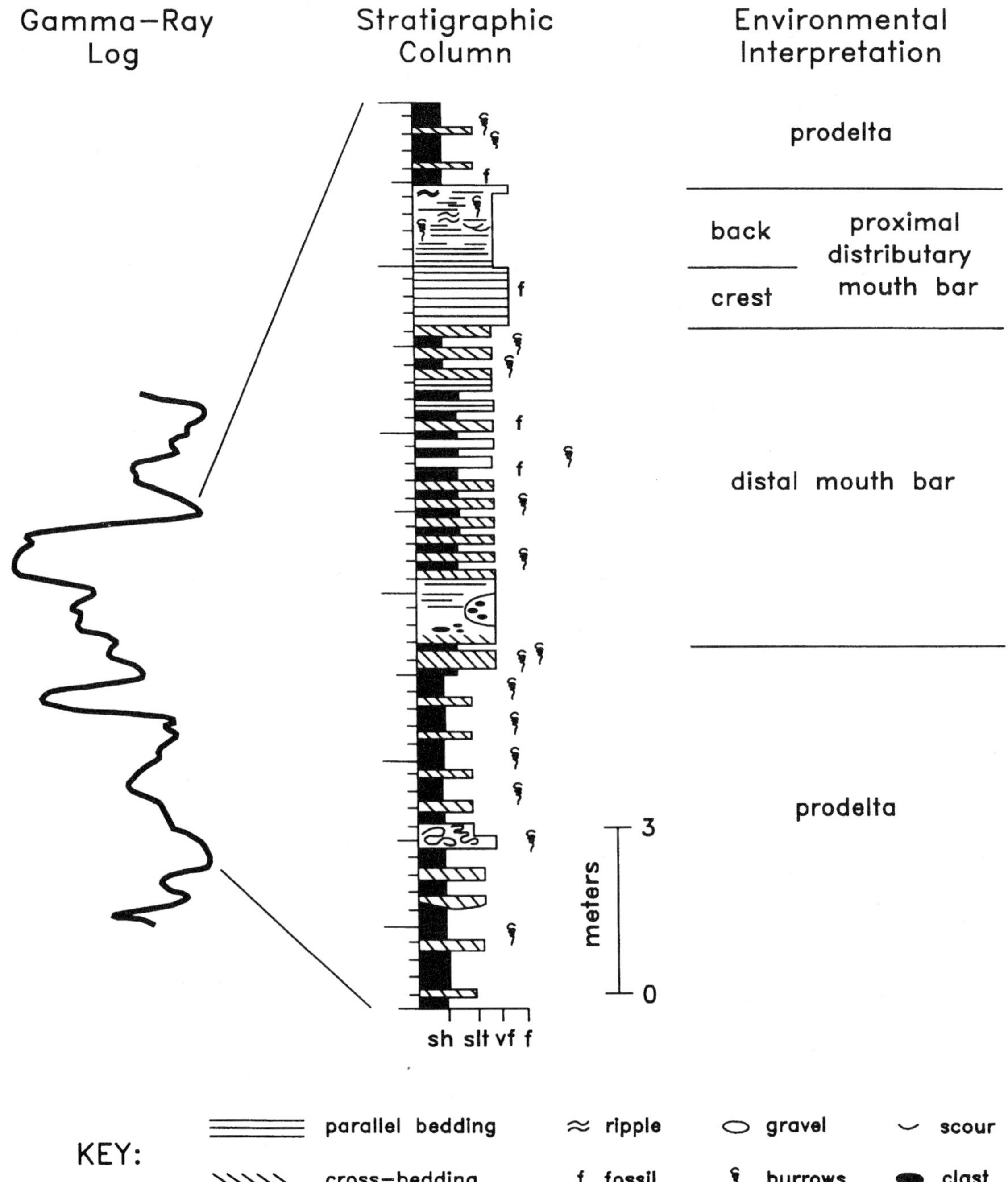

Figure 4. Columnar section of distributary mouth-bar sandstone illustrates the association of lithologies, textures, structures, and vertical facies associations. The gamma-ray log has a funnel shape.

Above the mouth-bar sandstone is a black shale and shaly siltstone. These rocks resemble the shales and siltstones beneath, and they attest to a rise of sea level or switching of distributary channel and a return of the prodeltaic environment.

Offshore Sand Ridge

Although subtle, the sand-ridge facies (8.7 m thick) displays a coarsening-upward texture and a funnel-shaped gamma-ray signature (Figure 5). The unit's base consists of siltstone with minor very fine sandstone and shale. Parallel lamination is the dominant structure, although ripple trough cross-laminae are also present. Upward in the core, very fine sandstone and siltstone with parallel laminae, planar cross-laminae, and ripple trough cross-laminae make up the middle section. Scour surfaces and shale clasts are present, as are a number of storm deposits, noted by their sharp basal contact, cross-laminae, and burrowed cap (Kreisa, 1981). Very fine sandstones with parallel laminae mark

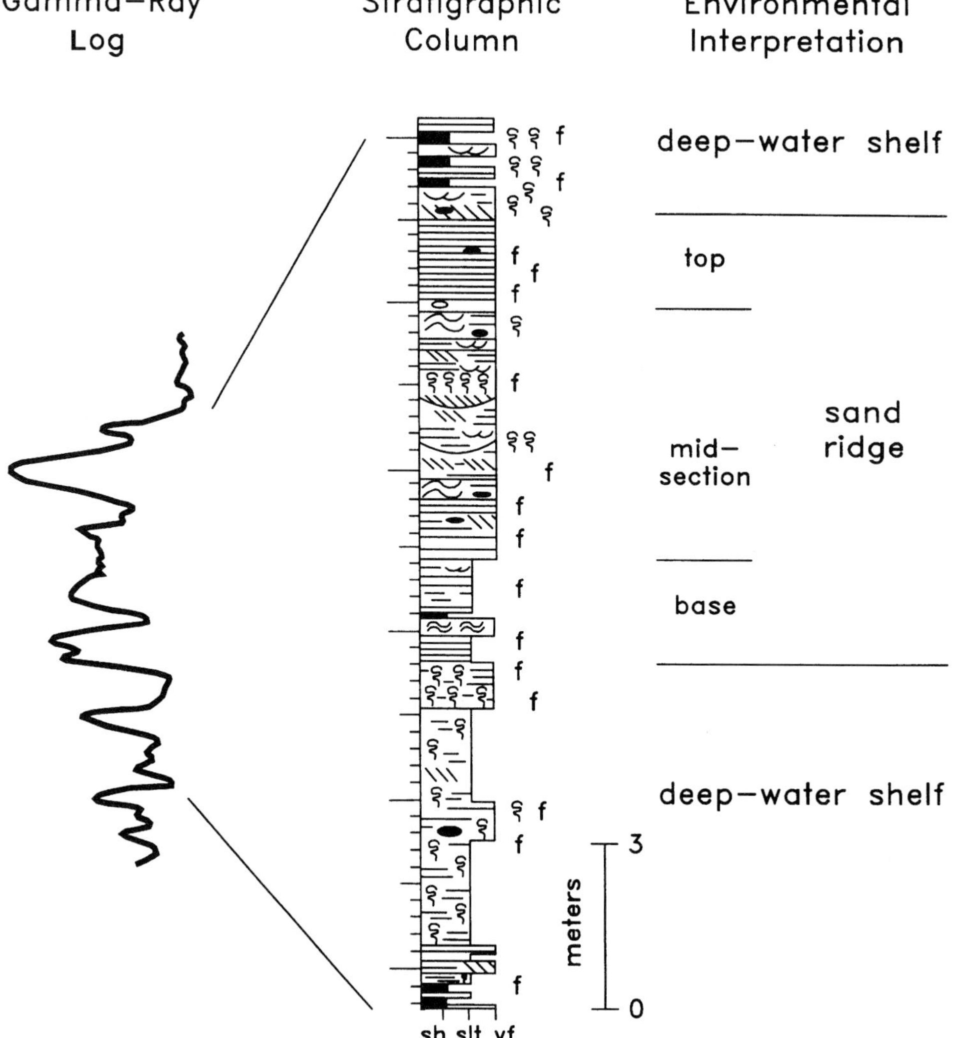

Figure 5. Columnar section of offshore sand-ridge sandstone. The sandstone is fossiliferous and parallel laminated; overlying and underlying siltstone and shale are bioturbated. Gamma-ray log shows a subtle funnel shape. The symbols are the same as in Figure 4.

the top of the unit. Zones of reworked fossils and shale clasts occur there. Skeletal grains, including brachiopods, crinoids, and gastropods, are common in the sandstones and siltstones from base to top, but burrows are confined to shale partings.

These fossiliferous sandstones formed as a sand ridge on the offshore Lock Haven shelf. As the ridge aggraded into shallower water, wave and current activity increased, and grain size coarsened slightly. Simultaneously, the predominant sedimentary structure changed from settle-out lamination (no flow) to cross-lamination (low-flow regime) to parallel lamination (upper-flow regime). Occasional periods of sea-floor erosion are indicated by the scour surfaces, rip-up clasts, thin tempestites, and fossil lags. Numerous epifaunal invertebrates inhabited the sand ridge and surrounding shelf; however, infaunal organisms burrowed into the sediment only during breaks in sand deposition (shale partings).

Underlying the sand-ridge sandstone is a burrowed siltstone and shale with minor sandstone, interpreted to have accumulated on the deep shelf below wave base. Near-absence of body fossils in the siltstones

with only moderate burrowing points to a stressed biological environment; the water chemistry was perhaps dysaerobic. Organic matter, pyritized fossil debris, and the green-gray color provide additional evidence for reducing conditions at the silt–water interface. Interbedded shales are well laminated and have few burrows, suggesting an even lower oxygen level in somewhat deeper water. Very fine sandstones, on the other hand, are fossiliferous (crinoids, brachiopods, bivalves, and tentaculitids) and moderately burrowed, reflecting deposition near wave base and under oxygenated conditions. Above the sand-ridge sandstone, the overlying facies suggests a drowning of the shelf by transgression and a return to the deposition of similar deep-water fine clastics.

Barrier Island

We divide the 11.6-m stratigraphic section into three parts (Figure 6). The middle part contains a distinctive parallel- and cross-laminated, very coarse sandstone. Particles include coarse rock fragments, feldspars and quartz, large skeletal debris, and phosphate grains. The basal contact is sharp. Overlying this

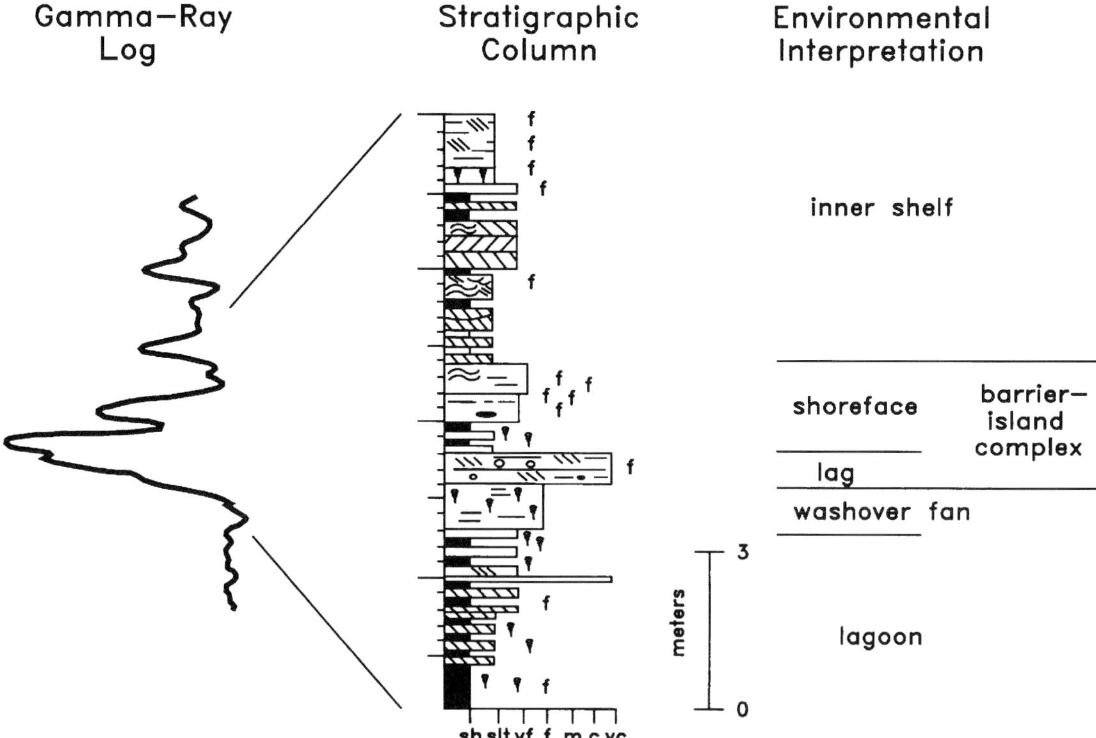

Figure 6. Columnar section of a transgressive barrier-island complex is marked by the coarse lag deposit of shoreface retreat. Underlying lagoonal sandstones are extensively bioturbated, whereas overlying sandstones of the lower shoreface and inner shelf are laminated and fossiliferous. Gamma-ray log has a relatively sharp base and bell shape. The symbols are the same as in Figure 4.

is a minor shaly, sandy siltstone and shale that is bioturbated and displays some deformation structures. Stratigraphically higher lies a fossiliferous sandstone: very fine to fine grained with abundant brachiopods, gastropods, crinoids, bivalves, tentaculitids, and bryozoans. Predominant structures are ripple bedding, burrows (especially *Asterosoma*), and shale partings.

The upper section of the core consists of interbedded siltstone, very fine sandstone, and shale. Sandstones and siltstones are somewhat fossiliferous, containing brachiopods, crinoids, and ostracods, and the primary structures include both parallel and cross-lamination. Also observed are a small-scale scour surface and a single thin bed of brachiopod valves with concave-down orientation. The shales may be laminated or burrowed. Some beds have been contorted by soft-sediment deformation, and many possess abundant organic matter.

The lower part of the core is marked by an extensively bioturbated fine sandstone. Coarse quartz sand is scattered throughout, and hints of laminae appear where burrowing is less intense. Very fine sandstone, siltstone, and shale alternate at the very bottom. The sandstones and siltstones occur in thin beds, laminae, and lenses, which show ripple bedding and contain a few small brachiopods. In contrast, the shales are well burrowed, and burrows illustrate a diverse assemblage of forms: sand- and silt-filled, small and horizontal, long and vertical, vertical U-shaped, and horizontal *Asterosoma*.

This sandstone body represents a transgressive barrier-island complex. We interpret an erosional ravinement to be present at the base of the middle section, having formed during a time of shoreface retreat (Swift, 1975). The barrier facies (that is, upper-shoreface and foreshore sands) were removed by erosion when sea level rose, and waves of the surf zone attacked the beach. Very coarse sandstone accumulated as a lag deposit atop the ravinement surface. Above the lag, rippled, fossiliferous finer grained sandstone with shale formed in the middle-lower shoreface environment as the transgression continued. The overlying, still-finer siliciclastics were laid down on the inner shelf, marking a transition from nearshore sand to offshore mud. Wave and current energy decreased systematically offshore, and grain size of the sediment concurrently decreased. This fining-upward sequence of transgressive beach–inner shelf sandstones is paralleled by the bell-shaped gamma-ray signature (Figure 6). Storms infrequently reworked the shelf sediment, as evidenced by the scour structure and resedimented brachiopods.

Lagoonal sediments lie beneath the barrier-island complex. Laminated, fine sandstone with scattered coarse quartz is thought to have been deposited by the action of overwash onto the leeward side of the barrier island. Storms breaking over the barrier transported sand into the lagoon; extensive bioturbation then developed during the interval following rapid deposition. Sediments of the lagoon proper included sparsely fossiliferous, very fine sand and thoroughly burrowed mud.

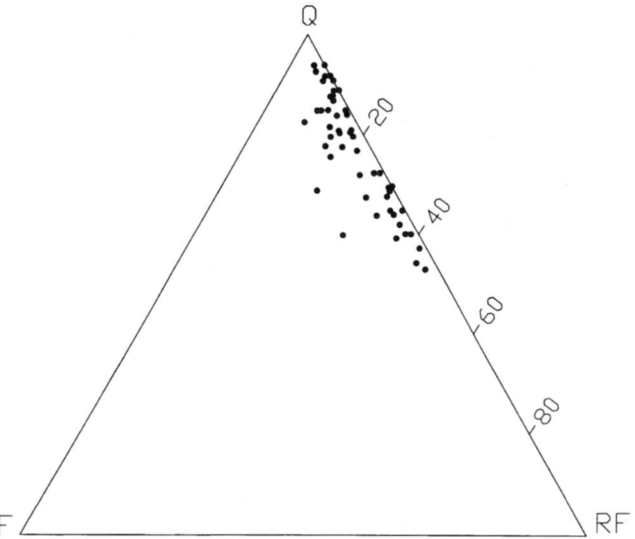

Figure 7. Quartz-feldspar-rock fragments ternary diagram plotting 51 Lock Haven sublitharenites (5% ≤ RF < 25%) and litharenites (RF ≥ 25%).

Streams

Fluvial sandstones are interpreted by their blocky or spiky pattern on the gamma-ray log, the shoestring geometry of sandstone on isopach maps, and an orientation perpendicular to the paleoshoreline (Boswell and Jewell, 1988). Sidewall cores show the rocks to be texturally mature and unfossiliferous.

FACIES AND LITHOLOGIES

Two thirds of the 51 sandstones are classified as sublitharenite, and the remainder as litharenite (Bruner and Smosna, 1994). Although the litharenites contain a greater number of rock fragments, they are otherwise quite similar to the sublitharenites. The two groups in fact form a petrographic continuum of lithic-rich rocks with a mean quartz-feldspar-rock fragment (Q/F/RF) ratio of 81/2/17 (Figure 7). Sorting is generally moderate to good. Mean grain size varies from 0.100 mm for shelf sandstones to 0.130 mm for mouth-bar sandstones, to 0.170 mm for fluvial sandstones, and to 0.240 mm for beach sandstones.

Quartz includes normal, undulose, polycrystalline, and stretched varieties, and many grains have inclusions of vermicular chlorite, kaolinite, and zircon, or intergrowths of feldspar. Rock fragments are overwhelmingly a mixture of phyllite and shale (Figure 8A). Phyllite clasts are composed of fine mica and chlorite with minor amounts of quartz, feldspar, apatite, and garnet; they display a marked foliation. Shale clasts consist of illite, quartz silt, and kaolinite. Rock fragments of dolomite, sandstone, and volcanics are, by contrast, very rare. Most of the feldspars appear fresh, but some are partly altered to sericite and illite, or replaced by dolomite. Muscovite, biotite, and chlorite constitute the main accessory minerals. All but six samples have a clay content <10% of the total rock volume.

During diagenesis, many rock fragments and feldspars were removed by dissolution, but secondary moldic pores can readily be differentiated in thin section, interpreted as to their origin, and point-counted. Rock fragments have also been squeezed into a pseudomatrix, but pseudomatrix can be identified and counted. These values for leached rock fragments, leached feldspars, and compacted rock fragments are then added to data for existing grain percents to determine the mineral composition before dissolution; that is, the original content of rock fragments and feldspars. Recalculating point-count data for all samples yields an original $Q_o/F_o/RF_o$ of 77/3/20 (Table 1).

The source area for these Upper Devonian sandstones must have contained a mixture of metamorphic and sedimentary rocks. Lithic-rich lithologies of the Lock Haven Formation match the composition predicted by Dickinson and Suczek (1979) and Dickinson (1988) for recycled sands from an orogenic provenance. This mix of sediments came from uplifted strata of the Acadian Mountains to the east, particularly Lower Cambrian low-grade quartzites, phyllites, and slates and their sedimentary cover. Continent-continent collision between North America, western Europe, and the Avalone terrane had just recently created the Acadian fold and thrust belt, and the Catskill deltaic complex constitutes the siliciclastic wedge that filled the adjoining foreland basin (Ettensohn, 1985).

The 51 samples obviously form a single population of sandstones, but the four sedimentary facies illustrate somewhat different mineral compositions (Figure 9). Despite an overlap, they have different contents of rock fragments (RF_o before leaching). Mineral composition was mainly controlled by mechanical destruction of rock fragments in the depositional environment (Davies and Ethridge, 1975; Mark, 1978; Espejo and López-Gamundí, 1994). Breakdown of shale and phyllite grains produced fine detritus of micas and clays; hence, muscovite, biotite, chlorite, illite, sericite, and smectite arc common components of both Lock Haven rock fragments and the sandstones' matrix. Other controls on sandstone mineralogy, such as tectonism, provenance, and climate, presumably remained constant in central Pennsylvania during Lock Haven deposition.

Meandering streams draining the Acadian Mountains transported and deposited sands rich in rock fragments (Figure 9). Resultant sandstones have a high, although variable, lithic content: RF_o ranges between 6% and 42% of the total rock volume and averages 24%. In between terms of compositional maturity, these are the most immature samples. Lithic composition not only varies considerably across this facies, as indicated by a large standard deviation, but the range in lithic content is also significant within any single fluvial sandstone. For example, the lithic content in one sandstone body varies between 11% and 42% within 1 m of section; another, between 6% and 32% in 3 m of core; and yet another, between 11% and 36% in 16 m. This extreme mineral variation most likely reflects drastic fluctuations in stream velocity and abrasive capacity; furthermore, no other facies exhibits such drastic mineral changes.

Figure 8. Photomicrographs of Lock Haven sandstones. Bar scales equal 0.250 mm. (A) Sandstone with abundant rock fragments of phyllite and shale (brown grains). Compaction has deformed these ductile lithics, squeezing them into intergranular pore space. (B) Large and small triangular primary pores between quartz grains have been lined and partly occluded by solid bitumen (black). (C) Oversized moldic pores >0.100 mm (center) were generated by the dissolution of rock fragments.

Abundant rock fragments were then delivered by streams to distributary mouth bars at Lock Haven delta fronts. Mouth-bar sandstones have a notably less variable lithic content, and RF_o of these sandstones ranges between 7% and 24%; the mean value is 15%. Presumably, these sediments were subjected to intense wave action and grain disintegration on the delta front. From the delta, lithic sands were swept seaward by storm currents to offshore sand ridges. Shelf sandstones have a mean RF_o close to that of mouth-bar sandstones, 14%; the range of lithics is somewhat lower (4%–21%). Storm action and offshore swells thus produced only minor destruction of labile rock fragments on the sand ridges. Barrier-island sands, worked continuously by longshore currents and breaking waves, possess the highest compositional maturity. Rock fragments are reduced to a mean value of 7%, and the narrow range of RF_o (4%–12%) indicates widespread breakdown of labile grains in this high-energy depositional setting. Differences in lithic content between the four facies do exist, depending to a large degree on the relative survivability of these unstable grains.

Results of this study compare favorably with other published investigations: mineral composition of siliciclastic sediments is sensitive to the depositional environment (Cameron and Blatt, 1971; Stonecipher and May, 1990). Fluvial sandstones are richest in metamorphic and sedimentary rock fragments; mouth-bar sandstones contain, on average, less than two thirds of the lithic content of fluvial facies; rock fragments are reduced but just a few percentage points between mouth-bar and shelf sandstones; the low content of rock fragments in barrier-island sandstones is half that of the mouth-bar facies. In total, two thirds of the rock fragments were destroyed between Lock Haven rivers and barrier islands. In like manner, Harper and Laughrey (1987), studying slightly younger Devonian Venago sandstones in Pennsylvania, reported a 20% decrease in rock fragments (plus pseudomatrix) between fluvial and mouth-bar sandstones, and a virtual elimination of rock fragments in foreshore sandstones. Davies and Ethridge (1975) observed a three-quarter reduction between fluvial and deltaic sandstones of the Eocene Wilcox Group in Texas.

FACIES AND PRIMARY POROSITY

Approximately one third of the porosity in the 51 sandstones is primary (Figure 8B), ranging up to a maximum value of 13%. Primary pores between rigid quartz grains are triangular in cross section, although when adjacent to ductile rock fragments, they can be quite irregular. Their size as measured in thin section varies between 0.005 and 0.080 mm, typically about one fourth to one third of the diameter of surrounding grains.

Because sedimentary environment controlled the content of rock fragments to a large extent, and because sandstone lithology in turn controlled the degree of compaction, samples of this study reveal a close association between facies, mineral content, compaction, and primary porosity (Figures 10, 11). Compaction deformed the ductile rock fragments of shale and phyllite and squeezed them into much of the intergranular pore space (Figure 8A). As a result, porosity systematically decreases in sandstones with an increasingly higher lithic content (Rittenhouse, 1971; Smosna, 1989). The graph in Figure 10 illustrates this

Table 1. Petrographic Data, Lock Haven Sandstones.*

Sample Number	Facies	Q_o	F_o	RF_o	ϕ_{mc}	ϕ_p	ϕ_{rf}
1	FL	40	1	21	0	0	0
2	FL	74	5	16	5	1	6
3	FL	69	7	15	2	0	8
4	FL	59	3	29	5	0	2
5	FL	63	7	6	23	3	0
6	FL	50	1	32	13	0	0
7	FL	69	4	11	8	2	4
8	FL	51	3	42	0	0	0
9	FL	65	5	11	13	2	6
10	FL	53	2	38	5	0	0
11	FL	46	3	36	9	0	0
12	FL	51	5	31	8	0	0
13	FL	57	3	32	3	0	0
14	FL	57	2	36	1	0	2
15	FL	59	5	28	2	0	1
16	FL	58	2	12	27	0	0
17	FL	67	2	15	15	0	0
18	FL	59	1	26	10	0	0
19	FL	52	2	27	16	0	0
20	MB	62	4	20	12	0	0
21	MB	70	3	18	8	2	13
22	MB	61	13	15	10	2	11
23	MB	54	2	24	4	0	1
24	MB	44	1	7	0	0	0
25	MB	68	7	12	13	6	2
26	MB	67	2	14	12	0	6
27	MB	66	1	11	11	3	5
28	MB	69	1	8	20	0	0
29	MB	57	0	24	8	0	0
30	MB	65	2	10	19	0	0
31	SR	63	9	21	6	2	4
32	SR	49	12	21	10	5	8
33	SR	60	3	11	17	0	0
34	SR	58	1	21	5	0	6
35	SR	51	1	21	10	0	1
36	SR	67	3	9	19	2	1
37	SR	67	2	5	25	0	0
38	SR	73	1	4	20	0	0
39	SR	35	1	12	5	0	0
40	B	34	2	4	33	0	0
41	B	71	0	6	20	0	0
42	B	75	0	9	12	3	3
43	B	65	3	12	13	0	2
44	B	71	1	8	15	0	0
45	B	81	0	5	11	1	1
46	B	74	2	4	12	0	0
47	B	69	1	5	17	7	3
48	B	72	0	7	24	3	2
49	B	59	1	7	32	13	1
50	B	56	1	10	32	1	6
51	B	62	1	8	11	0	0

*B = beach/barrier island, FL = fluvial, MB = distributary mouth bar, SR = offshore sand ridge, F = feldspar, Q = quartz, RF = rock fragments.

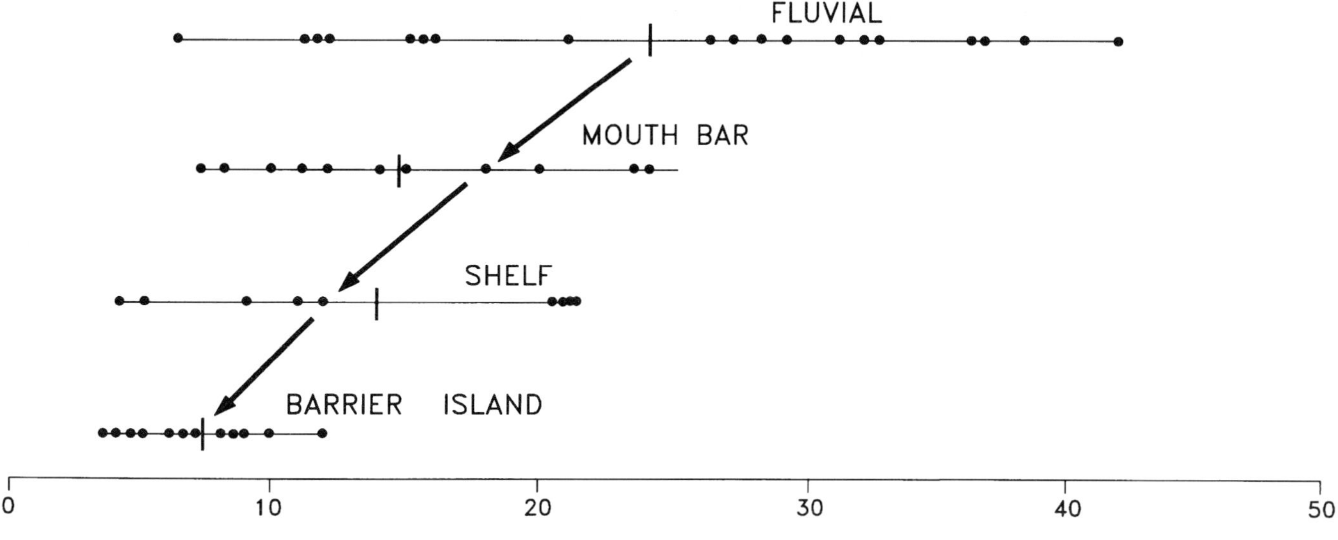

ORIGINAL ROCK FRAGMENTS, RF_o in percent

Figure 9. Graph of RF_o (content of rock fragments before chemical leaching) sedimentary environment; RF_o equals the volume percent of existing lithic grains plus pseudomatrix plus secondary porosity interpreted to have formed by dissolution of lithics. Vertical bars represent mean value for each facies. The data exemplify a facies dependency of lithic content among Lock Haven sandstones: lithic grains progressively decreased along the depositional systems tract.

correlation between lithic content and minus-cement porosity (sum of remaining primary porosity, quartz and dolomite cements, plus bitumen). Minus-cement porosity equals the volume of intergranular pore space before cementation but after the completion of mechanical compaction. Maximum minus-cement porosity of $\phi_{mc} = 33\%$ occurs in a sublitharenite with an original lithic content RF_o of 4%, and porosity falls to zero in a litharenite with a lithic content of 42%. (Two other sandstones with excessive clay matrix also possess a $\phi_{mc} = 0$.) The function $\phi_{mc} = 150/RF_o$ approximates the correlation between mineral content and original intergranular porosity.

In all four facies, most available pore space after compaction has been occluded by cement, matrix, and authigenic minerals (Bruner and Smosna, 1994). Although minus-cement porosity reaches 33%, final primary porosity ϕ_p does not exceed 13%, and two thirds of the samples have none. Syntaxial quartz cement has overgrown many detrital quartz grains. Ferroan dolomite commonly replaced feldspars and rock fragments, and in many places it extends beyond the boundary of framework grains to occupy adjacent pore space as a void-filling cement. Authigenic clays (illite, chlorite, and kaolinite) precipitated within intergranular pores and pore throats. Some clay matrix also occupies pore space. Finally, solid bitumen occurs as globs and interstitial stringers that coat detrital grains and line pores. Figure 10 also illustrates the present, reduced primary porosity. All but three samples plot below the curve $\phi_p = \phi_{mc}/3$; that is, more than two thirds of the original porosity has been occluded.

Minus-cement porosity can be related to sedimentary facies and mineral composition (Figure 11). From the fluvial environment to distributary mouth bar to sand ridge to barrier island, the lithic content of sandstones was continuously reduced as mechanical processes destroyed more and more shale and phyllite clasts. Fewer ductile rock fragments resulted in a lower degree of compaction. As a consequence, the average minus-cement porosity rises from 9% in fluvial sandstones to 11% in mouth-bar sandstones, to 13% in sand-ridge sandstones, and finally to 19% in barrier-island sandstones. Reduced porosity after cementation, however, is a different story, being very low in all but one Lock Haven facies. Present porosity averages 0.4%, 1.2%, and 1.0% in fluvial, mouth-bar, and shelf sandstones, respectively. Half of the barrier-island samples, on the other hand, have some measurable thin-section porosity; mean porosity for these 12 samples is 2.3% (two to six times more than in the other facies); and the maximum reaches 13%.

FACIES AND SECONDARY POROSITY

Two thirds of the porosity in the sandstones is secondary, and most of this is attributed to the leaching of phyllite and shale rock fragments (Figure 8C). Complete grain dissolution produced oversized molds of 0.100 to 1.700 mm. These molds often mimic compacted rock fragments in both size and shape, indicating that chemical leaching occurred after compaction. Intragranular pores of 0.025 to 0.050 mm resulted from partial leaching of rock fragments, whereas intragranular

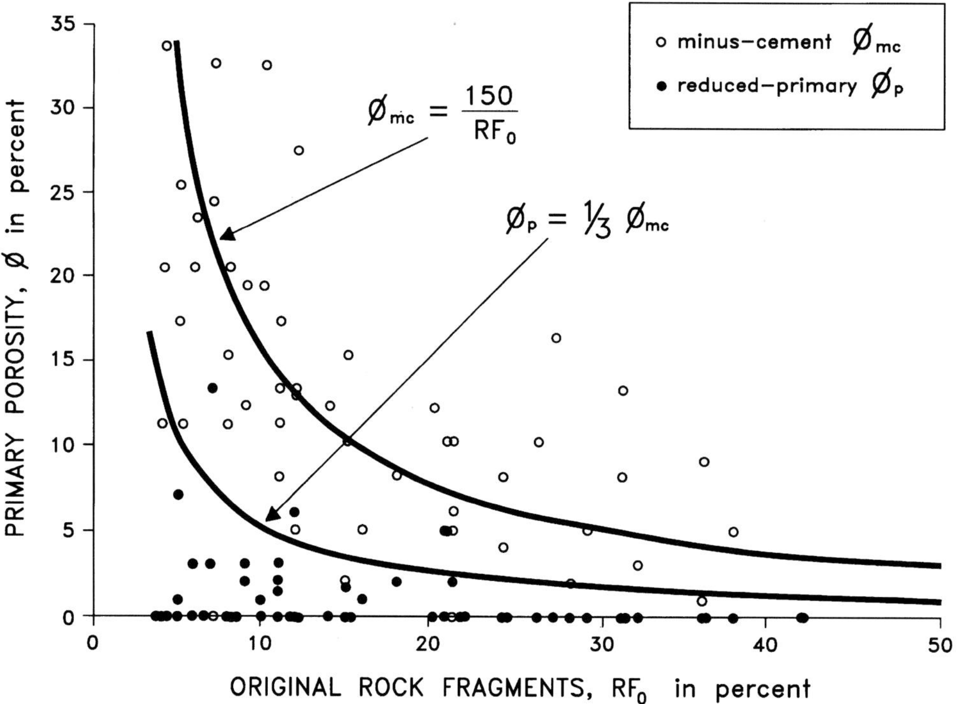

Figure 10. Graph of RF_0 minus-cement porosity and reduced primary porosity. ϕ_{mc} = existing primary porosity plus quartz and dolomite cements plus bitumen, and represents intergranular porosity before cementation. Porosity loss by compaction as a function of increasing lithic content is approximated by $\phi_{mc} = 150/RF_0$. ϕ_p = reduced porosity after cementation, and in most samples more than two thirds of the minus-cement porosity has been occluded.

micropores (<0.005 mm) illustrate a beginning stage of grain leaching. Partial and complete dissolution of feldspars has created additional secondary porosity.

The relationship between rock composition and secondary porosity is illustrated in Figure 12 (bottom part of graph), in which the upper limit of lithmoldic porosity defines a bell-shaped curve. Samples with a low content of rock fragments would, of course, contain few chemically unstable grains to remove by leaching. Two lithic-poor sandstones, for example, with original contents of 5% and 9% rock fragments, have 3% lithmoldic porosity. At a maximum, then, about half of the shale and phyllite rock fragments in these samples have dissolved. On the other hand, litharenites with >21% rock fragments never developed significant lithmoldic porosity. Compaction was extreme for sandstones with a high volume of lithics: mean values for primary porosity and permeability dropped to 0.4% and 0.04 md, and leaching fluids could not enter the rock to dissolve the unstable grains. Accordingly, a very small number of lithic grains were leached, and secondary porosity in lithic-rich sandstones does not exceed 2%.

Secondary porosity is greatest ($\phi_{rf} = 13\%$) in sandstones with an intermediate content of unstable grains. With an intermediate composition, compaction was moderate, and some primary porosity remained in the deep subsurface. Primary porosity not only contributed to reservoir quality, but allowed the introduction of leaching fluids, which dissolved rock fragments to create the secondary porosity. In these intermediate sandstones, as many as three quarters of the lithic grains have been dissolved.

Figure 12 also underscores the relationship between secondary porosity and sedimentary facies. Secondary porosity is greatest (ϕ_{rf} values of 6%–13%) in those sublitharenites with an original content of rock fragments between 10% and 21%. The four facies, however, have different mineral compositions, leading to differences in the distribution of lithmoldic porosity. Two thirds of the mouth-bar sandstones and two thirds of the shelf sandstones initially contained an intermediate amount of unstable grains ($10\% \leq RF_0 \leq 21\%$). In contrast, less than one third of the fluvial sandstones had an intermediate volume of unstable grains; this facies includes many samples with an overabundance of lithics. And very few of the barrier-island sandstones had an intermediate content; most contained too few lithics. Again, composition of the sandstones was chiefly a function of mechanical destruction of rock fragments in the depositional environment. Correspondingly, the mean value of secondary porosity is 3.5% in the mouth-bar sandstones, 2.2% in the sand-ridge samples, and 1.5% in both the fluvial and barrier-island sandstones.

In a recent analysis of Carboniferous reservoir rocks in Australia, Hamlin et al. (1996) reached comparable conclusions. They identified four sandstone facies in a nonmarine braid-delta system; the rocks are classified as sublitharenites, lithic grains range in abundance from 2% to 40%, and most of these are clasts of metamorphic and sedimentary rocks. Porosity is chiefly secondary, originating from the dissolution of rock fragments and rare feldspar. Their data show that as the lithic content increases from 13% to 21%, secondary porosity actually falls from 3.8% to 2.4%. Maximum porosity occurs in sandstone of an intermediate lithic composition.

Stonecipher and May (1990) also identified an analogous relationship between depositional facies and

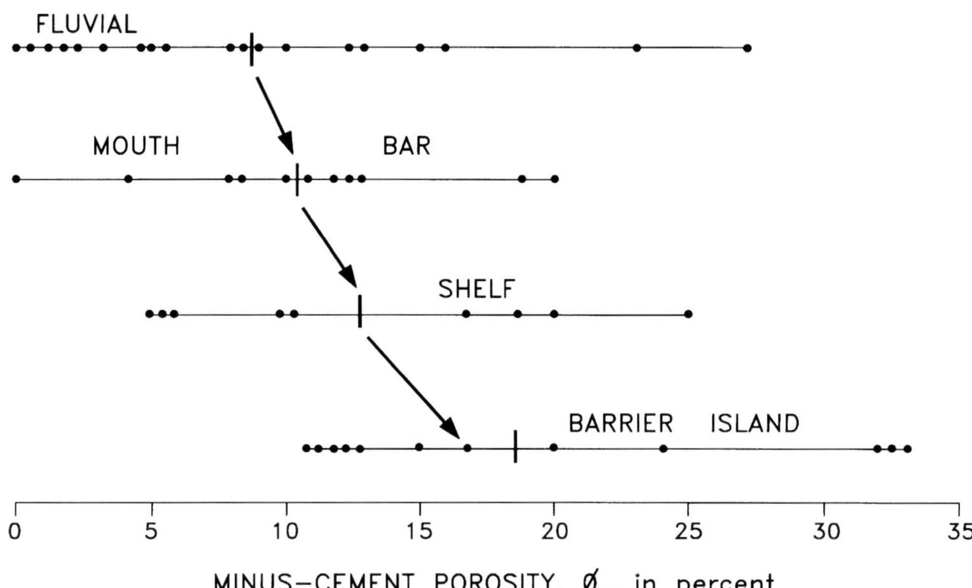

MINUS—CEMENT POROSITY, ϕ_{mc} in percent

Figure 11. Graph of minus-cement porosity vs. sedimentary environment. Sandstones with a smaller volume of ductile rock fragments underwent less compaction; consequently, they retain a higher primary porosity. Vertical bars represent mean value for each facies.

secondary porosity in the Wilcox Group of Texas. Although most porosity is primary in origin, secondary porosity in these sublitharenites (lithics range from 8% to 51%) depends in part on the content of rock fragments. Secondary porosity reaches highest values (3%–6%) in sandstones of distributary channel and mouth-bar facies with an intermediate content of rock fragments (17%–22%). Lithic-poor sandstones of the shoreface and tidal-flat facies, as well as lithic-rich sandstones of the tidal-channel facies, possess little or no secondary porosity.

Core Laboratories measured horizontal air permeabilities for 34 Lock Haven samples of this study, and 27 have a value <0.10 md; these are truly tight sandstones (unpublished report to Eastern States Exploration Company). But of the remaining samples—with poor to fair permeabilities (0.10–5.85 md)—fluid flow may be linked to porosity type and amount. Networks of lithmoldic pores are commonly observed in thin sections where several neighboring rock fragments have been leached. These networks, aligned parallel to bedding, must surely enhance the rocks' horizontal permeability. Also, the small intergranular primary pores may act as conduits between larger lithmoldic pores. The best reservoir sandstones, therefore, seem to possess a high total porosity (primary plus secondary) and an intermediate to high secondary porosity.

POROSITY PREDICTION

Subsurface exploration for Upper Devonian reservoirs in the Appalachian basin is based primarily on the mapping of sandstone members as identified by geophysical well logs. Prospects are then defined by the geometry of a sandstone body, its thickness, orientation, association with other sandstones, position with respect to shoreline or shelf margin, and structural configuration. Figure 13 depicts a composite isopach map of genetically related prospect sandstones: all four facies of this study have been arranged into an idealized Upper Devonian systems tract (modified from Boswell and Jewell, 1988).

The basic premise for our porosity predictions is that measurable differences in lithic content exist among the four sandstone facies. Each facies should consequently register different values for both primary and secondary porosities. Furthermore, within a sandstone member, the porosity values should change regularly along trends perpendicular to and parallel to the paleoslope.

The amount of rock fragments will decrease systematically along the systems tract of transitional-marine sandstones. Fluvial sandstones are compositionally immature litharenite with a high volume of lithic grains. Lithic content, however, can be extremely variable, and sublitharenite with comparatively few rock fragments may be closely interbedded with lithic-rich sandstone. Compared to this abundance of rock fragments in the fluvial facies, lithic grains are reduced by one third on the distributary mouth bar of the delta front, slightly more on the offshore shelf, and by two thirds on the barrier island.

Compaction of ductile metamorphic and sedimentary rock fragments in the litharenites and sublitharenites of a foreland basin will significantly reduce intergranular porosity. Minus-cement porosity may be approximated by the equation $\phi_{mc} = 150/RF_o$. Using mean values for Lock Haven samples, primary porosity before cementation of fluvial sandstones may have been 6%, of mouth-bar sandstones 10%, of shelf sandstones 11%, and of beach sandstones 21%. These values reflect a doubling and redoubling of minus-cement porosity along the systems tract.

Primary porosity, however, is almost everywhere greatly occluded by the introduction of cements, authigenic clays, and solid bitumen. In fact, almost nine tenths of the minus-cement porosity will eventually be destroyed. Again, using mean Lock Haven values,

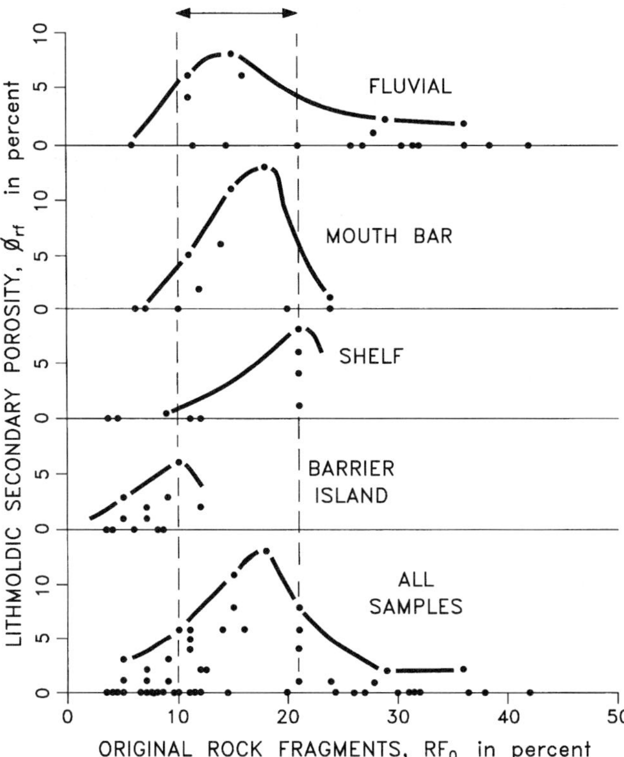

Figure 12. Graphs of sedimentary environment vs. content of rock fragments before leaching (RF$_o$) and lithmoldic secondary porosity (ϕ_{rf}). Secondary porosity is greatest for an original lithic content of $10\% \leq RF_o \leq 21\%$ (arrow at top).

reduced primary porosity will be 0.6% for fluvial sandstones, 1.1% for mouth-bar and shelf sandstones, and 2.1% for beach sandstones. Thus, only compositionally mature beach sandstones should possess any notable intergranular pore space.

Based on our analysis, secondary lithmoldic porosity should develop best in sandstones with a lithic content of $10\% \leq RF_o \leq 21\%$. Most of the mouth-bar sandstones have such an intermediate mineral content, with a mean RF$_o$ of 15%. This composition correlates to a maximum lithmoldic porosity of 11% on the bell-shaped plot of Figure 12. We expect, therefore, the mean value of lithmoldic porosity for this facies to be near 3.7% (a third of the maximum). Most of the offshore-shelf sandstones also have an intermediate composition with a mean RF$_o$ of 14%, which equates to a maximum lithmoldic porosity of 10% and points to a mean of 3.3% (a third of the maximum). In like manner, we predict that fluvial sandstones (RF$_o$ = 24%) will have a mean lithmoldic porosity of 1.5%, and beach sandstones (RF$_o$ = 7%) a mean lithmoldic porosity of 1.2%. Secondary porosity will be low in lithic-rich rocks near the source area, but should more than double along the systems tract to the delta front, where the volume of rock fragments becomes less abundant, and then decrease by two thirds as abrasion and mechanical breakdown remove still more lithics from the sediment.

CONCLUSIONS

To conclude, we rank the different sandstone facies with respect to reservoir quality. (1) Mouth-bar sandstones will have the best reservoir potential. Total porosity should be relatively good (mean of 4.8%) and lithmoldic porosity good, although primary porosity will be low. This combination of total and secondary porosity may also lead to a fair permeability. Moreover, half of the stratigraphic section is expected to be of reservoir quality; that is, having a thin-section porosity of ≥6% (half of the mouth-bar samples in our study have this much porosity). (2) Shelf sandstones will be similar to, but of a slightly lower quality than, mouth-bar sandstones. Total porosity should be fair (4.4%), and most of this is secondary. Only one third of the section may be of reservoir quality. (3) Beach sandstones will be of an even lower quality. Total porosity should be poor (3.2%), and most of this is primary. Only one third of the section may be of reservoir quality. (4) Fluvial sandstones will be the poorest reservoirs. Total porosity (2.1%), lithmoldic porosity, and primary porosity should all be low. Perhaps just one fifth of the section may be of reservoir quality.

Although differences in total porosities appear to be small, they become considerable in assessing marginally profitable reservoirs. In the comparative evaluation of drilling prospects, for example, an 8.0-m mouth-bar sandstone with mean porosity of 4.8% has the same gas-storage capacity (porosity thickness or $\phi \times h$) as an 8.7-m shelf sandstone with 4.4% porosity, a 12.0-m barrier-island sandstone with 3.2% porosity, and an 18.2-m fluvial sandstone with 2.1% porosity. Or from a different viewpoint, a mouth-bar sandstone (half of which is of reservoir quality) has 50% more gas-storage capacity than a shelf or beach sandstone of the same thickness (a third of which has reservoir quality), and 150% more storage capacity than a fluvial sandstone (a fifth of which is of reservoir quality). Our predictions should, of course, offer immediate benefit to the planning of drilling programs in the Appalachian basin, but we expect that the generalized relationships among porosity, lithology, and facies outlined by our study may be equally applicable to reservoir lithic sandstones of other basins.

ACKNOWLEDGMENTS

We thank Eastern States Exploration Company, especially Mike Canich and John Humphrey, and CNG Development Company for providing core samples and giving permission to publish the data. Neville Jones, Jon Gluyas, and an anonymous person reviewed the paper and made suggestions for its improvement. Alison Hanham and Debbie Benson drafted the illustrations.

REFERENCES CITED

Baldwin, B., and C.O. Butler, 1985, Compaction curves: AAPG Bulletin, v. 69, p. 622–626.
Bird, K.J., and C.M. Molenaar, 1992, The North Slope

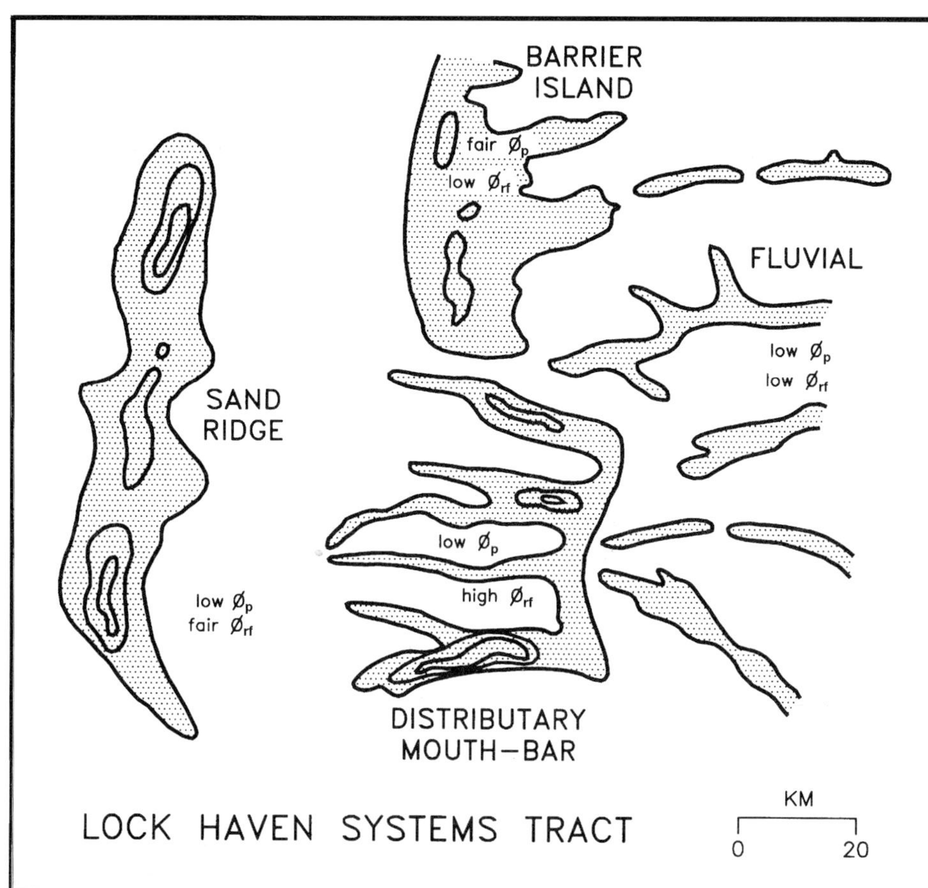

Figure 13. Composite isopach map of prospect sandstones in the Upper Devonian area of the Appalachian basin (modified from Boswell and Jewell, 1988). Fluvial, distributary mouth-bar, offshore sand-ridge, and barrier-island facies constitute a genetically related systems tract; each exhibits characteristic primary and secondary porosities. The contour interval equals 3 m.

foreland basin, Alaska, *in* R.W. Macqueen and D.A. Leckie, eds., Foreland basins and fold belts: AAPG Memoir 55, p. 363–393.

Björlykke, K., 1983, Diagenetic reactions in sandstones, *in* A. Parker and B.W. Sellwood, eds., Sediment diagenesis: Dordrecht, Reidel Publishing Co., p. 169–213.

Björlykke, K., 1984, Formation of secondary porosity: how important is it? *in* D.A. McDonald and R.C. Surdam, eds., Clastic diagenesis: AAPG Memoir 37, p. 277–286.

Boswell, R.M., and G.A. Jewell, 1988, Atlas of Upper Devonian/Lower Mississippian sandstones in the subsurface of West Virginia: West Virginia Geological and Economic Survey Circular C–43, 143 p.

Bouma, A.H., H.L. Berryhill, H.J. Knebel, and R.L. Brenner, 1982, Continental shelf, *in* P.A. Scholle and D. Spearing, eds., Sandstone depositional environments: AAPG Memoir 31, p. 281–327.

Bruner, K.R., and R. Smosna, 1994, Porosity development in Devonian lithic sandstones of the Appalachian foreland basin: Northeastern Geology, v. 16, p. 202–214.

Cameron, K.L., and H. Blatt, 1971, Durabilities of sand size schist and "volcanic" rock fragments during fluvial transport, Elk Creek, Black Hills, South Dakota: Journal of Sedimentary Petrology, v. 41, p. 565–576.

Chilingarian, G.V., 1983, Compactional diagenesis, *in* A. Parker and B.W. Sellwood, eds., Sediment diage-

nesis: Dordrecht, Reidel Publishing Co., p. 57–168.

Coleman, J.M., and D.B. Prior, 1982, Deltaic environments, *in* P.A. Scholle and D. Spearing, eds., Sandstone depositional environments: AAPG Memoir 31, p. 139–178.

Davies, D.K., and F.G. Ethridge, 1975, Sandstone composition and depositional environment: AAPG Bulletin, v. 59, p. 239–264.

Dennison, J.M., 1985, Catskill delta shallow marine strata, *in* D.L. Woodrow and W.D. Sevon, eds., The Catskill delta: Geological Society of America Special Paper 201, p. 91–106.

Dickinson, W.R., 1988, Provenance and sediment dispersal in relation to paleotectonics and paleogeography of sedimentary basins, *in* K.L. Kleinsphen and C. Paola, eds., New perspectives in basin analysis: New York, Springer-Verlag, p. 3–25.

Dickinson, W.R., and C.A. Suzcek, 1979, Plate tectonics and sandstone composition: AAPG Bulletin, v. 63, p. 2164–2182.

Elliott, T., 1978a, Deltas, *in* H.G. Reading, ed., Sedimentary environments and facies: New York, Elsevier, p. 97–142.

Elliott, T., 1978b, Clastic shorelines, *in* H.G. Reading, ed., Sedimentary environments and facies: New York, Elsevier, p. 143–177.

Espejo, I.S., and O.R. López-Gamundí, 1994, Source vs. depositional controls on sandstone composition in a foreland basin: the El Imperial Formation (Mid Carboniferous-Lower Permian), San Rafael basin, western Argentina: Journal of Sedimentary Research, v. A64, p. 8–16.

Ettensohn, F.R., 1985, The Catskill delta complex and the Acadian orogeny: a model, *in* D.L. Woodrow and W.D. Sevon, eds., The Catskill delta: Geological Society of America Special Paper 201, p. 39–49.

Galloway, W.E., 1984, Hydrogeologic regimes of sandstone diagenesis, *in* D.A. McDonald and R.C. Surdam, eds., Clastic diagenesis: AAPG Memoir 37, p. 3–13.

Hamlin, H.S., S.P. Dutton, R.J. Seggie, and N. Tyler, 1996, Depositional controls on reservoir properties in a braid-delta sandstone, Tirrawarra oil field, South Australia: AAPG Bulletin, v. 80, p. 139–156.

Harper, J.A., and C.D. Laughrey, 1987, Geology of the oil and gas fields of southwestern Pennsylvania: Pennsylvania Topographic and Geologic Survey, Mineral Resources Report 87, 166 p.

Heald, M.T., and R.E. Larese, 1973, The significance of the solution of feldspar in porosity development: Journal of Sedimentary Petrology, v. 43, p. 458–460.

Johnson, H.D., 1978, Shallow siliciclastic seas, *in* H.G. Reading, ed., Sedimentary environments and facies: New York, Elsevier, p. 207–258.

Kreisa, R.D., 1981, Storm-generated structures in subtidal marine facies with examples from the Middle and Upper Ordovician of southwestern Virginia: Journal of Sedimentary Petrology, v. 51, p. 823–848.

Loucks, R.G., M.M. Dodge, and W.F. Galloway, 1984, Regional controls on diagenesis and reservoir quality in Lower Tertiary sandstones along the Texas Gulf Coast, *in* D.A. McDonald and R.C. Surdam, eds., Clastic diagenesis: AAPG Memoir 37, p. 15–45.

Mark, G.M., 1978, The survivability of labile light-mineral grains in fluvial, aeolian and littoral marine environments: the Permian Cutler and Cedar Mesa formations, Moab, Utah: Sedimentology, v. 25, p. 587–606.

McCubbin, D.G., 1982, Barrier-island and strand plain facies, *in* P.A. Scholle and D. Spearing, eds., Sandstone depositional environments: AAPG Memoir 31, 247–279.

Miall, A.D., 1984, Deltas, *in* R.G. Walker, ed., Facies models, 2d. ed.: Geoscience Canada, Reprint Series 1, p. 105–118.

Moncure, G.K., R.W. Lahann, and R.M. Siebert, 1984, Origin of secondary porosity and cement distribution in a sandstone/shale sequence from the Frio Formation (Oligocene), *in* D.A. McDonald and R.C. Surdam, eds., Clastic diagenesis: AAPG Memoir 37, p. 151–161.

Parnell, J., 1987, Secondary porosity in hydrocarbon-bearing transgressive sandstones on an unstable Lower Paleozoic continental shelf, Welch Borderland, *in* J.D. Marshall, ed., Diagenesis of sedimentary sequences: Geological Society Special Publication 36, p. 297–312.

Pittman, E., 1979, Porosity, diagenesis and productive capability of sandstone reservoirs, *in* P.A. Scholle and P.R. Schluger, eds., Aspects of diagenesis: SEPM Special Publication 26, p. 159–173.

Potocki, D., and I. Hutcheon, 1992, Lithology and diagenesis of sandstones in the western Canada foreland basin, *in* R.W. Macqueen and D.A. Leckie, eds., Foreland basins and fold belts: AAPG Memoir 55, p. 229–257.

Potter, P.E., 1978, Petrology and chemistry of modern big river sands: Journal of Geology, v. 86, p. 423–449.

Pryor, W.A., 1973, Permeability-porosity patterns and variations in some Holocene sand bodies: AAPG Bulletin, v. 57, p. 162–189.

Reinson, G.E, 1984, Barrier island and associated strand-plain systems, *in* R.G. Walker, ed., Facies models, 2d ed.: Geoscience Canada, Reprint Series 1, p. 119–140.

Rittenhouse, G., 1971, Mechanical compaction of sands containing different percentages of ductile grains: a theoretical approach: AAPG Bulletin, v. 55, p. 92–96.

Schmidt, V., and D.A. McDonald, 1979a, The role of secondary porosity in the course of sandstone diagenesis, *in* P.A. Scholle and P.R. Schluger, eds., Aspects of diagenesis: SEPM Special Publication 26, p. 175–207.

Schmidt, V., and D.A. McDonald, 1979b, Texture and recognition of secondary porosity in sandstones, *in* P.A. Scholle and P.R. Schluger, eds., Aspects of diagenesis: SEPM Special Publication 26, p. 209–225.

Sclater, J.G., and P.A.F. Christie, 1980, Continental stretching: an exploration of the post-Mid-Cretaceous subsidence of the central North Sea basin: Journal of Geophysics Research, v. 85, p. 3711–3739.

Shanmugam, G., 1985, Significance of secondary porosity in interpreting sandstone composition: AAPG Bulletin, v. 69, p. 378–384.

Shanmugam, G., 1990, Porosity prediction in sandstone using erosional unconformities, *in* I.D. Meshri and P.J. Ortoleva, eds., Prediction of reservoir quality through chemical modeling: AAPG Memoir 49, p. 1–23.

Siebert, R.M., G.K. Moncure, and R.W. Lahann, 1984, A theory of framework grain dissolution in sandstones, *in* D.A. McDonald and R.C. Surdam, eds., Clastic diagenesis: AAPG Memoir 37, p. 163–175.

Smosna, R., 1989, Compaction law for Cretaceous sandstones of Alaska's North Slope: Journal of Sedimentary Petrology, v. 59, p. 572–584.

Stonecipher, S.A., and J.A. May, 1990, Facies control on early diagenesis: Wilcox Group, Texas Gulf Coast, *in* I.D. Meshri and P.J. Ortoleva, eds., Prediction of reservoir quality through chemical modelling: AAPG Memoir 44, p. 24–44.

Stonecipher, S.A., R.D. Winn, and M.G. Bishop, 1984, Diagenesis of the Frontier Formation, Moxa arch: a function of sandstone geometry, texture and composition, and fluid flux, *in* D.A. McDonald and R.C. Surdam, eds., Clastic diagenesis: AAPG Memoir 37, p. 289–316.

Swift, D.J.P., 1975, Barrier-island genesis: evidence from the central Atlantic shelf, eastern U.S.A.: Sedimentary Geology, v. 14, p. 1–43.

Walker, R.G., 1984, Shelf and shallow marine sandstones, *in* R.G. Walker, ed., Facies models, 2d edition: Geoscience Canada, Reprint Series 1, p. 141–170.

Mountjoy, E.W., and X.M. Marquez, 1997, Predicting reservoir properties in dolomites: Upper Devonian Leduc buildups, Deep Alberta Basin, *in* J.A. Kupecz, J. Gluyas, and S. Bloch, eds., Reservoir quality prediction in sandstones and carbonates: AAPG Memoir 69, p. 267–306.

Chapter 17

Predicting Reservoir Properties in Dolomites: Upper Devonian Leduc Buildups, Deep Alberta Basin

Eric W. Mountjoy
Xiomara M. Marquez[1]
Department of Earth and Planetary Sciences, McGill University
Montreal, Canada

ABSTRACT

Completely dolomitized Upper Devonian Leduc buildups at depths >4000 m have higher porosities and permeabilities than adjacent limestone buildups; dolostones are more resistant to pressure solution and tend to retain their porosity during burial. Distribution of pore types is controlled by depositional facies, whereas distribution of permeability is largely controlled by diagenetic processes, especially dolomitization.

In pool D3A of the Strachan reservoir, porosities and permeabilities are highest in the interior of the buildup where the strata are completely dolomitized. In the reef margin, porous and permeable dolomitized zones are interbedded with nonporous and nonpermeable limestone units. The presence of porous and permeable zones is closely related to the degree of dolomitization, with the greatest porosity and permeability occurring in completely dolomitized rocks.

The reservoir character in the Ricinus West buildup closely follows depositional units, despite complete dolomitization. At the reservoir scale, porosity and permeability have relatively similar values throughout the buildup. At the meter to tens of meters scale, the upper buildup interior is characterized by 1- to 2-m-thick, permeable and laterally continuous lagoonal strata. The lower reef interior consists of laterally discontinuous permeable zones. In the reef margin, permeability is controlled by fractures and interconnected vugs. At the millimeter scale, porosity and permeability are controlled by diagenetic processes.

Late cementation and dissolution processes have slightly decreased and increased porosity and permeability, mainly in the lower part of the reservoirs. Bitumen plugging decreased porosity and permeability in the upper part of the reservoirs. Although it is difficult to predict reservoir porosity and permeability trends, the secondary porosities in these deeply buried

---

[1]Department de Exploracion, Maraven S.A., Apartado 829, Caracas 1010-A Venezuela.

dolomites are mainly controlled by the primary porosity distribution and the depositional facies. The permeability is mainly controlled by diagenetic processes, especially dolomitization and various phases of cementation and bitumen plugging in the upper part of the reservoirs. Available data from the deep basin and the adjacent Rocky Mountains suggest that these porous dolomites are regionally extensive, and dolomite buildups elsewhere should have porosity and permeability variations similar to the Strachan and Ricinus West reservoirs. However, late-stage dolomite, anhydrite, and bitumen can locally partially to completely fill the pore spaces.

INTRODUCTION

The objective of this study is to better understand deeply buried limestone and dolomite reservoirs so that variations in reservoir porosity and permeability can be predicted in advance of drilling. Depositional facies and textures, diagenetic alteration, and burial effects are among the factors that can contribute to reservoir heterogeneity. We use examples from the deep basin of the Western Canada Sedimentary Basin as the focus of our study because they constitute prolific and widespread petroleum reservoirs.

Except for facies descriptions and the interpretation of depositional environments, few carbonate reservoirs in the Upper Devonian of the Western Canada Sedimentary Basin have been studied in detail for reservoir character. Leduc buildups (Upper Devonian, Frasnian) along the 320-km-long Rimbey-Meadowbrook reef trend in central Alberta (Figure 1) are prolific oil and gas producers. For example, the Ricinus West pool discovered in 1969 has produced 825 bcf, and the Strachan D3A pool discovered in 1967 has produced 911 bcf, with estimated original gas in place of 1.8 tcf and 1.4 tcf, respectively, for each field. The H_2S content varies from 3.2% to 10.8% in the Strachan D3A pool to 25% in the D3B pool, and 31% to 33.5% in Ricinus West. The Ricinus West field is larger (8×4 km) than Strachan D3A (6×3 km). The cumulative production compared to single section reserves of gas in place suggests that the dolomitized wells in Ricinus West are draining considerably more than one section and indicates that there is good reservoir continuity in this field (Podrusky et al., 1987).

The stratigraphy and geological setting of the Rimbey-Meadowbrook buildups have been outlined in Amthor et al. (1993, 1994), and their regional setting and depositional history have been described by Stoakes (1992). Information concerning the facies and depositional environments of Leduc buildups is mainly available from the undolomitized Redwater and Golden Spike buildups (Klovan, 1964; Wendte, 1974; McGillivary and Mountjoy, 1975; Reitzel et al., 1976; Burrowes, 1977; Jardine et al., 1977; Reitzel and Callow, 1977; Walls, 1978; Walls et al., 1979; Jardine and Wishart, 1982; Walls and Burrowes, 1985; Burrowes and Krouse, 1987; Carpenter and Lohmann, 1989; Wendte 1992a, b, 1994; Chouinard, 1993). A few studies discuss the extensively dolomitized part of the reef trend (Figure 1) (Layer, 1949; Waring and Layer, 1950; Andrichuk, 1958a, b; Illing, 1959; Barfoot and Rodgers, 1984; Barfoot and Ko, 1987; Machel and Mountjoy, 1987; McNamara and Wardlaw, 1991; Amthor et al., 1993; Drivet, 1993; Drivet and Mountjoy, 1993, 1997; Amthor et al., 1994; Marquez, 1994).

In terms of reservoir character in limestone buildups, only the Leduc Golden Spike (McGillivary and Mountjoy, 1975; Walls, 1978, 1983; Walls and Burrowes, 1985) and Swan Hills Judy Creek limestone reservoirs have been studied in detail (Wendte and Stoakes, 1982; Walls and Burrowes, 1990). In the case of dolomite reservoirs, only the Westerose buildup of the Rimbey-Meadowbrook reef trend has been studied (McNamara and Wardlaw, 1991). Mountjoy (1994) summarized the character of dolomitized reservoirs of the Devonian of western Canada. The present study and that of Drivet (1993) were designed to investigate the reservoir character of dolomite reservoirs in order to provide a suitable database that could be used for the prediction of reservoir quality. Not only was diagenesis and its modification of primary facies and porosity examined, but also the effects of diagenesis on reservoir quality with increasing depth. Distinguishing primary porosity and permeability from textures and fabrics that are overprinted by diagenesis and dolomitization (Mazzulo, 1992; Mountjoy, 1994) is difficult because primary facies and textures are often greatly modified or destroyed.

This chapter summarizes the variability of reservoir quality (porosity and permeability) in deeply buried carbonates, and addresses the potential controls of depositional facies and diagenesis on pore systems. Reservoir character was studied at three different scales: the entire reservoir, meter-scale depositional units, and the individual pore types (Weber, 1986). The objectives are: (1) to identify different pore types and to determine their distribution within Leduc reservoirs; (2) to determine reservoir continuity and variability, and porosity and permeability trends, relative to different depositional facies and diagenetic phases; (3) to compare reservoir characteristics of limestone and dolostone reservoirs in the deep basin; and (4) to determine how dolomitization, cementation, and reservoir bitumen have affected reservoir character.

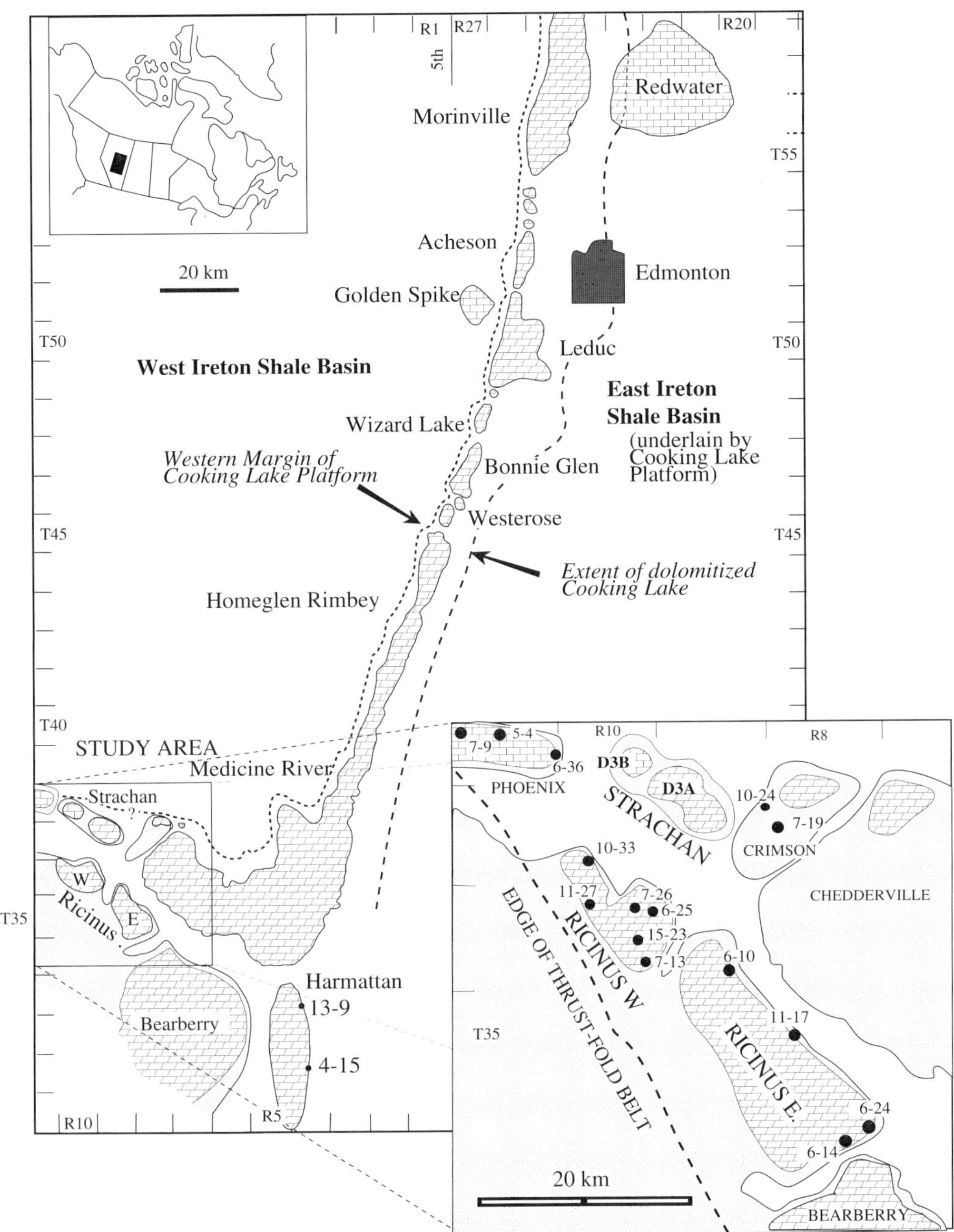

Figure 1. Study area in the deep basin at the southern end of the Rimbey-Meadowbrook reef trend in the center of the Alberta Basin. Inset map shows the distribution of the Strachan, Ricinus West, and East fields, and adjacent buildups.

The reservoir characteristics of these Leduc buildups have been partly documented in the shallow- (<2000 m) and intermediate-burial (2000 to ~3500 m) portion of the Rimbey-Meadowbrook reef trend (Reitzel and Callow, 1977; Barfoot and Rodgers, 1984; Barfoot and Ko, 1987; Hugo, 1990; McNamara and Wardlaw, 1991; Drivet, 1993; Amthor et al., 1994; Drivet and Mountjoy, 1994, 1997). In the deeper (>4000 m) part of the reef trend, there are few published data on reservoir properties and their variability within the Leduc buildups, except for a summary of regional-scale porosity and permeability variations by Amthor et al. (1994) and brief reports on the Strachan and Ricinus West gas reservoirs by Hriskevich et al. (1980) and Seifert (1990). Some information is also available in

theses (Laflamme, 1990; Drivet, 1993; Marquez, 1994). Consequently, little is known about the porosity and permeability variations within these Leduc reservoirs. The partly dolomitized Strachan and completely dolomitized Ricinus West buildups provide an ideal area for comparing the reservoir characteristics of deeply buried limestone and dolostone buildups. Information on the reservoir characteristics of the intermediately buried (2500 m) Homeglen-Rimbey dolomitized buildup is available in Drivet (1993) and Drivet and Mountjoy (1997).

On a regional scale, at shallow depth (<2000 m), porosity and permeability of limestones and dolostones have comparable values and distributions of porosity (Amthor et al., 1994). At these depths, dolomitization resulted in the redistribution and some decrease of porosity and a slight increase in permeability (Amthor et al., 1994). At deeper burial (>2000 m), however, dolostones are significantly more porous and permeable than adjacent limestones (Amthor et al., 1994; Drivet and Mountjoy, 1994; Mountjoy and Amthor, 1994), as also occurs in South Florida (Schmoker and Halley, 1982).

METHODS

All available cores from the Strachan and Ricinus West buildups in the Upper Devonian of central Alberta, between townships 34 to 39 and ranges 7–12W5 (Figures 2–5; Table 1) were logged and sampled from depths of ~3500 to 5000 m. Core parameters were observed and recorded systematically (Table 2). Core-derived horizontal permeabilities (K_h) and vertical permeabilities (K_v) and porosities were obtained from the Energy Resources Conservation Board (ERCB), Calgary. Permeability and porosity data from the gas-producing zone were measured from whole-core samples (full diameter, 8 cm; 0.3 m long). Because sample length is large relative to width, a potential bias exists because the horizontal to vertical ratio is too large (Lishman, 1969). Reservoirs can be considered stratified with respect to permeability because of vertical facies changes. Statistical analyses were obtained using the software STATVIEW II (1990). Permeability values >2000 md were excluded because they probably represent unrealistic values due to fracturing or large vugs. A few logs were analyzed using GEOGRAPHIX QLA2 log analysis software.

To compare porosity among different facies, an arithmetic mean of the core measurements is used, because porosity is a scalar quantity and commonly normally distributed. Permeability has a log normal or skewed distribution and commonly is highly variable over short distances, making a geometric mean a more appropriate choice (Wardlaw, 1990, 1992). McNamara and Wardlaw (1991) reported that the geometric mean of core-measured permeability provided the best correspondence, with permeability estimated from a pressure buildup test in the Leduc Westerose reservoir updip along the Rimbey-Meadowbrook reef trend (Figure 1). Permeability and porosity profiles within dolostones were obtained by plotting porosity, K_h, and K_v values for each sample depth for most of the wells studied in the Ricinus buildup (Marquez, 1994), with representative portions illustrated in this study. Arithmetic, geometric, and harmonic means of the permeability were calculated using the software STATVIEW II (1990).

Information from core observations and their correlation with porosity and permeability data are emphasized in this study for reservoir quality prediction. Maddox (1984) determined the correlation between core porosity and log porosity from neutron, density, and acoustic logs over the cored interval in the Ricinus West reservoir. Core analyses yielded similar to slightly higher porosities compared with those calculated from logs. Comparisons of porosities from cores and logs of Leduc dolomites along the Rimbey-Meadowbrook reef trend made by McNamara et al. (1991) and Drivet (1993) indicate that core porosities are generally comparable to porosities calculated from logs, when suitable logs are available. Furthermore, McNamara et al. (1991) showed that when a core contains pores larger than the core diameter of 8 cm, the value measured directly from cores underestimated the porosity by 3% or more. Thus, core analysis measurements can be considered reliable for the recognition of porosity and permeability trends, but will tend to underestimate porosity in coarse, vuggy carbonates.

Measurements of permeability at simulated reservoir conditions are one order of magnitude lower than similar measurements at ambient pressure (Vavra et al., 1991), so that measured values (ambient) represent maximum permeability. Which geologic parameters should be described and mapped to permit a reasonably accurate petrophysical quantification of carbonate reservoir models has been discussed and outlined by Lucia (1983, 1995), and Lucia and Conti (1987). Lucia (1995) advocates using rock-fabric units based on grain size and sorting, dolomite crystal size, separate-vug type and porosity, and total porosity.

FACIES AND DIAGENESIS

The Leduc Formation of the Strachan (Figures 2, 3) and Ricinus West fields (Figures 4, 5) are stromatoporoid-coral buildups of Upper Devonian age (Leduc Formation) with generally similar depositional facies (Marquez, 1994). The buildups consist of reef margin and reef interior environments (summarized in Figures 3, 5). The reef margin facies include coral rudstones, tabular stromatoporoid boundstones, and stromatoporoid-coral rudstones (Figure 6), indicating deposition in shallow, high-energy environments. The buildup interior comprises skeletal packstones/grainstones, skeletal wackestones, microbial laminites, and locally green shales (Figure 7) suggestive of shallowing-upward, peritidal conditions. The interior of the Ricinus West buildup is divisible into lower and upper parts. The lower portion consists of domal stromatoporoid floatstones, skeletal wackestones, and coral rudstones. The upper part (131 m) consists of six shallowing-upward parasequences (generally 8 m thick, but ≤27 m) and comprise from bottom to top: *Amphipora*-rich wackestones or packstones, locally with small domal stromatoporoids; skeletal grainstones; and peritidal microbial laminites. These cycles are two to three times as thick as outcrop equivalents (McLean and Mountjoy, 1994), which may be due to obliteration of some facies by dolomitization, or continuous deposition without obvious breaks (Marquez, 1994).

The limestone Strachan and dolomitized Ricinus West fields have undergone a complex diagenetic history and different diagenetic overprints (Figure 8) (Marquez, 1994). The Ricinus West buildup is similar to other dolomitized Leduc buildups along the Rimbey-Meadowbrook reef trend (Amthor et al., 1993, 1994; Drivet and Mountjoy, 1996). Near-surface sea-floor

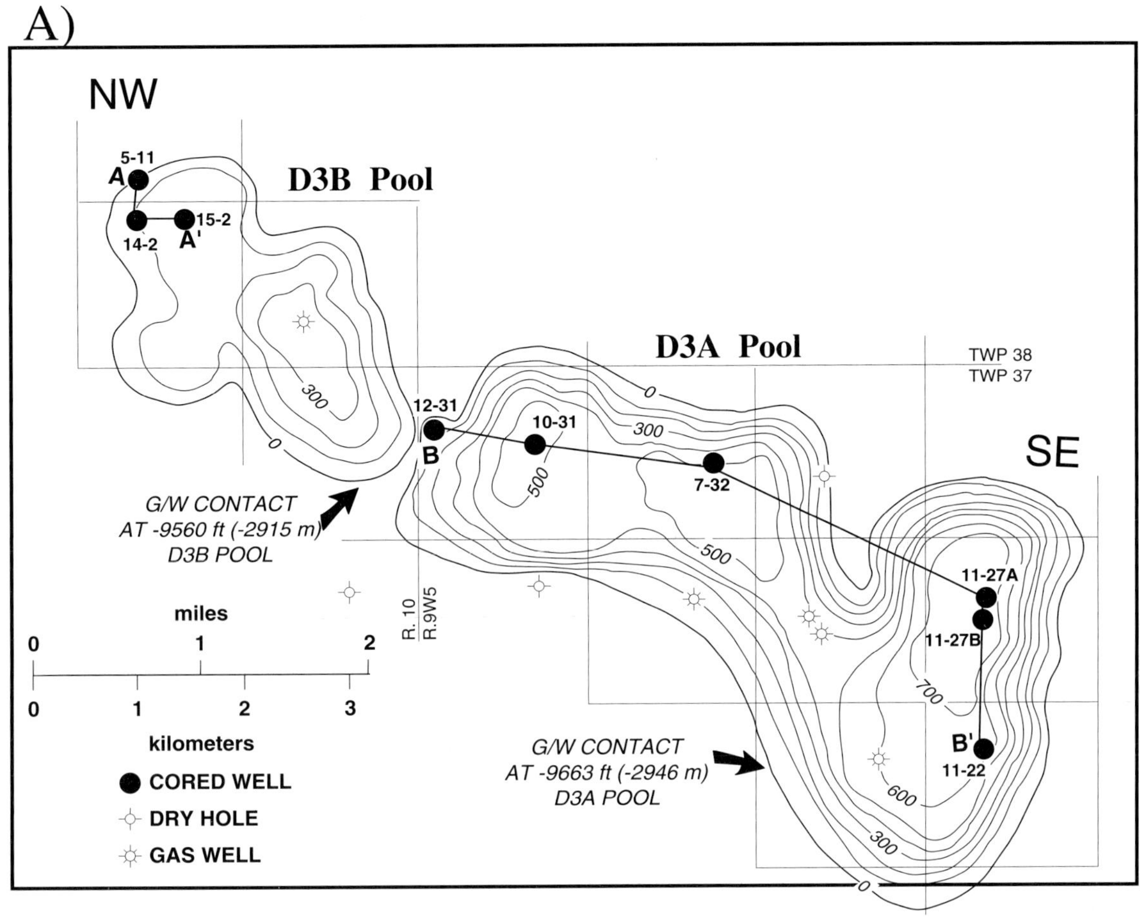

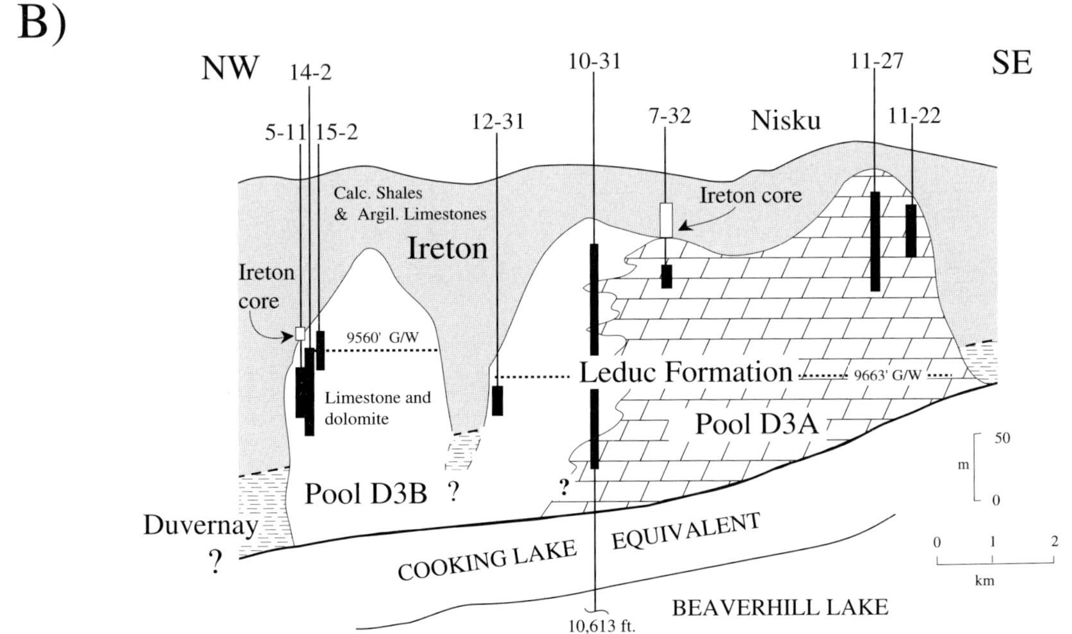

Figure 2. Leduc net pay isopach map of the Strachan buildup. (A) Map of well locations and net pay thickness (modified from Hriskevich et al., 1980). Wells 5-11, 14-2, 15-2, and 12-31 are nonproductive. (B) A NW-SE cross section shows distribution of limestone and partly dolomitized limestone (white), approximate gas/water (G/W) contact, and distribution of core.

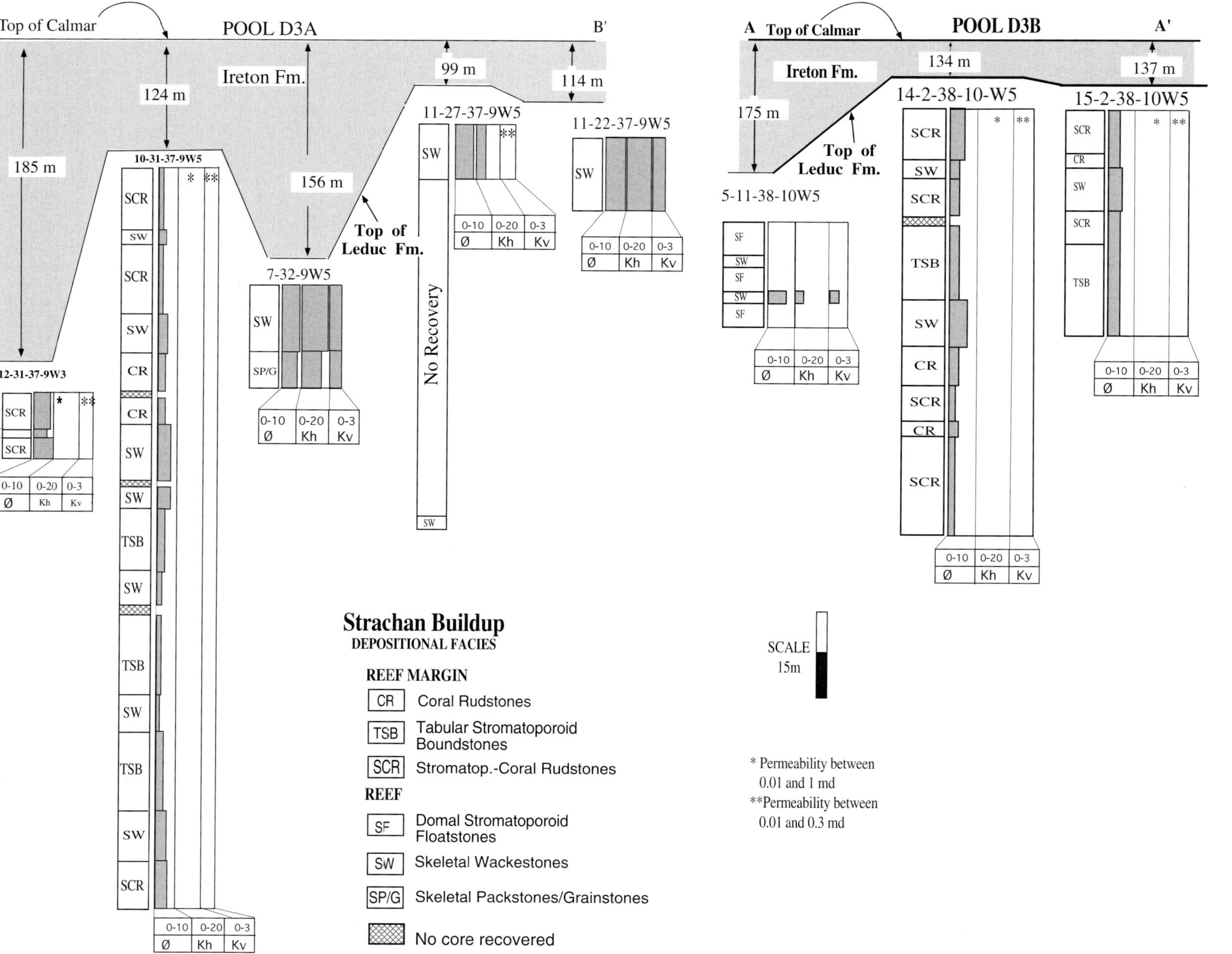

Figure 3. Distribution of depositional facies in Strachan D3A (section BB') and D3B pools (section AA'). See Figure 2 for well locations. Datum is top of Calmar Formation. To the right of each column are arithmetic average porosity in percent and geometric average permeabilities in millidarcies. Porosity and permeability data are based on core analyses from the Energy Resources Conservation Board (ERCB) in Calgary.

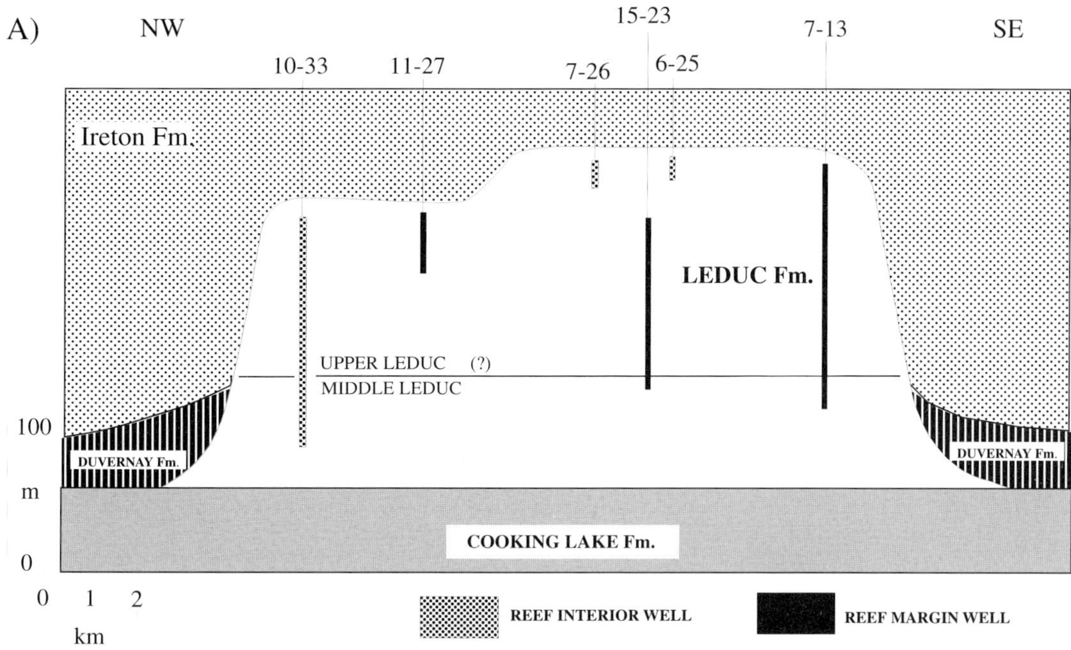

Figure 4. Ricinus West buildup. (A) A NW-SE cross section showing stratigraphy and location of cores. (B) Map of well locations and net pay thickness (modified from Hriskevich et al., 1980). G/W = gas/water.

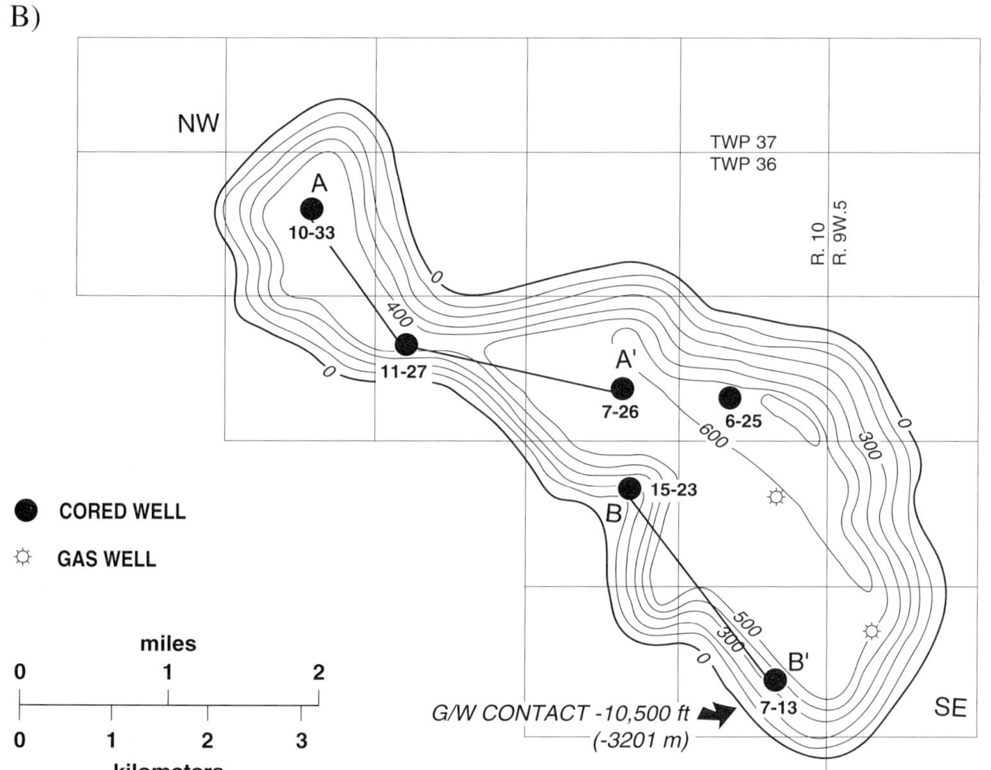

radiaxial fibrous calcites fill intergranular, shelter, and growth framework cavities in the buildup margins. Syntaxial overgrowths are rare and were followed by the precipitation of blocky calcite cements in skeletal pores. During intermediate burial, chemical compaction and partial to complete replacement dolomitization occurred. Complete dolomitization in the Ricinus West buildup obliterated many of the early paragenetic features. Most of the Leduc carbonates appear to have been replaced by dolomites at temperatures of ~40°–60°C and depths of 500–1200 m during the Late Devonian to Early Mississippian (Amthor et al., 1993; Mountjoy and Amthor, 1994; Drivet and Mountjoy, 1997). The sources and flow directions of the dolomitizing fluids are still uncertain. Replacement dolomitization was followed by fracturing, dissolution, and minor dolomite cement.

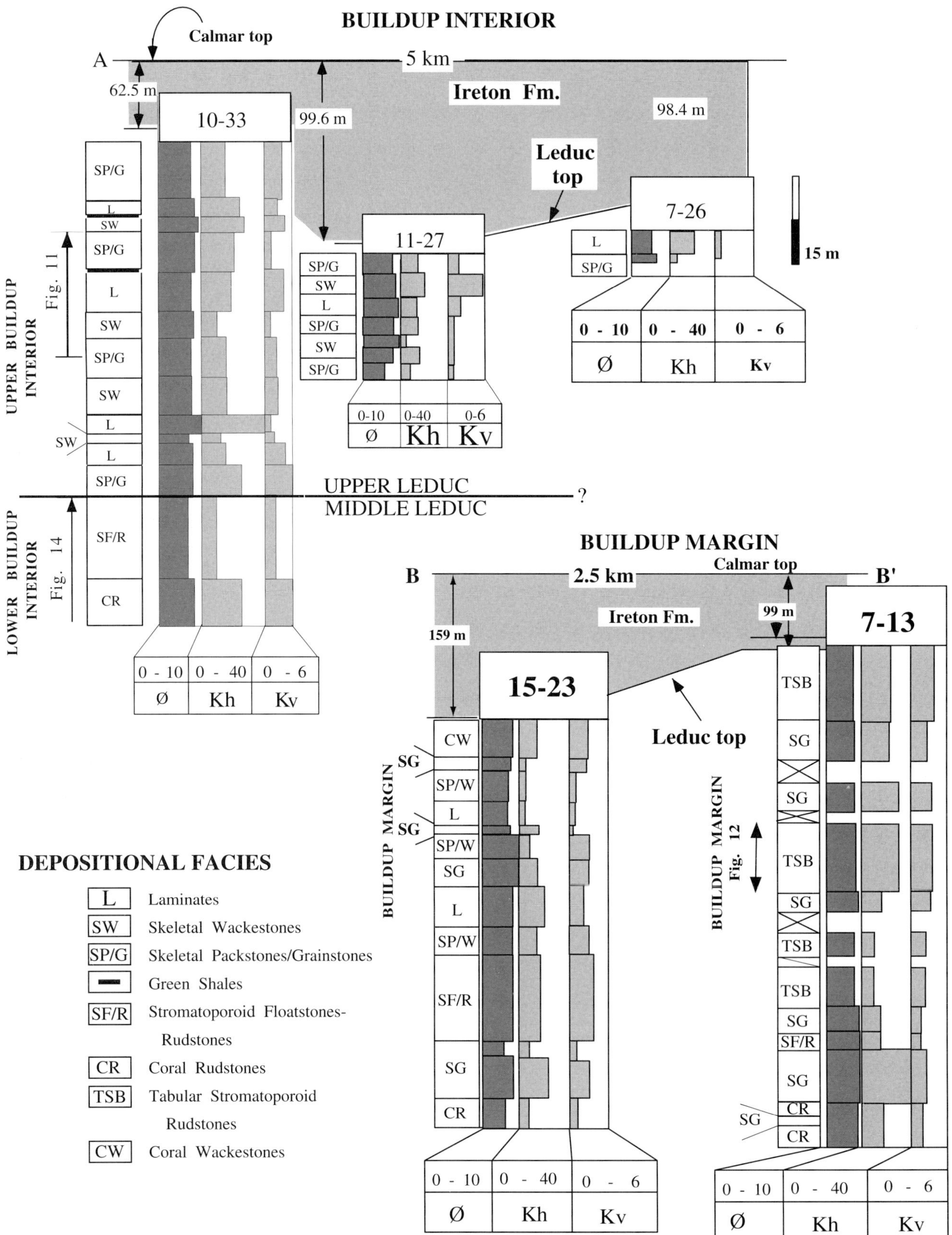

Figure 5. Depositional facies in the Ricinus West buildup interior wells 10-33, 11-27, and 7-26, and margin wells (lower right) 15-23 and 7-13 (see Figure 4 for well locations). Datum is top of Calmar Formation. To the right of each column are the arithmetic average porosity in percent and geometric average horizontal (K_h) and vertical (K_v) permeabilities in millidarcies. Porosity and permeability data are based on core analyses from ERCB.

Table 1. Wells and Cored Intervals from the Strachan, Ricinus, and Adjacent Buildups.

Well	Location	Cored Interval (ft)	Cored Interval (m)	Total (ft)	Total (m)	Field
Banff Aquitane	10-31-37-9W5	13,442–14,200	4098.2–4329.3	758.0	231.1	Strachan
Banff Aquitane	12-31-37-9W5	14,046–14,096	4282.3–4297.6	50.0	15.2	Strachan
Aquitane	15-2-38-10W5	13,485–13,605	4111.3–4147.9	120.0	36.6	Strachan
Chevron SOBC	14-2-38-10W5	13,466–13,526	4105.5–4123.8	60.0	18.3	Strachan
Aquit. Shell-Strachan	5-11-38-10W5	13,916–13,702	4242.7–4177.4	245.0	74.7	Strachan
Banff Aquitane	7-32-37-9W5	13,422–13,494	4092.1–4114.0	72.0	22.0	Strachan
B.A. et al.	11-27-37-9W5	13,187–13,197	4020.4–4023.5	10.0	3.0	Strachan
Gulf et al. Strachan	11-22-37-9W5	12,970–13,046	3954.3–3997.4	76.0	23.2	Strachan
Aquitane et al.	10-16-37-10W5	15,306–15,333	4666.5–4674.7	27.0	8.2	Strachan
Gulf et al. Strachan	10-24-37-9W5	13,322–13,378	4061.6–4078.7	56.0	17.1	Crimson
Archo Pacific Fina Cow Jk	7-33-37-8W5	12,133–12,168	3699.1–3709.8	35.0	10.7	Crimson
Husky et al.	11-28-37-8W5	12,133–12,178	3699.1–3712.8	45.0	13.7	Crimson
Gulf Poc et al.	7-19-37-8W5	13,436–13,493	4096.3–4113.7	57.0	17.4	Crimson
Imp. Chedderville	10-20-37-7W5	11,608–11,687	3539.0–3563.1	79.0	24.1	Chedderville
Pinn. Chedderville	16-19-37-7W5	11,642–11,692	3549.4–3564.6	50.0	15.2	Chedderville
Imp. Chedderville	10-29-37-7W5	11,991–12,016	3655.8–3663.4	25.0	7.6	Chedderville
BP Chedderville	6-30-37-7W5	11,585–11,595	3532.0–3535.1	9.8	3.0	Chedderville
Dome et al. Chedderville	14-9-37-7W5	11,634–11,713	3547.0–3571.0	79.0	24.1	Chedderville
Esso Mobil Ricinus	10-11-37-8W5	12,356–12,438	3767.1–3792.1	82.0	25.0	Chedderville
Imp. HB. Ricinus	12-28-36-7W5	12,452–12,518	3796.3–3816.5	66.0	20.1	Chedderville
Banff et al.	10-33-36-10W5	14,785–15,319	4507.6–4670.4	534.0	162.8	Ricinus
Banff Aquitane	11-27-36-10W5	14,664–14,724	4470.7–4489.0	60.0	18.3	Ricinus
Chevron Mobil	7-26-36-10W5	14,381–14,434	4384.5–4400.6	53.0	16.2	Ricinus
Banff Aquitane	15-23-36-10W5	14,616–15,067	4456.1–4593.6	451.0	137.5	Ricinus
Banff et al.	7-13-36-10W5	14,293–14,865	4357.6–4532.0	572.0	174.4	Ricinus
Banff et al. Ricinus	6-25-36-10W55	14,323–14,378	4366.8–4383.5	55.0	16.8	Ricinus
Mobil Ricinus	6-10-36-9W5	14,240–14,251	4341.5–4344.8	11.0	3.4	Ricinus
Mobil et al. Ricinus	11-17-35-8W5	13,814–13,921	4211.6–4244.2	107.0	32.6	Ricinus
Pan. Amer. Ricinus	6-24-34-8W5	13,414–13,475	4089.6–4108.2	61.0	18.6	Ricinus
Albany Amoco Ricinus	6-14-34-8W5	14,048–14,138	4282.9–4310.4	90.0	27.4	Ricinus
Banff Aquit. Ram River	7-9-37-10W5	15,404–15,510	4696.3–4728.7	108.0	32.9	Ram River
Husky Ram River	10-16-37-10W5	15,306–15,333	4666.5–4674.7	27.0	8.2	Ram River
Shell Canterra Ram R.	5-13-37-12W5	16,628–16,515	5069.5–5035.1	113.0	34.5	Ram River
Uno-Tex et al. Phoenix	5-4-39-11W5	14,220–14,280	4335.4–4353.7	60.0	18.3	Phoenix
RR Amoco et al. Ancona	7-9-39-12W5	15,539–15,553	4737.5–4741.8	14.1	4.3	Phoenix
C.S. et al. Phoenix	6-36-38-11W5	13,884–13,910	4232.9–4240.9	26.4	8.0	Phoenix

Table 2. Systematic Observations Made from Core.

1. Core Number, Box, Depth,
   Recovery (%)

2. Lithology
   a. Limestones
      Texture, fossils, contacts
      Depositional facies
      Degree of dolomitization

   b. Dolostones
      Crystal Size
      Fine: 30–62 µm
      Medium: 62–250 µm
      Very Coarse: >600 µm

3. Pore Type
   Intraskeletal
      Fossil type
   Intercrystalline
   Moldic
      Amphipora
      Thamnopora-like
   Vug
   Fracture
   Breccia

4. Pore Size
   Very large: >1.0 cm
   Large: 0.5–1.0 cm
   Medium: 2.0–5.0 mm
   Small: 0.1–1.0 mm
   Very small: >0.1 mm

5. Pore Shape
   Spherical
   Tubular
   Tabular
   Irregular
   Polyhedral

6. Pore Association Connection
   Matrix (intercrystalline)
   Fractures
   Touching vugs/molds

7. Cements
   Type
      Dolomite
      Anhydrite
      Sulfides
      Sulfur
      Quartz
      Reservoir bitumen
      Calcite
   Pore Type
   Degree of Filling
      Open
      Partly filled
      Filled

8. Fractures
   Orientation
      Subvertical
      Subhorizontal
   Intensity
   Width, Length
   Filling

9. Breccias
   Crackle
   Mosaic
   Rubble

Anhydrite cements postdate the dolomite cements. Microfractures filled with bitumen crosscut most diagenetic features. These hairline microfractures probably formed during overpressuring of a well-sealed reservoir during progressive burial by thermal cracking of crude oil to gas in conjunction with shearing related to tectonic compression (Marquez and Mountjoy, 1996). Calcite cements postdate microfracturing, with the latest phase related to thermochemical sulfate reduction (Krause et al., 1988; Marquez, 1994) (Figure 8).

Of critical importance to the rock fabrics, especially the permeability, is the size of the grains or dolomite crystals. Lucia (1983, 1995) grouped nonvuggy carbonates (both limestones and dolostones) into three porosity-permeability fields defined using particle and crystal-size boundaries of 20 and 100 µm. Fine to medium crystalline grainstone-dominated dolopackstones and medium crystalline mud-dominated dolostones plot in the 20–100 µm permeability field, and more coarsely crystalline rocks generally plot above the 100-µm boundary (Lucia, 1995). The replacement dolomites of the Strachan and Ricinus reservoirs consist predominantly of two types: (1) medium crystalline (60–250 µm), planar, euhedral to subhedral, forming dense and porous mosaics; and (2) fine crystalline (30–60 µm), planar, euhedral to subhedral, forming dense and porous mosaics. The medium crystalline dolomites are the most common, forming ≤90% of some wells (e.g., Stachan D3A 10-31, Crimson 10-24), and most would plot above the 100-µm boundary. The fine crystalline dolomites occur in the matrix of partially dolomitized packstones and rudstones (e.g., Strachan well 14-2) and would plot within the 20- to 100-µm boundaries. A third type, coarsely crystalline (250–600 µm), is locally abundant in Ricinus well 11-27. A fourth type, touching vugs with a matrix of intercrystalline dolomite, occurs in both the Strachan and Ricinus reservoirs.

PORE TYPES AND DEFINITIONS

The following sections summarize the pore types, permeabilities, and the resulting porosity network within these limestone and dolostone reservoirs (Tables 3–6). The Strachan reservoir has a weak water drive and

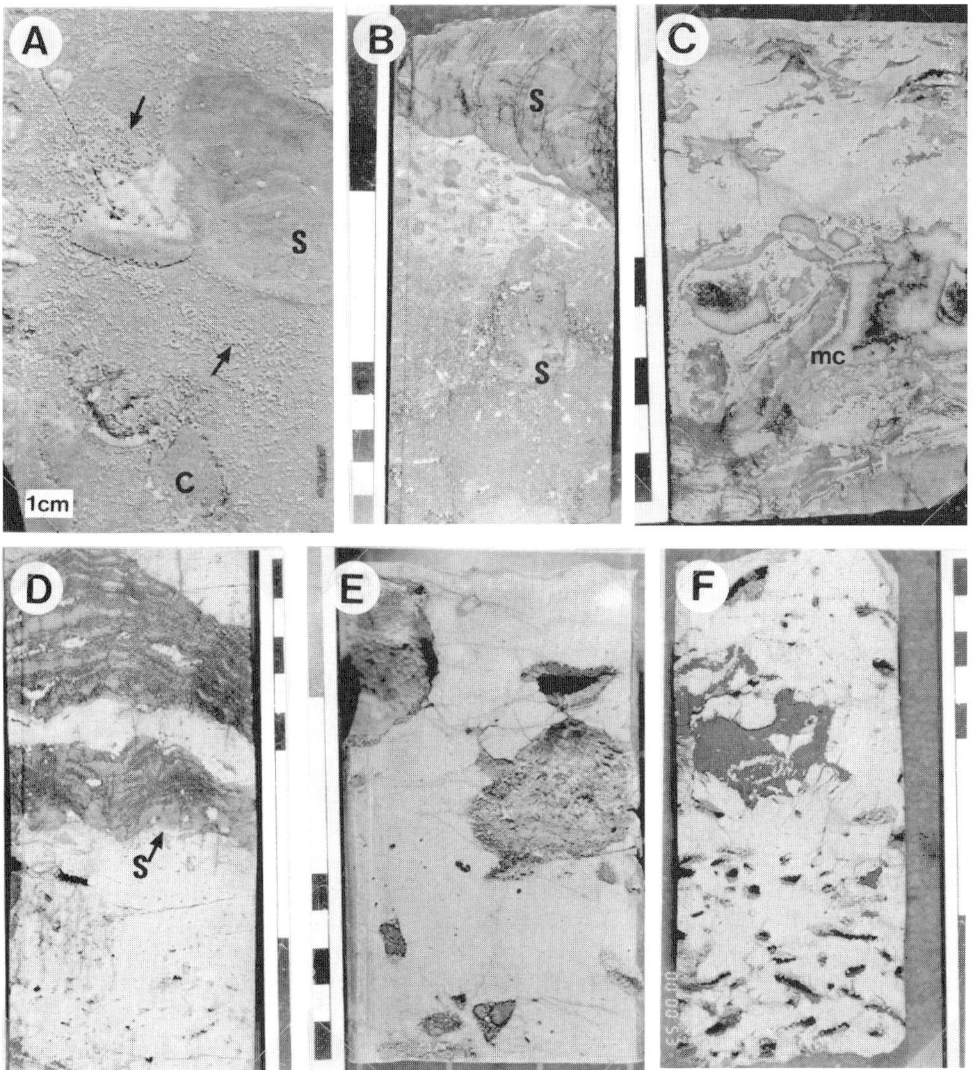

Figure 6. Core photographs showing distibution of dolomite in Strachan buildup. Scales for cores are in centimeters. (A) Selective replacement of lime mud matrix by isolated rhombs (110 μm) and patches (arrows) of dolomites in stromatoporoid (S) and coral (C) rudstones facies. Minor intercrystalline porosity, D3A 10-31-37-9W5, 4281 m. (B) Selective replacement of matrix and initiation of dolomite patches and mosaics (light gray). Compare with completely dolomitized sample (6F) only 0.5 m apart. Stromatoporoid (S) and coral rudstones facies, D3A 10-31-37-9W5, 4279.8 m. (C) Calcite marine cements (mc) and skeletal fragments surrounded by completely dolomitized matrix. Stromatoporoid and coral rudstones facies, D3A 10-31-37-9W5, 4181.4 m. (D) Replacement dolomite completely replaces matrix (light gray). Intraskeletal pores in stromatoporoid fragment (S) are partly filled with reservoir bitumen. Small vugs occur in the dolomitized matrix. Stromatoporoid and coral rudstones facies, D3A 10-31-37-9W5, 4265 m. (E) Replacement dolomitization of matrix results in a nonporous mosaic of dolomite crystals with some vugs. Partly leached domal stromatoporoid fragments. Reef margin stromatoporoid and coral rudstones facies, D3A 12-31-37-9W5, 4287.5 m. (F) Dissolution of skeletal fragments results in molds, solution-enlarged molds, and vugs. Pores are lined by reservoir bitumen. Stromatoporoid and coral rudstone facies, D3A 10-31-37-9W5, 4279.2 m.

is thus a partially "closed" system that underwent relatively rapid pressure decline during production (Hriskevich et al., 1980). A good understanding of the pore types and their distribution is essential for efficient recovery of the hydrocarbons in these reservoirs.

Pores types are characterized by different shape, size, orientation, and interconnection (Tables 3, 4), and are classified following Choquette and Pray (1970). They form three broad groupings with respect to their distribution within the buildup. Dense matrix refers to compact crystalline texture (Archie, 1952). Matrix intercrystalline porosity is defined as the spaces between dolomite crystals, except where there is evidence of dissolution, and larger pores (small vugs) are present (equivalent to the "pinpoint" porosity of McNamara and Wardlaw, 1991). Fracture porosity is subdivided into unfilled, partly filled, and filled, because fractures that appear to be filled may have permeability that is

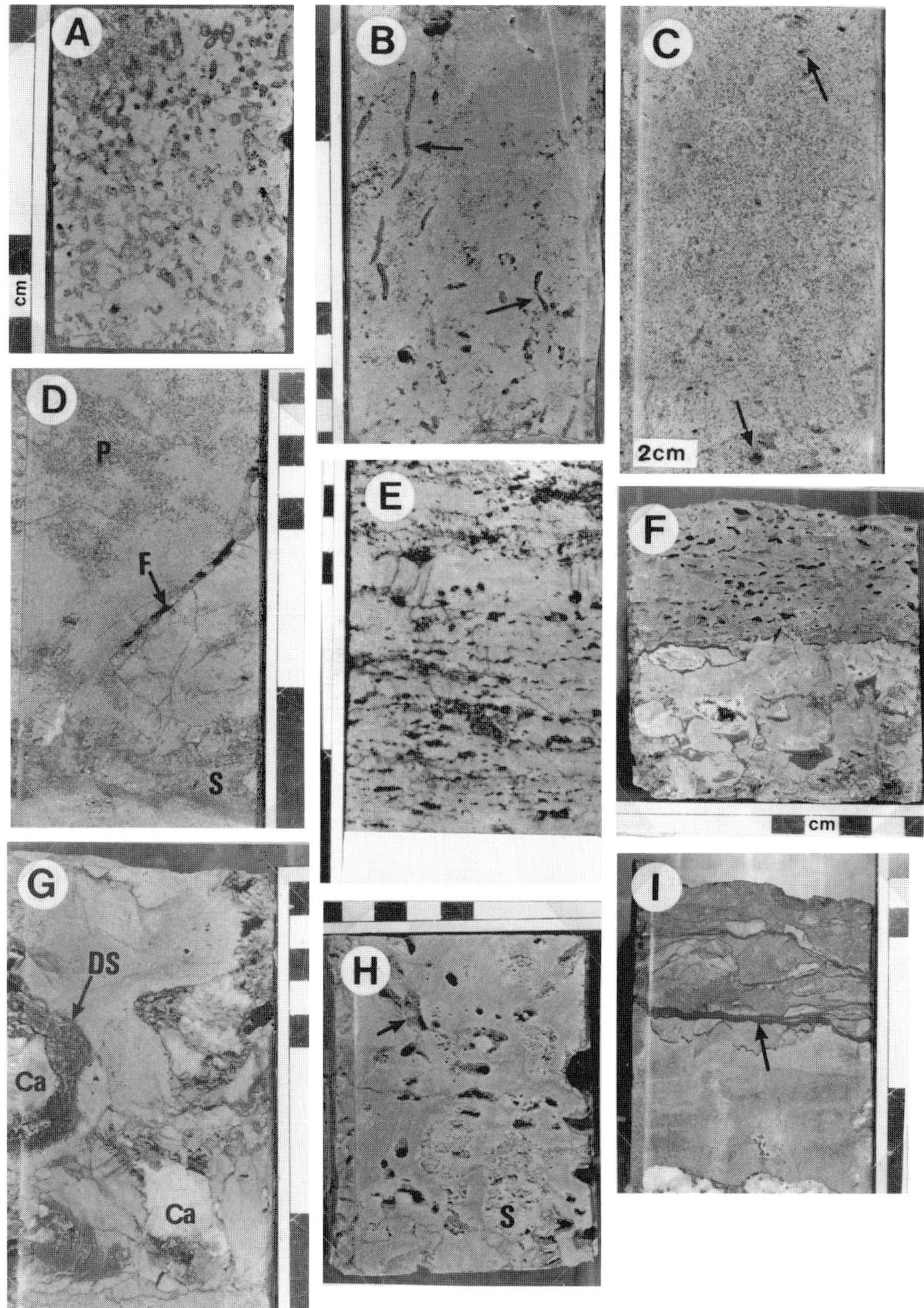

Figure 7. Core photographs showing depositional facies and associated pore types in dolomitized Ricinus West buildup interior well 10-33-36-10W5. Scales for cores are in centimeters. (A) Skeletal packstone facies, leached *Amphipora* moldic pores in a tight packstone matrix; 4601.5 m. (B) Skeletal wackestone facies, isolated tubular, moldic *Amphipora* pores (arrows) in a tight mudstone matrix; 4605.4 m. (C) Skeletal wackestone facies: intercrystalline pores partly filled with reservoir bitumen (black/staining), small isolated moldic pores (arrows), and vugs; 4601.2 m. (D) Skeletal wackestone facies: patches (P) of polyhedral intercrystalline pores partly filled with bitumen in an otherwise tight matrix. Stromatoporoid fragment (bottom) with solution-enlarged, tabular intraskeletal pores (S). Subvertical fracture (F) is partly filled with late calcite cement; 4579.2 m. (E) Microbial laminite facies: finely laminated mudstones with elongated, irregular, fenestral-like pores; 4608.5 m. (F) Stylolitic contact between microbial laminite facies and mudstones with pores filled with green shales; 4576.2 m. (G) Domal stromatoporoid floatstone facies: partly leached domal stromatoporoid (DS) with intraskeletal pores partly filled with reservoir bitumen (black). Late calcites (Ca) completely fill the remaining pore space; 4627.4 m. (H) Domal stromatoporoid rudstones: partly leached stromatoporoid (S) and coral (arrow) fragments in a dolomudstone matrix; 4664 m. (I) Green shaly laminations (arrow) and associated stylolites. Bottom half of core shows tight dolomudstone with vug completely filled with anhydrite; 4632.6 m.

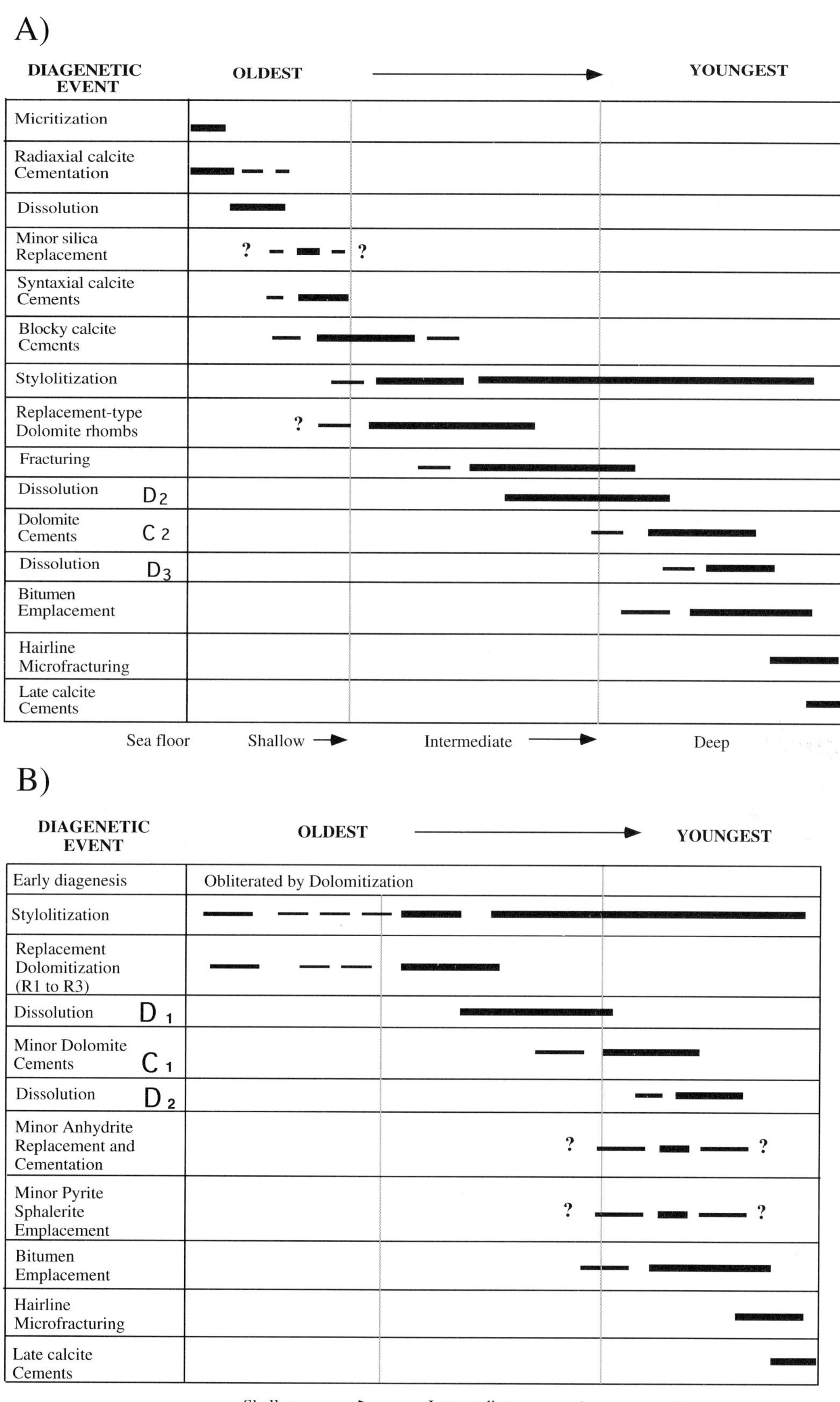

Figure 8. Paragenetic sequence for Strachan buildup: (A) limestone portion and (B) completely dolomitized portion, which also applies to Ricinus West and adjacent buildups. The onset of stylolitization (>500 m) designates the beginning of intermediate burial, and maturation of organic matter designates deep burial.

Table 3. Pore Types in Limestone and Partly Dolomitized Strata.*

Limestones

	Pore Type	Symbol	Shape	Size	Orientation	Connection	Degree of Filling*	Cement
FABRIC SELECTIVE	Intraskeletal: Corals		Spherical	50–100 µm		Connected pores within skeletal fragments	Partly filled	Bitumen
	Stromatoporoid		Tabular	50–100 µm	Parallel to growth structure		Partly filled	Bitumen
NONFABRIC SELECTIVE	Vug		Irregular	1–10 mm	Random	Intercrystalline pores in matrix fractures	Partly filled	Bitumen Calcite
	Fracture	F	Tabular	20–100 mm long	Subvertical	Fractures	Partly filled Filled	Dolomite C$_2$ Bitumen Calcite

Partly Dolomitized

	Pore Type	Symbol	Shape	Size	Orientation	Connection	Degree of Filling*	Cement
NONFABRIC SELECTIVE	Intraskeletal: Stromatoporoid		Tabular	50–100 µm	Parallel to growth structure	Connected pores within skeletal fragments	Partly filled	Bitumen
	Intercrystalline		Polyhedral	100 µm	Random pores in matrix	Intercrystalline	Partly filled	Bitumen
NONFABRIC SELECTIVE	Vug		Irregular	5–20 mm	Random	Intercrystalline pores in matrix Fractures	Partly filled	Bitumen Calcite
	Fracture	F	Tabular	10–60 mm long	Subvertical	Fractures	Partly filled Filled	Bitumen Calcite

*Filled = pore completely filled with cements; Partly filled = some cement present.

Table 4. Pore Types in Dolostones.

	PORE TYPE	SYMBOL	SHAPE	SIZE	ORIENTATION	CONNECTION	DEGREE OF FILLING*	CEMENT
FABRIC SELECTIVE	Intrasekletal		Tabular	50–100 µm	Parallel to growth structure	Connected pores in skeletal fragments	Partly filled	Bitumen
	Fenestral-like		Tabular	2–20 µm	Parallel to laminations	Intercrystalline pores in matrix	Partly filled	Bitumen
	Intercrystalline		Polyhedral	110–250 µm	Random	Fractures	Partly filled	Bitumen
	Moldic: Amphipora-like		Tabular	10–20 mm	Random to bedding parallel	Touching molds	Open	
	Stromatoporoid-like		Spherical	30–60 mm	Random	Touching molds intercrystalline	Partly filled	Anhydrite Bitumen
	Thamnopora-like		Tubular	20–40 mm	Random	Touching molds	Partly filled	Anhydrite Bitumen
NONFABRIC SELECTIVE	Vug		Irregular	1–70 mm	Random	Touching vugs	Partly filled fractures	Calcite Anhydrite Quartz
	Fracture		Tabular	40–80 mm long	Subvertical	Fractures	Partly filled Filled	Bitumen
	Breccia		Irregular	10–30 mm	Random	Intercrystalline fractures	Partly filled	Bitumen

*Filled = pore completely filled with cements; Open = no cement present; Partly filled = some cement present.

Table 5. Porosity and Permeability Values in Limestones, Partly Dolomitized Rocks, and Dolostones.*

Depositional Facies	Porosity			K_h (md)			K_v (md)			N
	Ar	Min	Max	Geo	Min	Max	Geo	Min	Max	
LIMESTONES										
STRACHAN BUILDUP MARGIN										
Stromatoporoid Floatstone	N/A			N/A			N/A			
Skeletal Wackestones	3.6	0.5	7.1	0.8	0.01	16.2	0.4	0.01	6.4	82
Tabular Stromatoporoid Boundstone	2.3	0.3	8.6	0.1	0.01	360	0.02	0.01	6.5	92
Stromatoporoid Coral Rudstone	2.6	0.3	15.4	0.1	0.01	2.1	0.01	0.01	74	179
Coral Rudstone	3.0	0.4	29.0	0.3	0.01	247	0.01	0.02	55	95
PARTLY DOLOMITIZED										
STRACHAN BUILDUP MARGIN										
Stromatoporoid Floatstone	N/A			N/A			N/A			
Skeletal Wackestones	4.91	2.0	8.0	0.21	0.01	20.7	0.02	0.01	11.5	8
Tabular Stromatoporoid Boundstone	3.5	0.3	9.2	1.1	0.01	715	0.1	0.01	9.8	74
Stromatoporoid Coral Rudstone	7.7	0.1	16.8	N/A	0.01	N/A	0.1	0.01	6.6	26
Coral Rudstone	4.5	0.5	9.9	2.2	0.03	82	1.0	0.01	147	41
DOLOSTONES										
STRACHAN BUILDUP INTERIOR										
Skeletal Wackestones	8.4	0.5	26	7.0	0.1	370	1.0	0.12	19	143
Skeletal Packstones/Grainstones	7.1	4.8	12.6	15.0	1.2	396	5.1	0.43	367	13
RICINUS WEST MARGIN										
Tabular Stromatoporoid Boundstone	6.3	1.6	12.4	16.7	0.2	384	3.7	0.01	129	195
Skeletal Wackestones	7.1	2.0	15.1	15.8	0.9	725	2.5	0.01	65	134
Stromatoporoid Floatstone	7.7	2.0	15.3	27.9	1.9	501	2.4	0.2	38	89
Coral Rudstone	6.6	3.0	12.4	18.5	2.3	262	2.0	0.2	28.4	30
Breccias	6.5	3.3	12.0	21.1	1.7	525	1.5	0.2	9.6	19
RICINUS WEST INTERIOR (UPPER PORTION)										
Microbial Laminites	6.2	0.9	13.8	15.6	0.7	900	2.8	0.01	96	154
Skeletal Wackestones	7.9	1.5	20.7	11.1	0.1	947	1.5	0.01	221	184
Skeletal Packstones/Grainstones	6.1	1.3	15.0	9.5	0.06	753	1.9	0.01	142	313
RICINUS WEST INTERIOR (LOWER PORTION)										
Stromatoporoid Floatstone	6.0	0.6	15.5	12.7	0.1	764	1.7	0.01	100	228
Coral Rudstone	5.6	1.0	15.2	11.9	1.6	418	1.5	0.01	41	70

*Ar = arithmetic mean; Geo = geometric mean; K_h = horizontial permeability; K_v = vertical permeability.

Table 6. Porosity and Permeability Ranges in Three Basic Pore Types, Ricinus West Reservoir.*

Main Porosity Types	Pore Types	Porosity (%)					K_h (md)			K_v (md)			Thickness (m)	Distribution Within Buildup
		Ar	σ	n	Min	Max	Geo	Min	Max	Geo	Min	Max		
Amphipora Molds	Moldic *Amphipora*-like													Upper
	Intercrystalline Fenestral-like	6.3	3.52	751	0.9	20.7	12.7**	0.06	2000	1.2**	0.01	861	131	Buildup Interior
Stromotoporoid Vugs	Moldic pores (Stromatoporoid-like)													Lower
	Intercrystalline Moldic (*Thamnopora*-like)	6.1	3.00	294	0.6	19.1	13.1**	0.1	1400	1.8**	0.01	834	61	Buildup Interior
Large Vugs and Fractures	Irregular vugs Breccias Intercrystalline	6.9	2.50	448	1.6	15.3	19.4**	0.2	2100	2.8**	0.01	141	259	Buildup Margin

*Ar = arithmetic mean, Geo = geometric mean, K_h = horizontal permeability, K_v = vertical permeability.
**n of the geometric mean not equivalent to n of porosity average (values above max were dropped).

significantly higher than associated matrix based on minipermeameter measurements (N.C. Wardlaw, 1992, personal communication). Interconnection between pores is described as fracture-connected, as interconnected vugs/molds, and as matrix intercrystalline porosity (McNamara and Wardlaw, 1991). Lucia (1995) suggests that separate-vug and touching-vug porosity should be distinguished. Separate vugs are typically fabric selective and include intrafossil, moldic, intragranular, and shelter porosity. Lucia classifies touching vugs as being typically nonfabric selective and includes cavernous, breccia, fracture, solution-enlarged fracture, intercrystalline, and fenestral porosity. However, in the case of these Upper Devonian buildups, most of the touching vugs are solution-enlarged molds (e.g., *Amphipora*) and are therefore fabric selective. Individual *Amphipora* appear to overlap and connect with each other in the horizontal direction. Specific pore types of the study area are discussed below.

Pore Types in Limestones

Limestones are present in pool D3B of the Strachan reservoir (wells 5-11, 14-2, 15-4) (Figure 2; Table 3). The facies distribution across pool D3B is illustrated in cross section AA' (Figure 3). Porosity and permeability plots are included with facies distribution to show the relationship between these reservoir parameters and the depositional facies. Skeletal wackestone facies have the highest porosity values (Table 5). Average K_h and K_v of limestones are very low overall (0.8 and 0.4 md, respectively). Pool D3B forms a poor reservoir.

The most common pore types in these limestones are, in decreasing abundance, intraskeletal (50%–100%), subvertical fractures (5%–15%), and vugs (5%). Intraskeletal pores have spherical shapes in corals (Figure 9A), tabular shapes in stromatoporoid fragments (Figure 9B), and are always lined with reservoir bitumen (a descriptive term for bitumen that lines and fills pore spaces to distinguish it from bitumen in source beds) (Rogers et al., 1974; Lomando, 1992). (In the remainder of the chapter, bitumen refers to reservoir bitumen.) Some interconnection is provided by the intraskeletal framework. Vugs are irregular in shape, small, and generally isolated. Minor subvertical fractures, with irregular surfaces that suggest dissolution, are filled with dolomite cement (C_2, saddle dolomite; Figure 8), bitumen, and late-stage calcite (Figure 9C, D).

Pore Types in Partly Dolomitized Limestones

Partly dolomitized limestones are present in pool D3A of the Strachan reservoir (wells 12-31 and 10-31) (Figure 2; Table 3). Facies distribution and relation to porosity and permeability are illustrated in cross section BB' (Figure 3). With 50 to 75 vol. % dolomite, all lime matrix is replaced, resulting in fine crystalline (30–60 µm) matrices with variable porosity ranging from porous intercrystalline to dense. Most skeletal fragments and marine cements remain as calcite (Figure 6C). In partly dolomitized limestones, average porosity is higher than in limestones (Table 5), but permeability is low (see section on diagenetic control on pore systems). Stromatoporoid-coral rudstone and skeletal wackestone facies have the highest average porosity values, 7.7% and 4.9%, respectively; average K_h and K_v are higher in dolomites than in limestones (K_h 7.0 vs. 0.8 md, and K_v 1.0 vs. 0.01, respectively).

Pore types (Table 3) include intraskeletal (40%–80%), intercrystalline (5%–20%), vugs (5%) and subvertical fractures (5%–15%). Intraskeletal pores are restricted to stromatoporoid fragments (Table 3). Intercrystalline pores are polyhedral in shape, ~60 µm in diameter. Small to medium, irregular vugs are common in the dolomitized matrix of all facies (Figure 6D, E).

Pore Types in Dolostones

Dolostones are present in pool D3A of the Strachan reef interior (wells 7-32, 11-27, 11-22) (Figure 3) and throughout the Ricinus West buildup. The facies distribution, porosity, and permeability values across the Ricinus West reservoir are illustrated in cross sections AA' and BB' (Figure 5). With >75 vol. % dolomite, most of the lime matrix is replaced, resulting in a dense dolomite mosaic (Figures 6, 7, and 9D). The size of the dolomite crystals ranges from 60 to 250 µm, with some muddy carbonates as fine as 30 µm. Most skeletal grains have been dissolved, forming slightly enlarged molds and irregular vugs, indicating a genetic relationship between the amount of dolomite and the moldic and vuggy pores (Figures 6, 9D–G; Table 4). These pores likely result from dissolution of calcite during or after replacement dolomitization. Porosity (≤8.4% average) and permeability are greater than in limestones and partly dolomitized limestones (Table 5).

These dolostones contain varying amounts of vugs (40%–100%), molds (10%–60%), fenestral-like (10%–30%), intercrystalline (5%–15%), fractures and breccias (5%–10%), and minor (<1%) solution-enlarged intraskeletal pores (Figure 7; Table 4). Vugs show a gradation from solution-enlarged molds to very large (3–7 cm), irregular pores that show little indication of their precursor fabric (Figure 9G). The most common moldic pores are those produced by selective dissolution of tubular *Amphipora* fragments (e.g., well 7-32, 4110.6 m) (Figures 6F, 7A, and 9F) in the upper part of the reef interior, referred to as *Amphipora*-like molds. In the lower part of the buildup interior, molds are common after dissolution of domal (spherical) stromatoporoids and tubuar *Thamnopora?* fragments (Figure 9E), stromatoporoid-like and *Thamnopora*-like, respectively. Molds of *Amphipora* and *Thamnopora* fragments are distinguished on the basis of their size, association with other pore types, and location within the buildup. Smaller pores that are tabular and aligned parallel to laminations (fenestral-like) are considered molds related to microbial laminite facies (Figures 7E, F, and 10D). However, some dolomites have a nonporous, interlocking, crystal mosaic (Figure 10B).

Intercrystalline pores in fine-crystalline replacement dolomite are rare and very small (10 µm; e.g., Ricinus West, well 10-33, 4633.8 m). In contrast, intercrystalline pores in the more abundant, coarsely crystalline replacement dolomite are larger (250 µm; e.g., Ricinus West, well 10-33, 4535 m; well 15-23, 4488 m, Figure 10C, D). Thus, these dolostones show a positive relationship between pore size and crystal size, as has been reported elsewhere for other dolomites (Lucia, 1983, 1995; Choquette et al., 1992).

Fractures are commonly oriented in a subvertical direction and are partly to completely filled by dolomite cement C_1 and bitumen or late calcite cement (Figures 7D, 10E). Fracture width varies 1–5 mm, and length ranges from 10 to 60 mm. Fractures are much more abundant in the buildup margins than in the buildup interior (compare Figures 11, 12, Ricinus wells 10-33 and well 7-13, respectively). Permeability is very sensitive to small changes in fracture porosity, which in turn is largely dependent on fracture width (Lucia, 1995). Dissolution along the fractures is indicated by associated solution-enlarged intercrystalline and vuggy porosity. Brecciated intervals occur in thin, isolated zones to zones several meters thick in the Ricinus West buildup margin (e.g., well 7-13; Figure 10F). Breccias (Figure 10F) are characterized by centimeter-size dolomite clasts with crackle, mosaic, and rubble textures (Beales and Hardy, 1980) that are likely to be the result of solution collapse (Amthor et al., 1993; Marquez, 1994). Matrices in the moldic breccias have intercrystalline porosity (Figure 10F). Similar pore types are recognized in dolomitized buildups (e.g., Westerose, Homeglen-Rimbey) along the Rimbey-Meadowbrook reef trend (Table 7) (McNamara and Wardlaw, 1991; Drivet, 1993). Bitumen-filled hairline microfractures and other voids are relatively abundant in the Strachan buildup. These microfractures are very late stage because they crosscut all diagenetic phases and extend subhorizontally, radially, and randomly away from vugs, molds, and fractures (Marquez and Mountjoy, 1996). Pore space occluded by bitumen is discussed under diagenetic controls and summarized in Table 8.

RESERVOIR CHARACTER IN DOLOSTONES

To evaluate the reservoir characteristics of dolostone buildups in the deeper part of the Alberta basin, the distribution and orientation of the different pore types in the completely dolomitized Ricinus West buildup were compared in three different parts of this buildup because of the excellent core available. The pore types were compared in three different parts of this buildup to determine the effect of facies on reservoir character. A number of different lithofacies (Figures 13, 14) occur in (1) the upper buildup interior (Upper Leduc Formation), (2) the lower buildup interior (Middle Leduc), and (3) the buildup margin (Figure 13).

Upper Buildup Interior: Moldic and Intercrystalline Porosity

Three basic pore types constitute this part of the buildup interior (in decreasing abundance) (Figure 11): (1) moldic (*Amphipora*-like), (2) intercrystalline, and (3) fenestral-like pores (Figure 7). Moldic (*Amphipora*-like) pores are tubular, 10 mm to >3 cm (Table 4), and randomly oriented in a dense matrix. Effective communication is provided by interconnected molds. Intercrystalline pores are small, polyhedral in shape, well connected, and commonly lined

with a thin (1- to 4-μm) coating of bitumen. Fenestral-like pores are small, tabular in shape, and oriented parallel to laminations. These pores are interconnected through intercrystalline pores in the matrix. Total porosity ranges 0.9% to 20.7% (mean: 6.3%); K_h varies 0.06 to 2000 md (geometric mean: 12.7 md); K_v ranges 0.01 to 861 md (geometric mean: 1.2 md) (Table 6).

Lower Buildup Interior: Poorly Connected Molds With Some Intercrystalline Porosity

Three basic pore types (Figure 14) also form the reservoir in the lower buildup interior: (1) moldic (stromatoporoid-like), (2) intercrystalline, and (3) moldic (*Thamnopora*-like). The lower buildup differs from the upper buildup interior because it contains larger and more abundant molds of skeletal fragments. Moldic (stromatoporoid-like) pores are very large, spherical, and randomly oriented in a dense matrix. These pores are interpreted to represent the dissolution of domal and bulbous stromatoporoids. However, these vugs do not change the permeability significantly, because intercrystalline pores are similar to those of the upper buildup interior. Additionally, >3-cm randomly oriented tubular pores, probably dissolved corals (*Thamnopora*-like), are common in a dense matrix. Total porosity varies from 0.6% to 19.1% (mean: 6.1%); K_h ranges 0.1 to 1400 md (geometric mean: 13.1 md); K_v from 0.01 to 843 md (geometric mean: 1.8 md) (Table 6).

Buildup Margin: Random Vugs, Breccias, and Intercrystalline Porosity

Vugs, breccias, and intercrystalline porosity (Figure 12) in decreasing abundance form the porosity types along the buildup margin (wells 7-13 and 11-27). Interconnection is commonly provided by fractures and some touching vugs. Total porosity ranges from 1.6% to 15.3% (mean: 6.9%); K_h ranges from 0.2 to 2100 md (geometric mean: 19.4 md); K_v from 0.01 to 141 md (geometric mean: 2.8 md) (Table 6).

SCALE, DISTRIBUTION, AND VARIABILITY OF DIFFERENT RESERVOIR TYPES

Large Scale (Tens of Meters to Kilometers)

The large-scale variations in porosity and permeability are shown in Ricinus West cross section, but overall the porosity and permeability are remarkably uniform (Figure 13). Although porosity and permeability are closely related to the depositional units and show varations in K_h from 4.9 to 26.7 md (Figure 13B), data from each of these parts of the Ricinus West reservoir indicate the lack of large-scale reservoir variability (Figure 13A; Table 6).

Medium Scale (Meters to Tens of Meters)

The differing arrangement of porosity and permeability in the three parts of the reservoir affects the petrophysical properties of the reservoir. The vertical and lateral arrangement of the dominant pore types

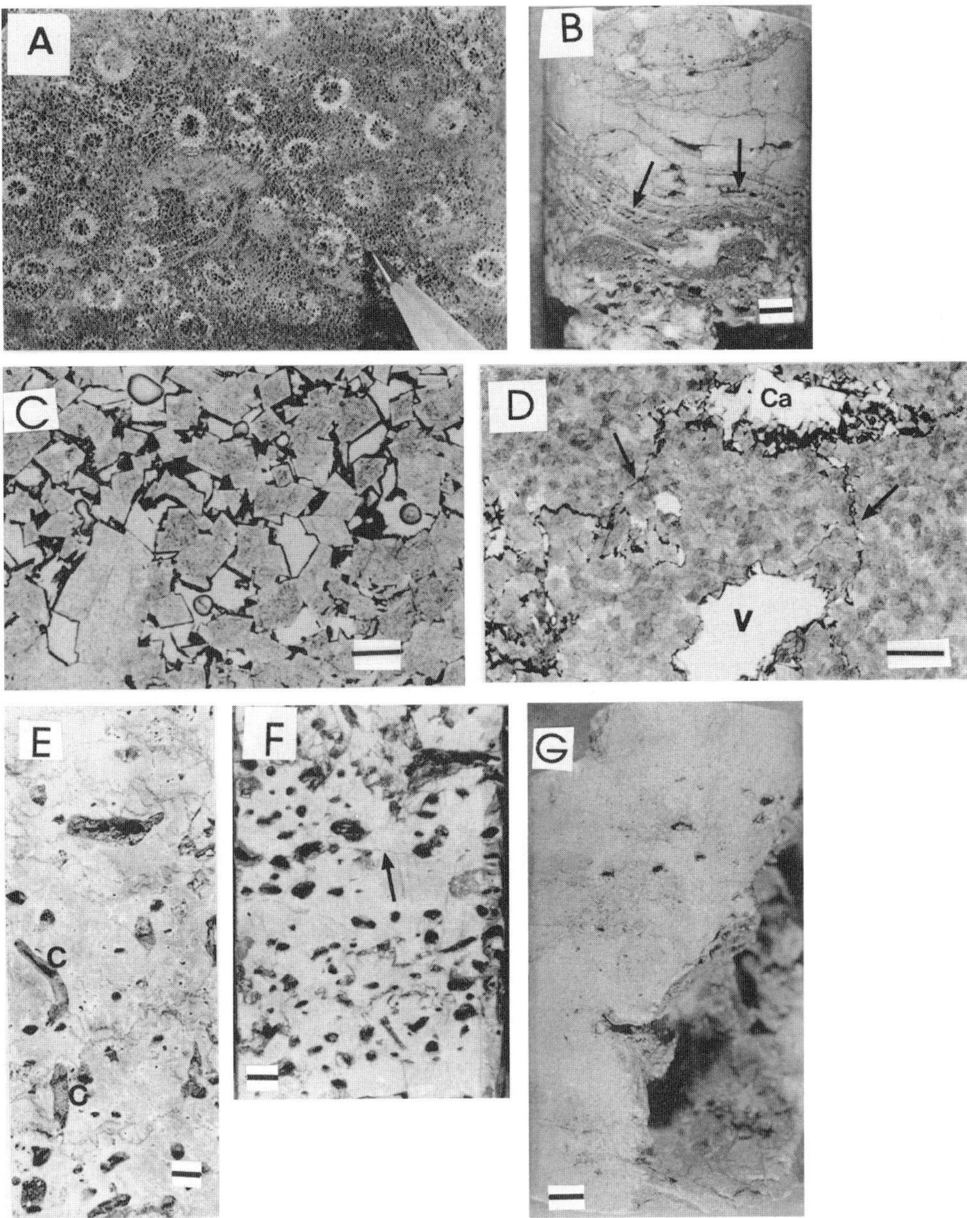

Figure 9. Core and thin-section photographs from Strachan and Ricinus buildups. Scales for cores are in centimeters. (A) Intraskeletal pores in calcite coral fragment. Strachan D3B 15-2-38-10W5, 4113.4 m. (B) Dolomitized tabular stromatoporoid boundstone facies with solution-enlarged intraskeletal pores (arrows) lined with reservoir bitumen. Some vugs filled with calcitized anhydrite. Ricinus West 7-13-36-10W5, 4621.3 m. (C) Porous, medium crystalline (62–250 µm), planar, euhedral replacement dolomite. Intercrystalline pores are lined with reservoir bitumen (black). Strachan D3A 10-31-37-9W5, 4287.8 m. Scale bar is 50 µm. (D) Nonporous replacement dolomite with irregular vugs (V). Stylolites (arrows) connect vugs, since they are filled with reservoir bitumen (black). Late calcite cement (Ca) partly fills vug. Ricinus West 10-33-36-10W5, 4601.5 m. Scale bar is 200 µm. (E) *Thamnopora?* molds (C) in a tight dolomite, coral rudstone facies. Some molds are solution enlarged. A few microfractures. Ricinus West, 7-13-36-10W5; 4514 m. (F) Tubular moldic pores probably after *Thamnopora* fragments in a tight matrix of replacement dolomite. Some small fractures (arrow) connect molds. (G) Large irregular vugs in replacement dolomite. Strachan D3A 12-31-37-9W5, 4295 m.

vary considerably within these areas, and cause variations at the medium scale in reservoir character. Thus, the K_h/K_v ratio is effectively much larger on this scale than it is on the core-measurement scale. Within the Ricinus West reservoir, several high- to medium-permeability intervals are separated by lower permeability zones (Figures 11–14; Marquez, 1994). In the upper buildup interior, laterally continuous, high K_h zones are associated with intercrystalline, moldic, and fenestral pores, whereas in the lower buildup interior, K_h zones are related to *Thamnopora*-like and stromatoporoid pores that are laterally discontinuous. In the buildup margin, K_v zones are common and are apparently related to subvertical fractures. Vertical permeability

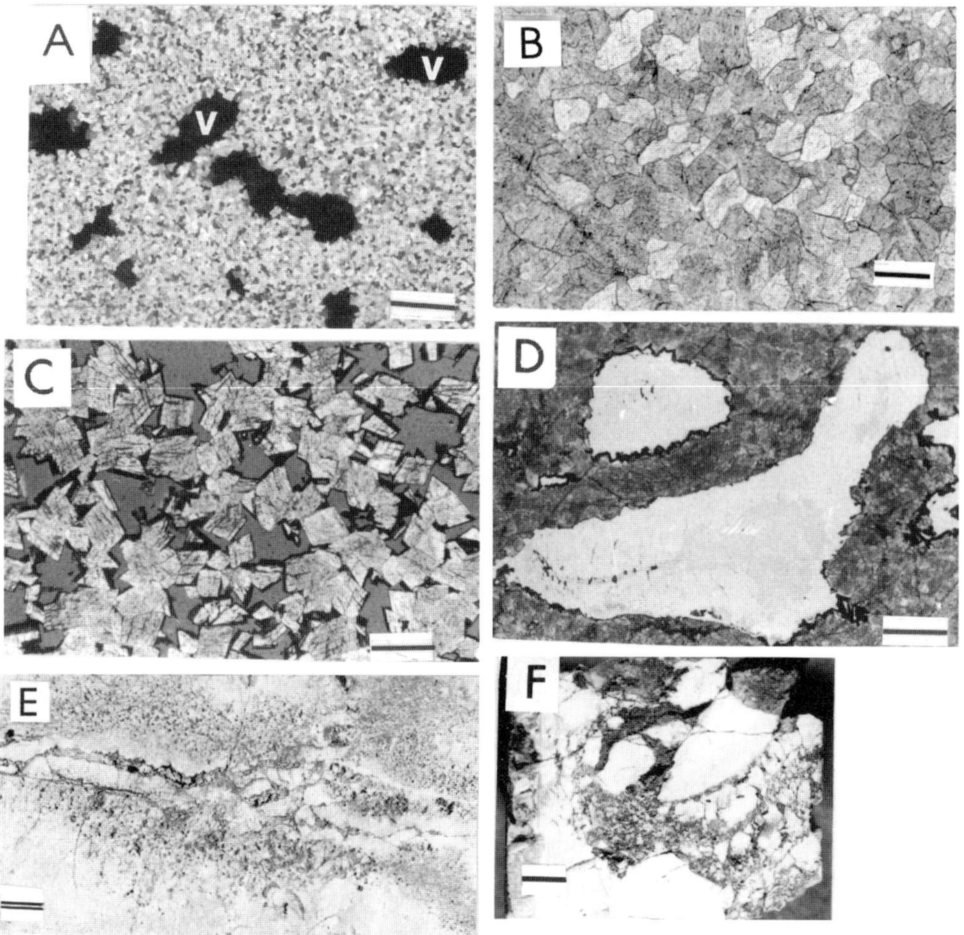

Figure 10. Core and thin-section photographs from dolomitized Ricinus buildup. Scales for cores are in centimeters. (A) Nonporous replacement dolomite with solution-enlarged pores (V), probably after coral fragments. Ricinus West 15-23-36-10W5, 4592.9 m. Scale bar 250 µm. (B) Nonporous mosaic of subhedral, medium crystalline, planar replacement dolomite. Crimson 10-24-37-9W5, 4032.3 m. Scale bar 100 µm. (C) Polyhedral intercrystalline pores (dark) in replacement dolomite lined by reservoir bitumen (black). Plane light. Ricinus West 15-23-36-10W5, 4488.4 m. Scale bar 50 µm. (D) Microbial laminite dolostone with irregular vugs in a tight matrix. Reservoir bitumen (black) lines vugs. Plane light. Ricinus West 11-27-36-10W5, 4490.8 m. Scale bar 250 µm. (E) Intercrystalline porosity in replacement dolomite adjacent to subvertical fracture partly dolomite cement. Ricinus West 10-33-36-10W5, 4621.6 m. (F) Dolomite breccia and fracture porosity within intercrystalline porosity. Pores are lined with bitumen. Ricinus West 7-13-36-10W5, 4393.5 m.

profiles (Figures 11, 12) indicate that the Ricinus West reservoir can be further subdivided into thinner facies and porosity-permeability slices as indicated in Figure 13B. In the reef interior, a lower sequence with laterally discontinuous permeable zones is overlain by a stacked sequence of thin, laterally continuous permeable zones with poor vertical interconnection. In the reef margin, thick, laterally discontinuous zones are characterized by higher K_v (Figure 12), apparently related to subvertical fractures. These data have implications not only for the understanding of reservoir heterogeneity and character, but also for the prediction of reservoir quality.

Small Scale (Meters to Millimeters)

At this small scale, porosity and permeability are controlled by the facies, individual pore types, and differences in diagenesis. In intercrystalline pores (Figure 11C) of the upper buildup interior, the presence of 10%–15% bitumen (percentage in terms of bulk volume) (Table 8) sufficiently lines most pore throats and thus greatly reduces the porosity and permeability. In stromatoporoid-like and *Thamnopora*-like molds in the lower buildup interior (Figure 14), bitumen is less abundant (~2%), and permeability variations are controlled by the presence of minor dolomite, anhydrite, and calcite cements. Bitumen-filled hairline microfractures are abundant in vugs in the upper part of the reef margin (above a paleo-oil-water contact); therefore, porosity and permeability are mainly controlled by the presence of subvertical fractures and vugs (Marquez and Mountjoy, 1996). In the lower part of the reef margin, touching vugs (with minor bitumen and cements) and some fractures provide K_v. The effects of bitumen, cementation, and dissolution are discussed below.

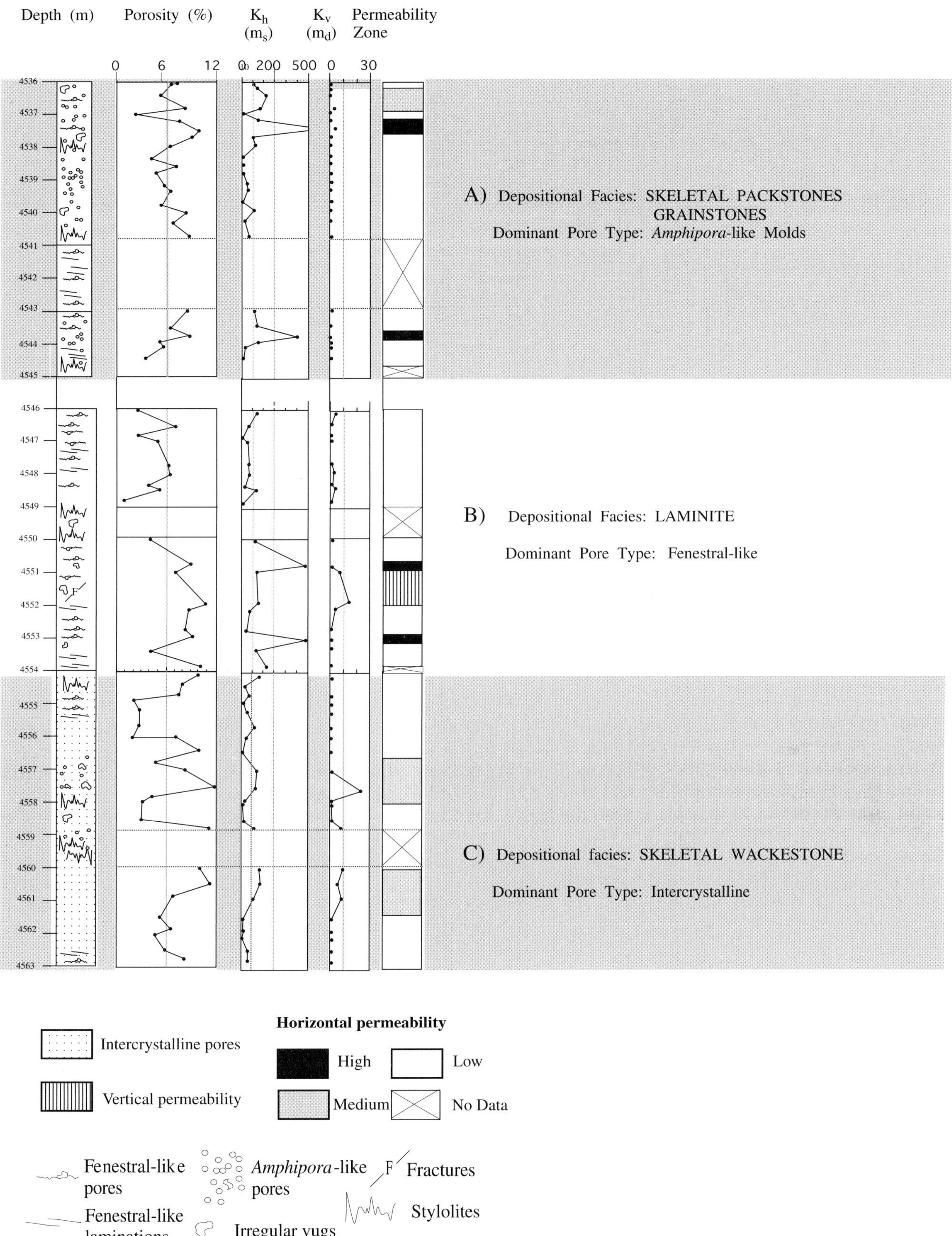

Figure 11. Examples of low-, medium-, and high-permeability zones in a 27-m-thick sequence in upper buildup interior, Ricinus West buildup interior (well 10-33; see Figure 5 for location of this interval). Thin, high-permeability zones are separated by thicker intervals of lower permeability. (A) Moldic (*Amphipora*-like) pores in skeletal packstone and grainstone facies. (B) Fenestral-like pores in microbial laminite facies. (C) Intercrystalline pores in skeletal wackestone facies. K_h = horizontal permeability, K_v = vertical permeability.

Dominant Pore Type: Vugs and/or Breccias
Depositional Facies: TABULAR STROMATOPOROID BOUNDSTONES

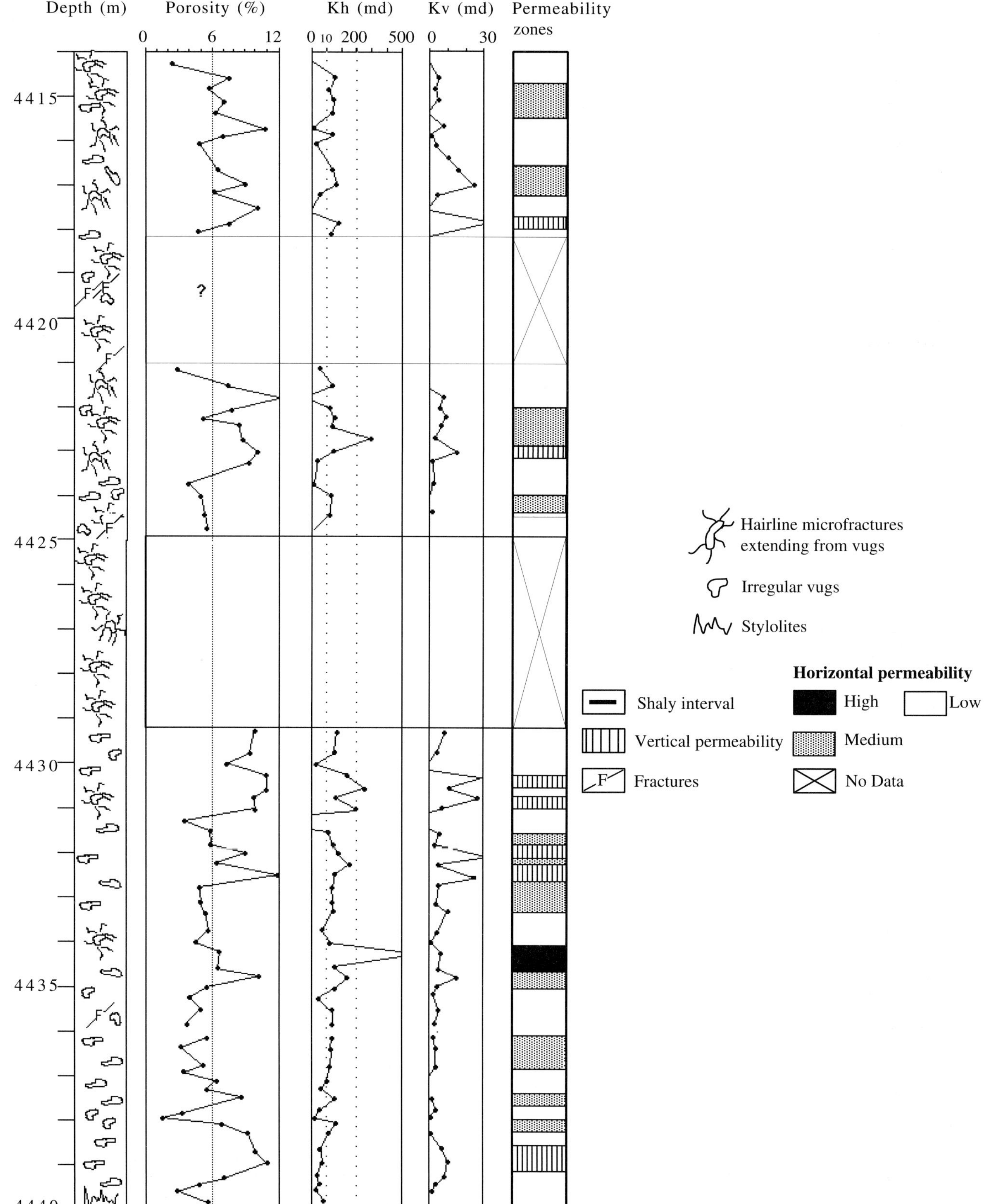

Figure 12. Examples of high-, medium-, and low-permeability zones in buildup margin, Ricinus West (well 7-13; see Figure 5 for location of this interval). In the lower 11 m, vertical permeability (K_v) is provided by connecting vugs and some fractures. In the upper 15 m, vertical permeability is dominated by partly open subvertical fractures. K_h = horizontal permeability.

Table 7. Porosity and Permeability in Limestone Buildups Along the Rimbey-Meadowbrook Reef Trend.

| Depositional Facies Reef Margin | Golden Spike* (<2000 m) | | Strachan (Pool D3B) (>4000 m) | | |
	Porosity Mean (%)	K_h (md)	Porosity Mean (%)	K_h (md)	Permeability K_v (md)
Coral Rudstones	4	1	3	0.3	0.01
Tabular Stromatoporoid Boundstones	5.5	10	2.3	0.1	0.02
Stomatoporoid Coral Rudstones	10	100	2.6	0.1	0.01

*From Walls and Burrowes (1985).
K_h = horizontal permeability, K_v = vertical permeability.

DEPOSITIONAL CONTROLS ON PORE SYSTEMS

Reservoir interconnection is the most important parameter in reservoir simulation studies (Van de Graff and Ealey, 1989), and is related to the presence and distribution of internal vertical and horizontal changes in permeability (Ghosh and Friedman, 1989). These changes in permeability are plotted in vertical profiles for the three parts of the West Ricinus reservoir (Figures 11, 12, and 14). Horizontal permeability data have a wide range of values and therefore have arbitrarily been grouped into high (>200 md), medium (10–200 md), and low (<10 md). Vertical permeabilities >10 md are considered high.

In the upper buildup interior, the vertical succession of *Amphipora*-like molds and fenestral-like and intercrystalline pores resemble the shallowing-upward depositional parasequences developed in the upper reef interior. These intervals range in thickness from 8 to 27 m (see well 10-33 in Figure 5). The vertical permeability profiles in Figure 11 show that in a 27-m-thick sequence containing *Amphipora*-like (Figure 11A), fenestral-like (Figure 11B), and intercrystalline pores (Figures 5, 8, and 11C), thin (1–2 m), high K_h zones are separated by thicker intervals (2–5 m) of lower permeability. This indicates that within each sequence, vertical interconnection is poor, and horizontal interconnection is provided by relatively thin permeable zones that would form fluid flow conduits. Slightly thicker permeability zones occur where intercrystalline porosity dominates. Based on cross sections AA' and BB' (Figure 5), depositional models for limestone buildups (McGillivray and Mountjoy, 1975) and outcrop studies of equivalent strata (McLean and Mountjoy, 1993a, b, 1994), these depositional sequences are extensive and laterally continuous. The upper buildup interior is characterized by six 8- to 27-m-thick intervals of different pore types that closely follow the depositional sequences, with poor vertical interconnection and thin, laterally continuous, highly permeable horizontal zones. Such thin, porous, and permeable laterally continuous strata would form conduits or flow units along which higher flows would occur.

In the lower buildup interior, vertical permeability is slightly higher (Figure 14). Horizontal permeability zones are slightly thicker where *Thamnopora*-like molds dominate (Figures 9F, 14B), with connectivity provided by touching molds. From previous studies (McGillivray and Mountjoy, 1975; McLean, 1992), facies associated with stromatoporoid and coral molds are known to be laterally discontinuous. This part of the buildup interior consists of thin, laterally discontinuous, porous, and permeable zones.

In the buildup margin, vertical permeability is common and associated with subvertical fractures (Figure 12, 4423 m) and connecting vugs (Figure 12, 4439 m). Horizontal permeability zones are thin (<1 m) and laterally discontinuous. Brecciated intervals that contribute to reservoir interconnection (see cross section BB', Figure 5) are common in the reef margin. Thus, the buildup margin is characterized by thick intervals where K_v predominates. This is similar for the Homeglen-Rimbey buildup margin (Drivet, 1993; Drivet and Mountjoy, 1997).

In summary, the buildup interior reservoir changes upward from laterally discontinuous *Thamnopora* moldic pores in a dolomite matrix with intercrystalline porosity to a similar matrix dolomite with *Amphipora* and fenestral-like pores in 1- to 2-m-thick units with high K_h separated by thicker intervals of dolomite with lower permeability. The buildup margin has better K_v due to subvertical fractures that connect the vugs and breccia zones. As noted above, the measured average matrix porosities and permeabilities in all three regions of the reservoir are similar and appear to be mainly a result of the dominant control on reservoir character by intercrystalline porosity in the matrix dolomites. However, considerable variation must occur in the porosity in different parts of the reservoir as a result of variations in moldic, vuggy, and breccia porosity.

DIAGENETIC CONTROLS ON PORE SYSTEMS

Dolomitization, cementation, dissolution, pressure solution, and bitumen plugging affect the pore systems and the relationship between porosity and permeability differently.

Dolomitization and Porosity

The degree of dolomitization in skeletal wackestones facies was investigated to examine the effects of dolomitization on porosity and other diagenetic effects (Figures

Table 8. Effect of Reservoir Bitumen on Porosity and Permeability in Dolomitized Skeletal Wackestone Facies, Ricinus West Reservoir.

Dominant Pore Types	Depth (m)	Sample Length (m)	A			B[†]	C = A + B	$D = \dfrac{(B \times 100)}{C}$	Comments
			ϕ^* (%)	K_h^* (md)	K_v^* (md)	Reservoir[†] Bitumen (%)	Prebitumen ϕ (%)	ϕ Reduction (%)	
Well 10-33-36-10W5									
Intercrystalline (10–250 mm)	4532.6	0.18	7.7	25.7	5.1	10	17.7	56.4	Optical textures
	4561.5	0.30	5.2	2.1	0.1	15	20.0	74.2	Some larger vugs (2 cm)
Small vugs (1–5 mm)	4601.0	0.18	2.9	1.6	0.1	10	12.9	77.5	Interbedded with microbial laminites
	4601.5	0.34	6.5	12.1	0.4	10	16.5	60.6	Pores oriented parallel to laminations
Some fractures (filled with bitumen)	4601.6	0.42	6.9	4.4	0.1	10	16.9	59.1	Interbedded with microbial laminites
	4628.9	0.36	7.8	9.6	1.5	2	9.2	15.2	Interbedded with domal stromatoporoid floatstones
Well 11-27-36-10W5									
Intercrystalline (10–250 mm)	4478.0	0.30	8.3	5.8	1.1	10	18.3	54.6	
	4478.3	0.25	11.3	45.8	24.8	1	12.3	8.1	1% bitumen in a patch in center of sample
Small vugs (1–5 mm)	4478.5	0/24	7.7	3.1	0.6	10	17.7	56.4	
Some fractures (filled with bitumen)	4478.8	0.16	7.5	3.7	0.1	10	17.5	57.1	
	4478.9	0.18	8.3	2.9	0.1	10	18.3	54.6	

*ϕ, K_v, K_h data are from ERCB files. K_v = vertical permeability, K_h = horizontal permeability
[†]Visual estimation.

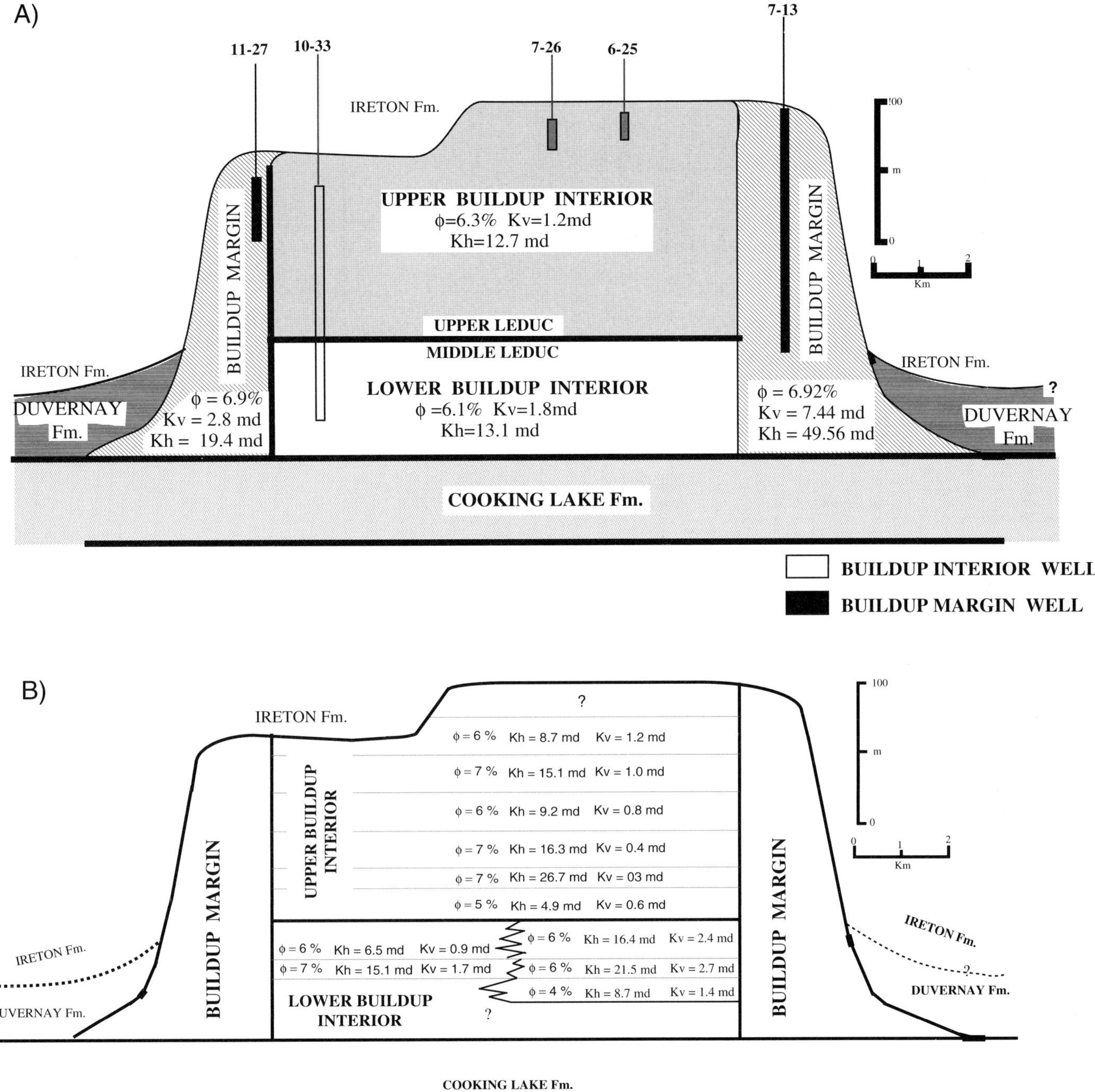

Figure 13. Subdivision of the Ricinus West reservoir in three large-scale regions; upper and lower buildup interior and buildup margin. (A) Average porosity (6%) is similar throughout the entire reservoir. Horizontal and vertical permeabilities are slightly higher in the buildup margin. (B) Medium-scale porosity and permeability subdivisions within the buildup interior. The upper buildup interior is subdivided into six zones, 8 to 27 m thick. Horizontal permeability in each zone is high, but vertical permeabilities are low. See Figure 11 for details of pore types and depositional facies. The lower buildup interior has laterally discontinuous porosity and permeability. The thick buildup margin has greater vertical permeability.

15–18). Skeletal wackestones were chosen because they have the most homogeneous depositional texture and are the most widespread facies throughout the buildups. In addition, except for the presence of ~10% bitumen, other late-stage diagenetic products such as dissolution vugs

and late calcite are minor components in these wackestones. Skeletal wackestones with <50 vol. % dolomite have porosities ranging from 1% to 5% (mean: 2.4%; Figure 16A, B). Completely dolomitized skeletal wackestones range in porosity from 0.5% to 14% (mean: 5.2%;

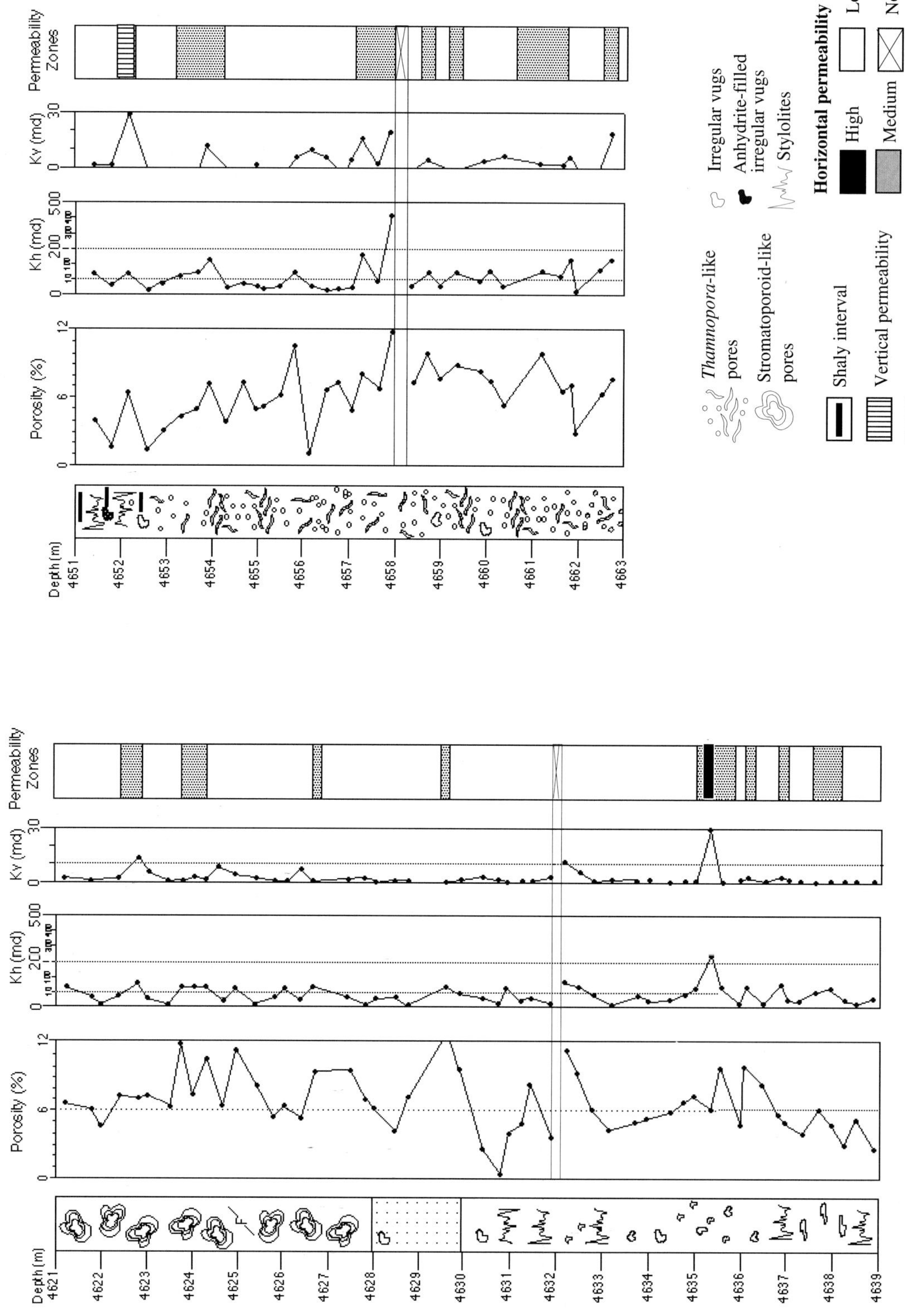

Figure 14. Examples of low-, medium-, and high-permeability zones in lower buildup interior, Ricinus West buildup (well 10-33; see Figure 5 for location of this interval). Higher permeability zones are thicker in the coral rudstone facies with abundant *Thamnopora*-like pores. Dominant pore types: (A) stromatoporoid-like molds in stromatoporoid floatstones/rudstones, (B) *Thamnopora*-like molds in coral rudstone facies. K_h = horizontal permeability, K_v = vertical permeability.

Figure 16C) and have considerably higher K_h ($\leq$100 md; Figure 16F). Thus, porosity increases from <5% in partially dolomitized limestones, on average, to $\leq$10% to 14% in completely dolomitized facies (Figure 16). Completely dolomitized skeletal wackestones in the reef interior of pool D3A (Figure 17A–C) have the highest porosities in the entire buildup, ranging from 3.4% to 25% (mean: ±9.8%) and much higher permeabilities (Figure 17D–E) than in partially dolomitized limestones. This is clearly shown by the log plot of percent dolomite and porosity [Figure 18, calculated from sonic and density logs using the Chaveroo method (Bateman, 1985) and assuming from core logging that the lithology was dominantly limestone and dolomite]. There is a satisfactory match with estimated amounts of dolomite and measured porosities from Strachan well 10-31, D3A pool (Figure 18). In the partially dolomitized intervals where limestone is $\geq$60%, the porosity is very low. Not all dolomites are porous, as shown by the low porosities in the completely dolomitized sections, especially in the interval 4179–4186 m. Porosity in this case is a combination of intercrystalline and vuggy and is best developed in strata with >50%–60% dolomite, with the highest porosities occurring in completely dolomitized rocks. These increases in porosity appear to result mainly from dissolution (D_l) of calcitic skeletal fragments and grains that were either associated with dolomitization or occurred later. Probably the development of intercrystalline porosity in the matrix also increased the porosity. The increased abundance of molds, solution-enlarged molds, and irregular vugs and porosity may not necessarily be related to dolomitization itself, but may in part be due to later solution events (Figure 8).

Dolomitization and Permeability

Dolomitization greatly affects permeability, but only when the rocks are 100% dolomite, which have highest absolute and average horizontal permeability (Figures 15–17D–F; geometric mean K_h: 27.9 md) and vertical permeability values [geometric mean K_v: 5.1 md (Table 5)]. As the percentage of dolomite increases, no significant trends occur in the permeabilities in these samples from the Strachan reservoir (Figures 6A–C; 15–17), but there is a large scatter in the data. Skeletal wackestones with <50 vol. % dolomite content tend to have very low horizontal permeabilities (0.01–0.4 md; Figure 16D, E). Completely dolomitized skeletal wackestones in the reef margin have greater horizontal permeabilities, ranging from 0.01 to 200 md (Figure 16F). Completely dolomitized skeletal wackestones in the Strachan buildup interior have high permeabilities from 0.01 to 450 md (Figure 17 D–F). Similar porosity and permeabilities occur in the completely dolomitized Ricinus reef margin 7-13 well (Figures 19, 20; compare Figure 17A with 19C). The crossplots of porosity with permeability from Ricinus 7-13 show considerable variation. In general, there is an overall increase in permeability with increasing porosity, which may be related to increasing crystal (or particle) size, as suggested by Lucia (1995). In Ricinus 7-13, below 4430 m there are several intervals with high K_v (Figure 12) that appear to be related to fracturing.

Moldic vugs can increase porosity, but if not connected will not contribute significantly to increased permeability.

Comparing the porosity-permeability plots with the three permeability classes of Lucia (1995) based on particle size indicates that the Devonian limestones and dolomites of the Strachan and Ricinus reservoirs fall within or above the >100-µm particle size (Figures 21, 22) [Lucia's (1995) rock-fabric/petrophysical class 1]. Both the dolomitized limestones (Strachan 10-31) and the 100% dolomitized intervals exhibit permeability values higher than expected. The limestones have permeabilities $\leq$10 md (Figure 21), whereas the dolomites have some permeability values slightly >1000 md (Figure 22). The reasons for permeabilities higher than expected is probably due to small and large vugs being well interconnected with each other via matrix intercrystalline porosity, and perhaps some fracturing. Interestingly, there is little difference in the porosity-permeability plots in the Ricinus buildup (Figure 22) between the buildup margin (7-13) and the buildup interior wells (10-33), even though there are more fractures in the buildup margin wells (cores with fractures tend to break, and it is difficult to measure their porosity and permeability). Thus, these dolomites and partially dolomitized strata with touching or interconnected vugs form excellent reservoirs with good horizontal permeabilities. Permeability appears to increase with dolomite crystal size, but this relationship has not been studied systematically in these reservoirs. Dolomitization of previously mud-dominated rocks tends to increase permeability and to some extent porosity.

Lucia (1995) considers that all touching vugs (fractures, breccia, caverns, fenestral) are diagenetic in origin and are unrelated to primary depositional features. Touching vugs in the Devonian strata of western Canada are abundant in all dolomitized reservoirs. Partial to complete dissolution of *Amphipora* and stromatoporoids formed a series of molds and solution-enlarged molds that make up touching or connected vugs where these fossils were abundant, especially along the buildup margins and in the adjacent lagoons. These vugs are interconnected where they touch adjacent vugs and via the medium to coarse intercrystalline porosity in the surrounding medium to coarsely crystalline matrix dolomites, forming a laterally extensive three-dimensional conduit for fluid flow. Hence, a moldic class related to primary textures needs to be added to Lucia's (1995, his figure 17) touching-vug category.

It is critical to determine whether such moldic vugs are physically separate or touching (connected). If separate, these vugs increase total porosity but do not significantly increase permeability. However, if they are touching, they increase permeability (Figure 22), and most of the porosity would be effective. Unlike the touching vugs classified by Lucia (1995) as being wholly diagenetic and unrelated to depositional fabrics, these Devonian moldic vugs are related to the depositional fabrics. Such fabrics are more readily characterized, and the resulting porosity and permeability are more easily mapped and extrapolated in these Devonian strata; in short, porosity and permeability are more predictable.

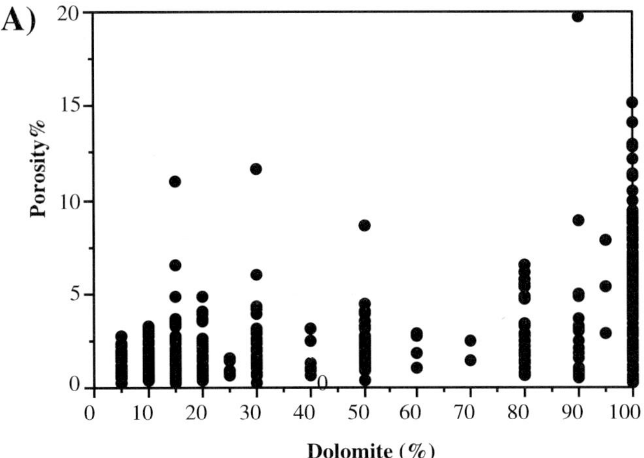

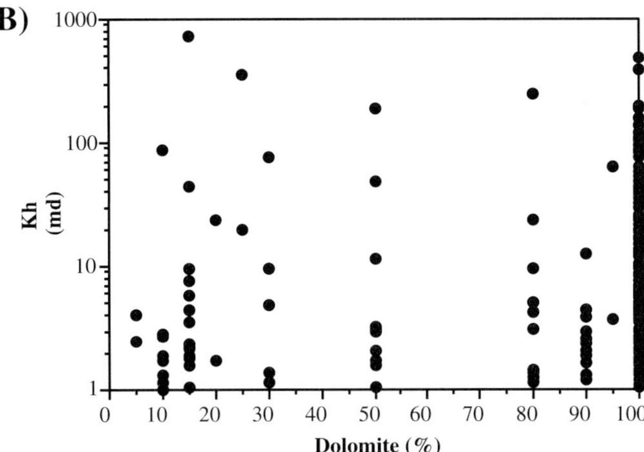

Figure 15. Core-measured porosity (A) and permeability (B) vs. dolomite content in all depositional facies of Strachan well 10-31 (pool D3A; see Figure 3 for location and facies distribution).

Thus, although all depositional facies have been modified by several diagenetic processes such as early cementation, chemical compaction, and later dissolution and cementation, replacement dolomitization and associated dissolution exert the most pronounced effect on reservoir fabrics (Figures 16–20). This effect is twofold: it results in significantly higher porosities and, more importantly, significant increases in permeabilities, thus enhancing their capacity to act as flow units.

Effect of Cementation and Dissolution

In Ricinus West, cementation occurred during more than one stage (Figure 8). The net effect of cementation is not only to reduce the pore size but to reduce or completely plug pore throats (Wardlaw, 1980), resulting in a drastic reduction of permeability. Anhydrite cementation is volumetrically important in the lower part of the reservoir. At least three dissolution events modified the Ricinus West reservoir rock. Dissolution D_1 (Figure 8) affected calcite skeletal fragments during or after replacement dolomitization, resulting in molds, solution-enlarged molds, and vugs in the entire reservoir, apparently improving the connectivity and permeability (Figures 6E, F, 7H, and 9B). Dissolution D_2 and minor

dissolution D_3 of matrix dolomite resulted in vugs and brecciated intervals that are more abundant in the reef margin. The distribution of these cements and occurrences and amounts of dissolution are difficult to predict.

Effects of Pressure Solution

The effects of stylolitization and other pressure solution phenomena in Leduc carbonates have been discussed by Mossop (1972) and Amthor et al. (1994), among others. The equivalent limestone facies in the Golden Spike buildup (1600 m present depth) have much higher porosities and permeabilities than limestones in the Strachan buildup (Table 7). This dramatic decrease of porosity and permeability is primarily the result of pressure solution (Figure 7F, I) during an additional 2 km of burial in the deeper basin, together with pore filling by later cements and bitumen. Even in the shallower part of the basin, pressure solution has caused considerable reduction in porosity in limestones and produced secondary effects such as the raised rim of the Redwater buildup (Mossop, 1972).

Effect of Bitumen on Porosity and Permeability

Bitumen has been reported as a common porosity-occluding phase in some reservoirs (McCaffery, 1977; Lomando, 1992). The precipitation of bitumen can cause wettability characteristics (the relationship between hydrocarbon-rock contact angles) to change from water-wet to mixed or strongly oil-wet (Lomando, 1992). For example, McCaffery (1977) reported that pore-lining bitumen in the Devonian Windfall D-3 pool made the rock intermediately wet for gas-water systems. Wettability strongly influences reservoir behavior and ultimate hydrocarbon recovery (Wardlaw, 1992). Therefore, the type, amount, and distribution of bitumen needs to be carefully evaluated.

In the Strachan and Ricinus West buildups, depositional facies with different pore types have variable amounts of bitumen with homogeneous to heterogeneous distribution (Figures 6D, F; 7C, D; 9B–D; and 10C, D). Skeletal wackestones have the highest percentages (15%) of bitumen and are characterized by abundant intercrystalline pores (Figures 9C, 10C), some small vugs, and, locally, fractures. All pore types are lined to partly filled with bitumen as thin (1–4 µm) coats and droplets (≤10 µm). Bitumen is most common in the upper part of reservoirs, mainly above the gas-water contact. The prebitumen porosity (visual estimates) and present (core-measured) porosity for this facies are shown in Table 8. In well 10-33, skeletal wackestones interbedded with microbial laminites in the upper reef interior of the Ricinus West reservoir have prebitumen porosity of 12.9% that has been reduced to 2.9% (a 70% reduction) after emplacement of an average bitumen content of 10% (at 4601 m). A similar porosity reduction occurs at 4561.5 m. In the lower reef interior, skeletal wackestones are interbedded with domal stromatoporoid facies and have a relative low bitumen content (2%; 4628.9 m), which reduced porosity by 22%. Similar reductions in porosity occur in well 11-27 (Table 8). Thus, precipitation of bitumen in the Ricinus West reservoir reduces porosity considerabiy.

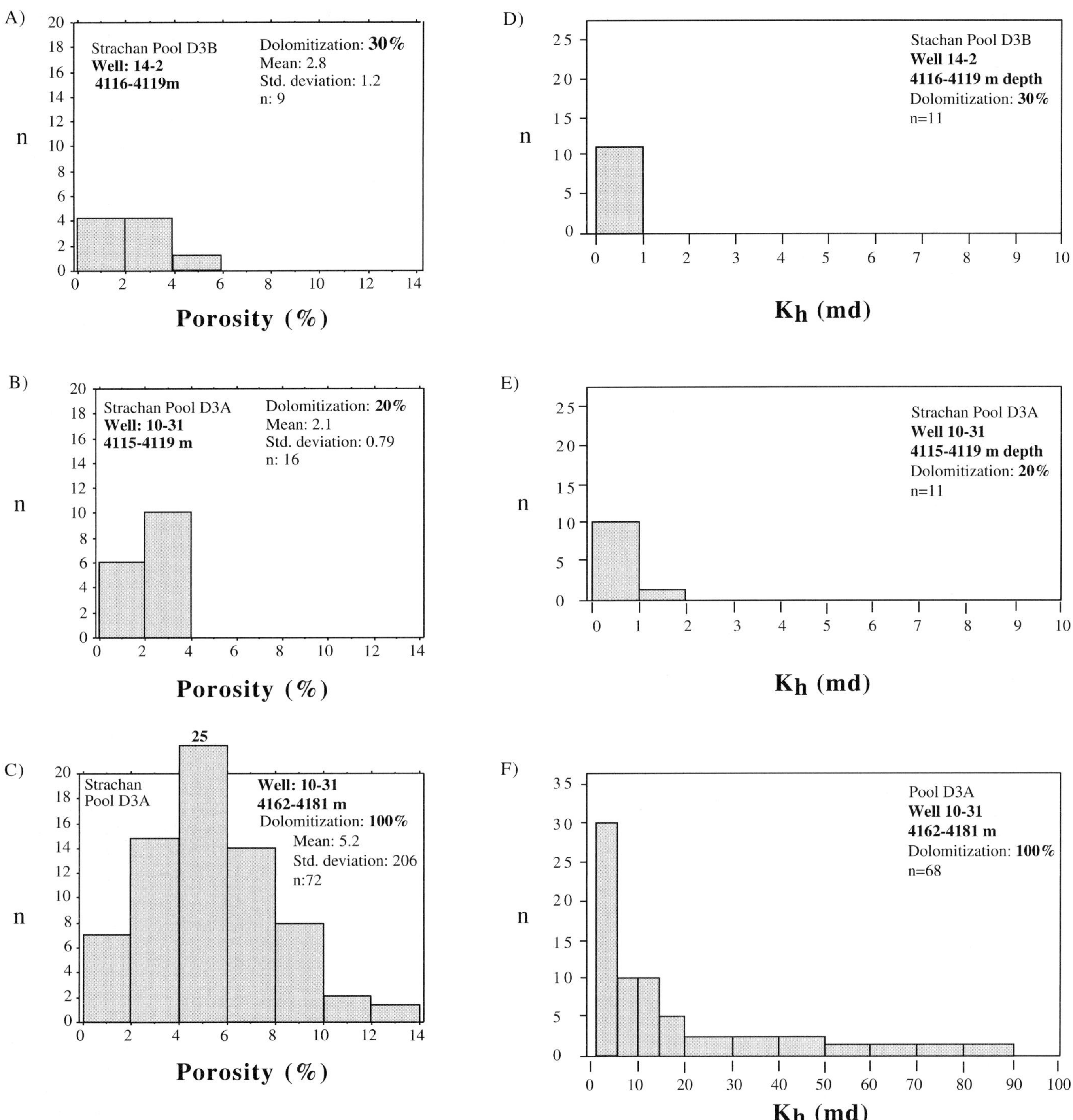

Figure 16. Porosity and permeability histograms measured from skeletal wackestone facies Strachan buildup showing variations in reef margin D3A and D3B pools: A–D, limestones (<50 vol. % dolomite content); E–F, dolostones (>75 vol. % dolomite content). See Figure 3 for well locations.

The greatest significance of bitumen precipitation on reservoir character is reduction of permeability and increase in reservoir heterogeneity. Comparison of core permeability in a continuous, 1-m-thick interval of skeletal wackestone facies (well 11-27, 4478 m; Table 8) illustrates the effect of bitumen plugging on permeability. With 10% bitumen, K_h averages 3.4 md and K_v averages 0.4 md, compared to 45.8 md and 24.8 md, respectively, in dolostones with 1% bitumen. This

reduction in permeability is caused by thin coats or droplets of bitumen that restrict or completely block pore throats. This bitumen was probably related to the deasphalting of the oil. This late-stage deep burial event imparts a component of reservoir heterogeneity that mostly is unrelated to depositional facies or prebitumen diagenesis. In addition, where overpressuring took place in an isolated or a sealed reservoir such as Strachan during the conversion of crude oil to gas beginning

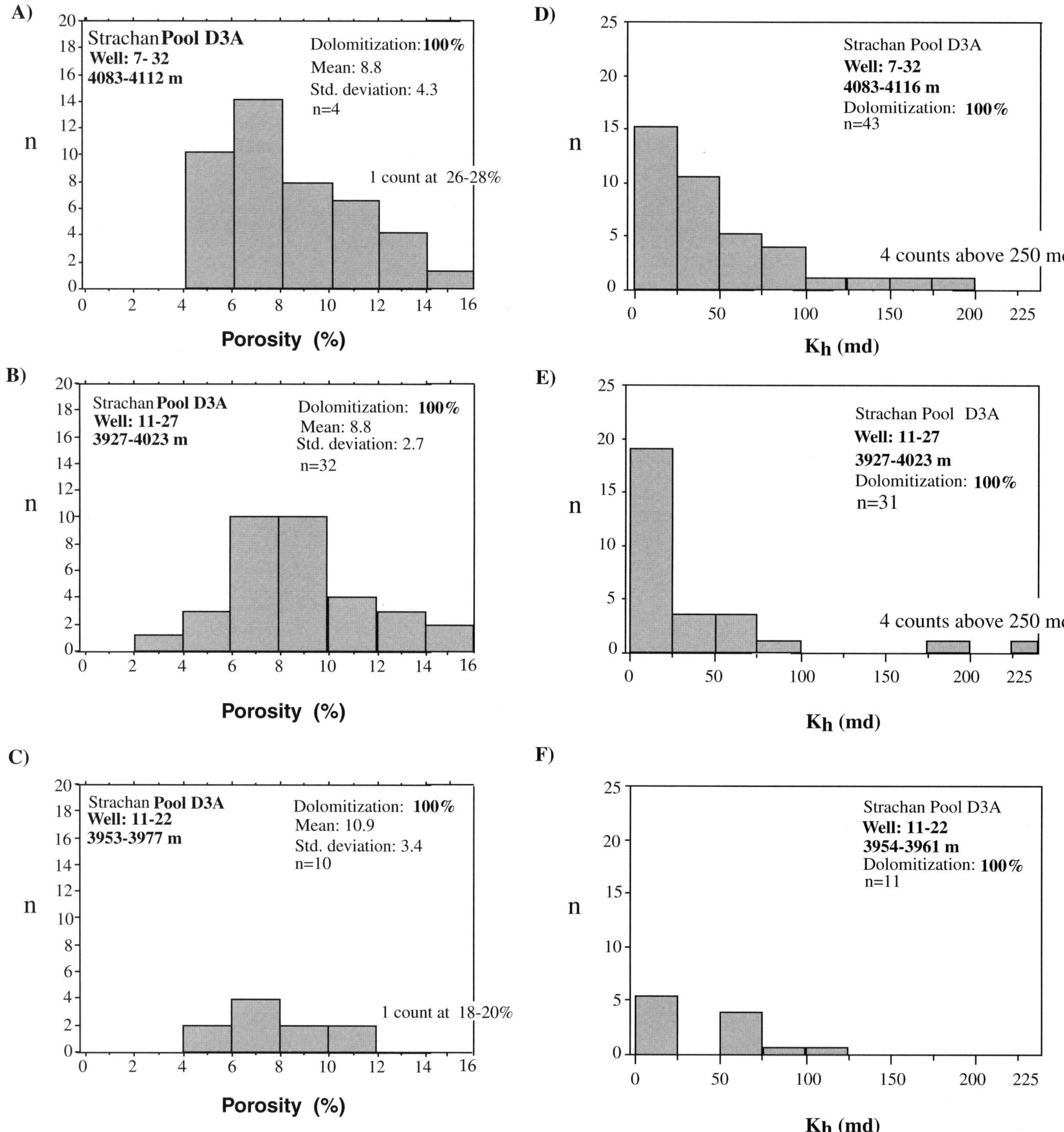

Figure 17. Porosity and permeability histograms measured from 100% dolomitized skeletal wackestone facies in wells 7-32, 11-27, and 11-22, interior of Strachan buildup pool D3A. See Figure 3 for well locations.

at temperatures above ~120°C, microfractures were formed and bitumen was forced into most pore throats and fractures (Marquez and Mountjoy, 1996).

Porosity Evolution in the Strachan and Ricinus West Buildups

Evolution of the pore systems is interpreted to have occurred in four major stages (Figure 23) that represent important changes in porosity and permeability of the reservoirs.

Stage 1: Deposition (Facies-Controlled Original Porosity)

The paleogeographic setting originally controlled depositional environments and, ultimately, the original porosity and permeability. Skeletal boundstones and rudstones with packstone matrices and marine

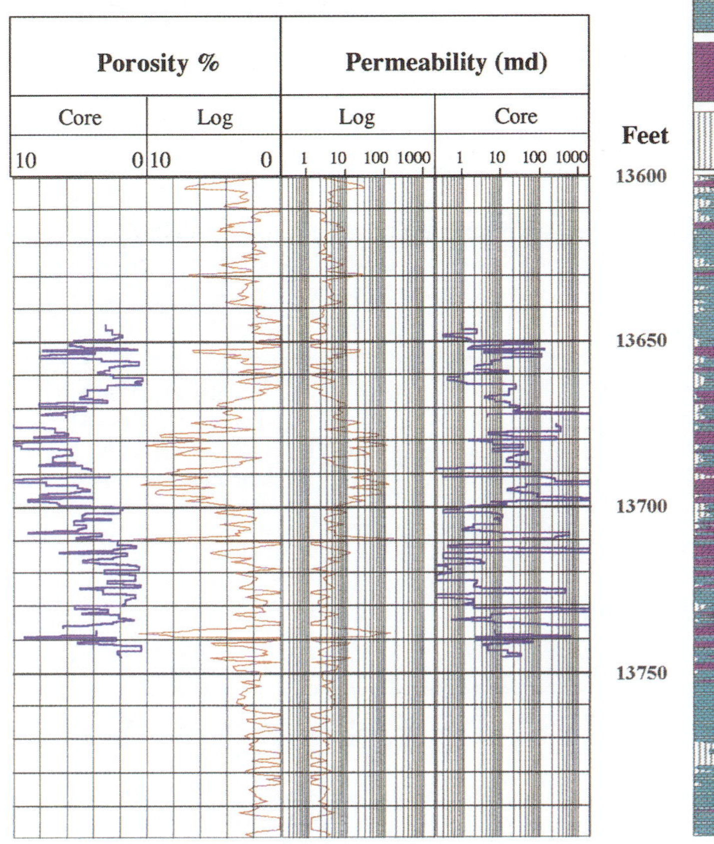

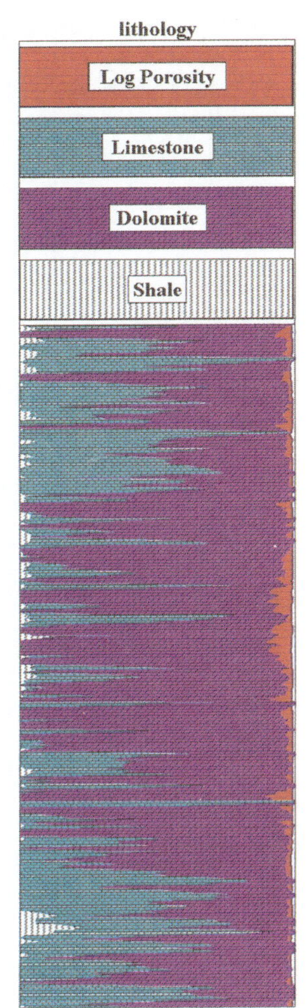

Figure 18. Partially dolomitized interval in Strachan well 10-31, D3A pool, showing (on the right) the relative amounts of limestone (blue), dolomite (magenta), and porosity (red), derived from equations (Chaveroo method) using simultaneous equations of sonic and density data (run in 1968) utilizing GEOGRAPHIX QLA2 log analysis software. On left, porosity and permeability curves; blue, measured core porosities and permeabilities (ERCB data), and brown, log calculated porosities and permeabilities. Porosity increases with increasing dolomitization. Highest porosities occur only in completely dolomitized strata. In general, there is a good match between the log-calculated porosity and lithology, except that the amount of dolomite calculated from logs is 20%–40% higher than that estimated from the cores.

cements were deposited at the reef margins. Skeletal rudstones, grainstones, packstones, wackestones, and mudstones were deposited in the reef interiors. Initial porosities were probably high and differed among the different facies. For example, modern packstones and grainstones from Florida have porosities from 40% to 67% and permeabilities from 1840 to 30,800 md, whereas wackestones and mudstones have higher porosities (68%) but lower permeabilities, 228 to 0.87 md (Enos and Sawatsky, 1981). Thus, depositional environments strongly control initial porosity and permeability. This primary porosity was reduced by near sea floor and early burial calcite cementation, especially along the buildup margins (Figures 3, 8). During burial, the sediments were subjected to mechanical and chemical compaction.

Stage 2: Replacement Dolomitization (Early Burial, Porosity Reduction)

During the early stages of replacement dolomitization (as observed in the Strachan buildup), isolated dolomite rhombs nucleated and grew in lime mud of all depositional facies (stage 2A, Figure 23). As dolomitization progressed, dolomite rhombs grew to form dolomite patches in the lime mud matrix (stage 2B) of rocks in the D3A pool, and also began filling some porosity. Less dolomite formed in the D3B pool (column 1, Figure 23). In places, advanced matrix

dolomitization formed a self-supporting framework that probably made these rocks more resistant to chemical compaction (stage 2C). Conversely, limestones were subjected to continuing pressure solution and porosity reduction (column 1, Figure 23). During or following replacement dolomitization, porosity was rearranged with dissolution (D_1) of allochems, forming vugs and increasing porosity (stage 2D).

Stage 3: Cementation, and Stage 4: Dissolution and Bitumen Emplacement

Either toward the end of fossil dissolution or afterward, minor amounts of dolomite cement (C_1) locally decreased porosity. Dolomite cement (C_2), anhydrite, and minor sulfides partly filled pores, mostly in the lower part of the reservoirs. The emplacement of bitumen during the Late Cretaceous notably reduced porosity and, more importantly, permeability, in the upper part of the reservoir.

DISCUSSION

Controls on Pore Types and Their Distribution

In carbonate reservoirs like the Leduc buildups, most depositional facies can be permeable or nonpermeable, depending on the type and extent of diagenesis. Porosity and permeability variations in the Strachan and Ricinus West buildups are controlled by a

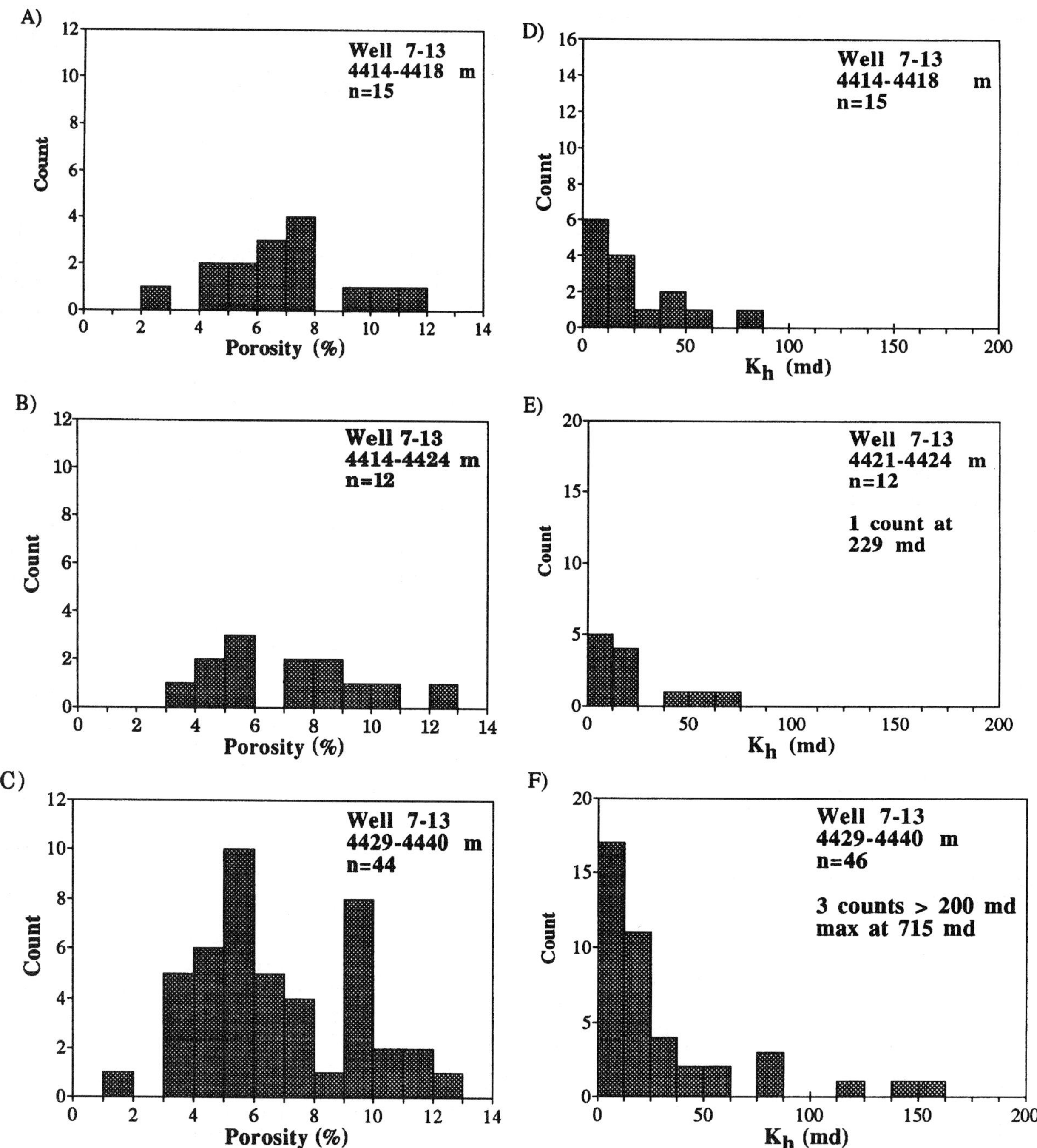

Figure 19. Porosity and permeability histograms from tabular stromatoporoid boundstone facies in Ricinus buildup margin well 7-13. Permeability increases slightly with increasing porosity (see Figure 12 for vertical plots of these data). Porosity is slightly bimodal.

series of complex interactions involving depositional environments and multiple diagenetic events during progressive burial, thermal maturation, overpressuring, and thermochemical sulfate reduction. The upper reef interior is characterized by porous, permeable, laterally continuous slices that range in thickness from 8 to 27 m (actual depositional cycles would have been 1 to 3 m thick), showing that arbitrary subdivision into layers of equal thickness, as was done by McNamara and Wardlaw (1991) in the analysis of Westerose, is not realistic. The lower Ricinus West reef interior appears to consist

of irregular, porous and permeable slices that range from 15 to 30 m thick, and the available data suggest that layers cannot be considered to be laterally continuous throughout the reservoir. The reef margin facies are thick and laterally discontinuous, with porous and permeable zones mainly controlled by subvertical fractures and brecciated intervals.

The distribution of porous and permeable zones within the dolomitized Ricinus West buildup differs from that of the Westerose buildup updip in the central part of the Rimbey-Meadowbrook reef trend (McNamara and

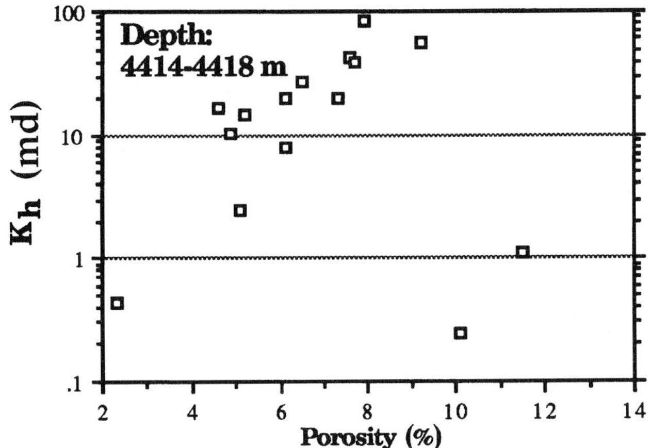

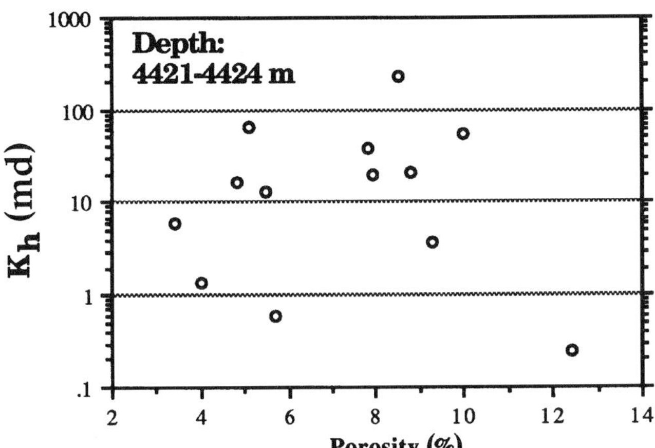

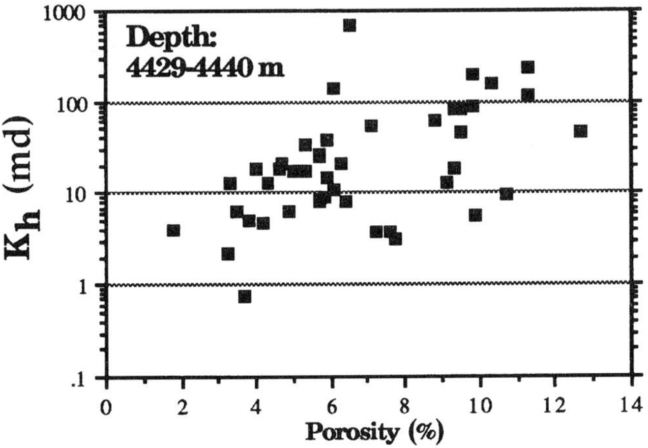

Figure 20. Crossplots of porosity and permeability data (from Figure 19) from tabular stromatoporoid boundstone facies in Ricinus buildup margin well 7-13. The interval 4429 to 4440 m shows a clearer trend of permeability increasing with porosity than do the overlying 4414 to 4418 or 4421 to 4424 m intervals.

Wardlaw, 1991), where the highest porosity and permeability rocks occur in the lower portion of the reef interior. This appears to be due to differences in primary facies distribution in the Westerose buildup because it is small (1.6 × 5.5 km and 214 m thick) compared to the Ricinus West

(4 × 9.5 km and 259 m thick). However, the main differences appear to be related in part to the different approaches used to map and group porosity and permeability data. In Ricinus West and the dolomitized part of Strachan (D3A pool), distribution of pore types closely follows the depositional sequences (Figure 5).

Predicting Reservoir Quality

It is difficult to extrapolate reservoir quality beyond the present database. However, comparing our data from the deep basin with information from intermediate and shallower parts of the basin and with what were once more deeply buried strata in Rocky Mountain outcrops reveals several important aspects. Thus, some generalizations and predictions can be advanced by comparing and integrating the above information with data from limestone and dolomite reservoirs elsewhere in the basin. Data from limestone buildups in the shallower parts of the Alberta basin suggest that the effects of sea floor and early diagenesis are relatively minor, except for (1) some buildup margins that contain extensive submarine cements (e.g., Golden Spike) (Mountjoy and Walls, 1977; Walls et al., 1979) and (2) subaerially exposed portions of buildup interiors that are locally strongly cemented (e.g., Golden Spike) (Walls and Burrowes, 1985, 1990). In the case of the older, isolated Swan Hills buildups (Havard and Oldershaw, 1976; Wong and Oldershaw, 1981; Wendte and Muir, 1995), subaerial exposure also has had a comparatively minor effect on reservoir quality; more than 90% of the porosity is of primary depositional origin.

The facies distributions in the buildups were predominantly controlled by sea level changes and thus are widespread and generally predictable (Wendte, 1992a, b), but with important variations caused by sediment supply and currents (McLean and Mountjoy, 1993b). Subaerial unconformities formed during sea level drops are also widespread and predictable (McLean and Mountjoy, 1994; Wendte and Muir, 1995). It is reasonable to assume that these nonporous cemented zones also are present locally in the dolomitized buildups. This, together with detailed observations from the partially dolomitized Miette buildup exposed in the Rocky Mountains (Mattes and Mountjoy, 1980) and reconnaissance observations from other dolomitized Rocky Mountain buildups (McLean and Mountjoy, 1993a, b, 1994), indicates that porous dolomites are regionally extensive and are largely controlled by the facies distribution and porosity of the original limestones. Thus, porosity is primarily facies controlled and the Leduc limestone buildups such as Redwater (Klovan, 1964) and Golden Spike (McGillivray and Mountjoy, 1975; Walls, 1978) can be used as general guides to predict the types and distribution of porosity and permeability, provided one takes into account porosity reduction with increasing burial.

Predicting the distribution of replacement dolomites and associated dissolution that took place during shallow burial is also difficult. However, most of the buildups in the Alberta basin are dolomitized, apparently because most were connected to conduit systems in the underlying platforms (e.g., Rimbey-Meadowbrook reef trend) (Amthor et al., 1993; Mountjoy and

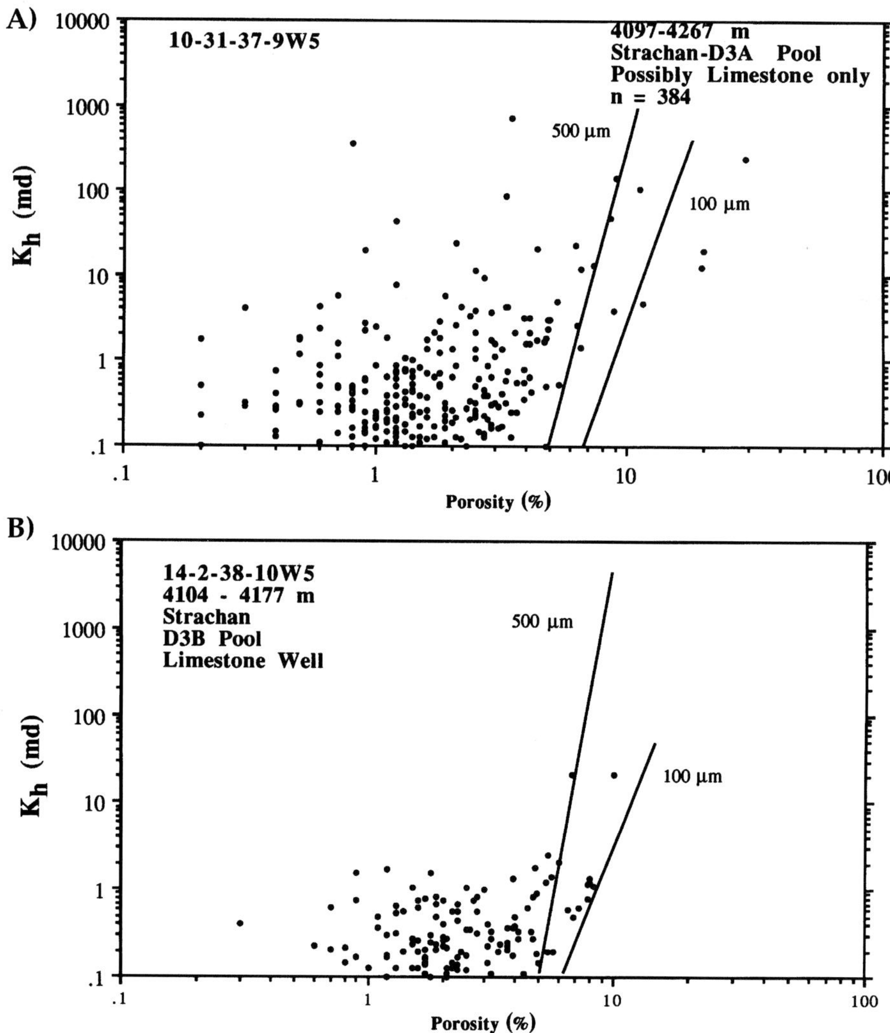

Figure 21. Crossplots of porosity-permeability (log scales) of mostly limestones in the Strachan D3A (A) and D3B (B) pools. Porosity scale is arithmetic and permeability scale is log. The two lines represent the 100-µm and 500-µm particle size boundaries used by Lucia (1983, 1995) to distinguish permeability fields, in this case rock fabric/petrophysical class 1. Rock fabric/petrophysical class 2 falls below the 100-µm line. The Strachan limestones have low porosities and permeabilities. The higher permeabilities from well 10-31 are due to permeability related to partial dolomitization.

Amthor, 1994) or were connected to fault and fracture conduit systems as in parts of the Peace River arch (Packard et al., 1990; Mountjoy and Halim-Dihardja, 1991) and elsewhere (D. Green, 1996, personal communication). Buildups away from these conduit systems such as Golden Spike, Redwater, and Miette are not dolomitized, or only partially dolomitized, because apparently they were not linked to a regional conduit system at the time of dolomitization. The four stages of porosity development relative to dolomitization (Figure 23) are representative of the diagenesis in different parts and stratigraphic levels of the basin (Walls and Burrowes, 1985, 1990; Mountjoy, 1994; Mountjoy and Amthor, 1994).

Fracturing, except for late-stage microfracturing (Marquez and Mountjoy, 1996), has not been studied sufficiently to determine its origin and overall distribution. The somewhat greater abundance of fractures in the buildup margins may be due to differential compaction between the buildup and the adjacent basin and to the buildup margins being more strongly cemented.

With increasing burial, limestones gradually lose their porosity due to pressure solution and cementation, which in the Alberta Basin has reduced or destroyed most of the primary porosity of limestones buried deeper than 3500 m (Drivet, 1993; Amthor et al., 1994; Marquez, 1994; this study). Also, plugging of pores and vugs in dolomite reservoirs by late-stage anhydrite, and to a minor extent carbonate cements, can locally reduce porosity and permeability, as observed in the Rimbey-Meadowbrook reef trend below depths of 2300 m immediately updip from the Strachan reservoir (Drivet, 1993; Mountjoy et al., 1997). Bitumen plugging caused by deasphalting will take place in those reservoirs in which crude oil has been cracked. Thus, except for these important modifications, it is reasonable to infer that dolomitized Leduc reservoirs elsewhere in the deep basin will have porosity and permeability variations similar to those observed in the Ricinus West and Strachan D3A reservoirs.

Although porosity and permeability are related to depositional facies and patterns of diagenetic overprinting, there is some variation within and between reefs, but this depends on the scale. At the large scale (tens of meters), the porosity and permeability are remarkably uniform (Figure 13); at the medium scale, the vertical and lateral arrangement of the dominant pore types varies considerably, causing variations in reservoir character. At the small (meter to millimeter) scale, porosity and permeability are controlled by facies, pore types, and differences in diagenesis. The approach used for characterizing reservoir properties and for classifying porosity, permeability, and pore systems is critical. Up to now, it has often been

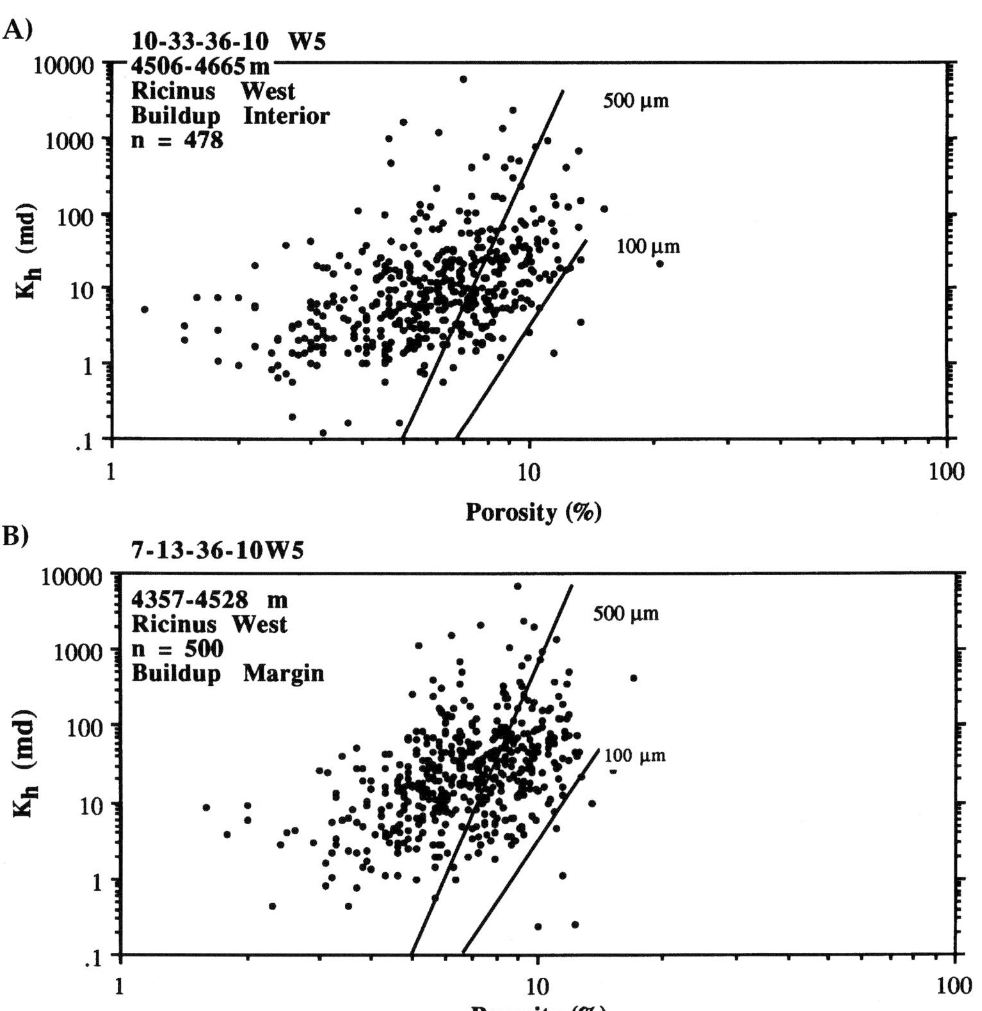

Figure 22. Crossplots of porosity-permeability (log scales) of the buildup interior (A) and margin (B) in the completely dolomitized Ricinus West reservoir. Lines and scales as in Figure 21. Data from well 10-33 (A) includes intervals plotted in Figure 11 and well 7–13 in Figure 12. The porosities and permeabilities are much higher than the values from limestones in Figure 21. There is essentially no difference in the distributions from the buildup margin and interior. In general, the permeabilities are higher than expected (compare with Lucia, 1995) and appear to be due to touching moldic vugs in a dolomite matrix with good intercrystalline porosity (see text).

one of arbitrary broad-scale lumping of horizontal slices and blocks having similar porosities. A reservoir description that closely follows the geology and depositional facies of the reservoir, and especially the diagenetic overprints, as outlined here for the Ricinus West buildup, provides a more realistic and accurate subdivision of reservoir properties. This cannot be performed unless the depositional facies, diagenesis, and pore types are thoroughly documented by means of careful core observations. In addition, grain and crystal size and sorting, and separate/connected-vug type and porosity (Lucia, 1995), are important in terms of describing rock fabrics.

CONCLUSIONS

The association and distribution of pore types and permeability within the Leduc Strachan and Ricinus West gas reservoirs in the deep Alberta basin indicate:

1. Depositional facies ultimately controlled the distribution of different pore types, whereas permeability is mainly controlled by diagenetic processes, especially dolomitization and dissolution, and various phases of cementation in the lower part of reservoirs below the paleo-oil/water contact and

bitumen in the upper part of reservoirs.
2. Completely dolomitized Upper Devonian Leduc buildups at depths >3000 m have higher porosities and permeabilities than limestones because dolostones are more resistant to pressure solution during burial.
3. The relationships between the proportion of dolomite and porosity are complex. In the Strachan buildup, a slight increase in porosity occurs with an increase in the amount of dolomite, which becomes more pronounced at >80% dolomite. This porosity increase appears to be related to the leaching of calcite and may be associated with dolomitization. In general, at burial depths >3000 m, porosity and permeability increase with increasing dolomitization, as in the partially dolomitized Strachan buildup (D3A pool), with the highest porosity and permeability occurring in completely dolomitized facies in the buildup interior. At the Strachan reef margin, porous and permeable dolostones are interbedded with nonporous and nonpermeable limestones, with strata that had higher primary porosities being preferentially dolomitized. These trends are modified by later cementation, dissolution, and bitumen emplacement. At these depths, limestones make

Stage 1: DEPOSITION

REEF MARGIN

REEF   INTERIOR

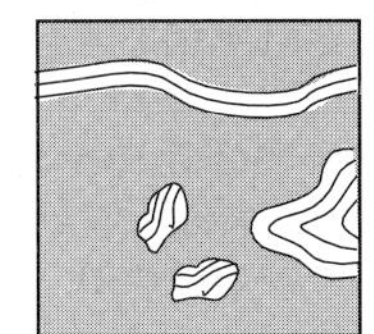

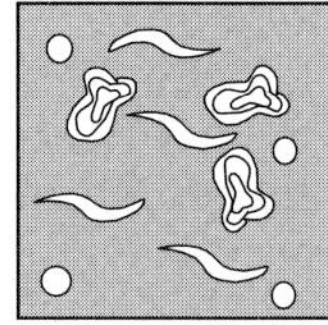

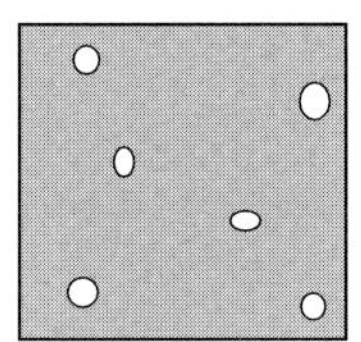

Boundstones, rudstones
with packstone matrices

Rudstones with
packstone and
grainstone matrices

Wackestones and
mudstones

Stage 2: PARTIAL TO COMPLETE DOLOMITIZATION

A) Nucleation and early growth of dolomite rhombs in the lime matrix.

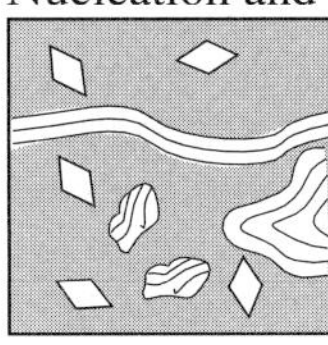 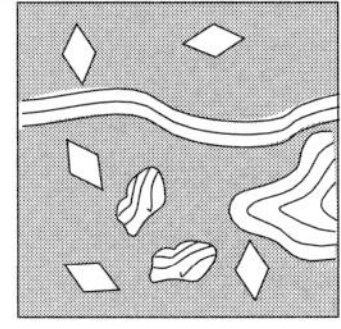 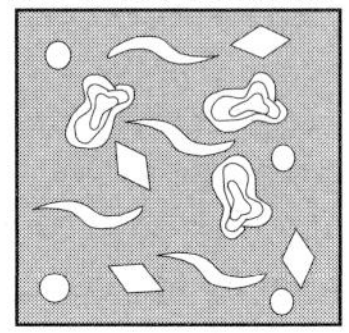 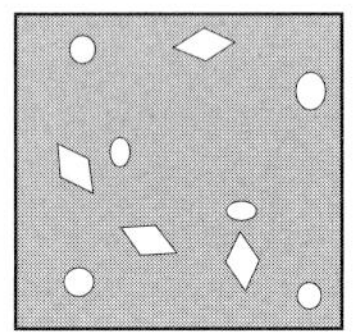

B) Further rhomb growth to form dolomite patches in lime matrix.

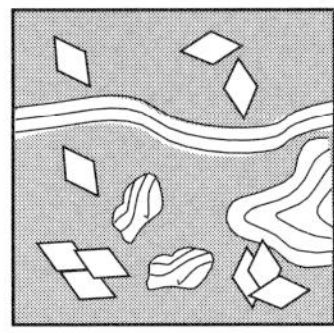 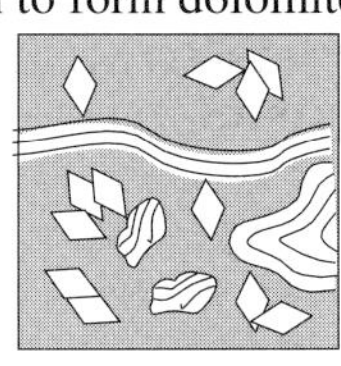 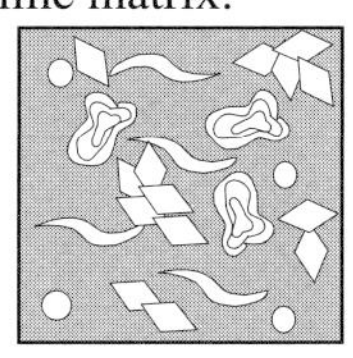 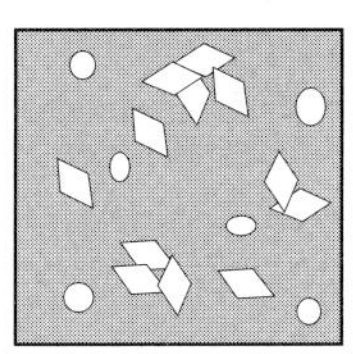

C) Advanced matrix replacement First dissolution of skeletal fragments and some lime matrix.

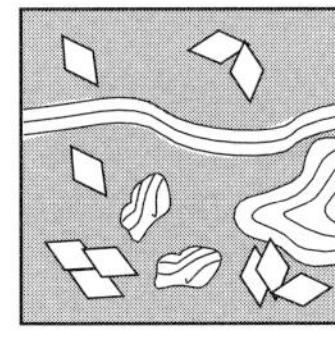   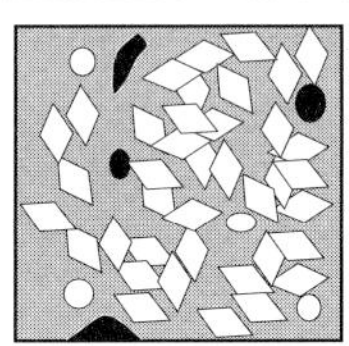

D) End of dolomite replacement. Partial to complete dissolution of calcite skeletal fragments.

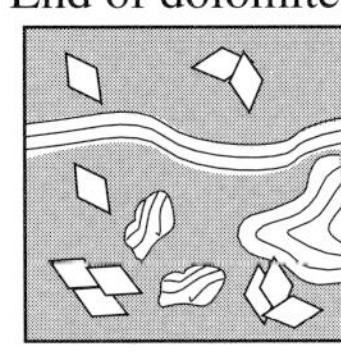 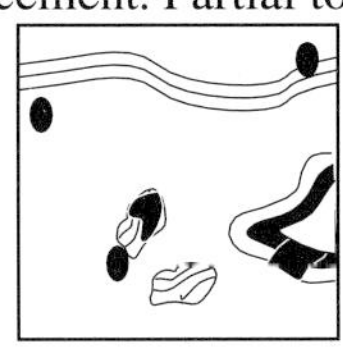 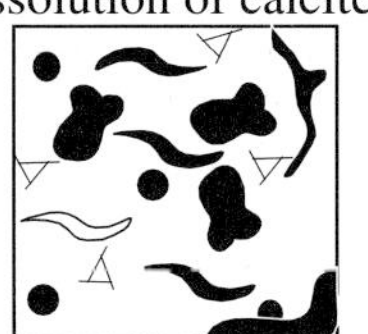 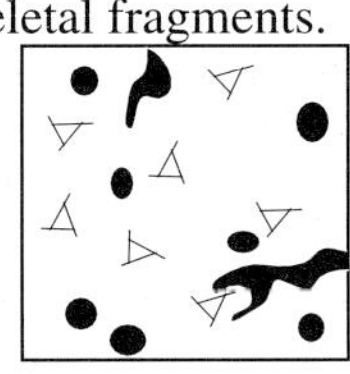

Stages 3 and 4: CEMENTATION, DISSOLUTION, MICROFRACTURING, AND BITUMEN PLUGGING

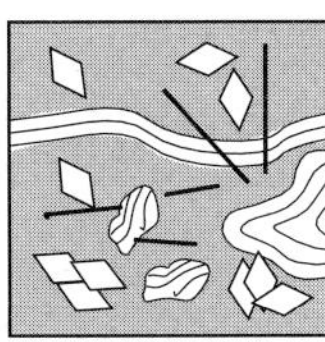 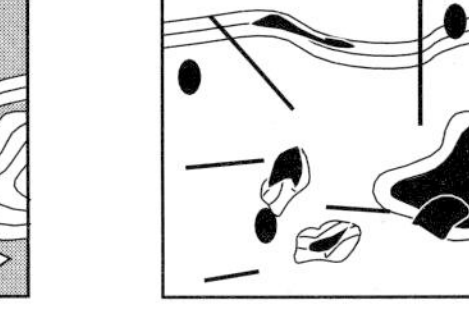  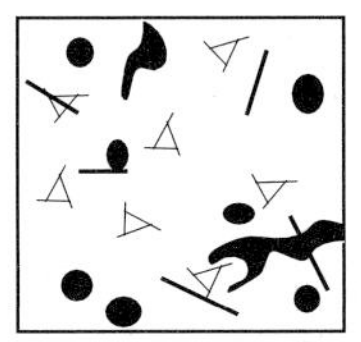

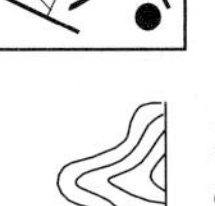

Figure 23. Inferred porosity evolution during replacement dolomitization of Leduc limestones and later diagenesis illustrating changes and modification of pore types. Stage 1: deposition and early porosity reduction; Stage 2: replacement dolomitization; Stages 3 and 4: late cementation, dissolution, microfracturing, and bitumen plugging.

Increasing Dolomitization

poor reservoirs because most of the primary porosity has been filled with cement or destroyed by pressure solution.

4. Pore types closely follow depositional facies in the Ricinus West buildup, despite complete dolomitization. In general, porosity and permeability are similar throughout much of this buildup. At the meter scale, the upper buildup interior reservoir is characterized by 1- to 2-m-thick, permeable, laterally continuous lagoonal zones separated by thicker intervals of nonporous, more finely crystalline dolomite. The lower reef interior has discontinuous permeable vuggy zones. The reef margin contains intervals of breccia, fractures, and connected vugs that provide vertical permeability. The best lateral permeability is related to laterally continuous skeletal wackestone to grainstone facies that have been dolomitized to a rock with good intercrystalline porosity in the upper buildup interior. In the buildup margin, the best vertical interconnections are provided by connected vugs and subvertical fractures in dolomites with intercrystalline porosity.

5. In dolomites, touching vugs of solution-enlarged molds of *Amphipora* and stromatoporoids form an excellent porous and permeable reservoir rock that is related to primary reef margin and reef flat facies. This moldic dolomite is an important reservoir rock that, when surrounded by dolomites with intercrystalline porosity, has permeabilities higher than expected. These dolomites appear to be largely responsible for the excellent productivity of these reservoirs.

6. At the millimeter scale, porosity and permeability are controlled by diagenetic processes. Postdolomitization processes, such as pressure solution, cementation, dissolution, and bitumen plugging, locally modified porosity and permeability. Anhydrite cementation, minor dolomite, calcite, and native sulfur cementation reduced porosity and, more significantly, permeability, producing a high degree of heterogeneity in the lower part of the reservoirs.

7. Bitumen reduces porosity and permeability by decreasing pore and pore-throat sizes, mainly in the upper part of the Ricinus West reservoir. Bitumen coating may cause wettability to change from water-wet to intermediate in water/gas systems. Bitumen can only be determined by core examination, petrographic analysis, and core porosity measurements. Errors in reservoir volumetric calculations from well logs can arise because of the lack of significant density contrast between crude oil and bitumen, making bitumen indistinguishable from oil.

8. Predicting reservoir fabrics is difficult. However, in this case, there is a general trend that Upper Devonian Leduc replacement dolomites form good to excellent reservoir rocks in the deep Alberta Basin, except where plugged by late-stage calcite, anhydrite, and bitumen. Early replacement dolomitization and associated calcite dissolution formed a vuggy reservoir rock connected by a matrix of intercrystalline porosity. The highest porosities and permeabilities occur in those facies that had the highest primary porosity and permeability; that is, the grainstones and framestones. The reservoir character was only slightly modified by later fracturing and cement and bitumen fillings.

Reconstruction of the depositional and diagenetic history from detailed studies of cores is essential for making realistic reservoir models. Differences in approach in grouping porosity and permeability data in reservoir characterization can lead to major differences in modeling and assessment of a reservoir. Critical to a realistic assessment is to describe the reservoir (depositional facies, sequences, and diagenesis, including grain and crystal size, sorting, separate-vug type, and porosity) as geologically faithfully as possible, rather than by means of arbitrary slices or blocks. Such studies demonstrate that carbonate reservoirs are very heterogeneous, and that discontinuous, relatively nonporous, and low permeability zones have the potential to impede fluid flow and greatly affect reservoir performance and production. More realistic models are essential for understanding reservoir production and for predictions of reservoir performance and ultimate recovery by enhancement techniques. Only after putting together such genuine reservoir models can the more detailed petrophysical information be incorporated in them and assessed.

Although it is difficult to predict reservoir porosity and permeability trends beyond these two reservoirs, it is clear that the secondary porosities in these deeply buried dolomites are mainly controlled by the primary porosity distribution, which in turn is controlled by the depositional facies. Observations elsewhere in the deep basin and the adjacent Rocky Mountains suggest that these porous Leduc dolomites are regionally extensive and should have porosity and permeability variations similar to the Strachan and Ricinus West reservoirs.

ACKNOWLEDGMENTS

This research was supported by Natural Science and Engineering Research Council (NSERC) operating grants and a strategic grant to E.W. Mountjoy and H.G. Machel, supplemented by funds from Chevron, Home Oil, Mobil Oil, Norcen, PanCanadian, Petro Canada, Shell Canada, and Imperial Oil. This paper has been modified and updated from Marquez (1994). X. Marquez received a scholarship from Maraven S.A. and a grant-in-aid from the American Association of Petroleum Geologists. We acknowledge the donation by Geographix of QLA2 log analysis software, and Brett Norris for his assistance and input with the log analysis. Individuals who shared their work and knowledge include: J. Amthor, G. Burrowes, G. Davies, W. Keith, B. Martindale, B. McNamara, H. Qing, J. Reimer, A. Rup, B. Scott, M. Teare, N. Wardlaw, B. Watt, and J. Wendte. Reviews of earlier versions and comments by E. Drivet, J. Duggan, D. Green, H. Machel, J. Paquette, N. Wardlaw, and S. Whittaker improved the manuscript. We appreciate the helpful editorial comments and suggestions of Rick Major and John Bloch and the helpful and thorough review by Julie Kupecz.

REFERENCES CITED

Amthor, J.E., E.W. Mountjoy, and H.G. Machel, 1993, Subsurface dolomites in Upper Devonian Leduc Formation buildup, central part of Rimbey-Meadowbrook reef trend, Alberta, Canada: Bulletin of the Canadian Society of Petroleum Geologists, v. 41, p. 164–185.

Amthor, J.E., E.W. Mountjoy, and H.G. Machel, 1994, Regional-scale porosity and permeability variations in Upper Devonian Leduc buildups: implications for origin and distribution of porosity in carbonates: AAPG Bulletin, v. 78, p. 1541–1559.

Andrichuk, J.M., 1958a, Stratigraphy and facies analysis of Upper Devonian reefs in Leduc, Stettler and Redwater areas, Alberta: AAPG Bulletin, v. 42, p. 1–93.

Andrichuk, J.M., 1958b, Cooking Lake and Duvernay Late Devonian sedimentation in Edmonton area of central Alberta, Canada: AAPG Bulletin, v. 42, p. 2189–2222.

Archie, G.E., 1952, Classification of carbonate reservoir rocks and petrophysical considerations: AAPG Bulletin, v. 36, p. 278–298.

Barfoot, G.L., and R.J. Rodgers, 1984, Leduc, a new life at 38: Journal of Canadian Petroleum Technology, May–June, p. 41–46.

Barfoot, G.L., and S.C.M. Ko, 1987, Assessing, and compensating for, the impact of the Leduc D-3A gas cap blowdown on the other Golden Trend pools: Journal of Canadian Petroleum Technology, July–August, p. 28–36.

Bateman, R.M., 1985, Open hole log analysis and formation evaluation: Boston, IHRDC Publishers, 647 p.

Beales, F.W., and J.L. Hardy, 1980, Criteria for the recognition of diverse dolomite types with an emphasis on studies on host rocks for Mississippi Valley-type ore deposits, in D.H. Zenger, J.B. Dunham, and R.L. Ethington, eds., Concepts and models of dolomitization: SEPM Special Publication 28, p. 197–213.

Burrowes, O.G., 1977, Sedimentation and diagenesis of back-reef deposits, Miette and Golden Spike buildups, Alberta: M.Sc. thesis, McGill University, Montreal, Quebec, 207 p.

Burrowes, O.G., and F.F. Krause, 1987, Overview of the Devonian System: subsurface Western Canada Basin, in F.F. Krause and O.G. Burrowes, eds., Devonian lithofacies and reservoir styles in Alberta: Calgary, 13th Canadian Society of Petroleum Geologists Core Conference and Display, p. 1–20.

Carpenter S.J., and K.C. Lohmann, 1989, $\delta^{18}O$ and $\delta^{13}C$ variations in Late Devonian marine cements from the Golden Spike and Nevis reefs, Alberta, Canada: Journal of Sedimentary Petrology, v. 59, p. 792–814.

Choquette, P.W., A. Cox, and W.J. Meyers, 1992, Characteristics, distribution and origin of porosity in shelf dolostones: Burlington-Keokuk Formation: Journal of Sedimentary Petrology, v. 62, p. 167–189.

Choquette, P.W., and L.C. Pray, 1970, Geologic nomenclature and classification of porosity in sedimentary carbonates: AAPG Bulletin, v. 54, p. 207–250.

Chouinard, H., 1993, Lithofacies and diagenesis of the Cooking Lake Platform carbonates, Alberta Basin subsurface, Canada: M.Sc. thesis, McGill University, Montreal, Quebec, 101 p.

Drivet, E., 1993, Diagenesis and reservoir characterization of Upper Devonian Leduc dolostones, southern Rimbey-Meadowbrook reef trend, central Alberta: M.Sc. thesis, McGill University, Montreal, Quebec, 115 p.

Drivet, E., and E.W. Mountjoy, 1993, Porosity variations in Upper Devonian Leduc dolomites, central Rimbey-Meadowbrook reef trend, Alberta (abs.): AAPG Bulletin, Annual Convention Abstracts, p. 93.

Drivet, E., and E.W. Mountjoy, 1994, Timing of dolomitization and secondary porosity in Upper Devonian Leduc dolostones, southern Rimbey-Meadowbrook reef trend, Alberta (abs.): Canadian Society of Exploration Geophysicists and Canadian Society of Petroleum Geologists Joint Annual Convention, Calgary, Alberta, p. 347–348.

Drivet, E., and E.W. Mountjoy, 1997, Burial dolomitization in the Leduc Formation (Upper Devonian), southern Rimbey-Meadowbrook reef trend, Alberta: Journal of Sedimentary Research, v. 67, p. 411–423.

Enos, P., and L.H. Sawatsky, 1981, Pore networks in Holocene carbonate sediments: Journal of Sedimentary Petrology, v. 51, p. 961–985.

Ghosh, S.K., and G.M. Friedman, 1989, Petrophysics of a dolostone reservoir: San Andres Formation (Permian), West Texas: Carbonates and Evaporites, v. 4, p. 45–119.

Havard, C., and A. Oldershaw, 1976, Early diagenesis in back-reef sedimentary cycles, Snipe Lake reef complex, Alberta: Bulletin of Canadian Petroleum Geology, v. 24, p. 27–69.

Hriskevich, M.E., J.M. Faber, and J.R. Langton, 1980, Strachan and Ricinus West gas fields, Alberta, Canada, in M.T. Halbouty, ed., Giant oil and gas fields of the decade 1968–1978: AAPG Memoir 30, p. 315–328.

Hugo, K., 1990, Mechanisms of groundwater flow and oil migration associated with Leduc reefs: Bulletin of the Canadian Society of Petroleum Geologists, v. 38, p. 307–319.

Illing, L.V., 1959, Deposition and diagenesis of some Upper Paleozoic carbonate sediments in western Canada: New York, Proceedings of the Fifth World Petroleum Congress, Section 1, p. 23–52.

Jardine, D., D.P. Andrews, J.W. Wishart, and J.W. Young, 1977, Distribution and continuity of carbonate reservoir: Journal of Petroleum Technology, v. 29, p. 873–885.

Jardine, D., and J.W. Wishart, 1982, Carbonate reservoir description: Dallas, Texas, Society of Petroleum Engineers, Paper 10010, 13 p.

Klovan, J.E., 1964, Facies analysis of the Redwater reef complex, Alberta, Canada: Bulletin of the Canadian Society of Petroleum Geologists, v. 12, p. 1–100.

Krouse, H.R., C.A. Viau, L.S. Eliuk, A. Ueda, and S. Halas, 1988, Chemical and isotopic evidence of thermochemical sulphate reduction by light hydrocarbon gases in deep carbonate reservoirs: Nature, v. 333, p. 415–419.

Laflamme, A.K., 1990, Replacement dolomitization in

the Upper Devonian Leduc and Swan Hills formations, Caroline area, Alberta, Canada: M.Sc. thesis, McGill University, 138 p.

Layer, D.B., 1949, Leduc oil field, Alberta, a Devonian coral reef discovery: AAPG Bulletin, v. 33, p. 572–602.

Lishman, J.R., 1969, Core permeability anisotropy: Petroleum Society of the Canadian Institute of Mining and Metallurgy, 20th Annual Technical Meeting, Edmonton, Alberta, Paper 6920.

Lomando, A.J., 1992, The influence of solid reservoir bitumen on reservoir quality: AAPG Bulletin, v. 76, p. 1137–1152.

Lucia, F.J., 1983, Petrophysical parameters estimated from visual descriptions of carbonate rocks: a field classification of carbonate pore space: Journal of Petroleum Technology, v. 35, p. 629–637.

Lucia, F.J., 1995, Rock-fabric/petrophysical classification of carbonate pore space for reservoir characterization: AAPG Bulletin, v. 79, p. 1275–1300.

Lucia, F.J., and R.D. Conti, 1987, Rock fabric, permeability, and log relationships in an upward-shoaling, vuggy carbonate sequence: University of Texas at Austin, Bureau of Economic Geology Geological Circular 87–5, 22 p.

Machel, H.G., and E.W. Mountjoy, 1987, General constraints on extensive pervasive dolomitization and their application to the Devonian carbonates of western Canada: Bulletin of Canadian Petroleum Geology, v. 35, p. 143–158.

Maddox, D.F., 1984, Reservoir simulation study—Ricinus West D-3A pool, Ricinus West field: Canterra Energy Ltd. Internal Report, Calgary, Alberta, 90 p.

Marquez, X., 1994, Reservoir geology of Upper Devonian Leduc buildups, deep Alberta Basin: Ph.D. thesis, McGill University, Montreal, Quebec, 285 p.

Marquez, X., and E.W. Mountjoy, 1996, Microfractures due to overpressures caused by thermal cracking in well-sealed Upper Devonian reservoirs, deep Alberta Basin: AAPG Bulletin, v. 80, p. 570–588.

Mattes, B.W., and E.W. Mountjoy, 1980, Burial dolomitization of the Upper Devonian Miette buildup, Alberta, in D.H. Zenger, J.B. Dunham, and R.L. Ethington, eds., Concepts and models of dolomitization: SEPM Special Publication 28, p. 259–297.

Mazzullo, S.L., 1992, Geochemical and neomorphic alteration of dolomite, a review: Carbonate and Evaporites, v. 7, p. 21–37.

McCaffery, F.G., 1977, Rock-fluid relationship studies on the Windfall D-3A reservoir and their application in evaluating gas cycling effectiveness: Journal of Canadian Petroleum Technology, January–March, p. 55–63.

McGillivray, J.G., and E.W. Mountjoy, 1975, Facies and related reservoir characteristics, Golden Spike reef complex, Alberta: Bulletin of the Canadian Society of Petroleum Geologists, v. 23, p. 753–809.

McLean, D.J., 1992, Upper Devonian buildup development in the Southern Canadian Rocky Mountains: a sequence stratigraphic approach: Ph.D. thesis, McGill University, Montreal, Quebec, 290 p.

McLean, D.J., and E.W. Mountjoy, 1993a, Stratigraphy and depositional history of the Burnt Timber Embayment, Fairholme Complex, Alberta: Bulletin of Canadian Petroleum Geology, v. 41, p. 290–306.

McLean, D.J., and E.W. Mountjoy, 1993b, Upper Devonian buildup, margin and slope development in the southern Canadian Rocky Mountains: Geological Society of America Bulletin, v. 105, p. 1263–1283.

McLean, D.J., and E.W. Mountjoy, 1994, Allocyclic control on Late Devonian buildup development, Southern Canadian Rocky Mountains: Journal of Sedimentary Research, v. B64, p. 326–340.

McNamara, L.B., and N.C. Wardlaw, 1991, Geological and statistical description of the Westerose reservoir, Alberta: Bulletin of Canadian Petroleum Geology, v. 39, p. 332–351.

McNamara, L.B., N.C. Wardlaw, and M. McKellar, 1991, Assessment of porosity from outcrops of vuggy carbonate and application to cores: Bulletin of Canadian Petroleum Geology, v. 39, p. 260–269.

Mossop, G.D., 1972, Origin of peripheral rim, Redwater reef, Alberta: Bulletin of Canadian Petroleum Geology, v. 20, p. 238–280.

Mountjoy, E.W., 1994, Dolomitization and the character of hydrocarbon reservoirs; Devonian of Western Canada, in A. Parker and B. Sellwood, eds., Quantitative diagenesis, recent developments and applications to reservoir geology: Dordrecht, Kluwer Academic Publishers, p. 33–94.

Mountjoy, E.W., and J.E. Amthor, 1994, Has burial dolomitization come of age? Some answers from the Western Canada Sedimentary Basin, in B. Purser, M. Tucker, and D. Zenger, eds., Dolomites, a volume in honour of Dolomieu: International Association of Sedimentologists Special Publication 21, p. 203–229.

Mountjoy, E.W., and M.K. Halim-Dihardja, 1991, Multiple phase fracture and fault-controlled burial dolomitization, Upper Devonian Wabamun Group, Alberta: Journal of Sedimentary Petrology, v. 61, p. 590–612.

Mountjoy, E.W., H. Qing, E. Drivet, X. Marquez, S. Whittaker, and A. Williams-Jones, 1997, Variable fluid and heat flow regimes in three Devonian dolomite conduit systems, Western Canada Sedimentary Basin: isotopic and fluid inclusion evidence/constraints, in I.P. Montanez, J.M. Gregg, and K.L. Shelton, eds., Basin wide fluid flow and diagenetic patterns: integrated petrologic, geochemical and hydrological considerations: SEPM Special Publication 57, p. 119–137.

Mountjoy, E.W., and R.A. Walls, 1977, Some examples of early submarine cements from Devonian buildups of Alberta: Miami, Rosentheil School of Marine and Atmospheric Science, University of Miami, Proceedings of the 3d International Coral Reef Symposium, v. 2, p. 155–161.

Packard, J.J., G.J. Pellegrin, I.S. Al-Aasm, I. Samson, and J. Gagnon, 1990, Diagenesis and dolomitization associated with hydrothermal karst in Famennian upper Wabamun ramp sediments, northwestern Alberta, in G.R. Bloy and M.G. Hadley, eds., The development of porosity in carbonate reservoirs: Canadian Society of Petroleum Geologists Continuing Education Short Course, Section 9, 25 p.

Podrusky, J.A., J.E. Barclay, A.P. Hamblin, L.P. Lee,

K.G. Osadetz, R.M. Procter, and G.C. Taylor, 1987, Conventional oil resources of Western Canada, Part I: Reservoir endowment: Geological Survey of Canada, Paper 87–26, 42 p.

Reitzel, G.A., and G.O. Callow, 1977, Pool description and performance analysis leads to understanding Golden Spikes miscible flood: Journal of Petroleum Technology, v. 29, p. 867–872.

Reitzel, G.A., G. Davidson, G.O. Callow, and D.R. Bates, 1976, Golden Spike D3-A pool oil depletion study, Calgary: Imperial Oil Ltd., Producing Department, Western Region, Report IPRC-5ME-76, 38 p.

Rogers, M.A., J.D. McAlary, and N.J. Bailey, 1974, Significance of reservoir bitumens to thermal maturation studies, Western Canada Basin: AAPG Bulletin, v. 58, p. 1806–1824.

Schmoker, J.W., and R.B. Halley, 1982, Carbonate porosity versus depth: a predictable relation for South Florida: AAPG Bulletin, v. 66, p. 2561–2570.

Seifert, S.R., 1990, Strachan Leduc gas pool, in M.L. Rose, ed., Oil and gas pools of Canada: Canadian Society of Petroleum Geologists, v. 1, variously paginated.

Stoakes, F.A., 1992, Woodbend megasequence, in J. Wendte, F.A. Stoakes, and C.V.Campbell, eds., Devonian–Early Mississippian carbonates of the Western Canada Sedimentary Basin: a sequence stratigraphic framework: SEPM Short Course 28, Calgary, p. 183–206.

Van de Graff, W.J.E., and P.J. Ealey, 1989, Geological modeling for simulation studies: AAPG Bulletin, v. 73, p. 1436–1444.

Vavra, C.L., M.H. Scheihing, and J.D. Klein, 1991, Reservoir geology of the Taylor sandstone in the Oak Hill field, Rusk County, Texas: integration of petrology, sedimentology, and log analysis for delineation of reservoir quality in a tight gas sand, in R. Sneider, W. Masssell, R. Mathis, D. Loren, and P. Wichmann, eds., The integration of geology, geophysics, petrophysics and petroleum engineering in reservoir delineation, description and management: AAPG Proceedings of the 1st Archie Conference, Houston, Texas, p. 130–158.

Walls, R.A., 1978, Cementation history and porosity development, Golden Spike Devonian reef complex, Alberta: Ph.D. thesis, McGill University, Montreal, Quebec, 307 p.

Walls, R.A., 1983, Golden Spike reef complex, Alberta, in P.A. Scholle, D.G. Bebout, and C.H. Moore, eds., Carbonate depositional environments: AAPG Memoir 33, p. 445–453.

Walls, R.A., and O.G. Burrowes, 1985, The role of cementation in the diagenetic history of Devonian reefs, western Canada, in N. Schneidermann and P.M. Harris, eds., Carbonate cement: SEPM Special Publication 36, p. 185–220.

Walls, R.A., and O.G. Burrowes, 1990, Diagenesis and reservoir development in Devonian limestone and dolostone reefs of western Canada: Canadian Society of Petroleum Geologists, Short Course Notes, Section 5, p. 5-1 to 5-18.

Walls, R.A., E.W., Mountjoy, and P. Fritz, 1979, Isotopic composition and diagenetic history of carbonate cements in Devonian Golden Spike reef, Alberta: Geological Society of America Bulletin, v. 90, p. 963–982.

Wardlaw, N.C., 1980, The effects of pore structure on displacement efficiency in reservoir rocks and in glass micromodels: Society of Petroleum Engineers Bulletin, no. 8843, p. 345–352.

Wardlaw, N.C., 1990, Characterization of carbonate reservoirs for enhanced oil recovery: Proceedings of the 1st Technical Symposium on Enhanced Oil Recovery, Tripoli, Libya, Paper 90-01-05, p. 85–105.

Wardlaw, N.C., 1992, Effects of carbonate rock-pore systems on oil recovery, in Subsurface dissolution porosity in carbonates: Recognition, causes and implications: AAPG Short Course, Calgary, p. 1–28.

Waring, W.W., and D.B. Layer, 1950, Devonian dolomitized reef, D-3 reservoir, Leduc field, Alberta, Canada: AAPG Bulletin, v. 34, p. 295–312.

Weber, K.J., 1986, How heterogeneity affects oil recovery, in L.W. Lake and H.B. Carroll, eds., Reservoir characterization: Orlando, Florida, Academic Press, p. 487–544.

Wendte, J.C., 1974, Sedimentation and diagenesis of the Cooking Lake platform and Lower Leduc reef facies, Upper Devonian Redwater, Alberta: Ph.D. thesis, University of California, Santa Cruz, 222 p.

Wendte, J.C., 1992a, Cyclicity of Devonian strata in the Western Canada Sedimentary Basin, in J. Wendte, F.A. Stoakes, and C.V. Campbell, eds., Devonian–Early Mississippian carbonates of the Western Canada Sedimentary Basin: a sequence stratigraphic framework: SEPM Short Course 28, Calgary, p. 25–40.

Wendte, J.C., 1992b, Platform evolution and its control on reef inception and localization, in J. Wendte, F.A. Stoakes, and C.V. Campbell, eds., Devonian–Early Mississippian carbonates of the Western Canada Sedimentary Basin: a sequence stratigraphic framework: SEPM Short Course 28, Calgary, p. 41–88.

Wendte, J.C., 1994, Cooking Lake Platform evolution and its control on Late Devonian Leduc reef inception and localization, Redwater, Alberta: Bulletin of Canadian Petroleum Geology, v. 42, p. 499–528.

Wendte, J.C., and I. Muir, 1995, Recognition and significance of an intraformational unconformity in Late Devonian Swan Hills reef complexes, Alberta, in D.A. Budd, A.H. Saller, and P.M. Harris, eds., Unconformities and porosity in carbonate strata: AAPG Memoir 63, p. 259–278.

Wendte, J.C., and F.A. Stoakes, 1982, Evolution and corresponding porosity distribution of the Judy Creek reef complex, central Alberta, in W.G. Cutter, ed., Canada's giant hydrocarbon reservoirs: Canadian Society of Petroleum Geologists, Core Conference Manual, CSPG-AAPG Convention, Calgary, Alberta, p. 63–81.

Wong, P.K., and A. Oldershaw, 1981, Burial cementation in the Devonian Kaybob reef complex, Alberta, Canada: Journal of Sedimentary Petrology, v. 51, p. 507–520

Index